10th International Conference on Materials and Mechanisms of Superconductivity

(M2S-X)

Journal of Physics: Conference Series Volume 449

Washington, DC, USA
29 July – 3 August 2012

ISBN: 978-1-62993-074-9
ISSN: 1742-6588

Printed from e-media with permission by:

Curran Associates, Inc.
57 Morehouse Lane
Red Hook, NY 12571

proceedings.com

Some format issues inherent in the e-media version may also appear in this print version.

Printed by Curran Associates, Inc. (2013)

Institute of Physics
Dirac House, Temple Back
Bristol BS1 6BE UK

Phone: 44 1 17 929 7481
Fax: 44 1 17 920 0979

techtracking@iop.org

Additional copies of this publication are available from:

Curran Associates, Inc.
57 Morehouse Lane
Red Hook, NY 12571 USA
Phone: 845-758-0400
Fax: 845-758-2634
Email: curran@proceedings.com
Web: www.proceedings.com

TABLE OF CONTENTS

PAPERS

A. COPPER OXIDE SUPERCONDUCTORS

B. IRON-BASED SUPERCONDUCTORS

C. HEAVY FERMION SUPERCONDUCTORS

D. NON-COPPER OXIDE AND NON-IRON-BASED SUPERCONDUCTORS

E. SPECIAL TOPICS SUCH AS NEW CLASSES OF SUPERCONDUCTORS AND TOPOLOGICAL SUPERCONDUCTORS

Journal of Physics: Conference Series **449** (2013) 011001 doi:10.1088/1742-6596/449/1/011001

Preface

The 2012 Materials and Mechanisms of Superconductivity Conference (M2S 2012), which occurs every three years, brought together world experts and young scientists to discuss open questions in the fundamental physics and applications of superconductors, and to disseminate the latest theoretical and experimental research results in superconductors and related novel materials. This conference of 600 participants acted as a valuable training ground in this technologically important area. We focused on key unanswered questions in high-temperature cuprate superconductors, high-temperature iron-based superconductors, topological superconductors, organic superconductors, and heavy-electron superconductors. The discovery of new materials and novel technological applications for electronic devices and for energy transmission and storage was emphasized. There were special sessions on superconductivity and energy, and outreach sessions, and an evening public lecture. There were also junior researcher symposia interspersed within the conference, thus providing an ideal environment for advanced graduate students and postdoctoral researchers to explore the latest theoretical and experimental methods used to investigate challenging questions in the physics of materials as it relates to both fundamental science and technological applications.

These proceedings are an archival testament to the excitement in the field and provide a valuable snapshot of the cutting-edge research of 2012. We hope this will be a valuable resource to active researchers in the field as well as an encouraging volume to excite new researchers to the ever-growing, multifaceted field of superconductivity.

We thank Bernd Lorenz and his Publications Committee for their tremendously creative and diligent work in putting this volume together. This Conference would not have been possible without the tireless work of our Program Committee, Chaired by Rick Greene and Co-Chaired by Mike Norman. Becky McDuffee, our Conference Secretary, deserves special mention for her Olympian efforts. And of course, many thanks to all of our ~600 participants, who made this entire conference such a success.

George Crabtree
Laura Greene
Peter Johnson

Organizing Committee

Conference Co-Chairs

George	Crabtree	Argonne National Laboratory - Argonne, IL & University of Illinois at Chicago and the Center for Emergent Superconductivity - Chicago, IL
Laura	Greene	University of Illinois at Urbana-Champaign and the Center for Emergent Superconductivity - Urbana, IL
Peter	Johnson	Brookhaven National Laboratory and the Center for Emergent Superconductivity - Upton, NY

Conference Treasurer

Peter	Johnson	Brookhaven National Laboratory - Upton, NY

Bernd T. Matthias Prize Chair

Paul	Chu	University of Houston - Houston, TX

H. Kammerlingh Onnes Prize Chair

Dirk	van der Marel	Université de Genève - Genève, Switzerland

John Bardeen Prize Chair

Eduardo	Fradkin	University of Illinois at Urbana-Champaign - Urbana, IL

Conference & Exhibit Management

Centennial Conferences	Louisville, CO

Journal of Physics: Conference Series **449** (2013) 011001 doi:10.1088/1742-6596/449/1/011001

Program Committee

Program Chair

Richard	Greene	University of Maryland - College Park, MD

Program Co-Chair

Michael	Norman	Argonne National Laboratory - Argonne, IL

Program Committee Members

Henri	Alloul	l'Université Paris-Sud - Orsay, France
Meigan	Aronson	Brookhaven National Laboratory - Upton, NY
Dmitri	Basov	University of California-San Diego - La Jolla, CA
Mac	Beasley	Stanford University - Stanford, CA
Collin	Broholm	The Johns Hopkins University - Baltimore, MD
Stuart	Brown	University of California-Los Angeles - Los Angeles, CA
Paul	Canfield	Ames Laboratory - Ames, IA
Paul	Chu	University of Houston - Houston, TX
John	Clarke	University of California-Berkeley - Berkeley, CA
Piers	Coleman	Rutgers State University - Piscataway, NJ
Andrea	Damascelli	University of British Columbia - Vancouver, Canada
Hong	Ding	Beijing Nat'l Lab for Condensed Matter Physics - Beijing, China
Peter	Hirschfeld	University of Florida - Gainesville, FL
Bernhard	Keimer	Max-Planck-Institut für Festkörperforschung - Stuttgart , Germany
Steve	Kivelson	Stanford University - Stanford, CA
Gabi	Kotliar	Rutgers State University - Piscataway, NJ
David	Larbalestier	National High Magnetic Field Laboratory - Tallahassee, FL
Andy	Mackenzie	University of St Andrews - St Andrews, Scotland
David	Mandrus	University of Tennessee - Knoxville, TN
Jochen	Mannhart	Max-Planck-Institut für Festkörperforschung - Stuttgart , Germany
Brian	Maple	University of California-San Diego - San Diego, CA
Igor	Mazin	Naval Research Laboratory - Washington, DC
Andy	Millis	Columbia University - New York, NY
Johnpierre	Paglione	University of Maryland - College Park, MD
Warren	Pickett	University of California-Davis - Davis, CA
Kosmas	Prassides	Durham University - Durham, United Kingdom
Maurice	Rice	ETH-Zürich - Zürich, Switzerland
Siddharth	Saxena	Cavendish Laboratory - Cambridge, United Kingdom
Zhi-Xun	Shen	Stanford University - Stanford, CA
Louis	Taillefer	Sherbrooke Université - Québec, Canada
Hidenori	Takagi	University of Tokyo - Tokyo, Japan
Joe	Thompson	Los Alamos National Laboratory - Los Alamos, NM
Shin-ichi	Uchida	University of Tokyo - Tokyo, Japan
Hai-Hu	Wen	Nanjing University - Nanjing, China
Fu-chun	Zhang	University of Hong, Kong - Pokfulam, Hong Kong

Journal of Physics: Conference Series **449** (2013) 011001 doi:10.1088/1742-6596/449/1/011001

Publication Committee

Publication Chair

Bernd	Lorenz	University of Houston - Houston, TX

Publication Committee Members

Eric	Bauer	Los Alamos National Laboratory - Los Alamos, NM
Sergey	Bud’ko	Ames Laboratory - Ames, IA
Erica	Carlson	Purdue University - West Lafayette, IN
Clarina	dela Cruz	Oak Ridge National Laboratory - Oak Ridge, TN
Xiaoli	Dong	Chinese Academy of Sciences - Beijing, China
Jim	Meen	University of Houston - Houston, TX
Johnpierre	Paglione	University of Maryland - College Park, MD
Haiyan	Wang	Texas A&M University - College Station, TX
Jian-Xin	Zhu	Los Alamos National Laboratory - Los Alamos, NM

Journal of Physics: Conference Series **449** (2013) 011001 doi:10.1088/1742-6596/449/1/011001

M2S 2012 Prize Winners

The 2012 Bernd T. Matthias Prize Winners:

Ivan Božović and **James N. Eckstein** for their pioneering and sustained contributions to the novel synthesis and engineering of superconducting materials.

Dirk Johrendt for his role in the discoveries of superconductivity in $(Ba,K)Fe_2As_2$ and related materials that have helped further the research of Fe-based superconductors.

The 2012 John Bardeen Prize Winners:

Steven A. Kivelson for his works on the role of phase fluctuations and on the interplay between unconventional superconductivity and electronic inhomogeneity.

James A. Sauls and **Chandra M. Varma** for their works on the identification of the pairing symmetry, pairing mechanism, and multiple superconducting phases in heavy-fermion superconductors.

The 2012 Heike Kamerlingh Onnes Prize Winners:

Øystein H. Fischer for leadership in magnetic superconductors and pioneering STM studies in the cuprate high-Tc materials.

Teunis M. Klapwijk for seminal experiments on the superconducting properties of superconductor-ferromagnet and superconductor-normal metal nano structures.

Herbert A. Mook, Jr. for several decades of important neutron spectroscopy and diffraction experiments on superconductors, especially those with magnetic tendencies.

Conference Photos

Address and presentation of the US Secretary of Energy Steven Chu

Presentation by Steve Kivelson, John Bardeen Prize Winner

Journal of Physics: Conference Series **449** (2013) 011001 doi:10.1088/1742-6596/449/1/011001

Kelly Reidy (Recipient of the NSF travel award for Graduate Students) explaining her poster presentation

Presentation by Ivan Božović, Bernd T. Matthias Prize Winner

Journal of Physics: Conference Series **449** (2013) 011001 doi:10.1088/1742-6596/449/1/011001

Winners of the Bernd T. Matthias, H. Kammerling Ones, and John Bardeen Prizes:

Ivan Božović, James N. Eckstein, James A. Sauls, Chandra M. Varma, Steve A. Kivelson, Dirk Jorendt, Herbert A. Mook Jr., Teunis M. Klapwijk, Øystein H. Fischer (from left to right)

Bob Cava (Princeton, left) and Ching-Wu Chu (University of Houston) in a question and answer panel at the ICAM/I2CAM – Sponsored Evening Symposium "The Foundations of High-Temperature Superconductivity: Twenty-five Years Young" directed towards junior researchers.

Speaker Piers Coleman (Rutgers) on stage at the back and student with levitating superconductor in front at the ICAM / I2CAM – Sponsored Public Lecture "From the Quantum to the Micron: The Mind-blowing Quest to Understand Matter and Life between the Angstrom and the Micron" directed towards the general public.

Journal of Physics: Conference Series **449** (2013) 011001 doi:10.1088/1742-6596/449/1/011001

Sponsors & Supporters

Advanced Light Source, Lawrence Berkeley National Laboratory

Air Force Office of Scientific Research

Argonne National Laboratory

Brookhaven National Laboratory

Center for Emergent Superconductivity

Council of the International Union of Pure and Applied Physics

Department of Energy, *Basic Energy Science*

Elsevier, Publisher of Physica C - Superconductivity and its Applications

ICAM-12CAM Institute for Complex Adaptive Matter

Institute of Physics, Chinese Academy of Sciences

Materials Research Society

National Science Foundation

Texas Center for Superconductivity at the University of Houston

University of Geneva

University of Illinois at Urbana-Champaign, Department of Physics

University of Illinois at Urbana-Champaign, Materials Research Laboratory

Conference Co-Chair Peter Johnson with the poster display honoring various Sponsors and Contributors to the M2S 2012 Conference

Journal of Physics: Conference Series **449** (2013) 011002 doi:10.1088/1742-6596/449/1/011002

Peer review statement

All papers published in this volume of *Journal of Physics: Conference Series* have been peer reviewed through processes administered by the proceedings Editors. Reviews were conducted by expert referees to the professional and scientific standards expected of a proceedings journal published by IOP Publishing.

Journal of Physics: Conference Series **449** (2013) 012001 doi:10.1088/1742-6596/449/1/012001

Twenty-five Years of High-Temperature Superconductivity – A Personal Review

Philip W. Anderson

Department of Physics, Princeton University, Princeton, NJ 08544, USA

E-mail: pwa@princeton.edu

Abstract. I present a personal review of the last twenty-five years of research on high-temperature superconductivity.

Twenty-five Years

Yes, folks, it has been 25 years and a few months since most of us first learned of the existence of the cuprate superconductors. The materials meeting in Boston of Dec '86 was the venue for the first few of a flood of confirmatory experimental papers; and the Bangalore meeting of January 1987 was the venue where theoretical ideas were first discussed among a substantial community. Is my talk there about RVB the first ever on theory of high T_c? I don't really know.

In that rather confused talk, and the Science paper which followed in March, I made two contentions which have turned out to be extraordinarily prescient (or lucky?). The first of these was that the CuO_2 planes which are the prominent feature of this class of material would turn out to be best described by the simple Hubbard Model

$$H = \sum_{i,j,\sigma} t_{ij} c^*_{i,\sigma} c_{j,\sigma} + U \sum_i n_{i\uparrow} n_{i\downarrow} \qquad [1]$$

which is often used to model the Mott metal-insulator transition. This model and this transition were a classic unsolved problem of theoretical physics—why should not this classic problem—the doped Mott insulator, we can call it—turn out to be the source of the new, puzzling behavior? It was not just guesswork, of course—I could see how the giant Jahn-Teller distortion which creates the planes would be expected to split out the single band of the Hubbard model—but still a bit lucky. Incidentally, electron, as opposed to hole, doping, in this view, does not lead to the simple Hubbard model, but has strong polaron effects. As far as I know, all of the papers which have made any headway with the problem work with the one-band Hubbard Model, and without fear of successful contradiction I shall take this as given.

The second contention was rather more of a leap in the dark. It had two parts: one was that the pairing was due to spin interactions via the antiferromagnetic superexchange implied by equation [1]; and the second was that the superconducting pairing could grow out of a novel insulating quantum state, a liquid of pair bonds or Resonating Valence Bond (RVB) state. In the first I was certainly not original:

Scalapino and Hirsch, and Schmidt-Rink et al, had presented similar ideas over a year earlier. But they spoke in terms of "spin fluctuations" and as will be one of my main points, spin fluctuations are by definition a weak coupling phenomenon in metals, not descriptive of what's going on here.

I surely wish that I could say "and starting from this it was all obvious!" Quite the opposite: as I have described elsewhere,[1] I almost immediately jumped off of my own bandwagon and abandoned the RVB, leaving it to be followed up by others—Maurice Rice and his group,[2] Patrick Lee et al,[3] Gabby Kotliar et al,[4] and others including Bob Laughlin, aka Hiawatha [5] -- and with my considerable group I spent a decade wandering off in the wrong direction and publishing a wrong book.[6] But it WAS obvious, though it was not until 2002, 15 years into the problem, that it began to seem so to me.

At this point I shall have to abandon the mode of personal reminiscence and go into a bit of technical discussion of the Hamiltonan in equation [1]. This is so simple that it basically has only two parameters aside from the temperature, the ratio $0 < U/t < \infty$ giving the coupling strength, and the doping $0 < x < 1$, defined as 1- the number of electrons per site.

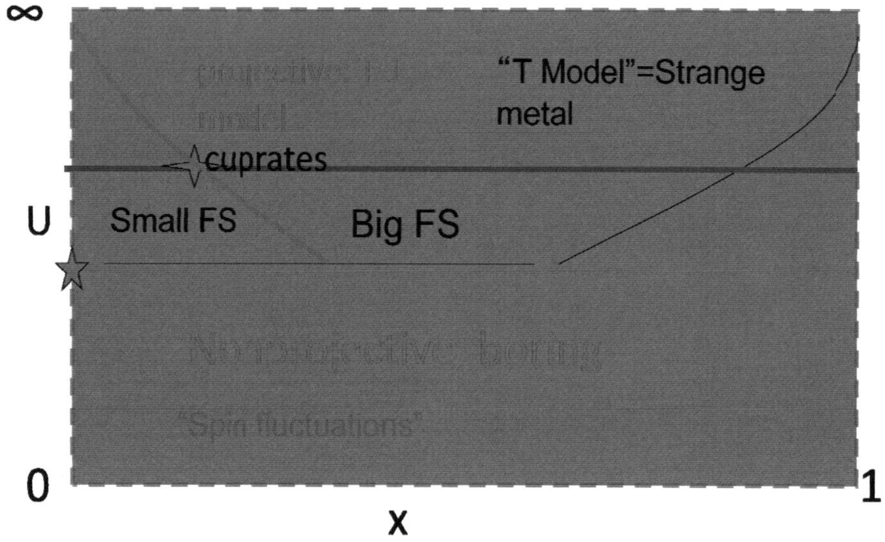

Fig 1. U vs x diagram for Hubbard model, showing the two fundamental separators: the Mott line where spin fluctuations diverge, and the Tallon separator between large and small Fermi surface physics.

In the figure 1 I plot U/t on a scale from 0 to ∞ vs x and try to delineate boundaries between broad classes of behaviors. If U is small we will just have a metal, even for small x; but as we increase U, especially at small x, the ladder diagrams for spin fluctuations, which I show in figure 2, begin to diverge, and the system can go unstable either to spin density waves or to singlet superconductivity. But it still remains a metal, especially above the transition temperature of whatever ordering happens.

For x=0, it is well-known that there is a critical U/t for the Mott transition. I like to think of the Mott transition as not a transition between states, like the superconducting transition, but between models. Above it, it no longer makes sense to start from electrons near a Fermi surface; instead one thinks of spins in a Heisenberg model. The other possible view of it is that it is where the spin fluctuations diverge: they become a cloud of bound states *outside* of the two-Fermion continuum, and acquire their own dynamics.

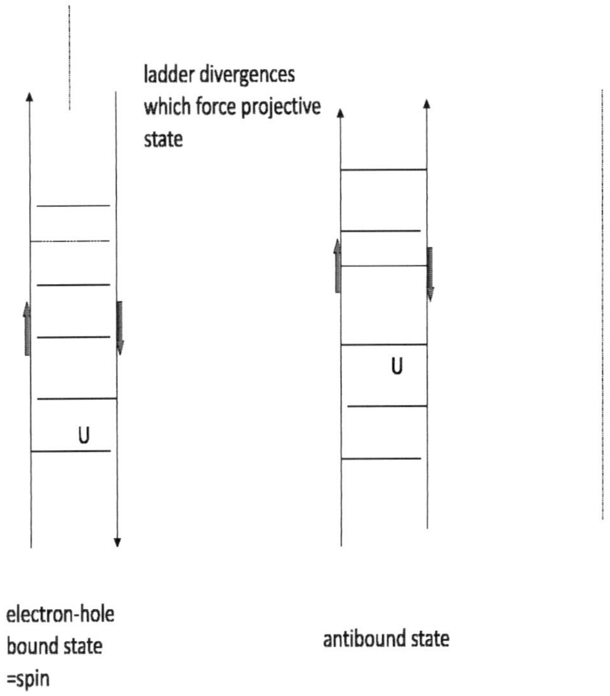

Fig 2. Spin fluctuation ladders the divergence of which occurs at the Mott line and separates off the antibound subspace.

But from that point of view this transition must extend into the continuum where x is finite; and I draw what to me seems the most likely trajectory for it in the U/t vs x space. The important point to make is that there is sound experimental evidence that the cuprates are where I have drawn them, always *above* this generalized Mott transition. Above this line, we must somehow eliminate the divergent bound states from our model, as was done formally by Walter Kohn in 1964 for the case x=0 (following my 1959 idea). What we did was to canonically transform away all of the matrix elements of the kinetic energy which led into doubly occupied hole states. This canonical transformation can be expanded in powers of 1/U, and this expansion replaces, in principle rigorously, the Hubbard model operating in the full space with a Heisenberg model acting in a space with spins only.

Maurice Rice and collaborators, as well as a number of others, generalized this transformation procedure to a derivation of the "t-J" Hamiltonian

$$\begin{aligned} H &= P\sum_{i,j,\sigma} t_{i,j} c^*_{i.\sigma} c_{j,\sigma} P + \sum_{i,j} J_{i.j} S_i \cdot S_j \\ &= T + J \quad \text{with} \\ \mathrm{P} &= \prod_i (1 - n_{i,\uparrow} n_{i\downarrow}) \end{aligned} \qquad [2]$$

P is the famous (or to some, notorious) Gutzwiller projector. But it is not here used as a calculational approximation, but as a rigorous constraint on the appropriate "effective low-energy theory" for the cuprate problem. Above the critical line in figure 1, in order to describe low-energy states of the Hubbard model one MUST confine them to a projected subspace if we are not to mix in very high-energy states into low-energy phenomena. If one does not do that one simply produces boring, misleading nonsense.

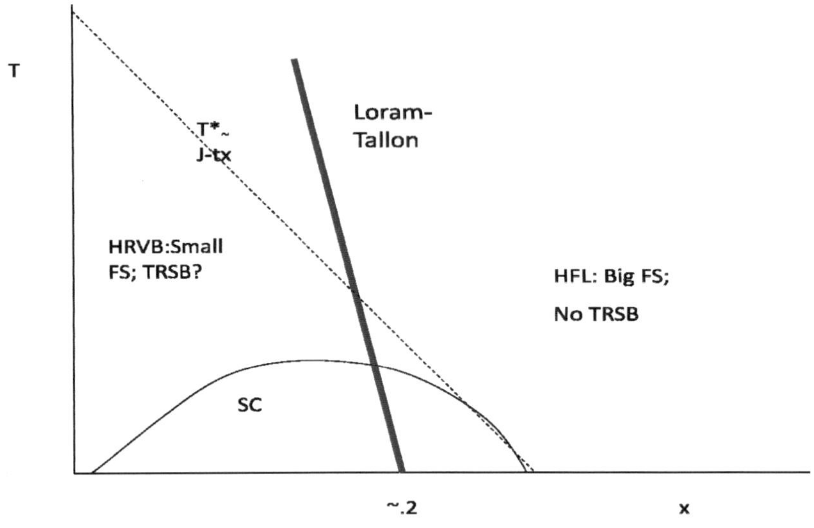

Fig 3. T – x diagram for cuprates showing Hidden FS to Hidden RVB separation at Loram-Tallon line. Also shown is T^*, where pairing occurs; a rough estimate is $T^* \sim$ J-tx from earlier argument. For the reader: is TRSB related to HRVB?

Therefore, when we confine ourselves to the cuprate line on this diagram, we are studying the states of the t-J Hamiltonian. We exhibit them in the familiar "generalized phase diagram" of the cuprates in the space of x vs T (Fig 3).

In this diagram there is again a very important boundary which is more fundamental than a mere phase transition or a critical point. To understand this, let us again look at the Mott limit x=0. In this limit the kinetic energy term T vanishes and the low-energy states are those of the Heisenberg model J. The most well-known of these is the antiferromagnet. The states of this with added holes were studied by Shraiman and Siggia in '89 and they have a Dirac-like zero at $k=\pi/2,\pi/2$. So added holes would have a small "pocket" Fermi surface around these nodes.

An alternative state which doesn't break translational symmetry is the RVB; as a number of us showed years ago this is not *competitive with* the antiferromagnet; rather the antiferromagnet can be thought of as *derivative from it*, a broken symmetry version of the same state. Therefore it is not at all surprising that it has the same Fermionic excitation spectrum for added holes, a set of four Dirac nodes at the $\pi/2,\pi/2$ points. Both of these insulating states, then, count their Luttinger's sum rules from x=0, not x=1 as one has for a simple metal. (It was Maurice Rice and collaborators who first emphasized this fact.) So in this region the sensible thing to do is to start by diagonalizing "J", and then add in "t" where the hole Fermions are in a *small* Fermi surface around or near the nodes.

On the other hand, when the doping is great enough we have all kinds of experimental evidence that there is a true Fermi surface, with area proportional to $n_h=1-x$. So for x large enough the state is based upon a *large* Fermi surface, which results from starting with T as one's H_0 and treating J as a perturbation.

So we have, as a function of x, another very fundamental boundary: between the states based upon a *large* Fermi surface, and those where the spins have already paired up at high energies and there remains only a *small* Fermi surface. This is marked as the heavy blue line in Fig 3. It is the locus of all kinds of physical changes, as was first noticed by Loram and Tallon.

The states to the right of the Loram -Tallon line have been studied in a sequence of papers by my student Casey and myself, and the situation there seems under control; I spoke about the explanation of the "strange metal", which occurs there, at the Boston APS meeting. It is not a conventional Fermi liquid, but can be related to transport properties and superconductivity in a "hidden" true Fermi liquid with a simple, sharp Fermi surface. It has little to do with critical fluctuations which are often adduced for it.

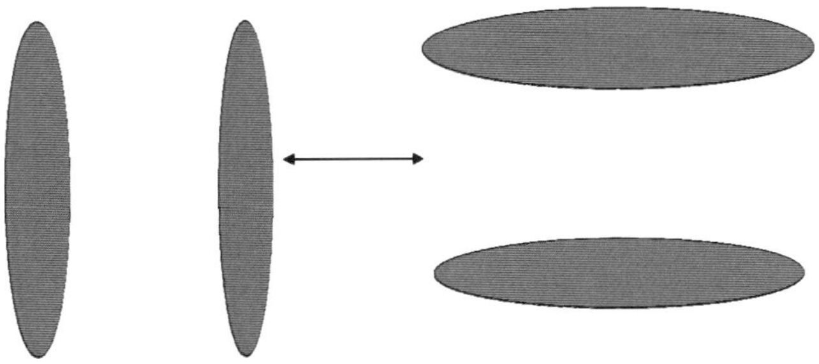

$= \text{“s+ id”} = (\cos k_x + \cos k_y) + i(\cos k_x - \cos k_y)$

Fig 4. "Portrait of an RVB" – to show extreme asymmetry of Green's function at x ~ 0. (Yazdani)

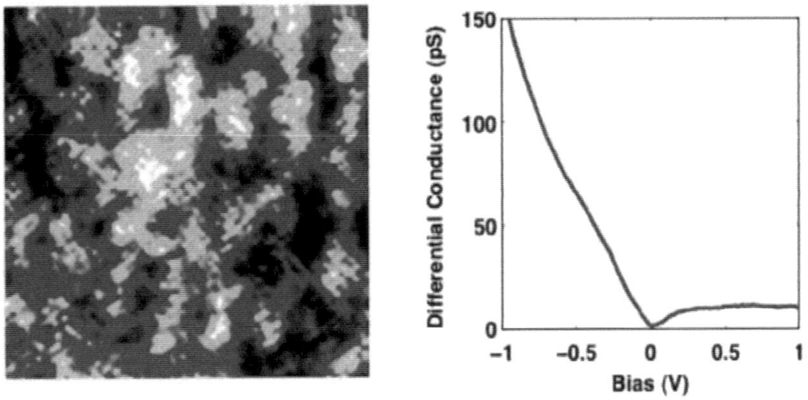

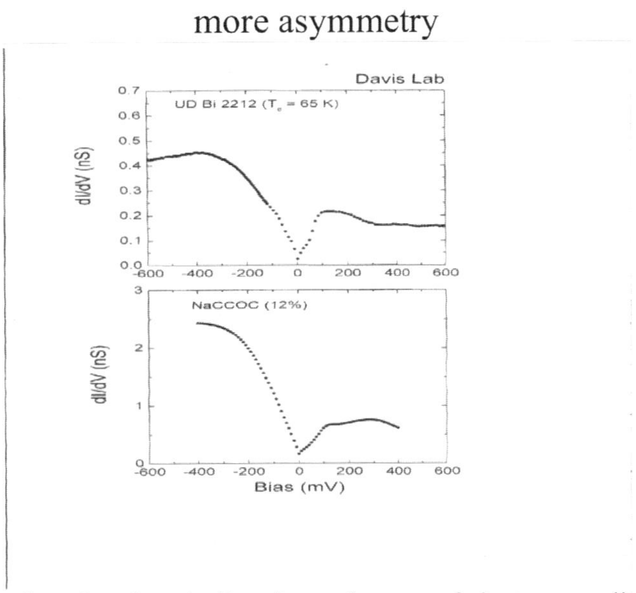

Fig 5. Early Davis tunneling data showing similar abrupt dropout of electron amplitude at end of "arc".

A second line, marked T*, is drawn in Fig 3. Empirically this is the locus of indications that spin pairing effects begin below T*. Theoretically, it agrees with the d-wave gap energy calculated in 1988 by Rice's group and by Kotliar and Liu from the t-J model, the Gutzwiller approximation and renormalized MFT.

But there are still many mysteries on the underdoped side, the famous "pseudogap" problem. Here we must first diagonalize "J", as I said, and at highish temperatures that means that the state will be an RVB: but which RVB? In the early days it was discovered, following Laughlin, that the RVB could be represented as a Gutzwiller -projected "flux phase", the one-electron state in which the wave-functions are required to experience a phase shift of π or $-\pi$ around each elementary plaquet in the square lattice. (this is completely equivalent, under projection, to my group's preference: an s+id superconductor) Some authors supposed that one could accommodate holes by simply allowing this π phase to change, but as Rice et al found, eventually, this phase is coupled to the lattice structure commensurably (they call this "umklapp") and does not move in k-space as holes are added. It is essentially an insulator because (also) at x=0 it has no electron excitations (see figure 4), only holes.

What happens at finite x? (see figure 6) If x is pinned at zero by keeping the chemical potential within the Mott gap there are only spinons, which can only be created in pairs and form a Dirac cone with its apex at ($\pi/2,\pi/2$). But as the chemical potential for holes reaches the edge of the Mott gap the spinon states become ordinary holes and the Dirac cone begins to accommodate a finite density of holes, making a small pocket of Fermi surface. If no superconductivity intervened, there would be an electron spectrum up to a finite energy E_F. But what seems to happen at low temperature is that the electrons are born paired with the holes, and that the spectrum is therefore necessarily perfectly symmetric up to E_F, where it breaks off in a cusp singularity on the electron side because all the states are gapped, but the gap retains the d-wave character of the RVB. I show you a number of instances of this behavior of the tunneling spectrum in figure 5, which has been emphasized by Ali Yazdani.

The further evolution of these pockets was recognized by Rice et al to lead to a crossover to the conventional Fermi surface; but because that group omitted the large contributions to the particle self-energies of the RVB, they made two unphysical predictions: at low doping, the pockets surround the nodes on the diagonals; and at near optimal doping, the n kinetic energy allows electron pockets, which are unphysical.

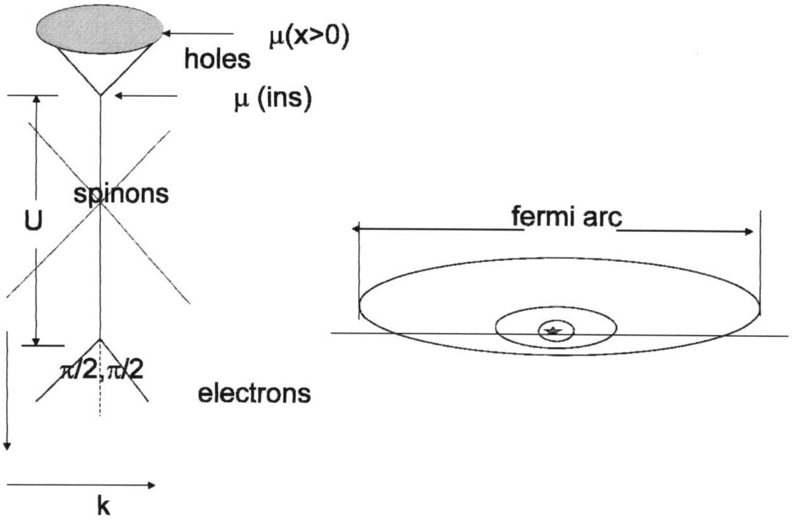

Fig 6. Sketch of energetics of "barely-doped" cuprate. Spinon Dirac cone becomes cone of hole states with, possibly, big variation of asymmetry factor. Pocket-arc Fermi surface surrounds node.

At this point the kinetic energy term begins to be finite because there are now a few empty sites for electron hopping. The Gutzwiller procedure refers *only* to the site occupancy and does not introduce in itself any correlations at all in momentum space, so that the kinetic energy, to a very good approximation, is simply renormalized by a factor g_t=2x/(1+x) proportional to the number of holes. That means that the phase rigidity of the superconductivity is very weak and the transition temperature very low (even 0 below x=.055). So the system undergoes a phase transition into a state where the electrons and holes **in the pockets <u>remain paired</u>** but lose their phase coherence. This is the "vortex liquid" phase discovered by Ong and whose limits have recently been beautifully defined by c-axis infrared spectroscopy of the Fribourg group (reported at this conference). See figure 7. This phase has an upper critical temperature where pairing in the pocket-arcs gives out.

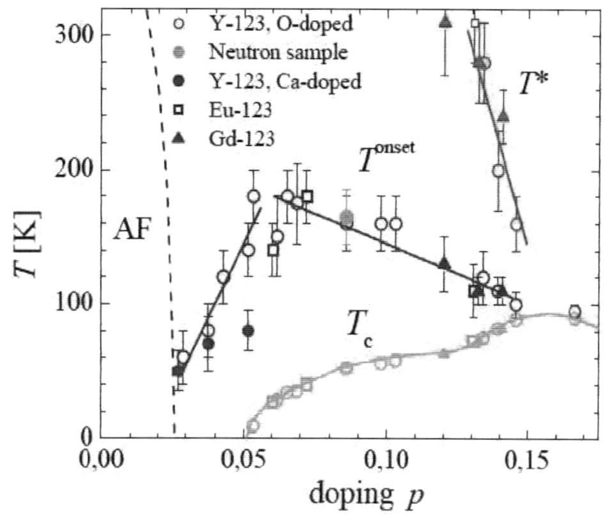

Fig 7. Delineation of 3 phases of pseudogap region: superconductivity in pocket-arcs, vortex liquid in pocket-arcs, RVB with no charge pairing. As measured by C. Bernhard group, Fribourg.

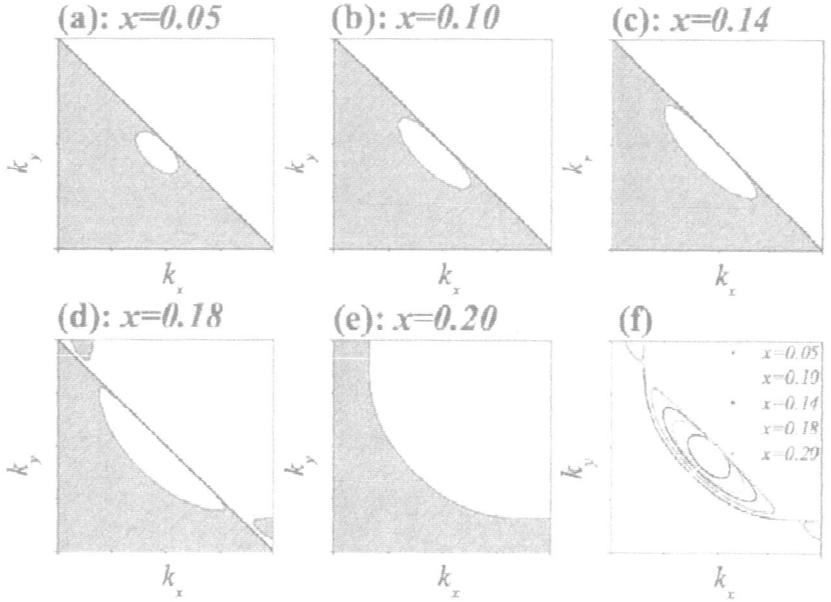

Fig 8. Pocket shapes as given by Rice et al, omitting spin pairing self energy.

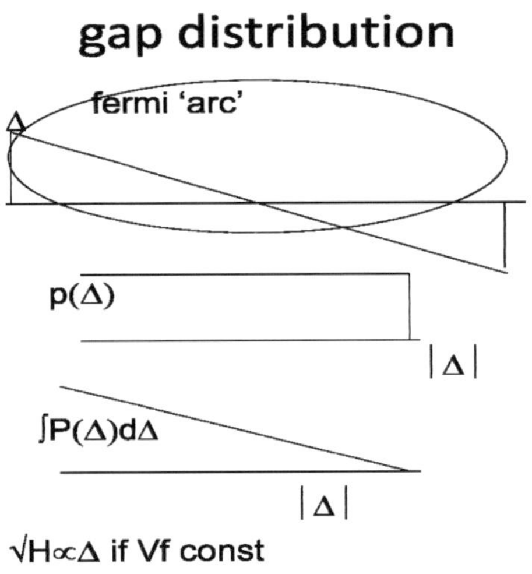

Fig 9. Gap distribution if arc gap is d-wave.

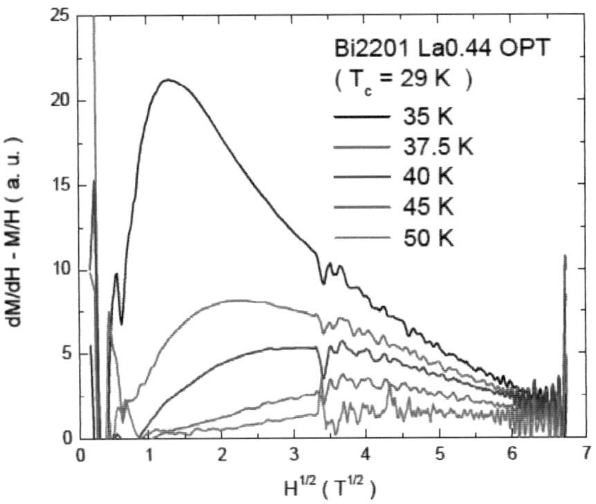

Fig 10. Gap distribution measured by nonlinear diamagnetism (Li): note amplitude T-dependent, NOT max gap.

But although this phase has an upper critical temperature because the superfluid density gives out, the gap as a function of momentum does NOT appear to be at all temperature dependent. This has been demonstrated in Lu Li's data on non-linear diamagnetism of, which can be shown to be a good measure of the gap distribution. The principle is illustrated in figure 8. We imagine the arc-pocket of Fermi surface as in the last figure, centered on the Brillouin zone diagonal where the gap is zero; so the gap as a function of k along the Fermi surface rises more or less linearly on both sides of zero as shown. If we convert that into the density of a given gap, P(Δ), that is constant as a function of delta, as shown in figure 9. Finally, we can convert gap into field by noting that the diamagnetism of electrons with a given gap disappears when $H > H_{c2}$, and H_{c2} may be defined as the field where

$$l_B = \sqrt{\Phi_0 / B} = \xi = hv_F / \Delta$$

$$\text{so} \quad \sqrt{H} \propto \Delta, \text{ since } v_F \text{ is nearly constant}$$

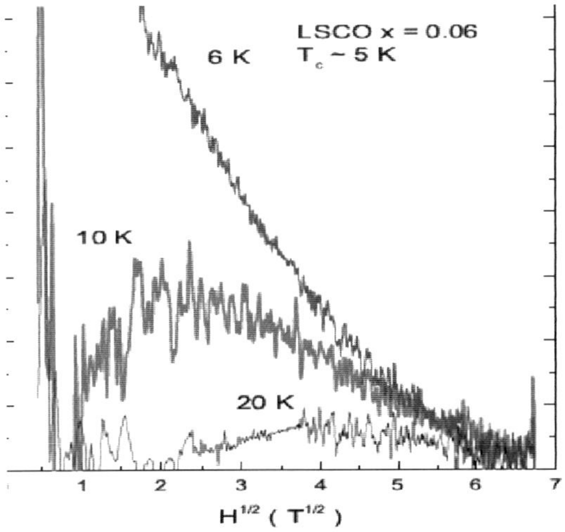

Fig 11. Same as Fig. 10 for a widely different x.

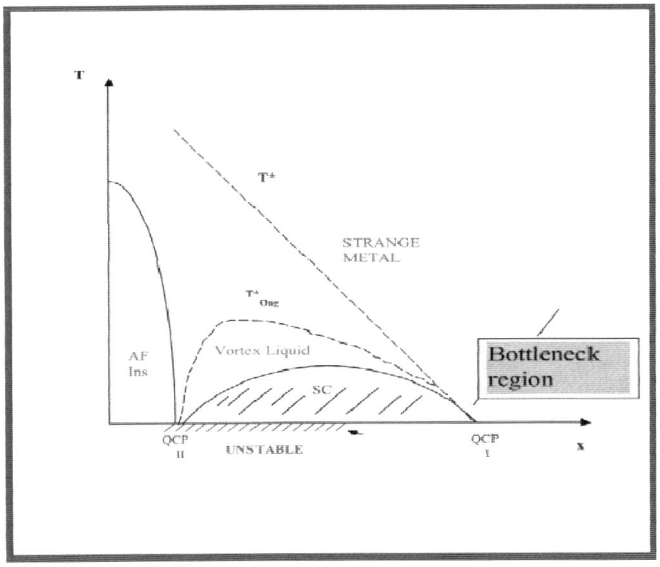

Fig 12. Summary of pseudogap physics.

A plot of ρ_s vs square root of H, then, which is available from Li's measurements for two different dopings (see figures 10 and 11) tells us the maximum value of the gap. What is noteworthy is that while the superfluid density decreases with T, the maximum field—and hence the maximum gap—does not. The gap is not self-consistently and cooperatively determined: it comes from outside the pocket-arc! We can conclude only that it is determined by the RVB spin pairing, which serves as a kind of Higgs-like mechanism to determine gaps for the real electrons in the pocket-arcs.

I present a summary in my final figure 12.

References:

[1] P. W. Anderson, Int Jl of Modern Physics B 25, 1(2011); cond-mat/1011.2736.
[2] F. C. Zhang, C. Gros, T. M. Rice and H. Shiba, J. Supercond. Sci. and Tech. 1, 36 (1988).
[3] P. A. Lee, Phys. Rev. Lett. 63, 680 (1989); P. A. Lee and X. G. Wen, , Phys. Rev. Lett. 78, 4111 (1997).
[4] G. Kotliar and L. Liu, Phys. Rev. B38, 5142 (1988).
[5] R. B. Laughlin, Annals of Improbable Research, May-June 2004, pp8-20.
[6] P. W. Anderson, "The Theory of High T_c Superconductivity", Princeton University Press, (1997).

Shape resonances in multi-condensate granular superconductors formed by networks of nanoscale-striped puddles

Antonio Bianconi

RICMASS Rome International Center for Materials Science
Superstripes, Via dei Sabelli 119A, 00185 Roma, Italy

E-mail: antonio.bianconi@ricmass.eu

Abstract. A characteristic feature of a superconductor made of multiple condensates is the possibility of the shape resonances in superconducting gaps. Shape resonances belong to class of Fano resonances in configuration interaction between open and closed scattering channels. The Shape resonances arise because of the exchange interaction, a Josephson-like term, for transfer of pairs between different condensates in different Fermi surface spots in the special cases where at least one Fermi surface is near a 2.5 Lifshitz topological transition. We show that tuning the shape resonances show first, the gap suppression (like a Fano anti-resonance) driven by configuration interaction between a BCS condensate and a BEC-like condensate, and second, the gap amplification (like a Fano resonance) driven by configuration interaction between BCS condensates in large and small Fermi surfaces. Shape resonances usually occur in granular nanoscale complex matter (called superstripes) because of the lattice instability near a 2.5 Lifshitz transition in presence of interactions. Using a new imaging method, scanning nano-X-ray diffraction, we have shown the generic formation in high temperature superconductors of a granular superconducting networks made of striped puddles formed by ordered oxygen interstitials or ordered local lattice distortions (like static short range charge density waves). In the superconducting puddles the chemical potential is tuned to a shape resonance in superconducting gaps and the maximum T_c occurs where the puddles form scale free superconducting networks.

Introduction

Understanding the mechanism that allows a quantum condensate to resist decoherence attacks of temperature is a major fundamental problem of condensed matter. The Bardeen Cooper Schrieffer wave-function [1-5] of the superconducting ground state has been constructed based on the theory of configuration interaction of all possible electron pairs (+k with spin up, and -k with spin down) on the Fermi surface in an energy window called the energy cut off of the interaction,

$$\left|\Psi_{BCS}\right\rangle = \prod_k (u_k + v_k c^+_{k\uparrow} c^+_{-k\downarrow})\left|0\right\rangle \qquad (1)$$

where $|0\rangle$ is the vacuum state, and $c^{+}_{k\uparrow}$ is the creation operator for an electron with momentum k and spin up. The many body BCS condensate with off-diagonal long range appears at the gap energy $\Delta(\kappa)$ below the Fermi level. The Schrieffer idea [5] came from the configuration interaction theory by Tomonaga involving a pion condensate around the nucleus [6]. The k-dependent structure of the interaction gives different values for the gap Δ (k) in different segments of the Fermi surface. The superfluid order parameter i.e., the superconducting gap is anisotropic in the k-space i.e., different in different locations of the k-space due different pairing strength. The k-dependent gap equation is given by

$$\Delta(\mu,k) = -\frac{1}{2N}\sum_{k'} \frac{V(k,k')\,\Delta(k'_y)}{\sqrt{(E(k')+\varepsilon_{k'}-\mu)^2+\Delta^2(k'_y)}} \qquad (2)$$

the critical temperature is given by

$$\Delta(k) = -\frac{1}{N}\sum_{k'} V(k,k') \frac{tgh(\frac{\xi(k')}{2T_c})}{2\xi(k')}\Delta(k') \qquad (3)$$

where the $\xi_n(k)=\varepsilon_n(k)-\mu$

These original BCS formulas describe the anisotropic superconductivity in the “clean limit”, where the single electron mean free path is larger than the superconducting coherence length. Real superconducting materials [7,8] have impurities and lattice disorder therefore the condition for the mean-free path $l > hv_F/\Delta_{av}$ where v_F is the Fermi velocity and Δ_{av} is the average superconducting gap was considered to be impossible to be satisfied [9,10]. In fact it is a very strict condition that implies that the impurity scattering rate $\gamma_{ab} << (1/2)(K_B/\hbar)T_c$ i.e., it should be smaller than few meV. The 40 years period from 1960 to 2000 was dominated by the *“dirty limit” dogma* [9,10] assuming an effective single Fermi surface for each superconductor. This dogma stated that all metals are in the *“dirty limit”* since impurity scattering and hybridization always mixes the wave functions of electrons on different spots in the same Fermi surface or in different Fermi surfaces where there are multiple bands crossing the Fermi level. The *“dirty limit”* dogma justifies the approximation in the theory of a k-independent V(k,k’) interaction i.e., an isotropic pairing interaction that is assumed to be a constant V_0 [9,10]. This approximation allows to derive the approximated simple universal BCS formula for the critical temperature T_C related to a single superconducting energy gap Δ_0 the energy needed to break the cooper pairs, averaged over many k-points in different bands:

$$\frac{2\Delta_0}{K_BT_C} = 3.52 \qquad (4)$$

$$\frac{T_c}{T_F} = \frac{0.36}{k_F\xi_0} \qquad (5)$$

$$K_BT_c \propto \hbar\omega_0 e^{-1/\lambda_{eff}} \qquad (6)$$

where T_F is Fermi temperature, $k_F=2\pi/\lambda_F$ is the wave-vector of electrons at the Fermi level, ξ_0 is the coherence length of the condensate, related with the size of the pair, $\hbar\omega_0$ is the energy if the boson meditating the pairing interaction, the effective coupling term $\lambda_{eff} = V_o N_{tot}$ is the product of the pairing strength V_0 and N_{tot} is the total density of states (DOS) at the Fermi level. Moreover using this approximation the BCS theory predicts that the critical temperature depends on isotope substitution with power law dependence $T_c \propto M^{-\alpha}$ where M is the atomic mass and α=0.5 is the isotope coefficient [10].

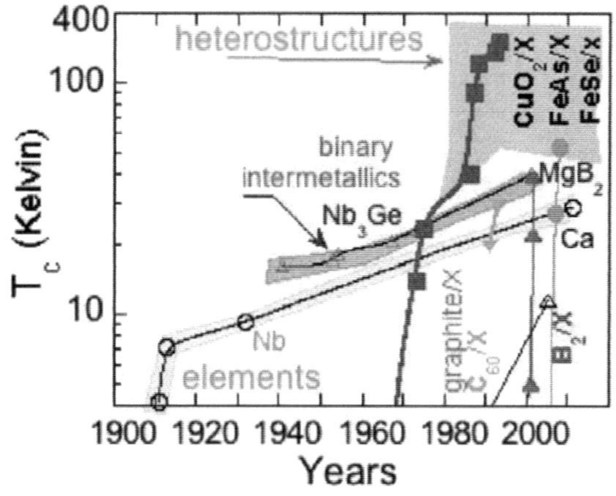

Figure 1: The evolution of the maximum superconducting critical temperature, as a function of the time of the discoveries of superconducting elemental metals (yellow), of binary intermetallics (blue) and of complex oxides (oxides).

These popular formulas are simple but they have completely washed out all interesting possible effects on the macroscopic superconducting parameters due to quantum interference effects in the configuration interaction between pairing channels in different points of the k-space that could appear in the *"clean limit"*. According with De Gennes this is not a problem since no spectacular effects have to be expected from the k-dependent V(k,k') interaction [10]. The last 40 years of the XX century can be called the period of the *"dirty limit"* dogma. The road map for the research of new high T_c materials based on the BCS approximated formulas was focused on metals showing peaks of total DOS, strong electron phonon coupling, and high energy phonon modes. The details of superconductivity in different materials have been described by the Migdal-Eliashberg approximation including the details if the attractive electron-phonon interaction and the Coulomb repulsion, keeping the assumption that the energy scale of the pairing interaction is much smaller than the Fermi energy $\hbar\omega_0 << E_F$ called the so called *"adiabatic limit"* i.e. the chemical potential in the metallic material is assumed to be far away from band-edge. The materials enter in the anti-adiabatic regime $E_F \approx \hbar\omega_0$ and the charge carriers become large or small polarons. Moving the chemical potential toward a band edge where the system approaches a metal-to-insulator transition. In the extreme antiadiabatic regime $E_F << \hbar\omega_0$ depending on the coupling one can have all possibilities from weak coupling BCS to BEC (Bose Einstein Condensate) that could be also below the band edge. Therefore near a band edge the system could be in the so called BEC-BCS crossover [11].

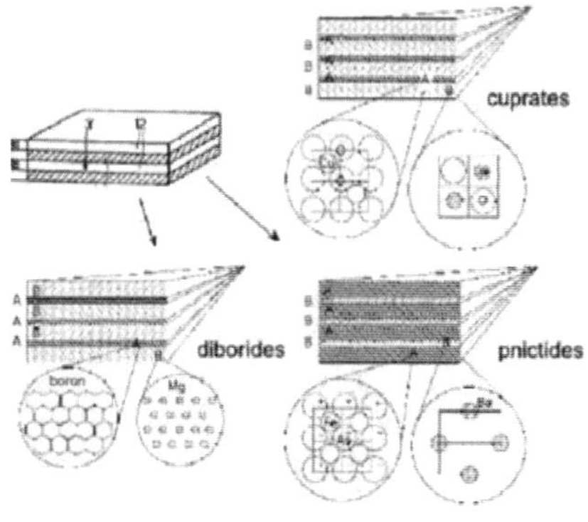

Figure 2: The lattice architecture for materials design of high temperature superconductors: superlattice of atomic layers intercalated by spacer layers.

The slow progress of materials science in the search for new high temperature superconductors in these last 100 years is shown in Fig. 1. The first discontinuity occurred by shifting from simple elemental metals to binary intermetallics in the 30's and the record was reached in metastable phases near a lattice instability. The second discontinuity in 1986 was the research shift from intermetallics to ceramics. In 1993 it was recognized that the cuprates are heterostructures at atomic limit therefore new high temperature superconductors should be made of superlattices of atomic superconducting units intercalated by spacers as the superlattice of superconducting atomic layers shown in Fig. 2 [12-15]. This theory has been confirmed in these last 20 years since all high temperature superconductors discovered so far are all made of similar heterostructures at atomic limit: diboride atomic layers [16-19], superlattices of iron based pnictides and chalcogenides layers [20-22] graphene, nanotubes [23] organic units intercalated by spacers. In this scenario the high T_c is controlled by the shape resonance [12-18] in superconducting gaps where one band is near an electronic topological transitions, that is a 2 1/2-order transition, called 2.5 Lifshitz transition [24-25].
The 2.5 Lifshitz transition was first introduced by Lifshitz in 1961 [24] and has been widely studied in solid state physics. The kinetic properties of metals at a 2.5 Lifshitz transition are well known [25-26]. The generic feature of the Fermi surface topology near a band edge in a multiband metal made of a superlattice of atomic layers is shown in Fig. 3. Two types of topological 2.5 Lifshitz transitions [18] appears by tuning the chemical potential near a band edge.
A clear practical realization of this scenario is MgB_2, a material known since 1953, where superconductivity was never measured for 46 years since no superconductivity was predicted by standard theories based on the effective single band "dirty limit" dogma but on the contrary high temperature superconductivity was predicted by the multiband theory if the chemical potential is driven near a band edge i.e., below the top of the boron $2p_{x,y}$ sigma band where it should form a multi-gap superconductor in the "clean limit" because of single electron hopping between π and σ bands are forbidden by symmetry [16-19]. The discovery of high T_c superconductivity in this material has shown that high temperature superconductivity appears where both the *"dirty limit dogma"* and the *adiabatic approximation* in BCS theory breakdown as first recognized in 2001 [16]. In fact the adiabatic condition breaks down near a 2.5 Lifshitz transition. However the anisotropic BCS theory [11] may works in the anti-adiabatic limit in multiband superconductors. This is shown by the fact that BCS seems to work pretty quantitatively for the Feshbach resonance in ultracold gases and for fullerenes despite the fact that the adiabatic condition is not at all well fulfilled.

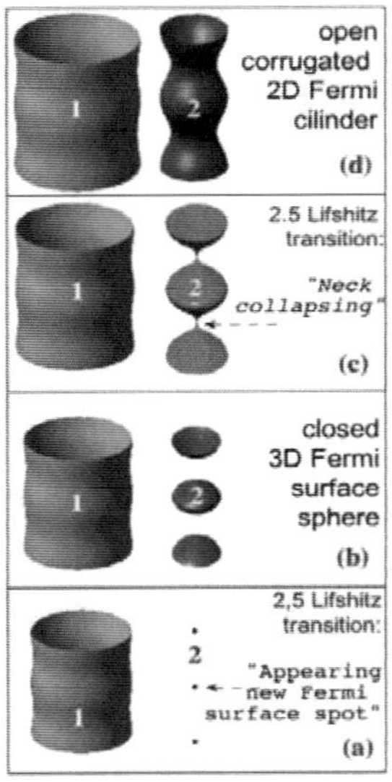

Figure 3: The evolution of Fermi surface topology in a generic two-band layered material where the chemical potential is moved near a band edge. A first type of 2.5 Lifshitz transition occurs here a new 3D Fermi surface appears (panel a). A second type of 2.5 Lifshitz transition occurs by moving further the chemical potential at the opening of a neck in a corrugated Fermi surface cylinder (panel c).

The presence of a 2.5 Lifshitz transition predicted also for cuprates in the two component scenario [27-30] was not accepted by the scientific community for many years but after 2003 many different experiments have observed 2.5 Lifshitz transitions in cuprates [31-37]. In cuprates a pseudo gap matter (probably made of polarons) where condensates should be in the antiadiabatic limit, was found to coexist with a Fermi liquid. The cuprates show beyond the insulator-to-metal transition two 2.5 Lifshitz transitions at doping 1/8 and near optimum doping as shown by the evolution of Hall effect [31] and of transport properties of the normal phase in high magnetic field [32,33], and quantum oscillations [35].
The presence of multiband superconductivity in proximity of band edges clearly appear in electron doped iron based superconductors in $Ba(Fe_{1-x}Co_x)_2As_2$ [21] and $a(Fe_{1-x}Ni_x)_2As_2$ [22] where the T_c dome occurs in the proximity of 2.5 Lifshitz transitions [22] as predicted by Innocenti et al [20].

2. The multiband anisotropic BCS theory

Multiband superconductivity [38,39] appears in anisotropic superconductivity, where the gaps are different in different large Fermi surfaces. The multiple Fermi surfaces cannot be reduced to a single effective band since the bands have different symmetry and/or are located in different spatial portions of the material. In this scenario the single particle hopping in presence of impurities and hybridization are forbidden. For example the condensate many body BCS wave-function for a two band superconductor made of a first a-band and a second b-band is given by:

$$|\Psi_{Kondo}\rangle = \prod_k (u_k + v_k a^+_{k\uparrow} a^+_{-k\downarrow}) \prod_{k'} (x_{k'} + y_{k'} b^+_{k'\uparrow} b^+_{-k'\downarrow}) |0\rangle \quad (7)$$

The term corresponding to the transfer of a pair from the "a"-band to the "b"-band and vice-versa appears with the negative sign [40] in the expression of the energy.

$$\sum_{k,k'} J(k,k')(a^+_{k\uparrow} a^+_{-k\downarrow} b_{-k\downarrow} b_{k\uparrow}) \quad (8)$$

where a^+ and b^+ are creation operators of electrons in the "a" and "b" band respectively and $J(k,k')$ is an exchange-like integral. This gain of energy is the origin of the increase of the transition temperature driven by this exchange-like [40] interaction between pairs in different points in the k-space. This is called also a Josephson-like pairing term, in fact the Josephson effect [41] describes the transfer of pairs between two different condensates in different spatial positions. This Josephson-like pairing interaction appears in the standard BCS multiband theory with a square exponent therefore it may be repulsive as it was first noticed by Kondo [40]. It is therefore different from the Cooper pairing process that is the conventional attractive intraband attraction for electron at the Fermi level in the single band BCS theory. In the case of repulsive Josephson-like interaction in multiband superconductivity the order parameter shows the sign reversal between the different bands or different points in the k-space. This well established theoretical result became popular in 2008 since it was proposed for the case of iron based superconductors with the name of $s\pm$ mechanism.

The k-dependent gap in each band n depends on the gaps in other bands in multi-condensate superconductivity

$$\Delta_n(\mu,k_y) = -\frac{1}{2N} \sum_{n' k'_y k'_x} \frac{V_{n,n'}(k,k')\,\Delta_{n'}(k'_y)}{\sqrt{(E_{n'}(k'_y) + \varepsilon_{k'_x} - \mu)^2 + \Delta^2_{n'}(k'_y)}} \quad (9)$$

where the k-dependent coupling is given by

$$V^o_{n,k_y;n',k'_y} = -V_o \int_S dx\,dy\,\psi_{n,-k}(x,y)\,\psi_{n',-k'}(x,y)\,\psi_{n,k}(x,y)\,\psi_{n',k'}(x,y) \quad (10)$$

3. Shape resonance in superconducting gaps: a type of Fano resonance.

In the standard theory of multiband superconductivity the Fermi level is assumed to far away from each band edge: i.e., the Fermi surfaces are large and intraband pairing in each band is in the adiabatic limit [38,39]. On the contrary some authors have proposed the idea of having a very small Fermi surface coexisting a large Fermi surface. The idea of having strong localized pairs coexisting with delocalized Copper pairs is quite old and goes back to the Bose-Fermi model [42-43] and the spin gap proximity effect mechanism of S. Kivelson, et al. [44].

The shape resonance model on the contrary considers the evolution of the Fermi surfaces tuning of the Fermi level from below (above) the bottom (the top) of a band to well above the

band edge of a second band with symmetry different from the large first one [46]. Therefore the shape resonance theory considers the evolution of superconductivity from the BEC like regime to the two bands BCS regime where both Fermi surfaces are large. The shape resonance theory [13-20,46] assumes that the anisotropic BCS theory with no approximations works in agreement with Leggett [11].
Spectacular effects similar to a Fano resonance [47-52] occur in the superconductivity mechanism in a multiband superconductor when the chemical potential is tuned near a band edge at a 2.5 Lifshitz transition. In this case the intraband pairing mechanism enters in the antiadiabatic limit in the particular band where the Fermi level is near the band edge.

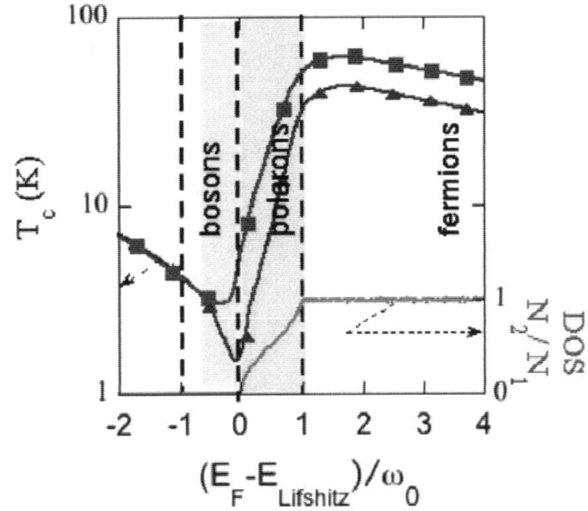

Figure 4: The critical temperature of a layered multi-band superconductor tuning the chemical potential near a 2.5 Lifshitz transition where the Fermi surface of a new appearing band evolves as shown in Fig. 3. The critical temperature in log-scale is plotted as a function of the Lifshitz energy parameter $z=(E_F-E_{Lifshitz})/\omega_0$ where ω_0 is the energy of the attractive intraband pairing interaction. The plot show two cases of Josephson-like pair exchange coupling ratio: a) c_{12}/c_{11}=-1.08 (*blue filled squares*) and c_{12}/c_{11}=-0.43 (*black filled triangles*). The colored regions indicate the BEC regime of the condensate (yellow) $-1<z<0$ and bipolaronic condensate (blue) $0<z<1$ in the new appearing Fermi surface.

The system can be treated in the BCS scheme [11] if the equation for the critical temperature near a 2.5 Lifshitz transition is solved together with the density equation as suggested by Leggett [5] taking into account that in this regime there is a large variation between the chemical potential in the superconducting phase and the normal phase

$$\Delta_n(k)=-\frac{1}{N}\sum_{n'k'}V_{n,n'}\left(k,k'\right)\frac{tgh\left(\frac{\xi_{n'}(k')}{2T_c}\right)}{2\xi_{n'}(k')}\Delta_{n'}(k') \qquad (11)$$

$$\rho = \frac{1}{S}\sum_{n}^{Nb}\sum_{k_x,k_y}\left(1 - \frac{\varepsilon_n(k_x,k_y) - \mu}{\sqrt{(\varepsilon_n(k_x,k_y) - \mu)^2 + \Delta_{n,k_y}^2}}\right) \quad (12)$$

We consider a two band superconductor where there is first large 2D cylindrical Fermi surface and the chemical potential is tuned across a 2.5 Lifshitz transition so that the Fermi surface topology of the new appearing Fermi surface evolves as shown in Fig. 2. The superconducting critical temperature as a function of the Lifshitz energy parameter $z = (E_F - E_{Lifshitz})/\omega_0$ where w_0 is the energy of the attractive intraband pairing interaction. measuring the energy distance form the energy $E_{Lifshitz}$ of the 2.5 electronic topological Lifshitz transition for the appearing of the a new Fermi surface as plotted in Fig. 4. This figure provides a typical example of shape resonance in the superconducting gaps.
The chemical potential crosses the first 2.5 Lifshitz transition, for the appearing of the new Fermi surface spot, at the value of the Lifshitz energy parameter $z = 0$, and the second 2.5 Lifshitz transition "opening a neck" at $z = 1$. The coupling in the first Fermi band is assumed to be in the standard BCS weak coupling regime $c_{11} = 0.22$. We consider here the case of strong coupling in the second band where the attractive coupling term $c22$ is about two times larger than the Cooper pairing coupling parameter in the first band $c_{22}/c_{11} = 2.17$. When the Lifshitz energy parameter is in the range $-1<z<0$ a *BEC condensate* is formed in the second band since all charges in the second band condense. At $z = 0$, we show evidence for the antiresonance due to negative interference effects between different pairing channels in the BCS condensate in the first band and the BEC in the new appearing band. This clearly shows is the quantum interference nature of shape resonances. The second curve shows the higher critical temperature for a larger c_{12} Josephson-like pair transfer value. T_c reaches the minimum value at the antiresonance for $z = 0$ in the limit of weak Josephson-like coupling, while for a strong Josephson-like coupling c_{12}, c_{21} the minimum due to the Fano-like antiresonance approach $z=-1$. Clearly we show in Fig. 4 that the shape resonance has a similar line-shape as a Fano resonance. Our minima correspond to the zero crossing which occurs near a Feshbach resonance in the ultracold Fermi gas problem.
These results provide a roadmap for material design of new room temperature superconductors with a lattice geometry that allow :

a) a multi-band superconductor in the clean limit with several bands crossing the Fermi level;
b) the different bands crossing the Fermi energy should have *different parity* and *different spatial locations* to avoid hybridization
c) the chemical potential is tuned near a 2.5 Lifshitz transition near a band edge.
d) the exchange-like or Josephson-like pair transfer integral should be as large as possible
e) a first condensate is in the adiabatic regime and a second condensate in the anti-adiabatic regime
f) The chemical potential can be tuned by pressure or gate voltage to get the Fano shape resonance in the superconducting gaps between pairing scattering channels in a BCS-BEC crossover
g) Tune the chemical potential by gate voltage tuning of using illumination or pressure so that the bipolaron condensate is converted into a BCS condensate in fact the maximum T_c occurs at a Lifshitz energy parameter z near 1.5 -2.

4. Optimum lattice inhomogeneity favoring high T_c.

The metals near a 2.5 Lifshitz transition are in a metastable phase in the verge of phase separation in the presence of interactions. It has been found that the metastability of the lattice form a network of nanoscale striped puddles of lattice, charge and orbital density waves called "superstripes" [53-56]. In each puddle the lattice and/or charge and/or spin and/or orbital 1D modulation gives multiple subbands crossing the Fermi level resulting in multiple Fermi surfaces in heterogeneous spots in the k-space.

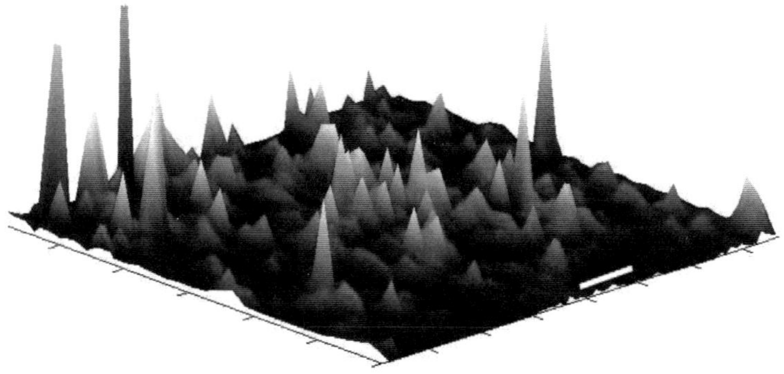

Figure 5: The landscape of nanoscale striped puddles of "local lattice distortions" in a cuprate superconductor La_2CuO_{4+y} measured by scanning nano x-ray diffraction [57] realizing the "superstripes" scenario [53-56]. The position dependence of the Q3-LLD superstructure intensity in the La_2CuO_{4+y} crystals with critical temperature, T_c, 37 K is plotted. The white bar indicates a length of 100 microns.

Indeed, in samples where a the dopants are frozen in a random distribution at a temperature lower than 200K, the local lattice distortions (LLD) detected by EXAFS XANES [29,30] and diffuse x-ray scattering show the formation of local lattice distortions that get self organized forming networks of nanoscale (10-100 nm) striped puddles [53-56] that recently have been visualized by scanning nano x-ray diffraction [57] shown in Fig. 5. The scanning XRD images show that a better self organization of LLD droplets favors higher T_c. In particular the density distribution of the nano-puddles follow a power-law distribution with a cut off. Investigating several sample it emerges that increasing the cut-off, i.e., going from an exponential to power law distribution the critical temperature increases as predicted by a recent theory for granular superconductors made of networks with power law distribution of superconducting nano grains connected with Josephson-like links.
Using a x-ray nanosized beam of 300 nm in diameter, we have obtained the imaging of the regions in La_2CuO_{4+y}, that contain incommensurate modulated local lattice distortions (LLD) shown in Fig. 5. The puddles are determined by self organization of local lattice distortions with superlattice wave-vector **q3** = 0.21 **b***+0.29 **c** therefore are called with the acronym Q3-LLD [57]. Fig. 4 shows the spatial distribution of Q3-LLD puddles for a superconducting sample. The Q3-LLD droplets form networks whose nature varies with superconducting critical temperature. We have used X-ray micro-diffraction apparatus at the ESRF to map the evolution of the Q3-LLD satellites for four single crystals of electrochemically doped La_2CuO_{4+y}, from the underdoped state to the optimum doping range, $0.06<y<0.12$.

The probability distributions, of the Q3-LLD XRD intensity for single crystals of electrochemically doped La_2CuO_{4+y} from the underdoped state to the optimum doping range, $0.06<y<0.12$ follow a power law distribution $P(x) \propto x^{-\alpha}\exp(-x/x_0)$ with a variable exponential cut-off x_0 with a constant power-law exponent $\alpha = 2.6 \pm 0.1$.
In Fig. 6 we have plotted the critical temperature of samples in the range 27-38 K associated with the droplet network of Josephson coupled nano-grains, as a function of the cut-off of the probability distribution x_0 of the intensity of x-ray intensities due to the variable density of Q3-LLD puddles. The critical temperature scales with the cut-off according to a power law with an exponent 0.4 ± 0.05. This result points again toward the importance of connectivity and an optimum inhomogeneity for high critical temperature. They are in qualitative agreement with the theoretical prediction of the increase of T_c in a granular superconductor on an annealed complex network made of Josephson coupled grain following a power law distribution with a finite cut-off [57]. In fact, for a power law distribution of links in a granular superconductor with an exponent α=2.6, the critical temperature is predicted to increase as a function of the cut-off with an exponent 3–α, as observed experimentally supporting the theory of quantum phase transitions in network of superconducting grains [58,59] and for bosonic scale free networks [60].

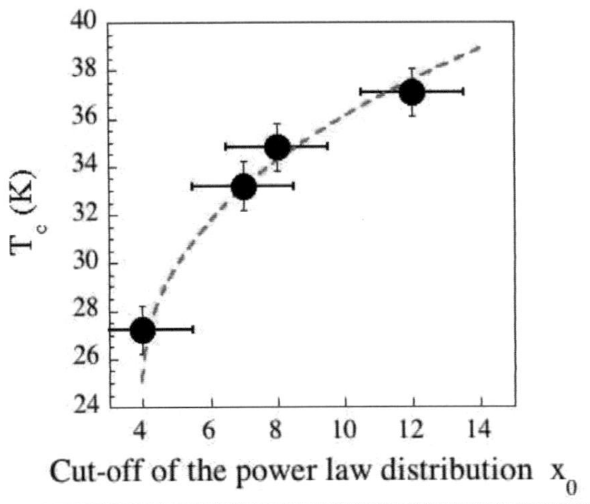

Figure 6: The critical temperature T_c in the range 25 K < T_c < 37 K for several super oxygenated La214 samples is plotted as a function of the cut-off parameter of the distribution of the LLD droplets density probed by the intensity distribution of the Q3-LLD superstructure satellites [57]. Error bars in the critical temperature are of ± 1 K. The dashed line is the fit with a power law curve with exponent 0.4 ± 0.05, in agreement with theoretical predictions in references [58-60] for granular superconductivity on a scale invariant network

In the superlattices the lattice complexity appears not only in the superconducting planes but also in the spacer layers. There the defect self organization forming striped structure will induce Fermi surface reconstruction in the superconducting layers that controls the Fermi surface topology so it can shift the 2.5 Lifshitz transitions.

The oxygen interstitials can be very inhomogeneous, even in "optimal" superconducting samples with a scale free distribution favouring high T_c as recently shown by scanning micro X-ray diffraction [61] and the oxygen interstitials self organization can be controlled by x-ray illumination [62-64].
The observed oxygen interstitals mobility is controlled by the tensile microstrain in the spacer layers due to the lattice misfit strain between the different units forming the super-lattices in Fig. 2 [65-69]. It is possible that the maximum Tc of 160K at a particular compressive 2% microstrain in cuprates is related with optimum self organization of the nanoscale puddles approaching the scale free distribution. Many authors are now reaching the conclusion that an optimum inhomogeneity in coplex matter favors high temperature superconductivity [70]. This idea is supported by the evidence that a similar nanoscale phase separation has been observed in iron based superconductors [71-72]. The emerging picture is that the first-order transition appears in the proximity of a 2.5 Lifshitz transit when the electron Coulomb interaction is switched on and recently interaction effects on the Lifshitz transitions have been systematically studied [73-75] while the topological change in the Fermi surface (called the 2.5 Lifshitz transition) was originally studied for non interacting electrons. The frustrated or arrested phase separation that appears in the experiments probing the structure fluctuations beyond the average structure point toward a key role of polaronic charge carriers in small Fermi surface [76].

5. Conclusions

We have presented the superstripes scenario for high temperature superconductors. These complex materials show k-dependent anisotropic pairing, superconductivity in the clean limit and multi-condensate superconductivity. The high T_c dome appears near a 2.5 Lifshitz transition near a band edge. The electrons in one of the multiple Fermi surfaces are in the BEC-BCS crossover regime. At list one of the borders of the high T_c dome is determined by the Fano anti-resonance of the shape resonance in the superconducting gaps where T_c drops toward zero at the 2.5 Lifshitz transition for the appearing of a new Fermi surface. At this particular regime there is a BEC-BCS crossover with coexisting BEC and BCS condensates that gives an anti-resonance in the configuration interaction controlled by the Josephson-like pair transfer term. This term gives an increasing high Tc amplification when the BEC is converted to a polaronic pair condensate and finally the maximum T_c is reached where the polaronic condensate moves toward a BCS condensate.
The materials near a 2.5 Lifshitz transition are in a metastable phase due to lattice instability and show networks of striped superconducting grains. It seem to us that the maximum T_c is reached at the particular condition where in each grain with a striped structure the critical temperature is amplified by the shape resonance in the superconducting gaps and the grains form a scale free network where T_c is pushed to the highest possible temperature.

Acknowledgments: The author thanks Tony Leggett, Lev Gorkov, Andrea Perali, Carlos Sa de Melo, Klim Kugel, Masatoshi Imada for useful discussions.

References.

1. Schrieffer JR 1964 *Theory of Superconductivity*, (for Josephson-like or exchange-like pair transfer see 300 pp.)
2. Blatt JM 1964 *Theory of Superconductivity* (Academic Press Inc,New York) for shape resonances see 362 pp and 215 pp
3. Leggett AJ 1980 *Modern Trends in the Theory of Condensed Matter* edited by A. Pekalski and R. Przystawa, Lecture Notes in Physics Vol. 115 (Springer-Verlag, Berlin).

4. Leggett A 1989 in *The New Physics* P. Davies ed., Cambridge University Press pp. 268-288
5. Schrieffer JR 1992 *Phys. Today* **45**, 46.
6. Tomonaga S 1947 Prog. Theo. Phys. (Kyoto) **2**, 6 .
7. Garland JW Jr 1963 *Phys. Rev. Lett.* **11** 111.
8. Geballe TH 1964 *Rev. Mod. Phys.* **36** 134.
9. Anderson PW 1959 *J Phys Chem Solids* **11** 26.
10. De Gennes PG 1966 Superconductivity of Metals and Alloys (Benjamin, Inc. New York, Amsterdam)
11. Leggett, A. J. 1980 Modern Trends in the Theory of Condensed Matter, edited by Pekalski A. and Przystawa R., Lecture Notes in Physics, Springer-Verlag, Berlin,Vol. 115, p. 13.
12. Bianconi A., 1994 Solid State Communications 89, 933.
13. Perali A, Bianconi A, Lanzara A, Saini N.L., 1996 Solid. State Commun. 100, 181.
14. Bianconi, A., Valletta, A.. Perali, A, Saini N.L. 1997 Solid. State Commun. 102, 369. DOI: 10.1016/S0038-1098(97)00011
15. Bianconi A, Valletta A, Perali A, Saini N L,1998 Physica C 296, 269. URL http://dx.doi.org/10.1016/S0921-4534 97 01825-X
16. Bianconi A, Di Castro D, Agrestini S, Campi G, Saini NL, Saccone A, De Negri S and Giovannini M 2001 J. Phys.: Condens. Matter **13**, 7383
17. Bussmann-Holder, A. and Bianconi, A. 2003 *Physical Review B* 67, 132509. doi:10.1103/PhysRevB.67.132509
18. Bianconi A, 2005 *Journal of Superconductivity* 18, 625.
19. Innocenti D, Poccia N, Ricci A, Valletta A, Caprara S, Perali A, Bianconi A 2**010** *Phys. Rev. B 82*, 184528.
20. Innocenti D, Caprara S, Poccia N, Ricci A, Valletta A, and Bianconi A, 2011 *Supercond. Sci. Technol.* **24**, 015012
21. Liu C et al 2011 *Phys. Rev. B* **84,** R020509
22. Ideta S et al 2012. arXiv:1205.1889
23. Bianconi A. 2006. *Phys. Stat. Sol. (a)* 203, 2950.
24. Lifshitz IM 1960. *Sov. Phys. JEPT* **11** 1130.
25. Varlamov A A, Egorov V S, and Pantsulaya A V 1989. *Advances in Physics* 38, 469.
26. Blanter Y M, Kaganov M I, Pantsulaya A V, and Varlamov A A 1994. *Physics Reports* 245, 159.
27. Bianconi A. and Missori M. 1994 *Solid State Communications* 91, 287.
28. Bianconi A., Di Castro, D., Bianconi, G., Pifferi, A., Saini, N. L., Chou, F. C., Johnston, D. C., and Colapietro M. 2000 *Physica C: Superconductivity* 341-348,1719.
29. Bianconi A, Saini N L, Lanzara A, Missori M, Rossetti T, Oyanagi H, Yamaguchi H, Oka K and Ito T 1996 *Phys. Rev. Lett.* **76,** 3412.
30. Bianconi A, Saini NL, Rossetti T, Lanzara A, Perali A, and Missori M 1996 *Phys. Rev. B* **54**, 12018
31. Balakirev F.F. et al. 2003 Nature 424, 913 doi:10.1038/nature01890.
32. Balakirev F. F. , Betts J. B., Boebinger G. S., Tsukada I., and Ando Y. 2006 New J. Phys 8, 194.
33. Balakirev F. F., Betts J. B. , Migliori A., Tsukada I., Ando Y., and G. S. Boebinger 2009 Phys. Rev. Lett. 102, 017004. URL http://dx.doi.org/10.1103/PhysRevLett.102.017004.
34. Laliberte F., Chang J., Doiron-Leyraud N., Hassinger E., Daou R., Rondeau M., Ramshaw B. J., Liang R., Bonn D. A., Hardy W. N., Pyon S., Takayama T., Takagi H., Sheikin I., Malone L., Proust C., Behnia K., and Taillefer 2011 L. *Nature Communications* 2, 432.
35. Ovchinnikov S G, Korshunov M. M., Shneyder E. I. 2009. Journal of Experimental and Theoretical Physics 109, 775. URL http://dx.doi.org/10.1134/S1063776109110077
36. Norman M R, Lin J., Millis A. J. 2010 Phys. Rev. B 81, 180513.
37. Perali A, Innocenti D, Valletta A and Bianconi A., 2012 Supercond. Sci. Technol. 25, 124002.
38. Moskalenko VA 1959 *Phys. Met. and Metallog.* **8**, 25.
39. Suhl H, Matthias BT and Walker LR 1959 *Phys. Rev. Lett.* **3** 552.
40. Kondo J 1963 *Prog. Theor. Phys.* **29** 1.
41. Josephson BD 1962 *Phys. Lett.* **1** 251.
42. Robaszkiewicz S, Micnas R, and Ranninger J. Phys. Rev. B 36, 180 (1987)
43. Micnas, R. Robaskiewicz S. in *"Theory Aspects of High-Temperature Superconductivity"* Proceedings of the NATO Advanced Study Institute, Delphi, Greece, 19-31 August 1996 Series: Nato Science Series E: Vol. 343, Kaldis, E.; Liarokapis, E.; Müller, K.A. (Eds.) 1997, 596 p. 61.
44. Emery V. J., Kivelson S. A., and Zachar O. 1997 *Physical Review B* 56, 6120.
45. Bussmann-Holder A., Micnas R., Robaszkiewicz S. 2005. in "Superconductivity in Complex Systems" Series: Structure and Bonding, Vol. 114 Müller, Karl Alexander; Bussmann-Holder, Annette (Eds.), XII, 394 p. 210.
46. Blatt JM and Thompson CJ 1963 *Phys. Rev. Lett.* **10** 332.
47. Fano U. , 1935 Il Nuovo Cimento (1924-1942) **12**, 154 URL http://dx.doi.org/10.1007/BF02958288.

48. Fano U, 1961, *Phys. Rev.* **124**, 1866
49. Bianconi A., 2003. Ugo Fano and shape resonances. *AIP Conference Proceedings* 652,13-18.
50. Vittorini-Orgeas, A. and Bianconi, A. 2009. *Journal of Superconductivity and Novel Magnetism* 22, 215.
51. Miroshnichenko A E, Flach S, and Kivshar Y S, 2010 *Reviews of Modern Physics* 82, 2257.
52. Poccia N, Ansuini A, and Bianconi A, 2011 *International Journal of Molecular Sciences* 12, 6810.
53. Bianconi A, 2000. *International Journal of Modern Physics B* 14, 3289.
54. Bianconi A., Di Castro, D., Saini, N. L., and Bianconi, G. 2002. *in* M. F. Thorpe and J. C. Phillips (eds.), Phase Transitions and Self-Organization in Electronic and Molecular Networks, Fundamental Materials Research, chapter 24. pp. 375-388. Kluwer Academic Publishers, Boston.
55. Bianconi A. and Poccia N., 2012. *Journal of Superconductivity and Novel Magnetism* 25,1403-1412.
56. Di-Castro D, Colapietro M, Bianconi G, 2000 *Int. J. Mod. Phys.* **14**, 3438
57. Poccia N, Ricci A, Campi G, Fratini M, Puri A, Di Gioacchino D, Marcelli A, Reynolds M, Burghammer M, Saini NL, Aeppli G, Bianconi A 2012 *Proc. Nat. Acad.* Sci USA **109**, 15685, URL http://dx.doi.org/10.1073/pnas.1208492109.
58. Bianconi G, 2012 *Physical Review E,* ***85***, 061113.
59. Bianconi G, 2012. *Journal of Statistical Mechanics: Theory and Experiment* **2012**, P07021
60. Halu A, Ferretti L, Vezzani A and Bianconi G 2012 *EPL (Europhysics Letters)* **99**,18001
61. Fratini M, Poccia N, Ricci A, Campi G, Burghammer M, Aeppli G, Bianconi A 2010 *Nature* **466**, 841
62. Poccia N, Fratini M, Ricci A, Campi G, Barba L, Vittorini-Orgeas A, Bianconi G, Aeppli G, Bianconi A, 2011 *Nature Materials* **10**, 733.
63. Poccia, N., Bianconi, A., Campi, G., Fratini, M., and Ricci, A. 2012. *Superconductor Science and Technology* 25, 124004
64. Campi, G., Castro, D. D., Bianconi, G., Agrestini, S., Saini, N. L., Oyanagi, H., and Bianconi, A. 2002. Photo-Induced phase transition to a striped polaron crystal in cuprates. *Phase Transitions* 75:927-933.
65. Bianconi A, Agrestini S, Bianconi G, et al. 2002 in "Stripes and Related Phenomena" Book Series: Selected Topics in Superconductivity, Vol. 8, pp 9-25, Springer doi:10.1007/0-306-47100-0_2 (Conference: 2nd International Conference on Stripes and High T_c Superconductivity (STRIPES 1998) Rome, Italy Date: Jun 02-06, 1998).
66. Di Castro D, Bianconi G, Colapietro M, et al. 2000 European Physical Journal B 18, 617 DOI: 10.1007/s100510070010
67. Bianconi, A., Saini, N. L., Agrestini, S., Di Castro, D., and Bianconi, G. 2000. *International Journal of Modern Physics B* 14, 3342. DOI:10.1142/S0217979200003812 (Conference: 3rd International Conference on Stripes and High Tc Superconductivity (STRIPES 2000) Location: Univ Rome La Sapienza, Rome, Italy Date: Sep 25-30, 2000)
68. Bianconi A, Agrestini, S., Bianconi, G., Di Castro, D., and Saini, N. L. 2001. *Journal of Alloys and Compounds* 317-318, 537. DOI: 10.1016/S0925-8388(00)01383-9
69. Garcia J, Bianconi, A, Benfatto M., and Natoli C. R., 1986. *Le Journal de Physique Colloques* 47, C8-49. DOI: 10.1051/jphyscol:1986807
70. Müller K A and Bussmann-Holder A 2005, Superconductivity in complex systems (Springer Berlin)
71. Ricci A, Poccia N, Campi G, Joseph B, Arrighetti G, Barba L, Reynolds M, Burghammer M, Takeya H, Mizuguchi Y, TakanoY, Colapietro M, Saini N., Bianconi A, 2011 *Phys. Rev. B* **84**, 060511.
72. Ricci A., Poccia N., Joseph B., Arrighetti G, Barba L., Plaisier J., Campi G., Mizuguchi Y., Takeya H., Takano Y., Saini N L, and Bianconi A, 2011. *Superconductor Science and Technology* 24, 082002.
73. Imada M, Misawa T and Yamaji Y., 2010 J. Phys.: Condens. Matter 22, 164206.
74. Yamaji Y, Misawa T and Imada M 2006 *J. Phys. Soc. Japan* 75, 094719
75. Kugel, K. I., Rakhmanov, A. L., Sboychakov, A. O., Poccia, N., and Bianconi, A. 2008. *Physical Review B* 78,165124.
76. Innocenti, D., Ricci, A., Poccia, N., Campi, G., Fratini, M., and Bianconi, A. 2009. *Journal of Superconductivity and Novel Magnetism* 22, 529.

In search for the pairing glue in cuprates by non-equilibrium optical spectroscopy

F Cilento[1], S Dal Conte[2,3], G Coslovich[4], F Banfi[2,3], G Ferrini[2,3], H Eisaki[5], M Greven[6], A Damascelli[7,8], D van der Marel[9], F Parmigiani[1,10] and C Giannetti[2,3]

[1]Sincrotrone Trieste S.C.p.A., Basovizza I-34012, Italy.
[2]I-LAMP (Interdisciplinary Laboratories for Advanced Materials Physics), Università Cattolica del Sacro Cuore, Brescia I-25121, Italy.
[3]Department of Mathematics and Physics, Università Cattolica del Sacro Cuore, Brescia I-25121, Italy.
[4]Materials Sciences Division, Lawrence Berkeley National Laboratory, Berkeley, CA 94720, USA.
[5]Nanoelectronics Research Institute, National Institute of Advanced Industrial Science and Technology, Tsukuba, Ibaraki 305-8568, Japan
[6]School of Physics and Astronomy, University of Minnesota, Minneapolis, MN 55455, USA.
[7]Department of Physics and Astronomy, University of British Columbia, Vancouver, BC V6T 1Z1, Canada.
[8]Quantum Matter Institute, University of British Columbia, Vancouver, BC V6T 1Z4, Canada.
[9]Département de Physique de la Matière Condensée, Université de Genève, Genève CH1211, Switzerland.
[10]Department of Physics, Università degli Studi di Trieste, Trieste I-34127, Italy.

E-mail: `claudio.giannetti@unicatt.it`

Abstract. In strongly correlated materials the electronic and optical properties are significantly affected by the coupling of fermionic quasiparticles to different degrees of freedom, such as lattice vibrations and bosonic excitations of electronic origin. Broadband ultrafast spectroscopy [1, 2] is emerging as the premier technique to unravel the subtle interplay between quasiparticles and electronic or phononic collective excitations, by their different characteristic timescales and spectral responses. By investigating the femtosecond dynamics of the optical properties of $Bi_2Sr_2Ca_{0.92}Y_{0.08}Cu_2O_{8+\delta}$ (Y-Bi2212) crystals over the 0.5-2 eV energy range, we disentangle the electronic and phononic contributions to the generalized electron-boson Eliashberg function [3, 4], showing that the spectral distribution of the electronic excitations, such as spin fluctuations and current loops, and the strength of their interaction with quasiparticles can account for the high critical temperature of the superconducting phase transition [5]. Finally, we discuss how the use of this technique can be extended to the underdoped region of the phase diagram of cuprates, in which a pseudogap in the quasiparticle density of states opens.

The microscopic modeling of the interaction of ultrashort light pulses with unconventional superconductors will be one of the key challenges of the next-years materials science, eventually leading to the full understanding of the role of the electronic correlations in controlling the dynamics on the femtosecond timescale.

1. Introduction

The generalized bosonic function

In conventional metals the scattering between quasiparticles (fermions) and lattice vibrations (bosons) is the microscopic mechanism that determines the transport and the optical properties. At low temperatures the electron-phonon interaction, denoted by $\alpha^2F(\Omega)$, provides the "glue" for the formation of the Cooper pairs, leading to the instability of the Fermi-liquid ground state upon the formation of the superconducting condensate. The critical temperature of the superconducting phase transition, that is of the order of a few degrees, is determined by the electron-phonon coupling constant, i.e. $\lambda_{e-ph}=2\int \alpha^2F(\Omega)/\Omega\, d\Omega$, through the McMillan's formula [6].

In the cuprate superconductors, the electronic correlations are responsible for the emergence of complex orders, influencing the phase diagram even far from the antiferromagnetic insulating phase at zero doping. These novel degrees of freedom, like paramagnon excitations [7], charge fluctuations [8] or loop currents [9] provide additional scattering channels to quasiparticles (QPs), strongly affecting their lifetime and dispersion over an energy range of the same order of the spectral distribution of the excitations. An "effective" description of the generic interaction between quasiparticles and excitations of both electronic and phononic nature, whose distribution at a given temperature follows the Bose-Einstein statistics, can be obtained by replacing the $\alpha^2F(\Omega)$ with a more general electron-boson coupling function [4], i.e., the *Bosonic Function* $\Pi(\Omega)$, defined as:

$$\Pi(\Omega) \equiv \alpha^2F(\Omega) + I^2\chi(\Omega) \tag{1}$$

where $I^2\chi(\Omega)$ accounts for the coupling with all the bosonic excitations of electronic origin.

The signatures of the QP-boson interaction usually manifest in most of the experiments that probe the electronic properties at equilibrium. The generalized bosonic function $\Pi(\Omega)$ can be extracted [4] by analyzing the kinks in angle-resolved photoemission data [10], the dip features in tunneling spectra [11, 12], the frequency-dependent scattering rate in optical spectroscopy [13, 3] and Raman spectra [14]. $\Pi(\Omega)$ shows some ubiquitous features, independent of the technique used, like a strong peak at 50-80 meV and a broad continuum that extends up to 300-400 meV. Although a main effort has been recently made [4], the determination of the relative weight of the electronic and phononic contributions to $\Pi(\Omega)$ remains elusive, since the electronic and phononic excitations coexist on the 0-90 meV energy scale. Solving this problem would constitute a step forward to unravel the puzzle of superconductivity in cuprates, while addressing the major question whether or not high-temperature superconductivity can be described in terms of a generalized Eliashberg formalism, in which the attractive interaction is 'retarded', i.e., mediated by virtual bosonic excitations of novel origin [15].

Non-equilibrium time-resolved spectroscopies are emerging as new fundamental tools to tackle these fundamental questions [1, 2] and directly probe the energy exchange between QPs and the coupled bosons, on their intrinsic timescales. The founding concepts of these techniques rely on the use of an ultrashort light pulse (pump) to prepare the system in a non-equilibrium state, i.e., with the distribution of fermionic QPs decoupled from the distribution of bosonic excitations, and a delayed probe that takes snapshots of the dynamics of the physical properties during the recovery of the equilibrium state. In the last years, many techniques have been developed to directly probe the different degrees of freedom relevant to the relaxation of photoexcited QPs. Time-resolved photoemission, which probes in real time the change of the QPs distribution with **k**-space resolution, has been used [16, 17] to measure the relaxation time of the non-equilibrium population and demonstrate the strong and selective coupling of QPs with a subset of phonons, most likely breathing and buckling modes. More recently, time-resolved electron diffraction, which is sensitive to the lattice displacement, has been applied [18] to directly investigate the structural dynamics of cuprates and demonstrate an anisotropic coupling with buckling lattice modes, already effective on the 50 fs timescale.

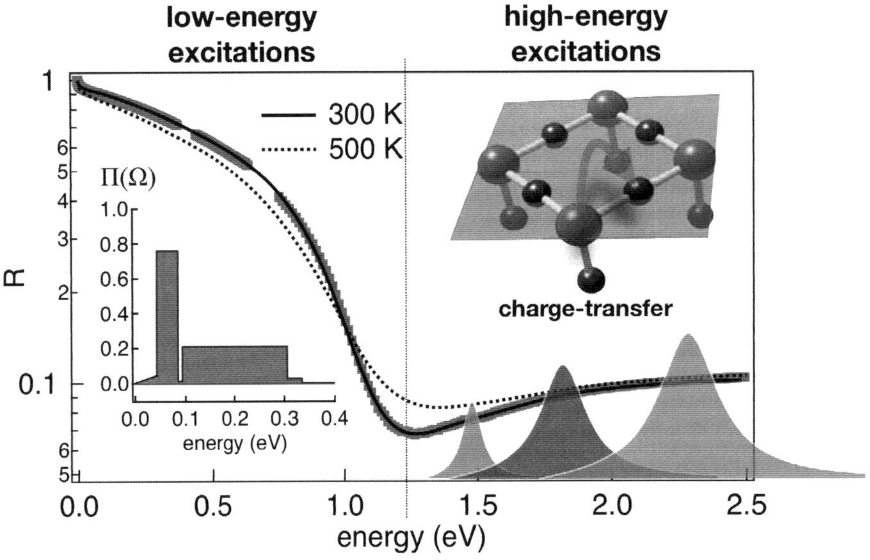

Figure 1. Reflectivity of the $Bi_2Sr_2Ca_{0.92}Y_{0.08}Cu_2O_{8+\delta}$ crystal at optimal doping (T_c=96 K), as measured by conventional spectroscopic ellipsometry (red dots). The reflectivity is well reproduced by an extended Drude model and a sum of Lorentz oscillators (black line). The total bosonic glue, $\Pi(\Omega)$, extracted from the data is shown in the left inset. The dashed black line represents the reflectivity calculated for the same system at T=500 K.

Among the different time-resolved techniques, we will focus on time-resolved optical spectroscopy, which probes the evolution of the dielectric function of the system after the excitation with a short light pulse. Although **k**-space-averaged, optics can be used to monitor the transient QPs scattering rate that is related to the coupling with bosons of electronic and phononic origin. In general, the pump-induced transient changes in the optical properties can be modeled by a two-steps process: i) the sudden photo-injection of fermionic QPs results in an effective increase of the plasma frequency of the Drude peak in the optical conductivity, without any change of the QPs scattering rate; ii) the subsequent heating of the bosonic excitations induces an increase of the scattering rate and a broadening of the Drude peak, without any change of the plasma frequency. Since changes in the plasma frequency and in the scattering rate affect the optical conductivity in qualitatively different ways, optical spectroscopy with femtosecond time-resolution can be used to probe the temporal evolution of the distribution of fermionic QPs and bosonic excitations, as a function of the delay from the pump pulse. Furthermore, the dynamics of bosonic excitations of electronic and phononic nature can be disentangled on the basis of their different timescales while exchanging energy with QPs.

In this work, we will briefly introduce the basics of equilibrium and non-equilibrium optical spectroscopy, we will review the main results on the determination of the relative electronic and phononic contributions to $\Pi(\Omega)$ and, finally, we will show how this technique can be extended to investigate the opening of a pseudogap in the quasiparticle density of states and possible anomalies in the temperature dependance of the QP self-energy.

2. Optical spectroscopy at equilibrium

Optical spectroscopy at equilibrium is a fundamental tool to investigate the electronic properties of strongly correlated materials [19] and unconventional superconductors [20]. Since the dielectric function contains direct information about the scattering rate of QPs and their dynamical effective mass, optical spectroscopies are intrinsically sensitive to the coupling of QPs to bosonic excitations.

2.1. The optical properties of a prototypical cuprate

In Figure 1 we report the *ab*-plane reflectivity of optimally-doped $Bi_2Sr_2Ca_{0.92}Y_{0.08}Cu_2O_{8+\delta}$ (Y-Bi2212) high-quality crystals [21] (T_c=96 K), measured by conventional spectroscopic ellipsometry [22] at 300 K. The normal-incidence reflectivity $R(\omega,T)$ is related to the dielectric function $\epsilon(\omega,T)$ by the relation:

$$R(\omega,T)=\left|\frac{1-\sqrt{\epsilon(\omega,T)}}{1+\sqrt{\epsilon(\omega,T)}}\right|^2 \tag{2}$$

The dielectric function is obtained by applying the Kramers-Kronig relations to the reflectivity for 6 meV$<\hbar\omega<$0.74 eV and directly from ellipsometry for 0.2 eV$<\hbar\omega<$4.5 eV. This combination allows a very accurate determination of $\epsilon(\omega)$ in the entire combined frequency range.

The reflectivity of Y-Bi2212 shows some general features common to most of cuprates:
i) below 1.25 eV the optical properties are dominated by a broad peak related to the optical response of the low-energy excitations in the conduction band. Interestingly, the dielectric function in this energy range cannot be simply reproduced by a Drude peak, in which a constant scattering rate τ and effective mass m^* of the QPs is assumed. The strongly frequency-dependent scattering rate and effective mass, resulting from the interaction with bosonic excitations, will be accounted for by the more general extended Drude model [20], presented in the next section;
ii) above 1.25 eV, the high-energy interband transitions dominate. In this energy range, the equilibrium dielectric function can be modeled as a sum of Lorentz oscillators at ~1.5, 2, 2.7 and 3.9 eV [2]. The attribution of these interband transitions in cuprates is a subject of intense debate. The ubiquitous charge-transfer (CT) gap edge (hole from the upper Hubbard band with $d_{x^2-y^2}$ symmetry to the O-$2p_{x,y}$ orbitals) in the undoped compounds is about 2 eV [20]. Upon doping, a structure reminiscent of the CT gap moves to higher energies, while the gap is filled with new transitions. This trend has been recently reproduced by Dynamical Mean Field Theory (DMFT) calculations of the electron spectral function and of the *ab*-plane optical conductivity for the hole-doped three-band Hubbard model [23]. The structures appearing in the dielectric function at 1-2 eV, that is, below the remnant of the CT gap at 2.5-3 eV, are possibly related to transitions between many-body Cu-O states at binding energies as high as 2 eV (for example singlet states) and states at the Fermi energy;
iii) the dressed plasma frequency, $\bar{\omega_p}$, defined through the relation $\mathrm{Re}\{\epsilon(\bar{\omega_p},T)\}=0$, is ~1 eV. In this energy range, the signatures of the pump-induced change of the QP distribution and of the heating of bosons can be more effectively disentangled, as will be discussed in Section 3.

2.2. The extended Drude model

In the extended Drude model (EDM) the physical processes responsible for the renormalization of the lifetime and effective mass of the QPs are accounted for in a phenomenological way, by replacing the frequency-independent scattering time τ with a complex temperature- and frequency-dependent scattering time $\tau(\omega,T)$:

$$\tau^{-1}\Rightarrow\tilde{\tau}^{-1}(\omega)=\tau^{-1}(\omega)-i\omega\tilde{\lambda}(\omega)=-iM(\omega,T) \tag{3}$$

where $1+\tilde{\lambda}(\omega)=\frac{m^*}{m}(\omega)$ is the mass renormalization of the QPs due to many-body interactions and $M(\omega,T)$ is the memory function.
In the EDM, the dielectric function $\epsilon_D(\omega,T)$ is given by:

$$\epsilon_D(\omega,T)=1-\frac{{\omega_p}^2}{\omega(\omega+M(\omega,T))}=1-\frac{{\omega_p}^2}{\omega(\omega(1+\tilde{\lambda}(\omega,T))+i/\tau(\omega,T))} \tag{4}$$

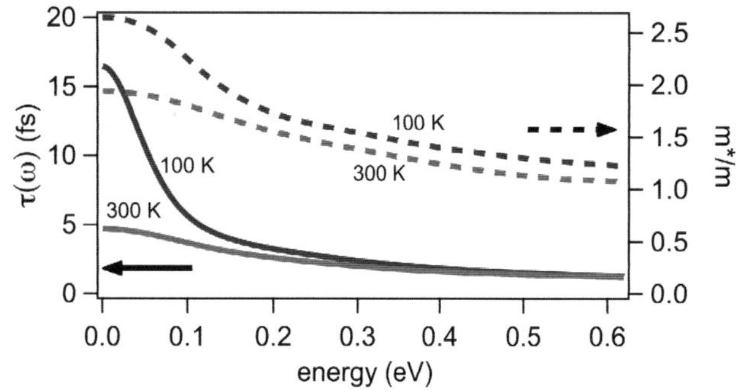

Figure 2. Frequency-dependent scattering rate (left axis) at 300 K (red line) and 100 K (blue line) and effective mass (right axis) at 300 K (red dashed line) and 100 K (blue dashed line) of the fermionic quasiparticles.

while the Drude optical conductivity $\sigma_D(\omega,T)$=$[1-\epsilon_D(\omega,T)]i\omega/4\pi$ reads:

$$\sigma_D(\omega,T) = \frac{i}{4\pi}\frac{{\omega_p}^2}{\omega + M(\omega,T)} = \frac{1}{4\pi}\frac{{\omega_p}^2}{1/\tau(\omega,T) - i\omega(1+\tilde{\lambda}(\omega,T))} \tag{5}$$

The renormalized scattering rate and effective mass can be directly extracted from the measured Drude optical conductivity, through the relations:

$$1/\tau(\omega,T) = \frac{\omega_p^2}{4\pi}\mathrm{Re}\left(\frac{1}{\sigma_D(\omega,T)}\right) \tag{6}$$

$$1+\tilde{\lambda}(\omega,T) = -\frac{\omega_p^2}{4\pi}\frac{1}{\omega}\mathrm{Im}\left(\frac{1}{\sigma_D(\omega,T)}\right) \tag{7}$$

Although this phenomenological version of the EDM does not provide any clue about the microscopic mechanisms responsible for the renormalization of the energy dispersion and lifetime of the QPs, it is very useful for directly extracting $\tau(\omega,T)$ and $m^*(\omega,T)/m$ from the optical data. In Figure 2, we report $\tau(\omega,T)$ (solid lines, left axis) and $m^*(\omega,T)/m$ (dashed lines, right axis) obtained by applying Eqs. 6-7 to the low-energy part of $\sigma_D(\omega,T)$, measured on optimally-doped Y-Bi2212 samples. At T=300 K (red lines), the value of the QP scattering time is ~5 fs at very low frequencies and decreases towards an asymptotic value of ~2 fs above 0.5 eV. These values of $\tau(\omega,T)$ imply that QPs scatters very quickly and exchange energy with bosons on the very femtosecond timescale. As the temperature decreases, the opening of the pseudogap prevents QPs from being scattered and an increase of the scattering time below 100 meV is measured, while the value of $\tau(\omega,T)$ above 100 meV is almost unaltered. A similar behavior is measured for the effective mass, that decreases towards the m^*/m=1 asymptotic value above 0.5 eV.

2.3. The extraction of the bosonic function from the optical conductivity

From the microscopic point of view, the Extended Drude formalism can be derived from the Holstein theory for normal metals [24]. Considering the Kubo formula and using complex diagrammatic techniques to evaluate the electron and boson thermal Green's functions and omitting vertex corrections (Migdal approximation), the Memory function $M(\omega,T)$ results:

$$M(\omega,T) = \omega\left\{\int_{-\infty}^{+\infty}\frac{f(\xi,T)-f(\xi+\omega,T)}{\omega+\Sigma^*(\xi,T)-\Sigma(\xi+\omega,T)+i\gamma_{\mathrm{imp}}}d\xi\right\}^{-1} - \omega \tag{8}$$

where f is the Fermi-Dirac distribution, $\Sigma(\omega,T)$ and $\Sigma^*(\omega,T)$ the electron and hole **k**-space averaged self-energies and γ_{imp} an intrinsic decay rate that accounts for the scattering by impurities. We pinpoint that, although the memory function $M(\omega, T)$ has the same analytical properties of the single-particle self-energy $\Sigma(\omega, T)$, it has a conceptually different meaning, since the optical transition at frequency ω involves a particle-hole excitation of the many-body system and provides information about the joint particle-hole density of states.

The electron self-energy $\Sigma(\omega, T)$ can be calculated as a convolution integral between the bosonic function $\Pi(\Omega)$ and a kernel function $L(\omega, \Omega, T)$:

$$\Sigma(\omega, T) = \int_0^\infty \Pi(\Omega) L(\omega, \Omega, T) d\Omega \tag{9}$$

The kernel function

$$L(\omega, \Omega, T) = \int \left[\frac{n(\Omega', T) + f(\Omega, T)}{\Omega - \omega + \Omega' + i\delta} + \frac{1 + n(\Omega', T) - f(\Omega, T)}{\Omega - \omega - \Omega' - i\delta} \right] d\Omega' \tag{10}$$

accounts for the distribution of the bosonic excitations through the Bose-Einstein distribution $n(\Omega, T)$, and can be calculated analytically:

$$L(\omega, \Omega; T_e, T_b) = -2\pi i \left[n(\Omega, T_b) + \frac{1}{2} \right] + \Psi\left(\frac{1}{2} + i\frac{\Omega - \omega}{2\pi T_e}\right) - \Psi\left(\frac{1}{2} - i\frac{\Omega + \omega}{2\pi T_e}\right) \tag{11}$$

where Ψ are digamma functions and the dependence of the different terms on the temperatures of the electronic QPs (T_e) and bosonic excitations (T_b) has been made explicit. In this formalism, the frequency-dependent scattering rate is a consequence of the microscopic interaction of QPs with a distribution of bosons at temperature T_b.

Close to optimal doping, the vertex corrections beyond Eliashberg theory can be reliably neglected and the EDM can be safely used to extract $\Pi(\Omega)$ from the optical conductivity, measured at the equilibrium temperature T. Although sophisticated maximum entropy techniques [13] have been developed to unveil the rich details of the bosonic function, the main features can be evidenced by simply fitting the model in (5) to the experimental reflectivity $R(\omega, T)$, reported in Figure 1, and assuming a simple histogram form for $\Pi(\Omega)$. The resulting bosonic function is reported in the left inset of Figure 1 and is characterized by:
i) a low-energy part (up to 40 meV) linearly increasing with the frequency. This part is compatible with either the coupling of QPs to acoustic [25] and Raman-active optical [26] phonons or the linear susceptibility expected for a Fermi liquid [27];
ii) a narrow, intense peak centered at ~60 meV, attributed to the anisotropic coupling to either out-of-plane buckling and in-plane breathing Cu-O optical modes [28] or bosonic excitations of electronic origin such as spin fluctuations [29];
iii) a broad continuum extending up to 350 meV, well above the characteristic phonon cutoff frequency (~90 meV) and usually attributed to the coupling with spin fluctuations [30, 31, 32, 7] or loop currents [33].

Using a sum of the EDM and four Lorentz oscillators at ~1.5, 2, 2.7 and 3.9 eV, the reflectivity of optimally-doped Y-Bi2212 can be satisfactorily reproduced up to 2.5 eV photon energy (see the black solid line in Figure 1). In the EDM the main role of the temperature is to change the density of the bosons and, as a consequence, the scattering rate of QPs. In Figure 1 we report $R(\omega, T)$ calculated for T=500 K (dashed line). The T-related increase of the scattering rate induces a broadening of the Drude peak, resulting in a decrease of the reflectivity below $\bar{\omega_p}$=1 eV and an increase of the reflectivity above $\bar{\omega_p}$.

Although very useful to determine the microscopic origin of the frequency-dependent electronic properties and capable of reproducing most of the features of the temperature-dependance of the dielectric function, the EDM does not allow to evaluate the relative contributions of the phononic and electronic contributions to the total $\Pi(\omega)$, since they spectrally coexist below 90 meV.

3. Non-equilibrium optical spectroscopy

Non-equilibrium optical spectroscopy permits to monitor the change of the dielectric function and, in particular, of the QPs scattering rate, during the relaxation process after the interaction with an ultrashort optical light pulse. Adding the temporal (t) dimension to the frequency (ω) dimension will be the key to disentangle the phononic and electronic contributions to $\Pi(\omega)$. A sketch of the technique is reported in Figure 3. A 100 fs infrared optical pulse (1.5 eV photon energy) is used to excite the sample. The dynamics of the dielectric function is probed by combining two ultrafast optical techniques: i) supercontinuum light generation in a photonic fiber and simultaneous detection of the 1-2 eV spectral region through a linear array, after dispersion in a prism; ii) Optical Parametric Amplification providing tunable and short (<100 fs) output pulses in the 0.5-1.1 eV range. The details of the experiment can be found in Refs. [1, 2, 5].

3.1. The extended Drude model out of equilibrium

During the interaction with the pump pulse, two physical processes are expected to take place on the sub-picosecond timescale:
-High-energy fermionic excitations are impulsively injected by the pump pulse. Since the lifetime of these excitations is very short, they rapidly relax toward the Fermi energy, producing a high-density of QPs at lower excitation energies. As far as no pseudogap is present, which would provide constraints for the QPs scattering processes, we assume that the excess of QPs can be described through an effective increase δT_e of the equilibrium electronic temperature T_e.
-The temperature T_b of the distribution of the bosonic excitations shows a δT_b increase because of the QP-boson coupling.
By noting that the Kernel function $L(\omega, \Omega; T_e, T_b)$ is the sum of separate terms depending on the electronic and bosonic temperatures, it is possible to use the EDM described in Section 2.3 to calculate the changes in the optical properties expected at t=0, i.e., soon after the excitation, in the two limits: a) Non-thermal scenario, i.e., $\delta T_e \gg \delta T_b$, in which the QPs are almost completely decoupled from the bosons. b) Quasi-thermal scenario, i.e., $\delta T_e \simeq \delta T_b$, in which the QPs and bosons are in thermal equilibrium at an effective temperature larger than the equilibrium temperature.

To facilitate the comparison with the experimental outcomes, we calculated the relative variation of the reflectivity at t=0, defined as the relative difference of the reflectivities out of equilibrium (R_{oeq}) and at equilibrium (R_{eq}):

$$\frac{\delta R}{R}(\omega, T) = \frac{R_{oeq}(\omega, T) - R_{eq}(\omega, T)}{R_{eq}(\omega, T)} = \frac{R(\omega, T_e + \delta T_e, T_b + \delta T_b) - R(\omega, T_e, T_b)}{R(\omega, T_e, T_b)} \tag{12}$$

The results are reported in Figure 4. In the case $\delta T_e \gg \delta T_b$ (yellow solid line), the variation of the reflectivity shows a positive, intense feature just below $\bar{\omega_p}$=1 eV, while it is negligible at higher energies. This behavior can be rationalized in terms of a small increase of the plasma frequency without any change in the scattering rate $\tau(\omega)$. For the case $\delta T_e \simeq \delta T_b$ (black solid line), the reflectivity variation is dramatically different, since it is dominated by the increase of the density of the bosons and of the QPs scattering rate, as evidenced by the broadening of the Drude peak. In the quasi-thermal scenario, $\delta R/R(\omega, T)$ turns from positive to negative at $\bar{\omega_p}$.

These results open the way for probing the dynamics of the dielectric function over a wide energy range, above and below $\bar{\omega_p}$, allowing to discriminate between the non-thermal ($\delta T_e \gg \delta T_b$) and quasi-thermal ($\delta T_e \simeq \delta T_b$) scenarios and directly measure the time necessary for the heating of the bosonic excitations.

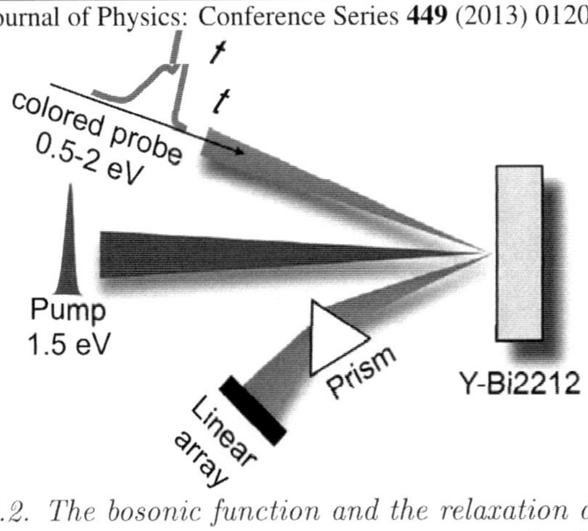

Figure 3. Sketch of the experimental setup for the 1.5 eV-pump and colored-probe optical measurements.

3.2. The bosonic function and the relaxation dynamics

The key point to analyze the time-resolved data is that the same bosonic function $\Pi(\Omega)$ obtained from the extended Drude model (see Eq. 9) controls the temporal dynamics of the energy exchange between QPs and bosons, as pointed out by P. Allen in Ref. [34]. This model, introduced for simple metals and referenced as the "effective-temperature model", can be extended to capture the more complex physics of cuprates, including the strong coupling with some Cu-O optical phonons. Formally, the total bosonic function can be written as $\Pi(\Omega)=I^2\chi(\Omega)+\alpha^2F(\Omega)_{SCP}+\alpha^2F(\Omega)_{lat}$, where $I^2\chi(\Omega)$ refers to the bosonic excitations of electronic origin at the effective temperature $T_{\rm be}$, $\alpha^2F(\Omega)_{SCP}$ to the small fraction of strongly-coupled phonons (SCPs), e.g., buckling and breathing Cu-O optical modes, at $T_{\rm SCP}$[16] and $\alpha^2F(\Omega)_{lat}$ to all other lattice vibrations at T_{lat}, including acoustic and Raman-active optical phonons.

A set of four coupled differential equations can be used to represent the following physical processes: a short laser pulse, with power density (absorbed) p, impulsively increases the effective electronic temperature of the QPs with a specific heat $C_e=\gamma_eT_e$ ($\gamma_e=\pi^2N_cN(\epsilon_F)k_b^2/3$, N_c being the number of cells in the sample and $N(E_F)$ the density of states of both spins per unit cell). T_e will then relax through the energy exchange with all the coupled degrees of freedom that linearly contribute to the total $\Pi(\Omega)=I^2\chi(\Omega)+\alpha^2F(\Omega)_{SCP}+\alpha^2F(\Omega)_{lat}$. The rate of the energy exchange among the different populations is given by [34]:

$$\frac{\partial T_e}{\partial t} = \frac{G(I^2\chi,T_{be},T_e)}{\gamma_eT_e}+\frac{G(\alpha^2F_{SCP},T_{SCP},T_e)}{\gamma_eT_e}+\frac{G(\alpha^2F_{lat},T_{lat},T_e)}{\gamma_eT_e}+\frac{p}{\gamma_eT_e} \quad (13)$$

$$\frac{\partial T_{be}}{\partial t} = -\frac{G(I^2\chi,T_{be},T_e)}{C_{be}} \quad (14)$$

$$\frac{\partial T_{SCP}}{\partial t} = -\frac{G(\alpha^2F_{SCP},T_{SCP},T_e)}{C_{SCP}} \quad (15)$$

$$\frac{\partial T_{lat}}{\partial t} = -\frac{G(\alpha^2F_{lat},T_{lat},T_e)}{C_{lat}} \quad (16)$$

where

$$G(\Pi_i,T_i,T_e)=\frac{6\gamma_e}{\pi\hbar k_b^2}\int_0^\infty d\Omega\Pi_i(\Omega)\Omega^2[n(\Omega,T_i)-n(\Omega,T_e)] \quad (17)$$

with $\Pi_i=I^2\chi$, α^2F_{SCP}, α^2F_{lat} and $n(\Omega,T_i)=(e^{\Omega/k_BT_i}-1)^{-1}$ the Bose-Einstein distribution at the temperatures $T_i=T_{be}$, T_{SCP}, T_{lat}. The specific heat (C_{SCP}) of SCPs is proportional to their density of states and is taken as a fraction f of the total specific heat, i.e., $C_{SCP}=fC_{lat}$.

Combining the four temperature model (4TM, Eqs. 13-17) with the extended Drude model (Eqs. 5,8,9,11), provides a full picture of the dynamics of $\delta R/R(\omega,T_{be},T_{SCP},T_{lat})$, in which $T_{be}(t)$, $T_{SCP}(t)$ and $T_{lat}(t)$ are functions of the time variable t and their dynamics is controlled by the three different ratios $G(\Pi_{\rm i},T_{\rm i},T_{\rm e})/C_{\rm i}$, where $C_{\rm i}=C_{\rm be}$, $C_{\rm SCP}$ and $C_{\rm lat}$.

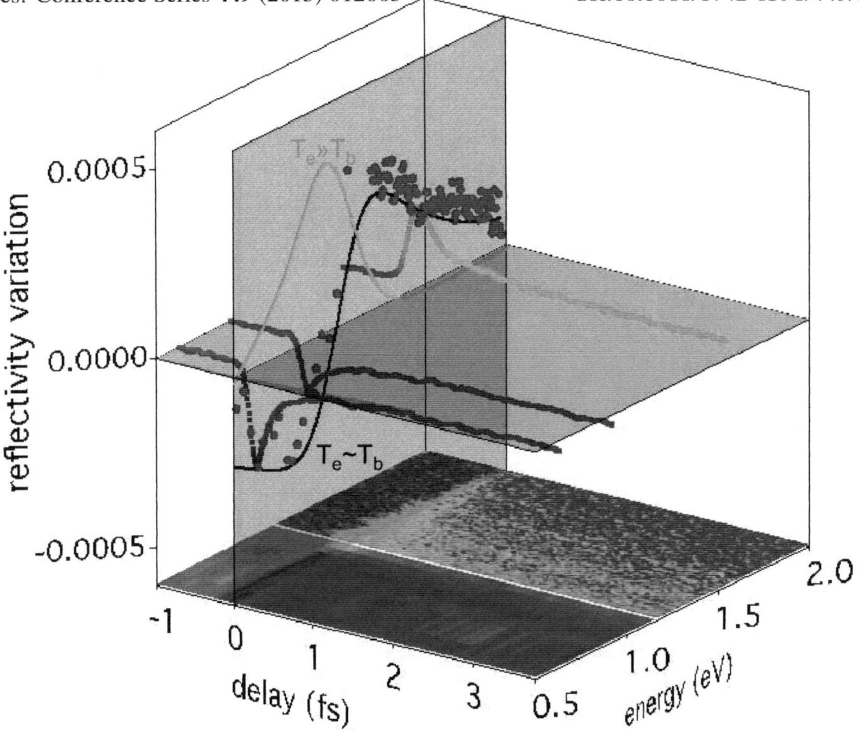

Figure 4. Relative reflectivity variation, i.e. $\delta R/R(\omega,t)$=($R_{exc}(\omega,t)$-$R_{eq}(\omega)$)/$R_{eq}(\omega)$, as a function of the probe photon energy and delay between the pump and probe pulses. The data have been taken on an optimally-doped Y-Bi2212 sample at room temperature. For further details, see Ref. [5]. The maximum $\delta R/R(\omega,t)$ at t=0 is calculated in the quasi-thermal (black line; $T_e \simeq T_b$) and non- thermal (yellow line; $T_e \gg T_b$) scenarios, using the parameters obtained from the fit to the equilibrium measurements. The red dots are the $\delta R/R(\omega,t$=0) measured by time-resolved optical spectroscopy.

4. Disentangling the electronic and phononic contributions to the bosonic function

Figure 4 reports the $\delta R/R(\omega,t)$, as measured through non-equilibrium spectroscopy on optimally-doped Y-Bi2212 samples at T=300 K, versus time (t, x-axis) and frequency (ω, y-axis). Some worthwhile features emerge from the data.

Already on the very short timescale (<100 fs), $\delta R/R(\omega,t)$ is negative below 1 eV and positive above 1 eV. The red dots are the values of $\delta R/R(\omega,t=0)$, i.e., the maximum reflectivity variation versus the probe wavelength. The data are well reproduced by the reflectivity variation calculated in quasi-thermal conditions ($\delta T_e \simeq \delta T_b$) through the EDM. It is worth noting that no signature of a change in the plasma frequency, characteristic of the non-thermal scenario ($\delta T_e \gg \delta T_b$), is detected within the time resolution of the experiment. Hence, it is possible to conclude that, already on the 100 fs timescale, the QPs are thermalized with some bosonic excitations participating to $\Pi(\Omega)$. The fast timescale of this thermalization implies a very large coupling and a relatively small specific heat. This finding strongly suggests that this process involves bosonic excitations of electronic origin and is consistent with the scattering time $\tau(\omega)\sim$1-5 fs estimated from the equilibrium optical conductivity measurements reported in Figure 2.

The temporal dynamics at fixed wavelength, i.e., $\delta R/R(t)$, is characterized by two decay times ($\sim$200 fs and $\sim$1 ps), universally measured in time-resolved experiments [35, 36]. The faster dynamics is related to the coupling of QPs to SCPs, controlled by the $\alpha^2 F_{SCP}$ part of the bosonic function and by C_{SCP}, while the slower dynamics is attributed to the coupling of QPs to all other lattice modes and it is controlled by $\alpha^2 F_{lat}$ and C_{lat}.

A quantitative analysis of the data can be carried out by using the combination of the four temperature model (4TM, Eqs. 13-18) and of the extended Drude model (Eqs. 5,8,9,11), previously discussed. The different contributions to $\Pi(\Omega)$ are extracted by fitting the calculated $\delta R/R(\omega,t)$ to the time- and frequency-resolved reflectivity data, reported in Figure 4. Considering that the energy distribution of phonons is limited to <90 meV, we assume that, for Ω>90 meV, $\Pi(\Omega)\simeq I^2\chi(\Omega)$. Within this assumption, the functional dependence of $\delta R/R(\omega,t)$ on $I^2\chi(\Omega)$, $\alpha^2F(\Omega)_{SCP}$ and $\alpha^2F(\Omega)_{lat}$ is simplified as a parametric dependence on three coefficients p_i, where $I^2\chi(\Omega)=p_1\Pi(\Omega<90\text{ meV})+\Pi(\Omega>90\text{ meV})$, $\alpha^2F(\Omega)_{SCP}=p_2\Pi(\Omega<90\text{ meV})$ and $\alpha^2F(\Omega)_{lat}=p_3\Pi(\Omega<90\text{ meV})$. Considering the constraint that $\Pi(\Omega)=I^2\chi(\Omega)+\alpha^2F(\Omega)_{SCP}+\alpha^2F(\Omega)_{lat}$ ($\Pi(\Omega)$ being the total glue function extracted from equilibrium optical spectroscopy and reported in Figure 1) and fixing the values $C_e/T_e=\gamma_e=10^{-4}$ $\text{Jcm}^{-3}\text{K}^{-2}$ and C_{lat}=2.27 $\text{Jcm}^{-3}\text{K}^{-1}$, the parameters phase-space of the model is significantly narrowed, allowing to unambiguously haul out the different contributions to $\Pi(\Omega)$ and to estimate C_{be} and C_{SCP}. For further details, see Ref. [5].

The analysis of the $\delta R/R(\omega,t)$ demonstrates that the entire high-energy part and ~46% of the peak at ~60 meV instantaneously thermalize with the QPs at a temperature $T_{\text{be}}\simeq T_{\text{e}}$. The spectral distribution and the estimated value of the specific heat of these excitations ($C_{\text{be}}<0.1C_{\text{e}}$) support their electronic origin. The $\alpha^2F(\Omega)_{SCP}$ is estimated to be ~34% of the peak at ~60 meV, corresponding to ~17% of the total $\Pi(\Omega)$, while $\alpha^2F(\Omega)_{lat}$ provides ~20% of the peak.

The measured values of the QPs-boson couplings can be used to estimate the upper bound for T_c, expected for each subset of the bosonic function. The maximal critical temperatures attainable are calculated assuming that $\Pi_b(\Omega)$ entirely contributes to the d-wave pairing and fixing the pseudopotential $\mu^*=0$. In the strong-coupling formalism, the critical temperature for d-wave pairing in a Fermi liquid with $\mu^*=0$, is approximately given by [6]:

$$T_c = 0.83\tilde{\Omega}\exp[-1.04(1+\lambda_i)/g\lambda_i] \tag{18}$$

where $\ln\tilde{\Omega}=2/\lambda_i\int_0^\infty \Pi_b(\Omega)\ln\Omega/\Omega d\Omega$, $\lambda_i=2\int\Pi_i(\Omega)/\Omega\, d\Omega$ is the electron-boson coupling constant and $g\in[0,1]$ is a parameter that accounts for the d-wave nature of the superconducting gap. The upper bound g=1 is used in the following.

We summarize the main results:

-The coupling to bosons of electronic origin is $\lambda_{e-be}=2\int I^2\chi(\Omega)/\Omega\, d\Omega$=**1.1±0.2**. This value corresponds to a maximal T_c=105-135 K.

-The coupling to strongly-coupled phonons (most likely breathing and buckling Cu-O optical modes) is $\lambda_{e-SCP}=2\int\alpha^2F(\Omega)_{SCP}/\Omega\, d\Omega$=**0.4±0.2**. This value is in complete agreement with the values measured on similar materials via different techniques, such as time-resolved photoemission spectroscopy [16], time-resolved electron diffraction [18] and single-color high-resolution time-resolved reflectivity [37]. The maximal critical temperature estimated considering only SCPs as the glue is T_c=2-30 K.

-The coupling to all other phonons is $\lambda_{e-lat}=2\int\alpha^2F(\Omega)_{lat}/\Omega\, d\Omega$=**0.2±0.2**, corresponding to T_c=0-12 K.

The large error bars in the values of λ_{e-be}, λ_{e-SCP} and λ_{e-lat} include: i) the experimental uncertainty in the pump fluence and other experimental parameters; ii) the possibility of adding to the 4TM a term that accounts for the anharmonic coupling of SCP to the lattice; iii) the possible overestimation of λ_{SCP} and λ_{lat} by a factor 8/5 [38], in the extreme case that the electron-electron scattering time (τ_{e-e}) is much larger than the electron-phonon scattering time and the energy exchange between the QPs and bosons begins before the establishment of a quasi-equilibrium fermionic population at $T_e+\delta T_e$.

Non-equilibrium optical spectroscopy allows disentangling the electronic and phononic contributions to the total bosonic function $\Pi(\Omega)$ of a cuprate superconductor. The strength of the coupling to bosons of electronic origin and its spectral distribution can account alone for the high-critical temperature (T_c=96 K) of the system. This supports the description of high-temperature superconductivity in cuprates in terms of a generalized Eliashberg formalism, in which the attractive interaction is mediated by virtual bosonic excitations of electronic origin, such as spin fluctuations or current loops.

5. Opening of a pseudogap and failure of the extended Drude model

In the calculation of the self-energy in Section 2.3, a constant density of states at the Fermi level has been considered. Although this approximation is reliable at T=300 K in optimally and overdoped systems, it fails dramatically as the temperature and the doping decrease and a pseudogap opens in the electronic density of states. A further evolution of the EDM, accounting for a non-constant electronic density of states, has been developed by Sharapov and Carbotte [39], and has been used to analyze spectroscopic data at equilibrium [40]. In this model, the imaginary part of the electronic self energy is given by:

$$\begin{aligned}\mathrm{Im}\Sigma(\omega,T) = -\pi\int_0^\infty \Pi(\Omega)\{\tilde{N}(\omega+\Omega,T)\,[n(\Omega,T)+f(\omega+\Omega,T)]+\\ +\tilde{N}(\omega-\Omega,T)\,[1+n(\Omega,T)-f(\omega-\Omega,T)]\}d\Omega\end{aligned} \tag{19}$$

while $\mathrm{Re}\Sigma(\omega,T)$ can be calculated through the Kramers-Kronig relations. To mimic the opening of a gap, the normalized density of states $\tilde{N}(\omega,T)$ can be written in the following form [40]:

$$\tilde{N}(\omega,T) = \begin{cases} \tilde{N}(0,T)+[1-\tilde{N}(0,T)]\left(\frac{\omega}{\Delta_{pg}}\right)^2 & for\ |\omega|\leqslant\Delta_{pg} \\ 1+\frac{2}{3}[1-\tilde{N}(0,T)] & for\ |\omega|\in(\Delta_{pg},2\Delta_{pg}) \\ 1 & for\ |\omega|\geqslant 2\Delta_{pg}\end{cases} \tag{20}$$

Where Δ_{pg} is the energy gap width, while $\tilde{N}(0,T)$ is the gap filling. In the top-right inset of Figure 5 we report $\tilde{N}(\omega,T)$ for the values Δ_{pg}=40 meV and $\tilde{N}(0,T)$=0.7.

Replacing Eqs. 9 and 11 with Eqs. 19 and 20 it is possible to calculate the $\delta R/R(\omega,t)$ induced by the pump pulse, even in the presence of a pseudogap.

In particular, three different relevant processes can be analyzed:

a) the impulsive heating of bosons at a temperature T_b higher than the equilibrium temperature T_0 of the system. A quasi-thermal scenario, with $T_e \simeq T_b$, and T_0=100 K is assumed. The calculated $\delta R/R(\omega,t=0)$, reported in Figure 5 (black line), is very similar to the reflectivity variation obtained in the case of a constant density of states at T=300 K (see Figure 4) and is related to the impulsive increase of the scattering rate and broadening of the Drude peak;

b) the impulsive filling of the pseudogap, as a consequence of the photoinjection of excess excitations. The $\delta R/R(\omega,t=0)$ is calculated assuming that the value $\tilde{N}$(0,T) is impulsively quenched from 0.7 to 0.9. The corresponding change in the density of states is shown in the top-right inset of Fig. 5. The result (see blue curve in Figure 5) is very different from the $\delta R/R(\omega,t=0)$ expected for an impulsive heating of the bosons. Very similar results are obtained by assuming the closing of the gap ($\Delta_{pg}<\Delta_{pg0}$) instead of its filling;

c) the impulsive decrease of the total $\Pi(\Omega)$. Although $\alpha^2F(\Omega)_{SCP}$ and $\alpha^2F(\Omega)_{lat}$ are expected to be doping- and temperature-independent, $I^2\chi(\Omega)$ could increase as the doping and the temperature decrease because either new magnetic excitations coupled with QPs emerge in the pseudogap phase [9, 41] or electronic correlations induce a non-Fermi liquid like increase of the self-energy at the antinodes. $\delta R/R(\omega,t=0)$ is calculated by assuming a small quench of the peak at 60 meV of the bosonic function (bottom-right inset of Fig. 5). The result (red solid line) is reported in Figure 5 and can be rationalized in terms of an impulsive decrease of

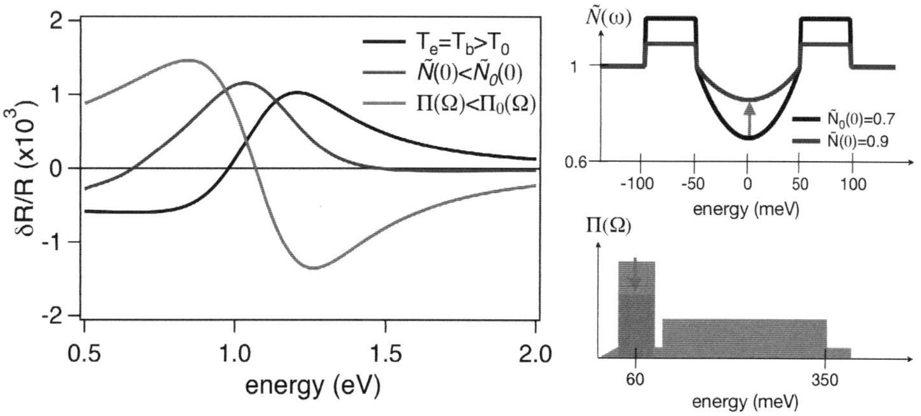

Figure 5. $\delta R/R(\omega, t{=}0)$ calculated in the case of: impulsive heating of the fermonic quasiparticles and bosonic excitations (black line), impulsive filling of the pseudogap (blue line), impulsive quench of the electron-boson coupling (red line). The right insets qualitatively show the change of the density of states, $\tilde{N}(\omega)$, and of the total glue, $\Pi(\Omega)$, assumed in the calculation of $\delta R/R(\omega, t{=}0)$. The equilibrium temperature T_0=100 K and the same $\Pi(\Omega)$ as the one determined at T=300 K have been used.

the coupling and consequent narrowing of the Drude peak, that is, the opposite effect to the transient heating of bosons.

In conclusion, we have shown that the use of non-equilibrium optical spectroscopy can be extended to doping and temperature regimes in which a pseudogap opens and electronic correlations strongly affect the electronic properties of the system.

6. Conclusions and Perspectives

Non-equilibrium optical spectroscopy is important and effective for investigating the physics of the cuprate superconductors and, more in general, of strongly correlated electron materials. Adding the temporal degree of freedom to the frequency resolution of equilibrium optical spectroscopy, allows disentangling the electronic and phononic contributions to the Eliashberg *Bosonic Function* $\Pi(\Omega)$ in a prototypical cuprate, i.e. optimally-doped Y-Bi2212. Our results demonstrate that, in principle, the strength of the coupling of quasiparticles to bosons of electronic origin can account for the high-critical temperature of the system (T_c=96 K). This result supports a description of high-temperature superconductivity in cuprates in terms of an attractive interaction, mediated by virtual bosonic excitations of electronic origin, where the most prominent candidates are spin fluctuations and loop currents. Furthermore, the analysis of the data obtained from non-equilibrium optical spectroscopy relies on a formalism (the Extended Drude model) that can be easily extended to more complex cases, e.g., systems with a gap in the density of states. This is very promising for further experiments aimed at investigating the elusive nature of the region of the cuprate phase diagram, in which a "pseudogap" in the density of states is present. Finally, the techniques described here can be easily extended to other complex materials, such as different classes of transition-metal oxides and iron-based superconductors.

These new experimental and theoretical tools, developed to investigate the electronic properties of correlated materials out of equilibrium, open intriguing perspectives for the next-years materials science. The modeling of the interaction of ultrashort light pulses with correlated materials is still at its infancy. The use of the "effective" temperatures and of the extended Drude models to describe the non-equilibrium dynamics on the sub-ps timescale has been proved to be a very effective tool to carry out a quantitative analysis of the data and single out the different

contributions to the pump-induced modification of the dielectric function. Nonetheless, these approximations cannot account for other important processes that play an important role in the femtosecond dynamics of strongly correlated materials. In particular:
-the origin of the interplay between high-energy Mott-like excitations at 1.5-2 eV and the onset of high-temperature superconductivity, recently demonstrated by non-equilibrium optical spectroscopy [2], is still an open question directly related to the nature of the superconducting phase in cuprates.
-the **k**-space distribution of the nascent population, after interaction with a femtosecond light pulse, can have strongly non-thermal features, such as an excess of excitations at the antinodes [17]. This feature could provide the access to novel non-thermal phases [42, 43, 44, 45, 46], in which superconductivity can be controlled by means of light pulses.

The joint effort of emerging theoretical tools, like non-equilibrium DMFT [47, 48] and time-dependent Gutzwiller approach [49, 50] and time-resolved experiments probing different physical properties, like time-resolved optical and photoemission spectroscopies and time-resolved electron diffraction, will provide the key to understand the role of the electronic correlations in controlling the ultrafast electronic properties in unconventional superconductors and transition-metal oxides.

Acknowledgments

We acknowledge discussions and suggestions from M. Capone and A. Chubukov. C.G., S. D.C. and F.P. have received funding from the European Union, Seventh Framework Programme (FP7 2007-2013), under Grant No. 280555 (GO-FASR project). F.C., G.C., and F.P. acknowledge the support of the Italian Ministry of University and Research under Grant Nos. FIRBRBAP045JF2 and FIRB-RBAP06AWK3. The crystal growth work was performed in M.G.'s prior laboratory at Stanford University, Stanford, CA 94305, USA, and supported by DOE under Contract No. DE-AC03-76SF00515. The work at UBC was supported by the Killam, Sloan Foundation, CRC, and NSERC's Steacie Fellowship Programs (A. D.), NSERC, CFI, CIFAR Quantum Materials, and BCSI. D. v.d.M. acknowledges the support of the Swiss National Science Foundation under Grant No. 200020-130052 and MaNEP.

References

[1] Cilento F, Giannetti C, Ferrini G, Dal Conte S, Sala T, Coslovich G, Rini M, Cavalleri A and Parmigiani F 2010 Ultrafast insulator-to-metal phase transition as a switch to measure the spectrogram of a supercontinuum light pulse *Appl. Phys. Lett.* **96** 021102
[2] Giannetti C *et al.* 2012 Revealing the high-energy electronic excitations underlying the onset of high-temperature superconductivity in cuprates *Nat. Commun.* **2**:353
[3] van Heumen E, Muhlethaler E, Kuzmenko AB, Eisaki H, Meevasana W, Greven M and van der Marel D 2009 Optical determination of the relation between the electron-boson coupling function and the critical temperature in high-T_c cuprates *Phys. Rev. B* **79** 184512
[4] Carbotte JP, Timusk T and Hwang J 2011 Bosons in high-temperature superconductors: an experimental survey *Rep. Prog. Phys.* **74** 066501
[5] Dal Conte S *et al.* 2012 Disentangling the Electronic and Phononic Glue in a High-T_c Superconductor *Science* **335** 1600
[6] Allen PB and Dynes R 1975 Transition temperature of strong-coupled superconductors reanalyzed. *Phys. Rev. B* **12** 905
[7] Le Tacon M *et al.* 2011 Intense paramagnon excitations in a large family of high-temperature superconductors. *Nature Phys.* **7** 725
[8] Ghiringhelli G *et al.* 2012 Long-range incommensurate charge fluctuations in (Y,Nd)$Ba_2Cu_3O_{6+x}$. *Science* **337** 821
[9] Li Y *et al.* 2010 Hidden magnetic excitation in the pseudogap phase of a high-T_c superconductor. *Nature* **468** 283
[10] Damascelli A, Hussaid Z and Shen Z-X 2003 Angle-resolved photoemission studies of the cuprate superconductors. *Rev. Mod. Phys.* **75** 473

[11] Lee J *et al.* 2010 Interplay of electron-lattice interactions and superconductivity in $Bi_2Sr_2CaCu_2O_{8+\delta}$. *Nature* **442** 546
[12] Ahmadi O, Coffey L, Zasadzinski JF, Miyakawa N and Ozyuzer L 2011 Eliashberg Analysis of Tunneling Experiments: Support for the Pairing Glue Hypothesis in Cuprate Superconductors. *Phys. Rev. Lett.* **106** 167005
[13] Hwang J, Timusk T, Schachinger E and Carbotte JP Evolution of the bosonic spectral density of the high-temperature superconductor $Bi_2Sr_2CaCu_2O_{8+\delta}$. *Phys. Rev. B* **75** 144508
[14] Muschler B, Prestel W, Schachinger E, Carbotte J P, Hackl R, Ono S and Ando Y 2010 An electron-boson glue function derived form electronic Raman scattering. *J. Phys.: Condens. Matter* **22** 375702
[15] Millis AJ, Monien H and Pines D 1990 Phenomenological model of nuclear relaxation in the normal state of $YBa_2Cu_3O_7$. *Phys. Rev. B* **42** 167
[16] Perfetti L, Loukakos PA, Lisowski M, Bovensiepen U, Eisaki H and Wolf M 2007 Ultrafast electron relaxation in superconducting $Bi_2Sr_2CaCu_2O_{8+\delta}$ by time-resolved photoelectron spectroscopy *Phys. Rev. Lett.* **99** 197001
[17] Cortés R, Rettig L, Yoshida Y, Eisaki H, Wolf M and Bovensiepen U 2011 Momentum-Resolved Ultrafast Electron Dynamics in Superconducting $Bi_2Sr_2CaCu_2O_{8+\delta}$. *Phys. Rev. Lett.* **107** 097002
[18] Carbone F, Yang D-S, Giannini E and Zewail AH 2008 Direct role of structural dynamics in electron-lattice coupling of superconducting cuprates. *Proc. Natl. Acad. Sci. USA* **105**:20161-20166
[19] Basov DN, Averitt RD, van der Marel D, Dressel M and Haule K 2010 Electrodynamics of correlated electron materials. *Rev. Mod. Phys.* **83** 471
[20] Basov DN and Timusk T 2005 Electrodynamics of high-T_c superconductors. *Rev. Mod. Phys.* **77** 721
[21] Eisaki H, Kaneko N, Feng DL, Damascelli A, Mang PK, Shen KM, Shen ZX, and Greven M 2004 Effect of chemical inhomogeneity in bismuth-based copper oxide superconductors. *Phys. Rev. B* **69** 064512
[22] van der Marel D *et al.* 2003 Quantum critical behaviour in a high-T_c superconductor. *Nature* **425** 271
[23] de' Medici L, Wang X, Capone M and Millis AJ 2009 Correlation strength, gaps, and particle-hole asymmetry in high-T_c cuprates: a dynamical mean field study of the three-band copper-oxide model. *Phys. Rev. B* **80** 054501
[24] Kaufmann HJ, Maksimov EG and Salje EKH 1998 Electron-phonon interaction and optical spectra of metals. *J. Superconduct.* **22** 755
[25] Johnston S *et al.* 2011 Evidence for forward scattering and coupling to acoustic phonon modes in high-T_c cuprate superconductors. *arXiv:1101.1302v1*
[26] Kovaleva NN *et al.* 2004 c-axis lattice dynamics in Bi-based cuprate superconductors. *Phys. Rev. B* **69** 054511
[27] Mirzaei SI *et al.* 2012 Evidence for a Fermi liquid in the pseudogap phase of high-T_c cuprates. *arXiv:1207.6704v1*
[28] Devereaux TP, Cuk T, Shen Z-X and Nagaosa N 2004 Anisotropic electron-phonon interaction in the cuprates. *Phys. Rev. Lett.* **93** 117004
[29] Dahm T *et al.* 2009 Strength of the spin-fluctuation-mediated pairing interaction in a high-temperature superconductor. *Nature Phys.* **5** 217
[30] Abanov A, Chubukov AV and Schmalian J 2001 Fingerprints of spin mediated pairing in cuprates. *J. Electron Spectrosc. Relat. Phenom.* **117** 129
[31] Chubukov AV and Schmalian J 2005 Superconductivity due to massless boson exchange in the strong-coupling limit. *Phys. Rev. B* **72** 174520
[32] Norman M and Chubukov A 2006 High-frequency behavior of the infrared conductivity of cuprates. *Phys. Rev. B* **73** 140501
[33] Varma C 2010 High-temperature superconductivity: Mind the pseudogap. *Nature* **468** 184
[34] Allen PB 1987 Theory of thermal relaxation of electrons in metals. *Phys. Rev. Lett.* **59** 1460
[35] Kabanov VV, Demsar J, Podobnik B and Mihailovic D 1999 Quasiparticle relaxation dynamics in superconductors with different gap structures: Theory and experiments on $YBa_2Cu_3O_{7-\delta}$ *Phys. Rev. B* **59** 1497
[36] Gedik N, Langner M, Orenstein J, Ono S, Abe Y and Ando Y 2005 Abrupt transition in quasiparticle dynamics at optimal doping in a cuprate superconductor system. *Phys. Rev. Lett.* **95** 117005
[37] Gadermaier C *et al.* 2010 Electron-phonon coupling in high-temperature cuprate superconductors determined from electron relaxation rates. *Proc. Natl. Acad. Sci. U.S.A.* **105** 257001
[38] Kabanov VV and Alexandrov AS 2008 Electron relaxation in metals: Theory and exact analytical solutions. *Phys. Rev. B* **78** 174514
[39] Sharapov SG and Carbotte JP 2005 Effects of energy dependence in the quasiparticle density of states on far-infrared absorption in the pseudogap state *Phys. Rev. B* **72** 134506
[40] Hwang J 2011 Electron-boson spectral density function of underdoped $Bi_2Sr_2CaCu_2O_{8+\delta}$ and $YBa_2Cu_3O_{6.50}$. *Phys. Rev. B* **83** 014507

[41] Li Y *et al.* 2012 Two Ising-like magnetic excitations in a single-layer cuprate superconductor. *Nature Phys.* **8** 404

[42] Kusar P, Kabanov VV, Demsar J, Mertelj T, Sugai S and Mihailovic D 2008 Controlled Vaporization of the Superconducting Condensate in Cuprate Superconductors by Femtosecond Photoexcitation. *Phys. Rev. Lett.* **101** 227001

[43] Giannetti C, Coslovich G, Cilento F, Ferrini G, Eisaki H, Kaneko N, Greven M and Parmigiani F 2009 Discontinuity of the ultrafast electronic response of underdoped superconducting $Bi_2Sr_2CaCu_2O_{8+\delta}$ strongly excited by ultrashort light pulses. *Phys. Rev. B* **79** 224502

[44] Coslovich G, Giannetti C, Cilento F, Dal Conte S, Ferrini G, Galinetto P, Greven M, Eisaki H, Raichle M, Liang R, Damascelli A and Parmigiani F 2011 Evidence for a photoinduced nonthermal superconducting-to-normal-state phase transition in overdoped $Bi_2Sr_2Ca_{0.92}Y_{0.08}Cu_2O_{8+\delta}$. *Phys. Rev. B* **83** 064519

[45] Stojchevska L, Kusar P, Mertelj T, Kabanov VV, Toda Y, Yao X and Mihailovic D 2011 Mechanisms of nonthermal destruction of the superconducting state and melting of the charge-density-wave state by femtosecond laser pulses. *Phys. Rev. B* **84** 180507

[46] Fausti D, Tobey RI, Dean N, Kaiser S, Dienst A, Hoffmann MC, Pyon S, Takayama T, Takagi H and Cavalleri A 2011 Light-Induced Superconductivity in a Stripe-Ordered Cuprate. *Science* **331** 189

[47] Eckstein M, Kollar M and Werner P 2009 Thermalization after an Interaction Quench in the Hubbard Model. *Phys. Rev. Lett.* **103** 056403

[48] Amaricci A, Weber C, Capone M and Kotliar G 2012 Approach to a stationary state in a driven Hubbard model coupled to a thermostat. *Phys. Rev. B* **86** 085110

[49] Schiró M and Fabrizio M 2010 Time-Dependent Mean Field Theory for Quench Dynamics in Correlated Electron Systems. *Phys. Rev. Lett.* **105** 076401

[50] André P, Schiró M and Fabrizio M 2012 Lattice and surface effects in the out-of-equilibrium dynamics of the Hubbard model. *Phys. Rev. B* **85** 205118

Generic strange-metal behaviour of overdoped cuprates

N E Hussey[1], H Gordon-Moys[1], J Kokalj[2] and R H McKenzie[2]
[1]H. H. Wills Physics Laboratory, University of Bristol, Bristol BS8 1TL, UK
[2]School of Mathematics and Physics, University of Queensland, Brisbane, 4072 Queensland, Australia

E-mail: n.e.hussey@bristol.ac.uk

Abstract. We present an analysis of the temperature dependence of the zero-field in-plane resistivity $\rho_{ab}(T)$ of overdoped $Tl_2Ba_2CuO_{6+\delta}$ (Tl2201). Taking our cue from earlier resistivity and angle-dependent magnetoresistance studies of Tl2201, as well as high-field measurements on $La_{2-x}Sr_xCuO_4$ (LSCO), we delineate $\rho_{ab}(T)$ into two T-dependent components below 200 K, one linear in temperature, the other quadratic. As found in LSCO, the T-linear component $\alpha_1(0)$ is finite for all superconducting samples, its magnitude scaling with the transition temperature T_c. By contrast, the T^2 coefficient $\alpha_2(0)$ is essentially doping independent. Such an extended regime (in doping) of T-linear resistivity at low T is at odds with conventional quantum critical scenarios involving the collapse of an ordered phase, possibly associated with the normal state pseudogap, to 0 K at a critical doping level. Its confirmation in Tl2201, whose electronic state (as revealed by quantum oscillation experiments) is highly homogeneous over hundreds of unit cells, appears to rule out phase separation or electronic inhomogeneity as the origin of this extended critical behaviour.

1. Introduction

The phrase "strange metal" has long been associated with the high-T_c cuprate superconductors. It was coined originally to describe the funnel-shaped region of the phase diagram above the superconducting dome in which the in-plane resistivity $\rho_{ab}(T)$ varies linearly with temperature and the inverse Hall angle $\cot\theta_H(T)$ varies quadratically with temperature [1]. In certain optimally doped cuprates, the T-linearity of the resistivity can extend up to extraordinarily high temperatures of order 1000 K [2]. This V-shaped region is defined by two temperature scales, T^* or the pseudogap temperature on the low-doping side and the coherence temperature T_{coc} on the other, both of which appear to merge at a critical doping level $p_{crit} \sim 0.19$ [3], close to optimal doping at which the superconducting transition temperature is a maximum. Its similarity to the temperature-doping phase diagram of some quantum critical metals and expectations from conventional quantum critical theory has prompted many in the field to associate the closing of the pseudogap at $p = p_{crit}$ to the existence of a zero-temperature quantum phase transition, possibly associated with the reconstruction of the Fermi surface [4].

Beyond this enigmatic region, it has been largely presumed that the electronic state of the cuprates evolves into a more conventional Fermi-liquid (FL) behaviour in the overdoped region. This presumption, however, is highly misleading. Only at very large doping, i.e. beyond the superconducting dome on the overdoped side, does the low-T resistivity follow a strictly T^2 dependence characteristic of a canonical FL [5,6]. Between optimal doping and the edge of the superconducting dome, there is a large swathe of the phase diagram in which $\rho_{ab}(T)$ remains anomalous and non-FL like. In this region, $\rho_{ab}(T)$ has been described either in terms of a single anomalous power law with an exponent n between 1 and 2 [7], or as a combination of two components, one linear-in-temperature, the other quadratic [8].

The latter description has gained experimental support from a combination of pulsed and persistent high magnetic field measurements carried out on two single-layer hole-doped cuprate families, LSCO and Tl2201 respectively [9-13]. In LSCO, pulsed fields were used to suppress superconductivity and to track the evolution of the normal state resistivity down to low temperatures (of order 1.5 K) [8]. These measurements revealed the persistence of a T-linear component $\alpha_1(0)T$ from optimal doping ($x = 0.17$) all the way out to $x = 0.29$, i.e. beyond the superconducting dome of LSCO. This unusual so-called 'extended criticality' contrasted markedly with expectations from conventional quantum critical scenarios in which non-FL-like resistivity survives only at a singular point in the phase diagram - the quantum critical point (QCP) - as seen, for example, in $YbRh_2Si_2$ [14] where the relevant tuning parameter is the strength of the magnetic field.

The origin of this extended T-linearity in LSCO is unknown at present. One possible explanation for the distinct behaviour of $\rho_{ab}(T)$ in LSCO is microscopic electronic phase separation into hole-rich (normal) and hole-poor (superconducting) regions, a phenomenon that has been well documented in the LSCO family [15]. In this scenario, the T-linear component is due to regions within the sample having a hole concentration close to optimal while the quadratic component originates from regions with a carrier density closer to 30 %, i.e. beyond the superconducting dome. This scenario is consistent with the observation that the coefficient $\alpha_1(0)$ scales with T_c reaching a maximum value of around 1 $\mu\Omega$cm/K at $p = p_{\rm crit}$ [8]. However, a similar phenomenology (i.e. coexistence of T and T^2 components in the resistivity with the T-linear term persisting over a wide region of the phase diagram) has also been seen in the Bechgaard salt $(TMTSF)_2PF_6$ [16]. Here $\alpha_1(0)$ and T_c display the same intimate correlation (in this case, T_c is varied through the application of hydrostatic pressure) yet as far as we are aware, there is no corresponding evidence for phase separation in these quasi-one-dimensional organic conductors and they have a well defined chemical stoichiometry [16].

An alternative scenario is one in which the two components in the resistivity arise from two independent scattering mechanisms being present in a homogeneous electronic state. Support for this picture comes from angle-dependent magnetoresistance (ADMR) measurements on overdoped Tl2201 where two distinct scattering rates were uncovered [10-14]; one that is isotropic over the Fermi surface and follows a T^2 dependence at low T and a T-linear term that is strongly anisotropic, peaking at the zone boundary and vanishing along the zone diagonal. The identification of a non-FL T-linear term in overdoped Tl2201 at low T is significant in that quantum oscillations (QO), the most definitive signature of fermionic quasiparticles, have been observed in crystals of similar doping [17-19]. Moreover, the observation of QO implies that the electronic state is highly homogeneous in overdoped Tl2201, with the carrier number changing by less than 0.2% over a distance equivalent to a cyclotron radius of order 1-2000 Å [18].

The observation of QO in Tl2201 would appear to rule out microscopic phase separation as the origin of the two T-dependent components in $\rho_{ab}(T)$ in overdoped cuprates. However, detailed ADMR measurements have only been carried out to date on samples with comparable T_c values (~ 15 K) and therefore similar carrier densities. Thus it has not been possible to confirm whether the low-T T-linear term in the scattering rate or resistivity persists over a wide doping range in Tl2201 and whether this extended critical behaviour is a generic feature of the high-T_c cuprate superconductors.

In this contribution, we reveal that the evolution of $\rho_{ab}(T)$ in Tl2201 with doping exhibits a remarkably similar phenomenology to that first reported in LSCO, at least in the normal state above T_c. This finding affirms that this strange-metal behaviour is indeed a generic property of overdoped cuprates, one that is intrinsic to the electronic state of these fascinating materials and one tied to the strong correlation physics inherent in their proximity to a Mott insulating state.

2. In-plane resistivity of Tl2201

The left panel of figure 1 shows $\rho_{ab}(T)$ data taken by Manako and co-workers on a non-superconducting (non-SC) Tl2201 single crystal [5] (labelled hereafter Tl-0K). As reported by Manako *et al.*, $\rho_{ab}(T)$ varies quadratically below approximately 50 K down to the lowest T measured (4.2 K), in a similar fashion to that found in overdoped non-superconducting LSCO (x = 0.33) [6]. As T increases, $\rho_{ab}(T)$ shows progressively less curvature and becomes almost T-linear near 300 K.

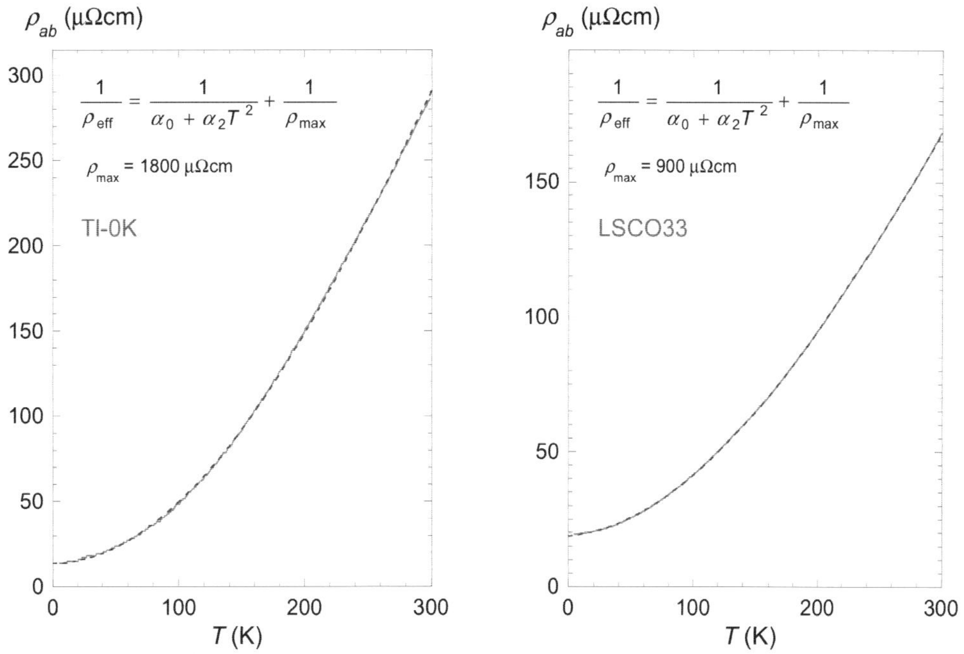

Figure 1. Left panel: in-plane resistivity $\rho_{ab}(T)$ (solid curve) of highly overdoped Tl2201, taken from Ref. [5].The dashed line is a fit to the expression quoted in the Figure. Right panel: Corresponding $\rho_{ab}(T)$ plot for highly overdoped non-superconducting LSCO (x = 0.33). Note the different ρ_{max} values.

A simple but elegant way to model the evolution of $\rho_{ab}(T)$ over the entire temperature range is to fit the data to the parallel resistor formula (PRF) first introduced by Weismann *et al.* to model the behaviour of saturating metals, i.e. metals whose resistivity tends to a constant (the so-called Mott-Ioffe-Regel or MIR limit) at high T once the mean free path ℓ has diminished to a value of order the lattice spacing a [20]. Using this fitting protocol, the two components in the PRF for both Tl-0K and non-SC LSCO33, reproduced in the left panel of figure 1, are the low-T component $\alpha_0 + \alpha_2 T^2$ and a maximum resistivity ρ_{max} which surprisingly, has a value consistent (to within a factor of 2) with the condition $\ell = a$ in both cases. Recall that for a quasi-2D metal, the Drude conductivity at the MIR limit corresponds to $\sigma_{MIR} = 1/\rho_{max} = (e^2/2\pi\hbar d)k_F a$. Thus for quasi-2D metals with comparable carrier densities and Fermi wave vectors k_F, ρ_{max} ought to scale with the interlayer spacing d. For Tl2201, d = 11.6 Å, while for LSCO, d = 6.4 Å.

The excellence of the two fits is surprising since it is now well established that in all cuprates, $\rho_{ab}(T)$ does not saturate at high T, but rather continues to increase, quasi-linearly with temperature, attaining values well in excess of the MIR limit. It has been shown however, both experimentally [21, 22] and theoretically [23], that once the dc resistivity of a non-saturating (bad) metal exceeds the MIR limit, the optical response becomes non-Drude like, with low-energy spectral weight being transferred to much higher energies of order the on-site Coulomb repulsion U. This transfer of spectral weight, a manifestation of strong electron correlation [23], implies that the escalation of the dc resistivity cannot be associated with an ever-increasing scattering rate (which would simply broaden the Drude response). Moreover, it has been argued previously [22] that below the MIR limit, the dc resistivity of bad metals like the cuprates may still evolve in a manner similar to that of saturating metals and thus the PRF is still an appropriate means of modelling the temperature evolution of $\rho_{ab}(T)$.

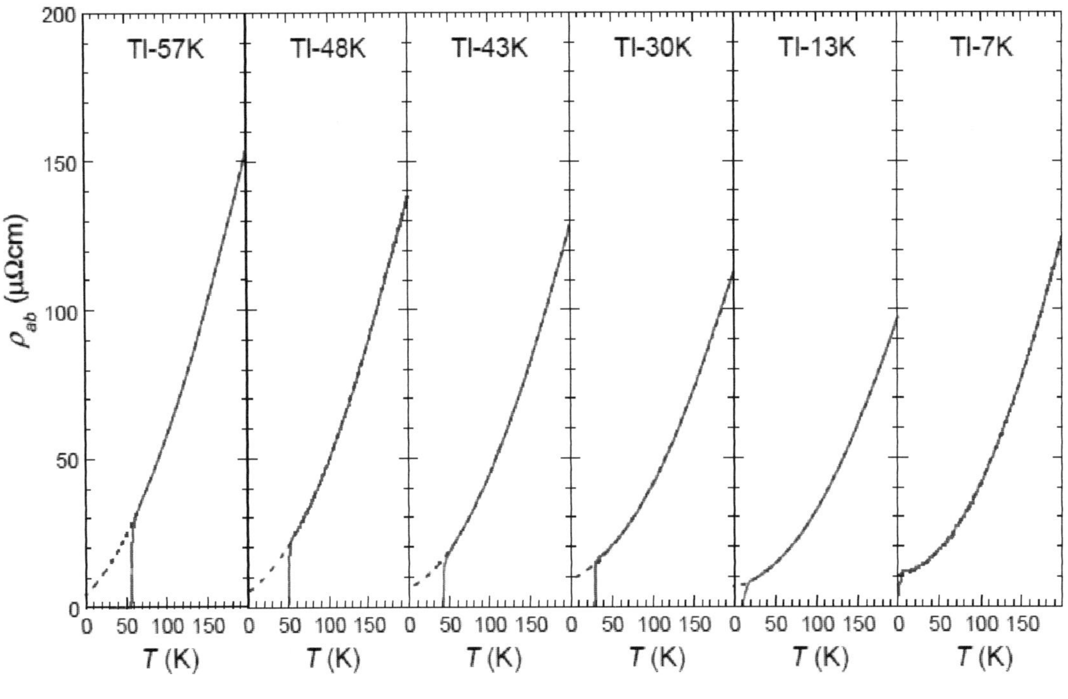

Figure 2. In-plane resistivity $\rho_{ab}(T)$ data (solid curves) for a series of Tl2201 crystals, taken from Ref. [5, 22,23].The dashed lines are fit to equation (1) between T_c and 200 K.

With this mind, we proceeded to model the $\rho_{ab}(T)$ data of a series of superconducting Tl2201 single crystals taken from the literature [5, 24, 25], having T_c values that varies between 0 K and 63 K, using the PRF. For all data sets below 200 K, the data could only be fitted satisfactorily with the inclusion of an additional T-linear term $\alpha_1 T$ such that the PRF read $1/\rho_{eff} = 1/[\alpha_0 + \alpha_1 T + \alpha_2 T^2] + 1/\rho_{max}$. Figure 2 shows the $\rho_{ab}(T)$ data together with the corresponding fits for a selected number of samples across the doping range. Note that for all fits, ρ_{max} is fixed at 1800 $\mu\Omega$cm. Note too that the residual resistivities are comparable across the range.

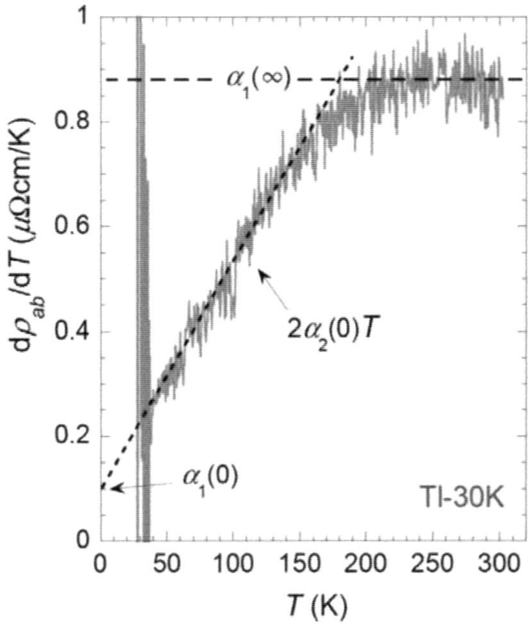

Figure 3. The temperature derivative $d\rho_{ab}/dT$ for a Tl-30K crystal [25] showing the various components of the $\rho_{ab}(T)$ curve, in particular the low-T T-linear term $\alpha_1(0)$, the low-T T^2 term $\alpha_2(0)$ and the high-T T-linear component $\alpha_1(\infty)$. The dashed lines are guides to the eye.

An alternative way to examine the T-dependence of $\rho_{ab}(T)$ is to plot the derivative $d\rho_{ab}/dT$. A representative example, for Tl-30K [25], is shown in figure 3. Above around 240 K, the derivative is essentially constant, illustrating the quasi-linear dependence of the high-T resistivity that was discussed above. As done previously [3], we label the coefficient of the high-T T-linear resistivity $\alpha_1(\infty)$. Between 40 K (i.e. above the fluctuation regime) and 150 K, $d\rho_{ab}/dT$ has a well-defined linear slope that extrapolates to a finite zero-temperature offset. These two features in the temperature derivative correspond respectively to the T^2 and T-linear terms included in the PRF fits of figure 2. The derivative plots therefore provide estimates of the two coefficients $\alpha_2(0)$ and $\alpha_1(0)$ that are independent of both the PRF and the residual resistivity α_0.

3. Discussion and conclusions

The resultant doping dependences of $\alpha_1(0)$ and $\alpha_2(0)$ are plotted in figures 4 and 5 respectively and compared directly with the results obtained from similar analysis carried out on the LSCO family [8]. The p values for Tl2201 were obtained from the recently revised T_c versus p diagram extracted from quantum oscillation experiments [18]. Apart from a factor of 2 difference in the magnitude of $\alpha_1(0)$ (note the different scales on the ordinate axes), the evolution of the linear and quadratic components of $\rho_{ab}(T)$ in LSCO and Tl2201 are almost identical. The key features to note here are the scaling of $\alpha_1(0)$ with T_c in both systems, the persistence of the T-linear component across the entire overdoped region of the phase diagram and the near constancy of the coefficient of the T^2 term as one approaches $p_{crit} \sim 0.19$ (marked by a vertical dashed line in figure 5). As mentioned in the Introduction, these two latter features are difficult to reconcile with conventional quantum critical scenarios in which the low T region of the Fermi–liquid-like resistivity would vanish on approaching the QCP. Indeed, the region over which $d\rho_{ab}/dT$ is linear, i.e. over which the T^2 term is robust, only decreases by of order 10 % (from 150 K to 135 K) between $p = 0.31$ and $p = 0.22$.

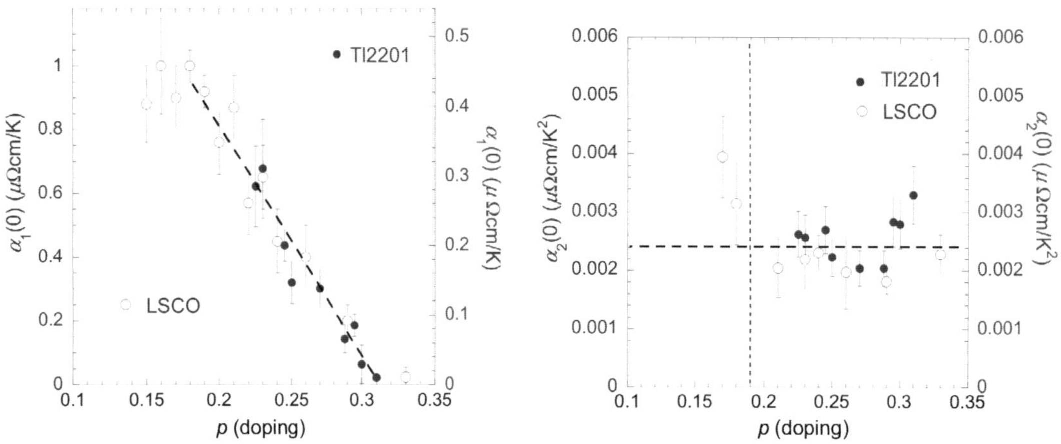

Figure 4. Left panel: Doping dependence of $\alpha_1(0)$, the coefficient of the T-linear component of the in-plane resistivity, in both LSCO (left ordinate axis) and Tl2201 (right ordinate axis). Note the left and right vertical scales differ by a factor of two. Right panel: Corresponding plot of the doping dependence of $\alpha_2(0)$, the coefficient of the T^2 component of $\rho_{ab}(T)$.

The confirmation of a near-identical phenomenology in the in-plane resistivity of two distinct cuprate families with different Fermi surface topologies, electronic anisotropies, disorder levels and electronic homogeneities provides strong evidence that this strange-metal behaviour is a generic property of the overdoped cuprates. The c-axis warping of highly overdoped Tl2201, as determined by both ADMR and QO experiments [18, 26], has a symmetry that is only consistent with the Fermi surface being located around the body-centred-tetragonal zone corners, implying that its Fermi surface topology remains hole-like up to $p = 0.30$. In LSCO, by contrast, the Fermi surface passes through a van Hove singularity at a doping level close to $p_{\text{crit}} = 0.19$ and becomes electron-like for $p > 0.19$ [27]. Thus, the phenomenology observed in $\rho_{ab}(T)$ in both systems beyond this doping level must be independent of Fermi-surface topology. The normal state resistive anisotropies of Tl2201 and LSCO differ by a factor of 10 or more, while their superconducting states are best described as two-dimensional (i.e. Josephson-coupled) and anisotropic but three-dimensional respectively [28]. The differences in the disorder levels of the two systems are self-evident in their respective residual resistivities.

Finally, but perhaps most significantly, the phenomenology summarized in figure 4 is observed in Tl2201 whose electronic state, as revealed by QO measurements [17-19], is extremely homogeneous, with the carrier number changing by less than 0.2% over a distance equivalent to a cyclotron radius of order 1-2000 Å [18]. It is therefore highly unlikely that this two-component resistive behaviour is due to microscopic phase separation. In the phase separation scenario one would expect similar T and doping dependence of the resistivity and Hall angle, but Hall angle shows, in contrast to the resistivity, T^2 dependence in the whole overdoped regime. This further suggests that the phase separation is not sufficient to describe transport properties in the overdoped regime, unless additional assumptions, e.g., anisotropic elastic scattering [29], are used. It also argues against overdoped LSCO being intrinsically inhomogeneous since it is, in effect, displaying the same transport behaviour as the highly homogeneous Tl2201 family. Indeed, a careful neutron diffraction study has shown that beyond optimal doping, overdoped LSCO is in fact much more homogenous and less susceptible to microscopic phase separation than its underdoped counterparts [30].

The presence of a T-linear term in $\rho_{ab}(T)$ of overdoped Tl2201 is consistent with the identification of an anisotropic T-linear term in the quasiparticle scattering rate by ADMR [10-13]. The scaling of $\alpha_1(0)$ with T_c is also consistent with the report of a correlation between the strength of anisotropic scattering (at a fixed temperature of 40 K) and T_c [11] and appears to be consistent with a connection between the low-T T-linear resistivity and the existence of a strongly k-dependent scattering process that is maximal at the zone boundaries and vanishes along the zone diagonals as well as the link between this scattering and superconductivity itself. The combined plot of $\alpha_1(0)(p)$, shown in the left panel of figure 4, demonstrates much more clearly that the T-linear component vanishes in both cuprate families at a critical doping level of $p_0 = 0.31$. This value of p_0 coincides precisely with the edge of the superconducting dome in Tl2201 and provides further strong support that the anisotropic (d-wave-like) T-linear scattering rate is linked to the pairing mechanism of high-T_c superconductivity.

In a recent set of papers [31, 32], we showed how the combination of an isotropic FL self-energy and an anisotropic marginal Fermi-liquid (AMFL) self-energy (having a d-wave form factor and a T- and ω-linear energy dependence) could explain not only the ADMR results, but also the dc resistivity, the Hall coefficient, the transverse magnetoresistance, the optical conductivity, the evolution of the quasiparticle mass and the specific heat coefficient of Tl2201 with overdoping. Thus, the form of the scattering would appear to reconcile a wide range of physical properties in the single-layer thallium system and given the similarity in the dc response of Tl2201 and LSCO, it would appear natural to apply the same phenomenology to LSCO and other cuprate families.

The AMFL model self energy [31, 32] includes a doping independent FL part which is capable of describing the weak doping dependence of $\alpha_1(0)$ shown in Figure 4 (right) and an anisotropic marginal Fermi liquid part whose strength increases with underdoping and captures the behavior of $\alpha_1(0)$ shown in Figure 4 (left). The success of the model again shows that the overdoped cuprates are not simple Fermi liquids. Specifically, the anisotropy over the Fermi surface, resembling the anisotropy of the underdoped (pseudogap) phase, starts to develop already in the overdoped side of the phase diagram. The isotropic scattering arises due to some local physics, presumably due to strong onsite Coulomb repulsion, and can be captured with, e.g., dynamical mean field theory (DMFT) or a Brinkman-Rice picture. The anisotropic part is a nonlocal effect, which needs to be described with more sophisticated methods, e.g., cluster DMFT [33]. Its origin is unclear and different fluctuations (antiferromagnetic, superconducting, and d-density wave), Fermi surface nesting and proximity to a QCP have been suggested to be relevant [32]. Considering the similarity between LSCO and Tl2201 discussed here the latter two seems unlikely. Furthermore, the success of the AMFL model shows, that there is no need to evoke more exotic theories, like two types of quasiparticles [e.g. spinons and holons with different scattering rates [34] or scattering mechanism connected to the charge conjugation properties of different currents [35], to simultaneously describe T-linear resistivity and T-quadratic Hall angle, at least in the overdoped regime.

In conclusion, the in-plane resistivity of both overdoped Tl2201 and LSCO have been shown to be composed of two T-dependent components below 200 K, one linear in temperature, the other quadratic. The T-linear component $\alpha_1(0)$ is finite for all superconducting samples with a magnitude that scales with T_c. This extended regime (in doping) of T-linear resistivity at low T is at odds with conventional quantum critical scenarios associating the pseudogap with the collapse of an ordered phase to 0 K at a critical doping level $p_{crit} \sim 0.19$. Its observation in Tl2201 appears to rule out phase separation or electronic inhomogeneity as the origin of this extended critical behaviour. A similar T-linear resistivity, with a magnitude scaling with T_c, has also been reported recently in the electron-doped cuprates [36]. The growing universality would appear to suggest that this strange-metal behaviour is both intrinsic and generic to all overdoped high-T_c cuprates.

Acknowledgments

This work was supported by the Engineering and Physical Sciences Research Council (U.K.) and the Australian Research Council. NEH acknowledges a Royal Society Wolfson Research Merit Award .

[1] Hussey N E 2008 *J. Phys. Cond. Matt.* **20** 123201
[2] Gurvitch M and Fiory A T 1987 *Phys. Rev. Lett.* **59** 1337
[3] Hussey N E, Cooper R A, Xu X, Wang Y, Vignolle B and Proust C 2011 *Phil. Trans. Roy. Soc. A* **369** 1626
[4] Doiron-Leyraud N and Taillefer L 2012 [incomplete]
[5] Manako T, Kubo Y and Shimakawa Y 1992 *Phys. Rev. B* **46** 11019
[6] Nakamae S, Behnia K, Yates S J C, Mangkorntong N, Nohara M, Takagi H and Hussey N E 2003 *Phys. Rev. B* **68** 100502
[7] Naqib S H, Cooper J R, Tallon J L and Panagopoulos C 2003 *Physica* **387C** 365
[8] Cooper R A *et al.* 2009 *Science* **323** 603
[9] Proust C, Boaknin E, Hill RW, Taillefer L and Mackenzie A P 2002 *Phys. Rev. Lett.* **89** 147003
[10] Abdel-Jawad M, Kennett M P, Balicas L, Carrington A, Mackenzie A P, McKenzie R H and Hussey N E 2006 *Nat. Phys.* **2** 821
[11] Abdel-Jawad M, Analytis J G, Balicas L, Carrington A, Charmant J P A, French M M J, Mackenzie A P and Hussey N E 2007 *Phys. Rev. Lett.* **99** 107002
[12] Analytis J G, Abdel-Jawad M, Balicas L, French M M J and Hussey N E 2007 *Phys. Rev. B* **76** 104523
[13] French M M J, Analytis J G, Balicas L, Carrington A and Hussey N E 2009 *New J. Phys.* **11** 055057
[14] Custers J *et al.* 2003 *Nature* **424** 524
[15] Tanabe Y, Adachi T, Omori K, Sato H, Noji T and Koike Y 2007 *J. Phys. Soc. Japan* **76** 113706
[16] Doiron-Leyraud N, Auban-Senzier P, de Cotret S R, Bourbonnais C, Jérome D, Bechgaard K and Taillefer L 2009 *Phys. Rev. B* **80** 214531
[17] Vignolle B, Carrington A, Cooper R A, French M M J, Mackenzie A P, Jaudet C, Vignolles D, Proust C and Hussey N E 2008 *Nature* **455** 952
[18] Bangura A F, Rourke P M C, Bensemann T, Matusiak M, Cooper J R, Hussey N E and Carrington A 2010 *Phys. Rev. B* **82** 140501(R)
[19] Rourke P M C, Bangura A F, Bensemann T, Matusiak M, Cooper J R, Carrington A and Hussey N E 2010 *New J. Phys.* **12** 105011
[20] Wiesmann H, Gurvitch M, Lutz H, Ghosh A K, Schwarz B, Strongin M, Allen P B and Halley J W 1977 *Phys. Rev. Lett.* **38** 782
[21] Takenaka K, Nohara J, Shiozaki R and Sugai S 2003 *Phys. Rev. B* **68** 134501
[22] Hussey N E, Takenaka K and Takagi H 2004 *Phil. Mag.* **84** 2847
[23] Merino J and McKenzie R H 2000 *Phys. Rev. B* **61** 7996
[24] Mackenzie A P, Julian S R, Sinclair D C and Lin C T 1996 *Phys. Rev. B* **53** 5848
[25] Tyler A W 1997 PhD Thesis (Cambridge)
[26] Hussey N E, Abdel-Jawad M, Carrington A, Mackenzie A P and Balicas L 2003 *Nature* **425** 814
[27] Yoshida T *et al.* 2007 *J. Phys. Cond. Matt.* **19** 125209
[28] Hussey N E, Cooper J R, Doyle R A, Lin C T, Liang W Y, Sinclair D C, Balakhrishnan G, Paul D. McK and Revcolevschi A 1996 *Phys. Rev. B* **53** 6752
[29] Abrahams E and Varma C M 2003 *Phys. Rev. B* **68** 094502
[30] Božin E S, Kwei G H, Takagi H and Billinge S J L 2001 *Phys. Rev. Lett.* **84** 5856
[31] Kokalj J and McKenzie R H 2011 *Phys. Rev. Lett.* **107** 147001
[32] Kokalj J, Hussey N E and McKenzie R H 2012 *Phys. Rev B* **86** 045312
[33] Gull E, Ferrero M, Parcollet O, Georges A and Millis A J 2010 *Phys. Rev. B* **82** 155101
[34] Anderson P W 1991 *Phys. Rev. Lett.* **67** 2092
[35] Coleman P, Schofield A J, and Tsvelik A M 1996 *Phys. Rev. Lett.* **76** 1324
[36] Jin K, Butch N P, Kirshenbaum K, Paglione J and Greene R L 2011 *Nature* **476** 73

Journal of Physics: Conference Series **449** (2013) 012005 doi:10.1088/1742-6596/449/1/012005

Theory of Pseudogap in Underdoped Cuprates

Masatoshi Imada[1,2], Shiro Sakai[1,2,3], Youhei Yamaji[1] and Yukitoshi Motome[1]

[1]Department of Applied Physics, University of Tokyo, [2]JST-CREST, Hongo, Bunkyo-ku, Tokyo, Japan, [3]Centre de Physique Theorique, Ecole Polytechnique, CNRS, 91128 Palaiseau, France

E-mail: imada@ap.t.u-tokyo.ac.jp

Abstract. Recent theoretical studies on the origin of the pseudogap emerging in underdoped cuprate superconductors are overviewed, based on insights obtained by a cluster extension of the dynamical mean field theory (cDMFT) for the doped two-dimensional (2D) Mott insulator. The pseudogap obtained in the cDMFT shows an *s*-wave-like full-gap structure, distinct from the *d*-wave superconducting gap. The zero-temperature electronic structure supports that a non-Fermi liquid phase exists and underlies the pseudogap. The non-Fermi liquid phase is separated from the larger-doped Fermi liquid by topological transitions of the Fermi surface and an emergence of zeros of Green's function. A coexisting evolution of the poles (Fermi surface) and zeros of the Green's function is a unique feature of the pseudogap phase. The spectra well reproduce the arc/pocket formation, together with basic experimental properties of the pseudogap phase in the cuprates. Furthermore, a full-gap structure is supported by a comparison with the results of Raman experiments. The overall feature supports the proximity of the Mott insulator and the significance of the quantum criticality of the Mott transition. These numerical results are further favorably interpreted by extending the exciton concept, known in semiconductors, to doped Mott insulators. In this composite fermion (CF) theory, the pseudogap emerges as a gap arising from a hybridization of the quasiparticle (QP) with the CF. The pairing channel opening between a QP and a CF solves the puzzle of the dichotomy between the *d*-wave superconductivity and the precursory insulating gap in the same antinodal region. A mechanism of superconductivity emerges from this pairing.

1. Introduction

Pseudogap ubiquitously found in the underdoped cuprates above the superconducting critical temperature is a puzzling and controversial phenomenon not well understood for more than 20 years after the discovery. Theoretically as well as experimentally, there exist a number of interpretations and proposals attempting to account for the origin of the pseudogap, while it is still under active debates [1, 2, 3, 4, 5, 6]. One scenario proposes that the pseudogap is a precursory gap of the *d*-wave superconductivity itself [7]. According to this scenario, the phase fluctuation existing even under a well-developed preformed pair destroys the superconducting order. Its simple picture has been challenged by the experimental observation of different structures between the pseudogap and the superconducting gap [8]. Another scenario proposes that the pseudogap emerges as a consequence of a real spontaneous symmetry breaking such as charge (stripe) or nematic orders [9, 10], *d*-density wave [11], and a phase with orbital current within a unit cell [12]. Symmetry broken states have been reported in some of the cuprates such as Y- and Bi-based compounds [13], while it is not clear at this moment whether all of the high-T_c cuprates show the same symmetry breaking. Other scenarios, such as the resonating valence bond [14] and the nodal liquid [15, 16] theories, propose strong fluctuations as the origin of the pseudogap. In these scenarios, the pseudogap is absent in the nodal point, with a d-wave symmetry as in the superconducting gap. This assumption is based on the experimental observation that the occupied part of the spectra indeed looks like d wave. However, since it has been difficult so far to observe the unoccupied part of the spectra at a sufficiently high resolution, the experimental pseudogap structure is in fact largely unknown. In any of the scenarios, the mechanism of the pseudogap formation is widely believed to be intimately correlated with the mechanism of the high temperature superconductivity itself.

Recent theoretical and computational progress, particularly cDMFT combined with the comparison with various experimental results has renewed the understanding of the pseudogap [17, 18, 19, 20, 21, 22, 23, 24, 25] and has revealed it as a distinct phase characterized by an unconventional metal. In this article, based on these numerical results without any *ad hoc* assumption, we examine various scenarios and conclude that the scenario by the preformed pair is not supported as a mechanism of the pseudogap. In addition, the numerical result shows that the pseudogap does not have the d-wave symmetry but shows a full-gap structure when we take into account the unoccupied part of the spectra. This contradicts the scenarios of the preformed pair, the nodal liquid, d-density wave and some of the resonating valence bond theories. The numerical results rather support that the pseudogap develops even without an explicit symmetry breaking such as the charge (nematic) order and the orbital current, whereas the pseudogap indeed grows solely as a consequence of the proximity (quantum criticality) to the Mott transition in the strong coupling region.

2. General remark on the origin of gap formation

Let us begin with a general remark on the mechanism of the (pseudo)gap formation from a theoretical viewpoint. In the mean-field picture of any symmetry breaking, including magnetic and superconducting origins, gap formations are understood always as consequences of hybridizations of the QP, where a pole in the Green's function of the hybridization partner causes the self-energy divergence of the QP Green's function. The divergence of the self-energy means the presence of a zero in the QP Green's function.

This mechanism is seen in the Hamiltonian given by the single-particle form as

$$H = \sum_i \xi_{\alpha_i} \hat{c}^\dagger_{\alpha_i} \hat{c}_{\alpha_i} + \sum_i \epsilon_{\beta_i} \hat{c}^\dagger_{\beta_i} \hat{c}_{\beta_i} + \sum_i \Delta_i (\hat{c}^\dagger_{\alpha_i} \hat{c}_{\beta_i} + \text{H.c.}), \quad (1)$$

where the QP with the creation (annihilation) operator $\hat{c}^\dagger_{\alpha_i}$ ($\hat{c}_{\alpha_i}$) hybridizes with the particle $\hat{c}^\dagger_{\beta_i}$ ($\hat{c}_{\beta_i}$) with the quantum numbers α_i and β_i, respectively. Then a hybridization gap proportional to Δ_i opens in the single-particle spectra after the diagonalization.

The antiferromagnetic gap in the mean field theory offers a typical example, where the gap is caused by the hybridization of electrons at a wave number k with the electrons at a different wavenumber $k+Q_0$, with Q_0 being the ordering vector. An $\uparrow$-spin electron with the momentum k hybridizes with a $\downarrow$-spin electron with the momentum $k + Q_0$, through the mean-field term $\Delta_{\text{AFM}}[\hat{c}^\dagger_{k\uparrow}\hat{c}_{k+Q_0\downarrow} + \text{H.c.}]$. Then, the self-energy for the $\uparrow$-spin electron with the momentum k diverges at the pole of the $\downarrow$-spin electron with momentum $k + Q_0$.

When the bare dispersion of $\hat{c}^\dagger_{k\sigma}$ in the absence of Δ_{AFM} is given by ξ_k and it hybridizes with the electron $\hat{c}^\dagger_{k+Q_0\bar{\sigma}}$, the Green's function G and its self-energy Σ of the electron $\hat{c}^\dagger_{k\sigma}$ are given as

$$G(k,\omega) = \frac{1}{\omega - \xi_k - \Sigma(k,\omega)}, \quad \text{and} \quad \Sigma(k,\omega) = \frac{\Delta^2_{\text{AFM}}}{\omega - \xi_{k+Q_0}}, \quad (2)$$

where ω is the energy. With the symmetry breaking, a zero surface appears in the single-particle Green's function. The zero surface is nothing but the self-energy pole, where $\Sigma = \pm\infty$ results in the breakdown of the perturbation expansion. The pole of the self-energy is nothing but the pole of the Green's function of the hybridization partner, namely, the electron at $k + Q_0$ for the above antiferromagnetic order. The gap formation and the destruction of the Fermi surface are caused by the zero surface which splits the pole surface and leads to the insulating gap.

3. Numerical methods

Details of the cDMFT method is given in the literature[26, 21]. The cDMFT calculation has been performed on the 2D Hubbard model defined by

$$H = -\sum_{\langle ij\rangle\sigma} t_{ij}(\hat{c}^{\dagger}_{i\sigma}\hat{c}_{j\sigma} + \text{H.c.}) + \sum_i U\hat{n}_{i\uparrow}\hat{n}_{i\downarrow} \tag{3}$$

under doping of carriers with the concentration δ measured from half filling. The model was solved with two choices of impurity solvers. One is the exact diagonalization for a 4-site cluster to clarify the properties at zero temperature. The other is the continuous-time Monte Carlo method [27] up to 16-site clusters for finite temperatures. To compare with experimental momentum-resolved spectra, it is necessary to introduce a method to bridge between the cluster calculation and the translationally invariant system, by periodizing the cluster results. Recently, for this purpose, it was shown that the cumulant periodization scheme is the best way to reduce finite-size effects to infer the thermodynamic limit within limited cluster sizes [21, 23]. The largest cluster size employed for the studies on the pseudogap so far is 16 sites, where the self-energy as well as the spectral weight near the Fermi level already appears to show a good convergence to the thermodynamic limit in the studied parameter and temperature region[23].

4. Evolution of zero and pole surfaces with topological phase transitions

We first overview what the cDMFT has revealed. In the results on the 2D Hubbard model, the zero surface indeed appears at half filling at large onsite interaction U even in the absence of the symmetry breaking [19, 21]. It separates the upper and lower Hubbard bands in the energy-momentum space and runs in the resultant Mott gap similarly to the symmetry broken cases (see Fig.1(a) of Ref. [19] or Fig.13(b) of Ref. [21]). The origin of the gap is not a simple mean field arising from the symmetry breaking but rather pure electron correlation itself.

In the lightly hole doped Mott insulator, the spectral weight transfers from the upper Hubbard to the top of the lower Hubbard band, at which the Fermi level is located. When U/t is large, the backbone Mott gap structure with a zero surface is still retained at a high energy. At the same time, an intricate small-energy structure emerges hierarchically near the Fermi level after the spectral weight transfer from the original upper Hubbard band upon hole doping. This generates a low-energy pole surface (Fermi surface at the Fermi level) of QPs which is split and considerably deformed by a low-energy zero surface emerging from a partial collapse of the Mott gap and the spectral weight transfer. Whether the weight transfer from the upper Hubbard band on the verge of the Mott transition occurs by a continuous shift of the weight through the Mott-gap-energy region or the weight appears near the Fermi level without the shift is not clear at the moment. The zero surface, coexisting and residing between the split pole surfaces, makes a small energy gap, which is identified with the pseudogap. The cDMFT with the solvers of the exact diagonalization at temperature $T = 0$ and the continuous-time quantum Monte Carlo method at $T > 0$ for the available clusters up to 16 sites with the cumulant periodization in the Brillouin zone consistently show the coexistence [28, 19, 21]. The coexistence is indeed a generic feature of lightly doped Mott insulators relevant to the pseudogap region in the cuprates.

Figure 1 shows how zero and pole surfaces evolve with doping in the 2D Hubbard model. In heavily-doped region ($\delta > 0.14$), the electronic structure has only a large electron-like Fermi

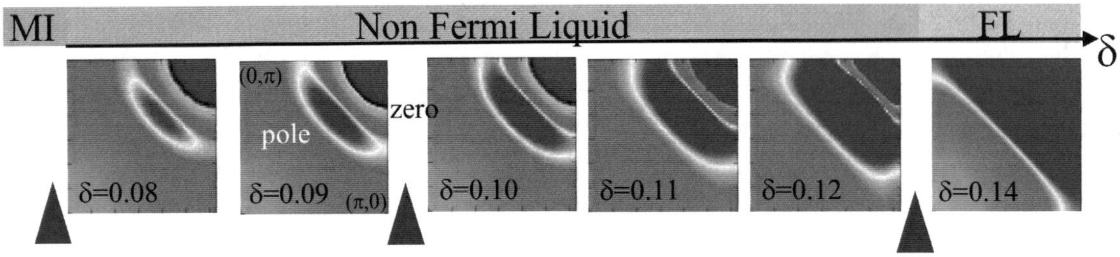

Figure 1. Evolution of pole (Fermi surface) and zero of the single-particle Green's function at the Fermi level $\omega = 0$. From the normal Fermi liquid (FL) with the electron-like Fermi surface to the Mott insulator (MI), several different quantum phase transitions caused by topological changes of the Fermi surface are designated by blue triangles. The red (light) and blue (dark) regions show negative and positive ReG, respectively. White boundaries indicate pole surfaces and dark boundaries indicate zero surfaces. The right triangle indicates a Lifshitz transition from an electron-like to a hole-like Fermi surface nearly simultaneously with the pair creation of the pole and zero surfaces. The middle triangle shows the location of a Lifshitz transition, where two open Fermi surfaces connect to a single closed Fermi pocket. The left triangle indicates the Mott transition where the Fermi surface pocket disappears with only the zero surface being left in the MI. The result was obtained by the cDMFT calculation at zero temperature with the 4-site cluster for the 2D Hubbard model only with the nearest neighbor transfer t and the interaction $U = 8t$. The computational details are found in Ref. [21]•

surface, characteristic to the Fermi liquid. With reducing doping, this Fermi surface changes its topology to hole-like. On top of that, between 12 and 14 % dopings, a zero surface emerges around (π, π) in pair with a small pole surface. The emergence of the zero surface signals the breakdown of the perturbation theory. Below this quantum phase transition with the topological change in the Fermi surface structure, non-Fermi liquid phase appears adiabatically disconnected with the overdoped Fermi liquid. The pseudogap phase appears further after another topological transition that is characterized as a Lifshitz transition transforming the two open Fermi surfaces into a single pocket. Since the outer part of the pocket is close to the zero surface, it is strongly damped and may not be visible in the angle resolved photoemission spectroscopy (ARPES), leaving a truncated arc-like structure in the inner part of the pocket [28]. The Mott insulator is realized after the final topological transition where the Fermi pocket vanishes. The evolution from the overdoped Fermi liquid to the Mott insulator may not be unique depending on the model parameters, even without symmetry breakings, and Fig. 1 illustrates one of the cases. Nevertheless, the existence of the topological phase transition (with emergence of the zero surfaces) separating the pseudogap phase from the conventional Fermi liquid is inevitable. Aside from details such as the quantitative accuracy of the result, if the evolution is continuous, the conventional Fermi liquid in the overdoped region with only the pole surface can be evolved by doping to the zero-surface-dominating Mott insulator, only when an intermediate phase exists: In this intermediate phase, the poles (Fermi surface) and zeros of the Green's function have to coexist to continuously bridge the pole-only Fermi liquid and zero-only Mott insulator.

This region of coexisting zero and pole surfaces with the emergence of the pseudogap cannot be adiabatically connected with the normal Fermi liquid in the largely-overdoped region, because, as we described above, the emergence of the zero surface implying the diverging self-energy of QP means the breakdown of the perturbation theory. Moreover, the emergence of the zero surface accompanies a simultaneous creation of a pole surface, which generates a quantum phase transition caused by a topological change in the Fermi surface. This quantum phase transition has nothing to do with a spontaneous symmetry breaking by itself, while it is categorized to a topological transition [29, 30, 31]. When interaction effects are involved in topological transitions, unconventional universality classes with a "soft" quantum criticality, characterized by the critical exponents such as $\beta > 1/2$ and $\delta < 3$, emerges in general around this marginal quantum critical point. This is in marked contrast with the conventional quantum criticality driven by the spontaneous symmetry breaking and described by the Landau-Ginzburg-Wilson scheme [29, 31]. Interaction effects may drive the topological quantum phase transition to the first-order one beyond the marginal quantum critical point. This first-order transition may continue at nonzero temperature and terminate at a finite-temperature critical point with the Ising universality class. When the transition occurs by the filling control, it occurs as the phase separation into different electron concentrations. However, such a phase separation is prohibited in the presence of the long-ranged Coulomb interaction. Then a meso-scale inhomogeneity or charge order becomes an intrinsic possibility as has been suggested in many experiments for the cuprates. Such competing orders observed in several types of cuprates indicate that the cuprates are close to the marginal quantum critical point driven by the topological transitions. Recently, the proximity of the first-order transition driven by the topological transition has been studied by the dynamical cluster approximation [32]. An intriguing future issue is how this soft criticality accompanying inevitable large quantum fluctuations induces instabilities toward competing orders such as superconductivity.

5. Full gap structure of pseudogap

A striking fact in the cDMFT result is that the pseudogap structure does not satisfy the d-wave symmetry and it appears to open a full s-wave-like gap [19, 21]. The result has clarified that the pseudogap is d-wave like only below the Fermi level, while it opens a full gap in the unoccupied part even in the nodal point. In the nodal region, the arc or pocket like structure of the Fermi surface becomes gapped in the unoccupied energy part. The s-wave structure of the pseudogap including the nodal point is robust against finite-size effect: It is consistently seen in the cluster sizes from 4 to 16.

The d-wave-like structure below the Fermi level well accounts for the ARPES results. The structure of a full gap with the isotropic s-wave-like amplitude can be confirmed only when we consider the total structure including the part above the Fermi level. The unoccupied part above the Fermi level has not been well probed so far because of the experimental difficulties such as the poor resolution in the inverse photoemission spectroscopy. The present theory and proposal can be critically tested by measuring whether the full-gap structure indeed exists in the nodal region above the Fermi level. The pseudogap structure in fact critically poses restrictions to the theories as well as to the mechanism of the superconductivity. It is desired to develop a high-resolution momentum-resolved probe for the unoccupied part.

The scanning tunnel microscopy (STM) in principle probes the unoccupied part, although it does not have a momentum resolution. One problem in the underdoped region is the inhomogeneity which may mix the responses in the superconducting and non-superconducting regions with different amplitudes of the gap in the temperature range normally accessible by the STM. Nevertheless, when we focus on a strongly underdoped region by ignoring the influence of the superconducting region at small ω, the STM appears to show a flatter bottom of the local density of states particularly in the large-gap (non-superconducting) region [33, 34, 35]. This result supports the existence of the full gap, if we eliminate the effects of the d-wave superconducting phase. It is an intriguing issue to investigate this region in more detail by examining the temperature dependence in the pseudogap phase and by using the Fourier-transform scanning tunneling spectroscopy (STS). This will serve for gaining more insight into the intrinsic momentum-resolved gap structure of the pseudogap phase particularly in the nodal region above the Fermi level.

Recent comparison of Raman intensity with the cDMFT result indeed supports that the full gap opens in the unoccupied part over the entire Brillouin zone including the nodal point [25]. By employing different light polarizations, Raman spectroscopy can separately probe particle-hole excitations in the nodal (B_{2g}) and antinodal (B_{1g}) regions in the Brillouin zone. It equally probes the energy region below and above the Fermi level. The Raman spectra show that the pseudogap grows more or less equally in the B_{2g} and B_{1g} polarizations in agreement with the prediction for the Raman spectra calculated from the cDMFT single-particle spectra that open the s-wave-like pseudogap. Furthermore, the Raman spectra show that the superconducting gap has an energy scale much smaller than the pseudogap and a peak outside the pairing gap appears in the pseudogap energy scale, suggesting entirely different origins of these two gaps.

6. Composite fermion theory

The pseudogap formation has been established in the numerical results on the doped Mott insulator. In this section, we discuss a generic mechanism of the pseudogap formation in lightly doped Mott insulators, by extending the exciton concept that has played a significant role in understanding excitations of semiconductors and insulators [36, 37].

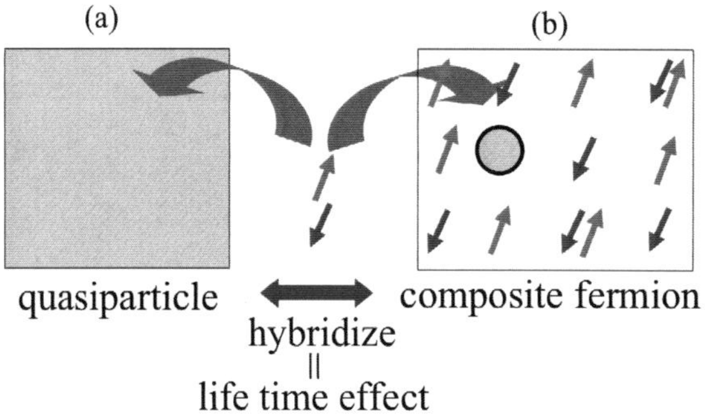

Figure 2. Schematic illustration of two ways of adding electrons to lightly doped Mott insulators: (a) QP added in the sea of nearly half-filled electrons depicted as light-blue (gray) background (b) CF as a weakly bound state of the added electron and an existing hole. These two excitations both have 1/2-spins depicted by arrows. They also hop and become extended in real space, giving their dispersions. Gray circle in (b) represents a hole.

In the underdoped cuprates, the Mott gap is still a robust underlying structure, where a zero surface persists below the upper Hubbard band. At the same time, a low-energy-scale structure near the gap edge including the QP excitations also evolves with doping. In this circumstance, we have two possible ways of low-energy excitation of doped carriers. One is QP and the other is composite fermion (CF) excitations as are illustrated in Fig. 2. The QP is spatially extended as a single-particle state. This becomes a low-energy excitation because of the kinetic energy gain. On the other hand one may add an electron near a holon site, to form a weakly bound state of the added electron and the existing holon similarly to the case of excitons. In the Mott insulating state, such an exciton is a tightly bound state with the binding energy of the Mott gap. This exciton-type excitation is the only one allowed in the large U limit. It is a localized excitation in the Mott insulator and observed at the Mott gap (more precisely, the charge transfer gap) edge of the optical conductivity [38]. At large but finite U, even in the lightly doped systems, this excitonic excitation may still be robust and distinguished from QPs when the added electron forms a state bound with the holon as one naturally expects. However, the binding energy must be dramatically weakened, because the doublon-holon attractive force (that generates the binding into an exciton) is screened by other doped carriers. We note that the screening is expected to be efficient even with a tiny doping concentration especially in two dimensions because of a constant density of states at the gap edge for the normal quadratic dispersion. Then the bound state becomes extended over several sites. This bound state is called CF[36]. The binding energy of the CF is originally scaled by the Mott gap in the Mott insulator while it is scaled by t in the lightly doped insulator as is shown in Ref. [24].

The two types of excitations may physically represent "two sides of a coin": The QP represents the coherent part whereas the CF picks up the incoherent part of the carrier. These two excitations are not perfectly the eigenstate of the Hamiltonian and they have finite life times and nonzero damping amplitudes. It means that the QP decays into the CF after a certain life time, while the CF mutually decays into the QP again. This process is nothing but the hybridization of the QP and CF as we will see below.

These two faces of excitations are successfully expressed by the slave boson theory originally formulated by Kotliar and Ruckenstein [39], by extending and improving it. The original Kotliar-Ruckenstein formalism is equivalent to the Gutzwiller approximation, where the band narrowing is represented by the decreasing amplitude of the slave boson condensate in the mean-field framework. Castellani *et al.* [40] extended it by including the Gaussian fluctuation of the boson condensate, which enables to describe the upper and lower Hubbard bands. Further improvement has been achieved by considering the fluctuation beyond the Gaussian level by incorporating the excitonic bound state formation of the QP with a holon [36, 37].

More concretely, in the slave boson formalism, an electron c is described as a particle composed of a boson (slave boson representing holon e, doublon d and singly occupied up- and down-spin bosons $p_\uparrow$ and $p_\downarrow$) and a fermion representing the QP f_σ with spin σ in the form $\hat{c}^\dagger_{i\sigma} = (\hat{p}^\dagger_{i\sigma}\hat{e}_i + \hat{d}^\dagger_i \hat{p}_{i\sigma})\hat{f}^\dagger_{i\sigma}$. Then the transfer term $t_{ij}\hat{c}^\dagger_{i\sigma}\hat{c}_{j\sigma}$ in the original Hubbard model is rewritten in the summation of terms such as $t_{ij}\hat{p}^\dagger_{i\sigma}\hat{p}_{j\sigma}(\hat{\Upsilon}^\dagger_{i\sigma} - \hat{C}^\dagger_{i\sigma})(\hat{\Upsilon}_{i\sigma} - \hat{C}_{i\sigma})$, where $\hat{C}^\dagger_{i\sigma} = \hat{e}_i \hat{f}^\dagger_{i\sigma}$. A Stratonovich-Hubbard field Υ introduced here plays the role of the CF. The present formalism naturally takes into account the CF substantiated by this Stratonovich-Hubbard field [36, 37]. Now it becomes clear that the above rewritten transfer contains the hybridization of a QP f with a CF Υ after averaging out fluctuating charge bosons and spin bosons as $\langle \hat{p}^\dagger_{i\sigma}\hat{p}_{j\sigma}\rangle$ in the decoupling approximation. The hybridization is scaled by t_{ij} and represents the physical process that the CF is dissolved into and recombined from a QP and a holon (or doublon) in binding-unbinding fluctuations. The effective Hamiltonian reads

$$H = \sum_k \xi(k)\hat{f}^\dagger_k \hat{f}_k + \sum_k \alpha(k)\hat{\Upsilon}^\dagger_k \hat{\Upsilon}_k + \sum_k \Delta(k)\hat{f}^\dagger_k \hat{\Upsilon}_k + \text{H.c.}, \tag{4}$$

where $\alpha(k)$ is the CF dispersion, while the QP dispersion $\xi(k)$ is also a renormalized one in the absence of the CF. Here, the CF has been defined so far as the bound state of a holon and an added electron, while another CF can equally be formed from the bound state of a doublon and an added hole. The formulation for the latter type is obtained straightforwardly from the former type, and for simplicity of the description, we do not explicitly introduce the latter type of the CF here.

Now the hybridization of QP and CF generates a hybridization gap, which is identified with the pseudogap in the cuprates. This hybridization gap is entirely different from the Mott gap precursor and from the superconducting gap as well. In fact, the pseudogap requires the underlying remnant of the Mott gap to exist but is generated as a lower-energy hierarchy developed with the CF.

The CF theory reproduces a number of puzzling experimental results as well as remarkable numerical results obtained by the cDMFT [21] and quantum Monte Carlo simulations [41]. For instance, in addition to the consistent doping dependence of the pseudogap with the full-gap structure, the whole structure of the spectral weight $A(k, \omega)$ with upper and lower Hubbard bands and the QP dispersion, the hole pocket/arc structure of the Fermi surface at $\omega = 0$, asymmetry of the density of states with respect to the $\omega > 0$ and $\omega < 0$ parts, and growth of the density of states with increasing doping concentration, indicated in the specific heat, are very consistent with the experimental and cDMFT indications[36].

7. Dichotomy and superconductivity

In this section, we discuss plausible consequences inferred from the results based on the cDMFT and the CF theory in the previous sections. Since QP and CF represent the coherent and incoherent part of the single-particle excitation, respectively, they may both play roles in pairing. The QP pairing leads to the conventional Cooper pair, while the CF may also form a bound state with another CF or with a QP. This is regarded as an incoherent pair, which contributes to the superconducting gap without the phase coherence. In this sense, the CF offers a novel mechanism of superconductivity. The pairing of the CF and QP is already considered explicitly in a perturbative treatment[37]. We here note that the CF dynamically has a large dipole moment consisting of the hole and the electron (doublon) because of the spatially extended structure of their bound state, and the pairing may occur through this dynamical dipole fluctuation as in the case of the attraction through the van der Waals interaction, but with a much larger energy scale. Polarizable CF fluctuations are a natural source of the glue for the pairing that has a conceptual similarity to the exciton mediated pairing. Internal degrees of freedom of a CF have not been explored in detail in the literature, although the CF self-energy has to some extent been considered[37].

The CF degrees of freedom give a solution to the puzzle of the dichotomy known in the cuprates. Here the dichotomy means that experimentally the insulating pseudogap coexists with the superconducting d-wave gap in the superconducting phase both mainly in the antinodal region[42]. Since the pseudogap in the antinodal region enhances the insulating behavior rather than the pairing precursor, it must severely compete with the superconducting d-wave gap in this antinodal region. Then the question is why such a d-wave gap having the largest gap in the antinodal region grows under the insulating pseudogap in the same region without destroying each other.

Our numerical and theoretical findings solve this puzzle: In the antinodal region, the QP dispersion is gapped and the coherent pairing is killed. However, this gapped region is nothing but the region where the CF poles run. This means that the polarizable CF fluctuations most efficiently couple with QPs and offer glues for QP pairings particularly in the nodal region. The pairing with the CF involved may also most efficiently occur in the antinodal region. In the antinodal region, the Cooper pairs of the QPs and the incoherent pairs of the CFs appear strongly mixed and it is not meaningful to distiguish them. In this way, the CF pair and QP pair are mutually transferable as the coherent and incoherent parts of the same pair, and the CF pair reinforces the QP Cooper pair. Despite the reinforcement, we expect that the QP gap appears following a conventional d-wave symmetry and is still smaller than the CF incoherent gap.

Spatially inhomogeneous gap has been observed in the STS [34, 43]. Two distinct energy scales of the gap have also been pointed out in ARPES[8, 42]. Since the pairing mechanism discussed here has a dual character of coherent and incoherent contributions, respectively from QPs and CFs, it is tempting to assign the large gap region observed in STS as that arising from the CF pair presumably trapped near the dopant position thus contributing to an incoherent and large gap. The large gap in the antinodal region not necessarily correlated with the superconductivity with the d-wave form may also be accounted for by the contribution from the incoherent CF pairing. The incommensurate charge modulation found in STS[44] may also represent localized pairs involving CFs or their real-space condensation into the charge order.

8. Concluding Remark

We have discussed recent progress made for the understanding of the pseudogap observed in the cuprate superconductors. The pseudogap phase is identified as the phase with the coexistence of the poles and zeros of the QP Green's function. This phase is separated from the conventional Fermi liquid by the topological transitions and distinguished by the existence of the zero surface at the Fermi level. The quantum criticality of the topological transitions is described by an unconventional universality class beyond the scheme of Landau, Ginzburg and Wilson. The full s-wave-like gap structure of the pseudogap revealed in the cDMFT and supported by the Raman experiments imposes a severe constraint on the origin of the pseudogap and high-T_c superconductivity as well. The pseudogap as a generic phenomenon of the lightly doped Mott insulator has a consistent interpretation as the hybridization gap of the QP and the excitonic CF. The pairing of the CF solves the dichotomy of the pairing in the antinodal region coexisting with the insulating pseudogap in the same region. Experimental detection of the CF by an appropriate probe is a real challenge, because it is charge neutral. Studies on the unoccupied part of the spectra also provide an experimental challenge. These hidden parts are crucial to understand the pseudogap and the mechanism of the superconductivity in the cuprates. Further detailed studies on the hidden CF physics is a challenging issue for future.

Acknowledgments

The authors thank useful discussions with T. Hanaguri and A. Georges. This work was supported by Grant-in-Aid for Scientific Research (Grant No. 22340090), from MEXT, Japan. We acknowledge the Strategic Programs for Innovative Research (SPIRE), MEXT, and the Computational Materials Science Initiative (CMSI), Japan for their financial support.

References

[1] Imada M, Fujimori A and Tokura Y 1998 *Rev. Mod. Phys.* **70** 1039
[2] Damascelli A, Hussain Z and Shen Z X 2003 *Rev. Mod. Phys.* **75** 473
[3] Basov D N and Timusk T 2005 *Rev. Mod. Phys.* **77** 721
[4] Norman M R, Pines D, Kallin C 2005 Adv. Phys. **54** 715
[5] Lee P A, Nagaosa N, Wen X G 2006 *Rev. Mod. Phys.* **78** 17
[6] Hufner S, Hossain M A, Damascelli A and Sawatzky G A 2008 *Rep. Prog. Phys.* **71** 062501
[7] Emery V J and Kivelson S A 1995 *Nature* **374** 434
[8] For example, see Kondo T, Takeuchi T, Kaminski A, Tsuda S and Shin S 2007 *Phys. Rev. Lett.* **98** 267004
[9] Daou R, Chang J, LeBoeuf D, Cyr-Choiniere O, Laliberte F, Doiron-Leyraud N, Ramshaw, B J, Liang R X, Bonn D A, Hardy W N, Taillefer L 2010 *Nature* **463** 519
[10] Vojta M 2008 *Adv. Phys.* **58** 699
[11] Chakravarty S, Laughlin R B, Morr D K, and Nayak C 2001 *Phys. Rev. B* **63** 094503
[12] Varma C M 2006 *Phys. Rev. B* **73** 155113
[13] For example, see Kaminski A, Rosenkranz S, Fretwell H M, Campuzano J C, Li Z, Raffy H, Cullen W G, You H, Olson C G, Varma C M, Hochst H 2002 *Nature* **416** 610
[14] Rice T M, 2005 *Prog. Theor. Phys. Suppl.* **160** 39
[15] Balents L, Fisher M P A, and Nayak C 1998 *Int. J. Mod. Phys. B* **12** 1033
[16] Franz M and Tesanovic Z 2001 *Phys. Rev. Lett.* **87** 257003
[17] Civelli M, Capone M, Kancharla S S, Parcollet O and Kotliar G 2005 *Phys. Rev. Lett.* **95** 106402
[18] Kyung B, Kancharla S S, Senechal D, Tremblay A-M S, Civelli M and Kotliar G 2006 *Phys. Rev. B* **73** 165114
[19] Sakai S, Motome Y and Imada M 2009 *Phys. Rev. Lett.* **102** 056404
[20] Liebsch A and Tong N-H 2009 *Phys. Rev. B* **80** 165126
[21] Sakai S, Motome Y and Imada M 2010 *Phys. Rev. B* **82** 134505
[22] Sordi G, Haule K and Tremblay A-M S 2010 *Phys. Rev. Lett.* **104** 226402
[23] Sakai S, Sangiovanni G, Civelli M, Motome Y, Held K and Imada M 2012 *Phys. Rev. B* **85** 035102
[24] Imada M, Yamaji Y, Sakai S and Motome Y 2011 *Ann. Phys. (Berlin)* **523** 629

[25] Sakai S, Blanc S, Civelli M, Gallais Y, Cazayous M, Measson M-A, Wen J S, Xu Z J, Gu G D, Sangiovanni G, Motome Y, Held K, Sacuto A, Georges A, Imada M 2012 *arXiv*:1207.5070
[26] Kotliar G, Savrasov S Y, Palsson G and Biroli G 2001 *Phys. Rev. Lett.* **87** 186401
[27] Gull E, Mills A J, Lichtenstein A I, Rubtsov A N, Troyer M, and Werner P 2011 *Rev. Mod. Phys.* **83** 349
[28] Stanescu T D and Kotliar G 2006 *Phys. Rev. B* **74** 125110; Stanescu T D, Civelli M, Haule K and Kotliar G 2006 *Ann. Phys. (N.Y.)* **321** 1682
[29] Imada M 2005 *Phys. Rev. B* **72** 075113.
[30] Misawa T and Imada M 2007 *Phys. Rev. B* **75** 115121
[31] Imada M, Misawa T and Yamaji Y 2010 *J. Phys. Condens. Matter* **22** 164206
[32] Chen K-S, Meng Z Y, Pruschke T, Moreno J, and Jarrell M 2012 *arXiv:* 1207.0796
[33] see Supplementary Figure 1 of Ref. [44].
[34] McElroy K, Lee, D-H, Hoffman J E, Lang K M, Lee J, Hudson E W, Eisaki H, Uchida S and Davis J C 2005 *Phys. Rev. Lett.* **94** 197005
[35] Kohsaka Y, Hanaguri T, Azuma M, Takano M, Davis J C and Takagi H 2012 *Nat. Phys.* **8** 534.
[36] Yamaji Y and Imada M 2011 *Phys. Rev. Lett.* **106** 016404
[37] Yamaji Y and Imada M 2011 *Phys. Rev. B* **83** 214522
[38] Uchida S, Ido T, Takagi H, Arima T, Tokura Y and Tajima S 1991 *Phys. Rev. B* **43** 7942
[39] Kotliar G and Ruckenstein A E 1986 *Phys. Rev. Lett.* **57** 1362; Lechermann F, Georges A, Kotliar G and Parcollet O 2007 *Phys. Rev. B* **76** 155102
[40] Castellani C, Kotliar G, Raimondi R, Grilli M, Wang Z and Rozenberg M 1992 *Phys. Rev. Lett.* **69** 2009
[41] For instance, see Preuss R, Hanke W, Grober C and Evertz H G 1997 *Phys. Rev. Lett.* **79** 1122
[42] Tanaka K, Lee W S, Lu D H, Fujimori A, Fujii T, Risdiana, Terasaki I, Scalapino D J, Devereaux T P, Hussain Z, and Shen Z-X 2006 *Science* **314** 1910; Terashima K, Matsui H, Sato T, Takahashi T, Kofu M, and Hirota K 2007 *Phys. Rev. Lett.* **99** 017003
[43] Lang K M, Madhavan V, Hoffman J E, Hudson E W, Eisaki H, Uchida S, and Davis J C 2002 *Nature* **415** 412
[44] Hanaguri T, Lupien C, Kohsaka Y, Lee D H, Azuma M, Takano M, Takagi H and Davis J C 2004 *Nature (London)* **430** 1001

Journal of Physics: Conference Series **449** (2013) 012006 doi:10.1088/1742-6596/449/1/012006

Magnetism and Superconductivity in the Pseudogap Phase of Underdoped Cuprates

A. J. A. James[1], R. M. Konik[1], K. Huang[2], W.-Q. Chen[3,2], T. M. Rice[1,2,4] and F. C. Zhang[2]

[1]Condensed Matter Physics and Materials Science Department, Brookhaven National Laboratory, Upton, NY 11973, USA

[2]Center for Theoretical and Computational Physics and Department of Physics, The University of Hong Kong, Hong Kong SAR, China

[3]Department of Physics, South University of Science and Technology of China, Shenzhen, Guangdong, China

[4]Institut fuer Theoretische Physik, ETH Zurich,CH-8093 Zurich, Switzerland

Abstract. The theoretical description of the anomalous properties of the pseudogap phase in the underdoped region of the cuprate phase diagram lags behind the progress in spectroscopic and other experiments. A phenomenological ansatz, based on analogies to the approach to Mott localization at weak coupling in lower dimensional systems, for the single particle propagator in the pseudogap phase has been proposed by Yang, Rice and Zhang. This ansatz has had success in describing a range of experiments, especially spectroscopies such as ARPES and aspects of STM results. Recently this approach has been extended to successfully interpret the magnetic excitations in the spin response response. In the charge channel, d-wave cooperon excitations accompany the opening of the pseudogap and they generate an additional d-wave attraction for quasiparticles at the remnant nodal Fermi surfaces.

1. Introduction

The phase diagram of the cuprates is by now quite well established. The two limits of a standard Landau Fermi liquid metallic state at strong overdoping and an AF ordered Mott insulator in the parent stoichiometric compound, are well understood. In between the full Fermi surface at overdoping shrinks as the doping is reduced, before finally disappearing at zero doping. In an intermediate density range four disconnected Fermi arcs, centered on each nodal direction, are observed in ARPES experiments [1]. These Fermi arcs lead to only a partial gapping of the Fermi surface, hence the name pseudogap. This pseudogap phase at underdoping is the most puzzling and anomalous region of the phase diagram. One much debated question is whether this partial gapping of the Fermi surface is a consequence of some form of superlattice order in this underdoped region. Certainly there are a number of examples of superlattice instabilities in underdoped cuprates e.g. the static stripe phase in $La_{2-x}Ba_xCu_2O_4$ at $x = 1/8$ [2] with antiphase modulated superconductivity coexisting with charge and spin density waves. However no superlattice instabilities have been reported in the cleanest and best ordered examples of underdoped cuprates in the pseudogap phase at zero magnetic field, e.g.the double chain cuprate $YBa_2Cu_4O_8$ [3] and O-chain ordered Ortho II- $YBa_2Cu_3O_{6.5}$[4]. This absence leads us to argue that the partial truncation Fermi surface is not driven by a superlattice instability. Rather, the appearance of superlattices in certain underdoped cuprates like the d-wave superconductivity at lower temperatures reflect instabilities of the underlying pseudogap phase. Some years back Anderson[5] advocated the interpretation of the pseudogap phase as a well defined, but ultimately unstable, fixed point, similar to the Landau Fermi liquid in standard metals. The key issue then is to understand how such a highly anomalous phase can come about as the cuprates evolve from the full Fermi surface Landau Fermi liquid at overdoping, to a Mott insulator at stoichiometry. Although the basic model that describes the cuprates, namely a single band Hubbard model on a 2D square lattice, looks deceptively simple a direct 'a priori' approach is very challenging. We are looking for an anomalous phase stablized by short range order, in a system of strongly interacting fermions. Some years ago Yang, Rice and Zhang[6] proposed a phenomenological approach which seeks to give a consistent description of the key anomalous electronic properties that characterize the pseudogap phase. The YRZ approach was based on a generalization of the well studied case of multi-leg Hubbard ladders to the case of a 2D square lattice. In particular, in two model ladder systems, a partial truncation of the Fermi surface is found at light doping already the weak coupling - namely, wider Hubbard ladders with an even number of legs[7] and a 2D array of 2-leg Hubbard ladders with long range interladder hopping[8]. In both systems there only short range correlations predominantly in the antiferromagnetic (AF) and d-wave pairing (dSC) channels. Their general properties are consistent with the original proposal for a resonant valence bond (RVB) phase proposed by Anderson very soon after the discovery of the cuprate superconductors[9]. We will not go into details here about the considerations that led to the phenomenological form for the single particle propagator and refer the reader to a recent review[10] which covers both the reasoning behind the proposed form and the comparison to a range of experiments. It is worth noting however that an alternative derivation starting from the strong coupling coupling t-J model was recently given by Lee[11]. Following on an earlier paper by Ng[12], he pointed out that in the factorization approach with spinons and holons, if the hole hopping term is treated as a source of attraction which leads to spinon-holon bound state, a form similar to YRZ follows for the propagator. For more details see the recent paper by James *et al*[13]. The YRZ ansatz [6]for the coherent part of a single particle Green's function $G^{YRZ}(\mathbf{k},\omega)$ in a doped RVB spin liquid takes the form

$$\begin{aligned} G^{YRZ}(\mathbf{k},\omega) &= \frac{g_t(x)}{\omega - \xi(\mathbf{k}) - \Sigma^{RVB}(\mathbf{k},\omega)}; \\ \Sigma^{RVB}(\mathbf{k},\omega) &= \Delta_R^2(\mathbf{k})/(\omega + \xi_0(\mathbf{k})) \end{aligned} \tag{1}$$

where

$$\begin{aligned} \xi(\mathbf{k}) &= \xi_0(\mathbf{k}) - 4\bar{t}'(x)\cos k_x \cos k_y - 2\bar{t}''(x)(\cos 2k_x + \cos 2k_y) - \mu_p \\ \xi_0(\mathbf{k}) &= -2\bar{t}(x)(\cos k_x + \cos k_y) \\ \Delta_R(\mathbf{k}) &= \Delta_0(x)(\cos k_x - \cos k_y) \end{aligned} \tag{2}$$

The renormalized hopping coefficients $\bar{t}(x), \bar{t}'(x)$ and $\bar{t}''(x)$ contain Gutzwiller factors and the RVB gap function decreases linearly with increasing hole density x as $\Delta_0(x) = \Delta_0(1 - x/0.2)$. These approximate strong coupling forms are taken from the earlier renormalized mean field theory for the the $t - J$ model The key ingredient in $G^{YRZ}(\mathbf{k},\omega)$ is that the RVB pairing gap opens not on the Fermi surface as in a BCS superconductor, but is pinned by Umklapp scattering processes to the fixed surface in $\mathbf{k}$ space where $(\cos k_x + \cos k_y)$ vanishes - the U- surface which coincides with reduced zone in the case of simple AF ordering. At stoichiometry the RVB gap on a fixed surface in reciprocal space leads to an insulating state. Finite doping causes 4 closed hole pockets to appear centered on the nodal directions and lying inside this U-surface. The spectral weight of the quasiparticles varies strongly around the pockets which causes them to appear as the 4 nodal arcs observed in ARPES experiments. Recent ARPES experiments by the Brookhaven group on underdoped BSCCO samples in the normal pseudogap phase are a stronger test of this Green's function.[14, 15]. They used the finite temperature broadening of the signal to extract the quasiparticle dispersion not only below, but also up to an energy roughly 3kT above the Fermi energy. They observed clear evidence of a coherent quasiparticle dispersion through the Fermi level but with particle-hole asymmetry arising from back bending of the quasiparticle dispersion at positive energy to form a closed pocket in agreement with the YRZ form. More details on these very interesting and innovative experiments can be found in the talk by Peter Johnson. Additionally, earlier spectroscopic STM experiments reported quasiparticle Fermi arcs consistent with the pocket front surface, i.e the surface nearest the zone center, where the spectral weight is largest[16]. All told these experiments agree nicely with the quasiparticle dispersion and the interpretation of the 4 disconnected Fermi arcs proposed by YRZ. These spectroscopic experiments are the most explicit test of the RVB form for the propagator. The predictions that follow for a range of other experiments were worked out by a number of groups, especially by Carbotte, Nicol and their collaborators [17, 18, 19, 20] and by Valenzuela and Bascones[21] . These range from measurements of the specific heat, infrared conductivity, Raman scattering, to the temperature and doping dependence of the superconducting penetation depth. Again the overall agreement between the phenomenological theory, when generalized to describe also a d-wave superconducting state, and experiment is quite satisfactory. One feature that these experiments test is the relative weak dependence of a number of low temperature properties on the hole density. In the theory there is not much change in properties that are sensitive to the near nodal parts of the Fermi arcs, such as the low temperature specific heat and the low temperature variation of the penetration depth. More discussion on this topic can be found in the recent review[10]and the papers quoted therein. Here we concentrate on two recent studies of the collective modes in the spin and charge sectors in the pseudogap phase.

2. Magnetic Excitation Spectrum

We start with the interpretation of the magnetic excitation spectrum that is observed in the pseudogap phase. A series of neutron scattering experiments on several cuprates have found an unusual spectrum which, in keeping with the enhanced AF response, is centered at the wavevector (π, π). It has been named as a hourglass spectrum consisting of two components: a triplet mode centred at (π, π) which disperses upward symmetrically around (π, π), and lower frequency legs which stretch down to zero energy at a ring of wavevectors displaced slightly away from the central wavevector (π, π). These have an incommensurate intercept at zero energy [22]. There has been much discussion in the literature on the interpretation of this unusual spectrum. Here we concentrate on a recent series of calculations by James, Konik and Rice [13] based on the YRZ ansatz for propagator in the pseudogap phase. They used a RPA scheme to calculate the spin response

$$\langle \mathbf{S} \cdot \mathbf{S} \rangle_{YRZ}(\omega, \mathbf{k}) = -\frac{3}{\pi} \frac{\chi_{YRZ}(\omega, \mathbf{k})}{1 - J(\mathbf{k})\chi_{YRZ}(\omega, \mathbf{k})}, \tag{3}$$

where χ_{YRZ} is simply a particle-hole bubble obtained using the YRZ propagators. To describe the spin response in the superconducting state an additional d-wave pairing amplitude is included in the Green's function. The parameters in the propagator have values appropriate for the description of underdoped $La_{2-x}Sr_xCuO_4$. The energy scale of the magnetic response is largely determined by the RPA exchange interaction J. An appropriate choice of J allows one to obtain a common magnetic response for several underdoped cuprates as illustrated in Fig. 1.

The hourglass in both theory and experiment has strong incommensurate response at lowest energies concentrated at four points displaced slightly from the wavevector (π, π) along the axes at wavevectors, $(\pi, \pi \pm \delta)$ and $(\pi \pm \delta, \pi)$. As the energy is initially increased this response evolves inwards towards (π, π) and simultaneously becomes more isotropic in its distribution

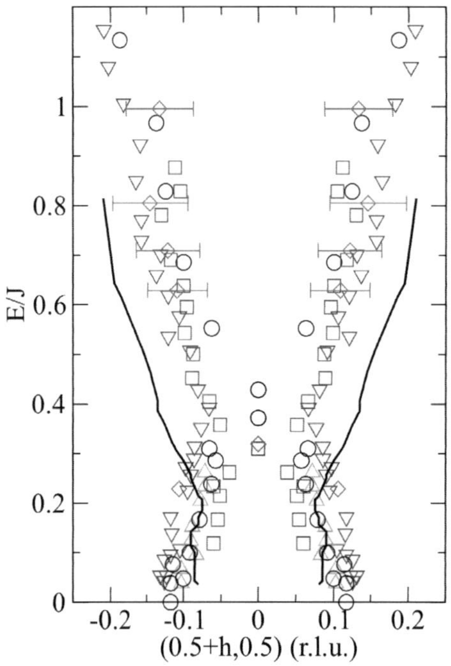

Figure 1. Hourglass dispersion of the resonance near the central wavevector (π, π). The thick black line is the position of the maximum intensity peak in the RPA calculations[13]. Experimental data points (appropriately rescaled) are taken from experiments on a number of underdoped 214 and YBCO cuprates - for details see James *et al* [13]

around (π, π). Whether this inward dispersion reaches (π, π) depends in the observations on the particular cuprate that is being examined. At an energy of order $J/5$ the dispersion reverses and moves outwards from (π, π) and is more isotropically distributed about (π, π). The low energy part of the spectrum in this theory is caused by quasiparticle-hole excitations between opposite near-nodal pockets. This interpretation of the low energy part of the spectrum as particle-hole excitations agrees with neutron scattering experiments performed on $YBa_2Cu_3O_{6.5}$-OrthoII samples. This cuprate is among the best ordered underdoped cuprates and has the advantage that large single crystals are available. The experiments of Stock *et al* [23] show a clear suppression of the low energy spectrum as T drops below T_c consistent with a suppression of the low energy particle-hole excitations as the superconducting gap opens up on the nodal Fermi arcs. Note, the calculations of James *et al* are for clean materials. The presence of impurities,even nonmagnetic impurities such as Zn, introduce local moments [24], so that a mixture of itinerant and local moments will occur in most cuprates. Recently the spin response at higher energies has been measured by RIXS experiments with enhanced resolution by Le Tacon*et al*[25]. Interestingly the calculations of James*et al* also reproduce the key features of these experiments[13].

3. Superconductity in the Pseudogap Phase of the Cuprates

Turning to the charge response, we can divide theories of the pseudogap phase that do not rely on some form of superlattice into two classes. One proposes that the pseudogap results from an extended region of enhanced superconducting fluctuations above a transition temperature which is lowered by strong phase fluctuations. These are a consequence of reduced phase stiffness, which in an underdoped Mott insulator is determined by the hole doping rather than the electron density. This theory implies that when thermal fluctuations are suppressed in the superconducting groundstate, a standard d-wave gap opens over the full Fermi surface,as in overdoped cuprates. However as pointed out by Deutscher[26], the large difference in the Andreev tunneling signal between overdoped and underdoped cuprates is difficult to rationalize in this scenario. The YRZ proposal for the pseudogap phase belongs to the second class which assumes an insulating pseudogap appears as a precursor to Mott insulating behavior at stoichiometry - a proposal which agrees with the Andreev results[27]. Some years back Norman,Pines and Kallin[28] raised the interesting question of whether the associated reduction in the Fermi surface area simply weakens the d-wave superconductivity. There are however a number of strong arguments in favor of a more nuanced effect of the pseudogap on superconductivity. This can occur if a low energy cooperon resonance accompanies opening of the antinodal pseudogap. The analogy between the physics underlying the pseudogap and Hubbard ladder discussed earlier is a reason to explore this possibility.

Konik, Rice and Tsvelik [29] examined an array of coupled 4-leg Hubbard ladders near half-filling. In wider ladders the bands pair up on a hierarchy of energy scales with the pairing of the outermost and innermost bands having the largest gap [7]. This band pair exhibits enhanced responses similar to the half-filled 2-leg Hubbard ladder. For a suitable parameter choice, the larger energy gap causes this band pair to be half-filled and insulating in an initial finite density range but supporting a cooperon collective excitation. Under these conditions, a pairing attraction is generated on the Fermi pockets in the central band pair through coupling in the Cooper channel to the cooperon of the insulating band pair. Although this is clearly an artificial model, Konik et al [29] proposed it as a tractable model which contains the essential features of the cuprates, namely *d*-wave pairing on remnant Fermi pockets supported by coupling to the cooperon excitation living on the remaining gapped bands. There have been a number of reports of enhanced, even giant, penetration lengths for d-wave superconductivity into nonsuperconducting phases of underdoped cuprates [30, 31, 32].Bozovic *et al* [31] reported a giant proximity effect in uniform trilayer junctions of high temperature superconductors at temperature T above the superconducting transition temperature T_c of the spacer underdoped cuprate layer in the pseudogap phase. These reports motivated Huang and coworkers[33] to model this system using the YRZ form for the normal state propagator of the lower Tc underdoped spacer cuprate layer. They found that this normal phase when modeled by YRZ propagators with a d-wave attraction in line as expected at low energies from renormalization results for the Hubbard model[10] , had a cooperon pole in the pairing propagator. The excitation energy of this cooperon goes unstable at Tc. A cooperon excitation in the barrier leads to an enhanced penetration of the superconducivity into the spacer layer in the trilayer junction. Choosing input parameters that reproduced the transition temperatures in the trilayer, Huang *et al* [33]could reproduce the giant proximity effect reported by Bozovic and coworkers[31] quite accurately as illustrated in Fig. 2. The contribution of the cooperon excitation can be deduced from the distribution of the Josephson current in momentum space of the underdoped sample (doping $x \approx 0.1$). A detailed examination of origin of the enhancement shows that the contribution from the cooperon excitation is significant and may be comparable with that from the quasiparticles in the nodal regions of k-space. Note that an alternative explanation of the enhanced penetration depth[31] based on a model with strong phase fluctuations has been given by Marchand *et al*[34]. There are also experiments on $YBCO/Y_xPr_{1-x}Ba_2Cu_3O_7/YBCO$ superstructure, which also report a very large proximity length at $x < 0.55$[30, 32]. However these strong proximity effects need a different explanation.

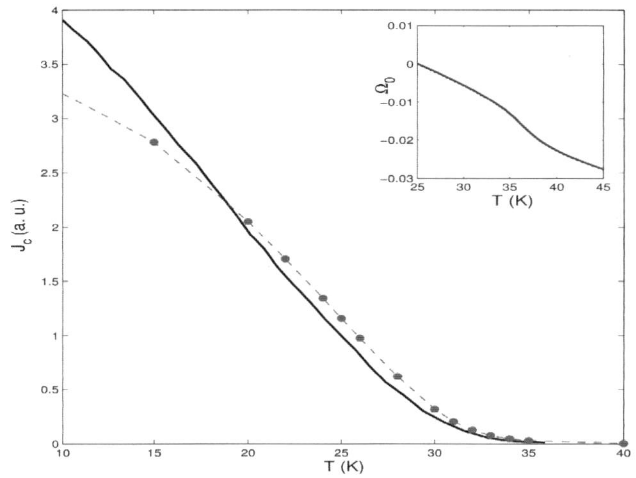

Figure 2. The critical Josephson current in a trilayer junction through a underdoped spacer layer. The black curve is the experimental data of Bozovic *et al* [31]. The red dots are the results of Huang *et al*[33] using the YRZ propagator for the spacer layer. The inset shows the calculated temperature dependence of the cooperon excitation energy above the Tc of the underdoped spacer layer.

4. Conclusions

At present, in the study of the cuprate superconductors there is a lack of consensus on the theory side, to put matters politely. On the experimental side there is continuing strong progress in improving resolution, samples, experimental techniques etc. This situation calls for a phenomenological approach which seeks to develop a theory which consistently explains all, or at least many of the anomalous data taken on the cuprates . The approach described above is motivated by a desire to fill this gap. It certainly has had a number of clear successes but there a still some open issues. The most glaring and most puzzling issue is the explanation of the quantum oscillations observed in underdoped cuprates at high magnetic fields and low temperatures.

The work at Brookhaven National Lab. received support from the Center for Emergent Superconductivity, an Energy Frontier Research Center funded by the DOE Office of Science, Office of Basic Energy Science under award DEAC0298CH1088. In addition the work at Hong Kong University received support in part from Hong Kong RGC GRF grant HKU701010, and at ETH Zurich from the Swiss Nationalfond and MANEP network.

References

[1] Norman M R, Ding H, Randeria M, Campuzano J C, Yokoya T, Takeuchi T, Takahashi T, Mochiku T, Kadowaki K, Guptasarma P, Hinks D G 1998 *Nature* **392** 157.
[2] Li Q, Huecker M Gu G D, Tsvelik A M and Tranquada J M 2007 *Phys. Rev. Lett* **99** 067001.
[3] Tomeno I, Machi T, Tai K, and Koshizuka N, Kambe S, Hayashi A, Ueda Y, Yasuoka H 1994 Phys. Rev. B **49** 15327.
[4] Stock C, Buyers W. J L, Tun Z, Liang R, Peets D, Bonn D, Hardy W N, and Taillefer L 2002 *Phys. Rev. B* **66** 024505.
[5] Anderson P W 2002 Physica B **318** 28.
[6] Yang K-Y, Rice T M, and Zhang F C, 2006 *Phys. Rev. B* **73**, 174501
[7] LeHur K and Rice T M 2009 Ann. Phys. **324** 1452.

[8] Konik R M, Rice T M and Tsvelik A M 2006 *Phys. Rev. Lett* **96** 086407
[9] Anderson P W 1987 Science **235** 1196.
[10] Rice T M, Yang K-Y and Zhang F C 2012 *Rep. Prog. Phys.* **75**, 016502
[11] Lee P A (private communication)
[12] Ng T K, 2005 *Phys. Rev. B* **71**, 172509
[13] James A J A, Konik R M and Rice T M 2011 arXiv 1112.2676
[14] Yang H-B, Rameau J D, Johnson P D, Valla T, Tsvelik A, Gu G D.2008 *Nature* **456** 77
[15] Yang H-B, Rameau J-D, Pan Z-H, Gu G D, Johnson P D, Claus H, Hinks D G, and Kidd T E 2011 *Phys. Rev. Lett.* **107**, 047003
[16] Kohsaka Y, Taylor C, Wahl P, Schmidt A., Lee J Fujita K, Alldredge J W, McElroy K, Lee J, Eisaki H, Uchida S, Lee D-H, Davis J C 2008 *Nature* **454** 1072.
[17] Illes E, Nicol E J, Carbotte J P 2009 *Phys. Rev. B* **79** 100505R
[18] LeBlanc J P F, Nico E J, Carbotte J .P 2009 *Phys. Rev. B* **80** 060505.
[19] Carbotte J P, Fisher K. A. G, LeBlanc J P F, Nicol E J 2010 *Phys. Rev. B* **81** 014522.
[20] LeBlanc J P F, Carbotte J P, Nicol E J, *Phys. Rev. B* 2010 **81** 064504.
[21] Valenzuela B, and Bascones E 2007 *Phys. Rev. LetT* **98** 227002.
[22] For reviews see Tranquada J 2007 chapter in Handbook of High- Temperature Superconductivity: Theory and Experiment, ed. J R Schrieffer, Springer and Fujita M*et al*, J Phys. Soc. Jpn. (in press); arXiv:1108.4431.
[23] Stock C, Buyers W. J L, Liang R, Peets D, Tun Z, Bonn D, Hardy W N, and Birgeneau 2004 *Phys. Rev. B* **69** 014502.
[24] Suchaneck A, Hinkov V, Huag D, Scultz L, Berhhard C, Ivanov A, Hradli K, Lin C TT, Bourges P and Keimer B 2010 *Phys. Rev. Lett.* **105**, 037207
[25] Le Tacon M *et al* 2011 *Nature Physics* **454** 1072.
[26] Deutscher G 2005 *Rev. Mod. Phys.* **77** 109
[27] Yang K-Y, Huang K, Chen W-Q, Rice T M and Zhang F C 2010 *Phys. Rev. Lett* **105** 167004
[28] Norman M, Pines D and Kallin K 2005 *Adv. Phys.* **54** 715
[29] Konik R M, Rice T M and Tsvelik A M 2010 *Phys. Rev. B* **82**, 054501
[30] Fivat P, Triscone J-M, and Fischer O 1996 J Low Temp Phys **105**, 1319
[31] Bozovic I, Logvenov G, Verhoeven M A J, Caputo P, Goldobin E, and Beasley M R 2004 *Phys. Rev. Lett* **93**, 157002
[32] Wojek B M, Morezoni E, Eschenko D G, Suter A, Prokscha T, Keller H, Koller E, Fischer O, Malik V K, Bernhard C and Doebeli M 2012 *Phys. Rev. B* **85**, 024505
[33] Huang K, Chen W-Q, Rice T M and Zhang F C 2012 *EPL***99**, 24002
[34] Marchand D, Covaci L, Berciu M, and Franz M 2008 *Phys. Rev. LetT* **101**, 097004

Photoemission Studies of the Pseudogap Regime in the High T_c Cuprate Phase Diagram

P.D. Johnson[1], H.B Yang[1], J.D. Rameau[1], G.D. Gu[1], T.E. Kidd[2], H. Claus[3] and D. Hinks[3]

[1] Condensed Matter Physics and Materials Science Department, Brookhaven National Laboratory, Upton, NY
[2] Physics Department, University of Northern Iowa, Iowa
[3] Materials Science Division, Argonne National Laboratory

E-mail: pdj@lbnl.gov

Abstract. By examining thermally excited states above the Fermi level in angle resolved photoemission spectroscopy, we show that the disconnected Fermi arcs observed in the pseudogap state of the cuprate phase diagram are actually components of fully enclosed hole pockets. The spectral weight of these pockets is vanishingly small at the Luttinger surface, coincident with the antiferromagnetic zone-boundary formed by the underlying spin lattice, creating the illusion of the Fermi arcs. The area of the pockets as measured in this study is consistent with the doping level, and hence the carrier density, of the samples measured. Furthermore, the shape and area of the pockets is well reproduced by phenomenological models of the pseudogap phase that reflect its proximity to a Mott insulator.

1. Introduction

High T_c superconductivity remains a challenge for the materials science community. Both the cuprates [1] and the more recently discovered Fe-based superconductors [2] present new and exciting phenomena that do not lend themselves to orthodox treatment. In the case of the cuprates, the superconducting phase emerges from a Mott insulator upon selective doping with charge carriers. However, in the case of the Fe-based materials, the superconductivity emerges from an underlying metallic state. It is therefore important to obtain a correct understanding of the normal states from which high temperature superconductivity emerges. Recognizing this, the pseudogap phase of the underdoped cuprates has been heavily studied. This region of the phase diagram presents its own challenges, including a gap in the spectral response even in the normal state [3], and a Fermi surface that seemingly consists of disconnected Fermi arcs.[4] The presence of this gap, or pseudogap, has promoted considerable discussion as to whether there exists some form of pairing of the carriers at temperatures above that at which the long range phase coherence associated with superconductivity develops [5], or whether there exists another ground state that competes with the superconductivity.[6,7] The disconnected Fermi arcs present another puzzle. Indeed, to date, there is no precedent in any other condensed matter system for such a phenomenon. Two candidate descriptions of the arc have emerged. In one picture superconducting fluctuations above Tc result in a d-wave gap with the finite arc reflecting closure of the gap in the nodal region due to classical thermal fluctuations.[8] In the other picture, as discussed later, the arcs represent one side of a hole-pocket associated with the proximity to the Mott insulating state.[9-13] These are commonly referred to as one-gap or two-gap models, respectively.[14]

Quantum oscillation experiments appear to indicate the presence of just such small Fermi pockets in the underdoped regime and a large Fermi surface in the overdoped regime.[15] Within the quantum oscillation community, there have been several different models proposed for the nature of the pockets. However it is important to recognize that those experiments are carried out in high magnetic fields. A recent study has reported that such high magnetic fields can in themselves lead to a spurious reconstruction of the Fermi surface resulting in the formation of pockets.[16] In the present paper we discuss, in more detail, our recent Angle-Resolved Photoemission Spectroscopy (ARPES) studies of the underdoped Bi2212 cuprates, with a particular emphasis on the possibility of nodal hole pockets. As discussed in detail elsewhere, the existence of such pockets is consistent with the fact that cuprates are strongly correlated systems or doped antiferromagnetic Mott insulators.

2. Experimental Procedure

Photoemission, by its nature, requires an electron in the initial state. Thus, it is seen as one of the primary probes of the occupied electronic states below the chemical potential. However, techniques have been developed to enable some level of study of the states above the chemical potential, E_F. These include the recognition that the occupation probability of the states immediately below and above E_F is determined by the finite width in energy of the Fermi-Dirac (FD) distribution function. Thus, access is provided to states above E_F simply by dividing the ARPES spectrum by the FD function.[17,18] Another technique that has become very prevalent with studies of the high T_c materials is the so-called symmetrization method.[4] Here one effectively forces a particle-hole symmetry onto the system and symmetrizes the data around the chemical potential. In the studies reported here, the spectra are analyzed by first removing any broadening due to experimental resolution, using the so called Lucy-Richardson technique,[19] and then dividing by the appropriate temperature dependent Fermi-Dirac distribution. As discussed elsewhere,[19] rather than simply dividing by a resolution broadened FD function, it is important to follow such a procedure, if one wants to avoid analysis induced binding energy shifts and false intensity variations in high resolution data.

The photoemission studies reported in this paper were carried out on underdoped cuprate samples, both Ca doped and oxygen deficient varieties. The Ca-rich crystal was grown from a rod with $Bi_{2.1}Sr_{1.4}Ca_{1.5}Cu_2O_{8+\delta}$ composition using an arc-image furnace with a flowing 20% O2-Ar gas mixture. The maximum T_c was 80K for Ca doped samples. The sample was then annealed at 700°C giving a 45K Tc with a transition width of ~2K. The oxygen-deficient $Bi_2Sr_2CaCu_2O_{8+\delta}$ (Bi2212) crystals were produced by annealing optimally doped Bi2212 crystals, at 450°C to 650°C for 3–15 days. The spectra shown in this paper were all recorded on beam line U13UB at the NSLS using a Scienta SES2002 electron spectrometer. Each spectrum was recorded in the pulse-counting mode with an energy and angular resolution of 15 meV and 0.1° respectively.

3. Results

Figure 1 shows spectra obtained after the analysis described above from the optimally doped Bi2212 (T_c=91K) and underdoped Bi2212 (T_c=65K) both in the normal state.[20] In both cases, the spectra were initially recorded with a photon energy of 16.5 eV and at a temperature of 140K. The spectra, after analysis, from the optimally doped materials show bands moving clearly through E_F as one moves away from the nodal direction. This is consistent with the loss of superconductivity in the normal state. However the spectra, after analysis, from the underdoped material show a gap in the spectral function that moves to deeper binding energies as one moves away from the nodal direction. The latter gap is consistent with the findings of phenomenological models of the pseudogap phase such as the so-called YRZ model, which predict the presence of Fermi hole pockets with x-dependent area in the pseudogap phase.[10-12] The predicted band dispersions from the YRZ model [10] are shown in fig. 2 for a doping level of p = 0.1 or $T_c \sim$ 65K. The figure shows the calculated spectral function $A(\underline{k},\omega)$ moving away from the nodal direction using the expression $A(\underline{k},\omega) = -(1/\pi)\mathrm{Im}\, G(\underline{k},\omega)$

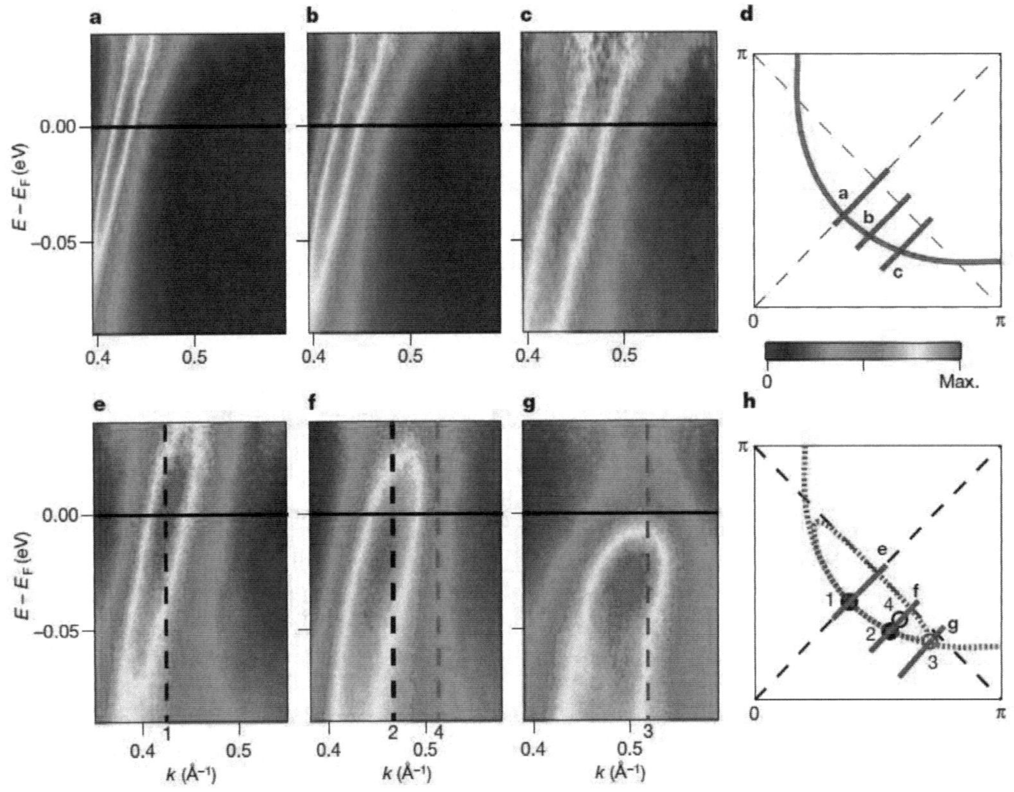

Fig 1. Reproduced from ref. [20], spectra, after analysis as described in the text, obtained as a function of angle α away from the nodal direction for optimally doped Bi2212 (upper panels) and the underdoped Bi2212 (lower panels) $T_c = 65K$. The location of the different cuts in momentum space are indicated in the schematic diagrams on the right.

where the Green's function G(k,ω) is given by the form

$$G(k,\omega) = \frac{g_t}{\omega-\xi(k)-\Delta_R^2/[\omega+\xi_o(k)]+i\Gamma} + G_{inc} \qquad (1)$$

with the various terms defined in ref. [10]. We have introduced a broadening term Γ to allow for finite lifetime effects. We note here that the expression

$$\Delta_R^2/[\omega + \xi_o(\underline{k})] \qquad (2)$$

is effectively a real self-energy term reflecting a hole moving through the background of paired singlets characterizing the RVB state.[21] It is this term that is responsible for the vanishing back side of the pocket. Each band in fig. 2 is characterized by two Fermi-surface crossings, one corresponding to a pole in the Green's function and one corresponding to a zero value. Within the context of the YRZ ansatz, the latter zeros occur wherever

$$\xi_0(\underline{k}) = -2t(\cos k_x + \cos k_y) = 0 \qquad (3)$$

Here t is the nearest neighbor hopping term and x is the doping level. Examination of equation 3 shows that it is satisfied along the magnetic zone boundary running from (π,0) to (0,π). As can be seen in fig. 2(b), the two Fermi surface crossings associated with the dispersing bands map out the boundary of a hole pocket with an area that approximately scales with the doping level.

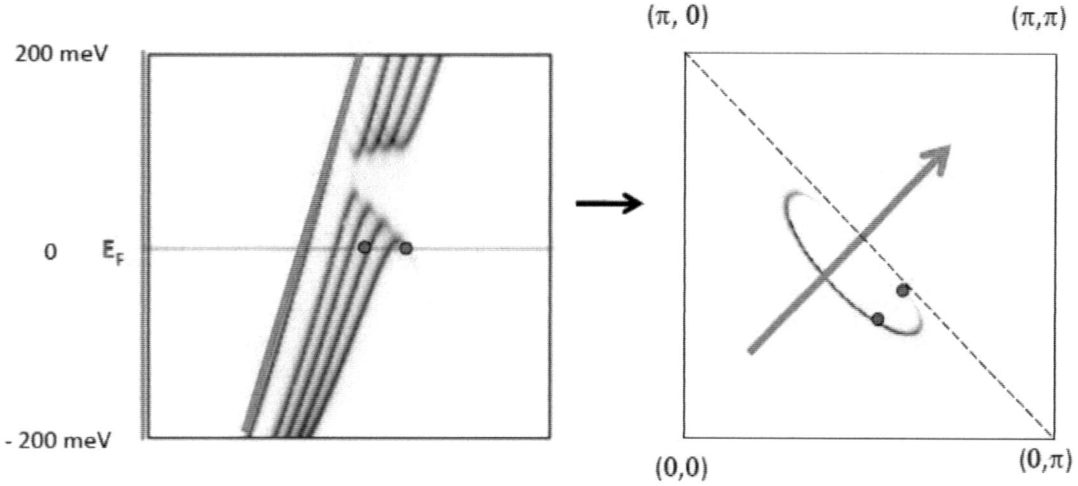

Fig 2. Dispersions calculated within the framework of the YRZ model on moving away from the nodal direction as defined by the green dispersing band. The red points indicate the Fermi surface crossings. The right panel shows the pocket defined by the crossings. The diagonal dashed line in the right panel defines the Luttinger surface which in this case coincides with the anti-ferromagnetic zone boundary.

That such a pocket can exist with one side's FS crossings defined by the poles of the Green's function and the backside defined by its zeros, is, as noted above, consistent with the observation that these materials are close to the Mott insulating state. Fig. 3 shows more finely detailed data recorded from the end of the arc region of a T_c = 65K sample.[22] The figure shows both the raw data and the same data after analysis as described earlier. Fig. 3(o) holds the key to the analysis. It is assumed that the dispersion of a given band will follow that implied by fig. 2. Then, experimentally, one makes two key observations, point A, the initial FS crossing and, point B, the maximum travel of the band before loss of intensity. Then, based again on fig. 2, we make the assumption that a third point, C, defines a second Fermi surface crossing or "Ghost" crossing. The k-space symmetry, or bending back, of the band reflected across X as shown in the figure is consistent with the pair-like structure of the pseudogap assumed by the YRZ model.

Fig. 4(a) shows the results of such an analysis for three samples of different doping levels as indicated. Fig 4(b) shows the results of calculations using the YRZ ansatz for comparison. The agreement between experiment and calculation for the FS is quite remarkable. Again we note that other theoretical approaches that recognize the strongly correlated nature of the cuprates will also produce pockets. The inset to figure 4(a) shows the area of the pockets as a function of doping, versus the nominal doping level deduced from the Presland formula relating doping to T_c.[23] It would appear that at higher doping levels there is some deviation from the 1:1 correspondence observed at lower levels. It is unclear whether this represents a breakdown of the Luttinger sum rule concept [24] or the formation of electron pockets in the anti-nodal region. The latter would not be observable in the experiment because of the presence of the pseudogap. However, we also note that the presence of electron pockets at doping levels much below optimum is inconsistent with the results of the YRZ ansatz.[10] We note that the pockets reported in the present study are in relatively good agreement with the area bounded by the locus of superconducting

Bogoliubov band minima $k_B(E)$ extracted from spectroscopic imaging scanning tunneling spectroscopy (SISTS) studies of a sample with a similar doping level.[25]

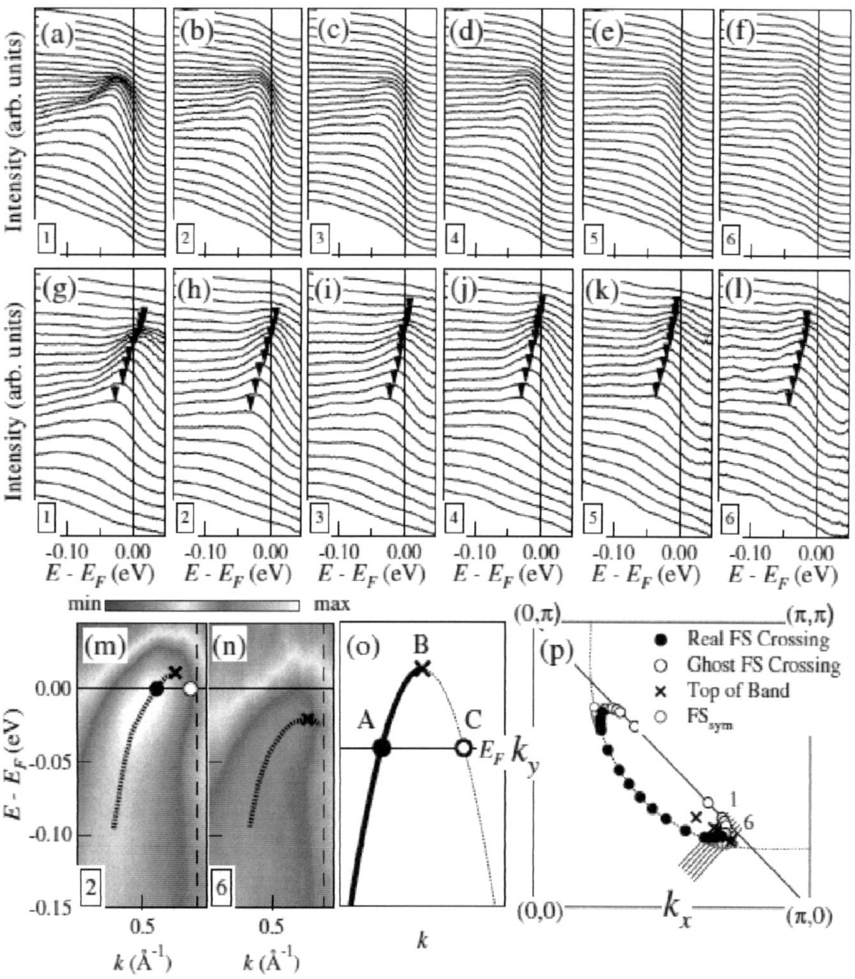

Fig 3. The upper panels (a-f) show spectra as recoded in the vicinity of the end of the arc. The center panels (g-i) show the same spectra following analysis as described in the text. Panel (m) shows a spectrum on the arc and (n) a spectrum off the arc. Panel (o) shows the basis of the analysis that results in the pocket shown in panel (p). Reproduced from Ref. [20].

4. Discussion

The calculated Fermi pockets in fig. 4(b) reflect different values of the pseudogap parameter Δ_R where

$$\Delta_R = \Delta_0(\cos k_x - \cos k_y) \tag{4}$$

represents a pseudogap having the same underlying symmetry as the superconducting gap for these materials. In fig. 5 we plot those values as a function of doping. The plot suggests a critical point in the vicinity of x=0.2 doping at which point the Fermi surface switches from being doping-dependent nodal

Fermi pockets to a large Fermi surface centered on the (π, π) point. Such a transition should be evident in, for instance, transport measurements and indeed it is, as found in the studies of Tallon et al.[26]

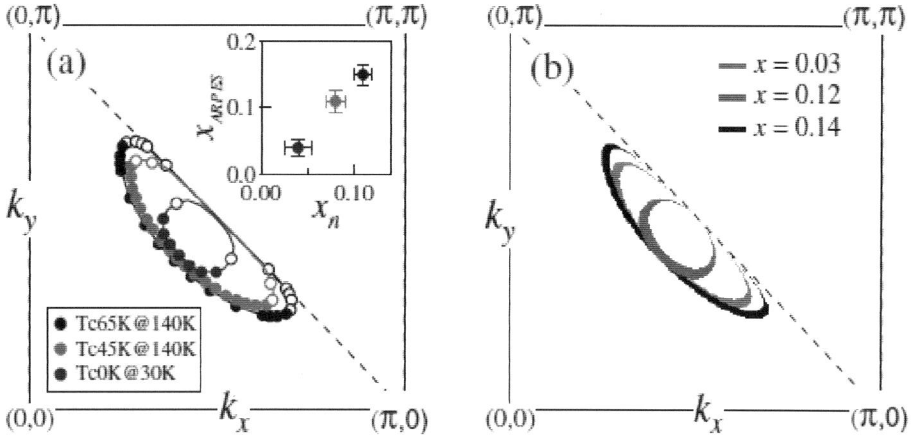

Fig. 4. (a) Experimentally pockets determined in the manner outlined in figure 3 for a series of doping levels. (b) Nodal Fermi pockets calculated within the framework of the YRZ ansatz for the same doping levels. Reproduced from ref [20].

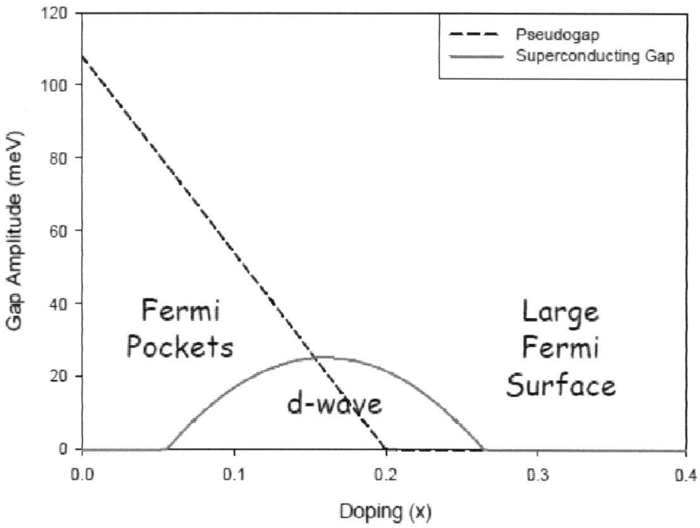

Fig. 5 Plot of the pseudogap parameter Δ_0 as a function of doping x. The plot would indicate the presence of a possible critical point in the vicinity of 0.2 as suggested in ref [10].

The studies and analysis reported in this paper provide strong support for the possibility that the disconnected Fermi arcs observed in the Bi2212 system represent one side of nodal hole pockets with area scaling in proportion to the doping level. As noted earlier, the development of such pockets is entirely consistent with the fact that the cuprates emerge from the Mott-insulating state.

Although not reproduced here, we have shown elsewhere that in the anti-nodal region the spectral response in the normal state reveals a gap symmetric with respect to the chemical potential. A symmetric gap is an indication of some form of pairing. What exactly is the nature of the pairing? While some authors have suggested the possibility of pre-formed Cooper pairs,[8] others have suggested some form of singlet pairing.[21] Here we note again that the symmetric gap is observed only along the copper-oxygen bond directions. The gap is not symmetric away from these directions, implying that the pairing is not showing the full d-wave symmetry observed for the superconducting state as it exists below T_c. This would suggest that the pairing observed near the zone boundary is in fact real-space singlet pairing reflecting the underlying spin lattice of the CuO_2 planes.

Acknowledgements
The authors would like to thank Seamus Davis, Mike Norman, Philip Phillips, Maurice Rice, John Tranquada, Alexei Tsvelik, Subir Sachdev, Tonica Valla, and Ali Yazdani for useful discussions. The work at Brookhaven is supported in part by the U.S. DOE under Contract No. DE-AC02-98CH10886 and in part by the Center for Emergent Superconductivity (CES), an Energy Frontier Research Center funded by the U.S. DOE, Office of Basic Energy Sciences. The work at Argonne is partially supported by the U.S. DOE under Contract No. DE-AC02-06CH11357 and partially by the same CES. T. E. Kidd acknowledges support from the Iowa Office of Energy Independence Grant No. 09-IPF-11.

References:
[1] J.G. Bednorz and K.A. Muller, Zeitschrift fur Physik B – Condensed Matter, 64, 189 (1986)
[2] Y. Kamihara, T. Watanabe, M. Hirano, H. Hosono, JACS 130, 3296 (2008)
[3] T. Timusk and B. Statt, Rep. Prog. Phys. 62, 61 (1999)
[4] M. R. Norman et al., Nature (London) 392, 157 (1998).
[5] V. J. Emery and S. A. Kivelson, Nature (London) 374, 434 (1995).
[6] A.V. Chubukov and D. K. Morr, Phys. Rep. 288, 355 (1997).
[7] S. Chakravarty, C. Nayak, and S. Tewari, Phys. Rev. B 68, 100504 (2003).
[8] M.R. Norman,, A. Kanigel, M. Randeria, U. Chatterjee, and J.C. Campuzano, Rev. B 76, 174501 (2007)
[9] T.D. Stanescu, G. Kotliar, Phys Rev B, 74, 125110, (2006).
[10] K.Y. Yang, T. M. Rice, and F. C. Zhang, Phys. Rev. B 73, 174501 (2006).
[11] K.Y. Yang et al., Europhys. Lett. 86, 37 002 (2009)
[12] S. Sachdev, M. Metlitski, M. Punk, J. Phys.-Condensed Matter, 24, 294205 (2012)
[13] Seungmin Hong and Philip Phillips, arXiv:1110.0440 (2012)
[14] S Hufner, M A Hossain, A Damascelli and G A Sawatzky, Rep. Prog. Phys. **71** 062501 (2008)
[15] Doiron-Leyraud et al., Nature (London) 447, 565 (2007)
[16] Wu, T. et al. Nature 477, 191194 (2011).
[17] Kisker, E. et al., Phys. Rev. Lett. 58, 1784–1787 (1987).
[18] Matsui, H. et al., Phys. Rev. Lett. 90, 217002 (2003).
[19] J. D. Rameau, H.-B. Yang, and P. D. Johnson, J. Electron Spectrosc. Relat. Phenom. 181, 35 (2010).
[20] H.-B. Yang et al., Nature (London) 456, 77 (2008).
[21] P.W. Anderson, Science, 235, 1196 (1987)
[22] H.-B. Yang et al., Phys. Rev. Lett., 107, 047003 (2011)
[23] M.R. Presland, J.L. Tallon, R.G. Buckley, R.S. Liu and N.E. Flower, Physica C 176, 95 (1991)
[24] K.B. Dave, P.W. Phillips, C.L. Kane, arXiv:1207.4201 (2012)
[25] Y. Kohsaka et al., Nature (London) 454, 1072 (2008).
[26] J. L. Tallon, J. W. Loram, J. R. Cooper, C. Panagopoulos, and C. Bernhard, Phys Rev B, 68, 180501 (2003).

Large in-plane Elastic Scattering due to Cu-vacancies in an Electron Doped Superconductor

S. Kunwar[1], F. C. Niestemski[2], V. Madhavan[3]

[1]Department of Physics, King Fahd University of Petroleum and Minerals, Dhahran, 31261, KSA
[2]Department of Physics, Stanford University, Stanford, California 94305, USA
[3]Department of Physics, Boston College, Chestnut Hill, Massachusetts 02467, USA

Email: kunwar@kfupm.edu.sa

Abstract. High transition temperature (high-Tc) superconductivity arises in cuprate materials when sufficient holes or electrons are doped into the CuO_2 layers of their antiferromagnetic (AF) insulating parent compounds. While hole doping immediately turns the material superconducting, mere electron doping is not enough to induce superconductivity in some of the cuprates which need to be annealed in a low oxygen environment in order to induce superconductivity. In this work we analyse STM spectra on an electron doped superconductor, $Pr_{1-x}LaCe_xCuO_{4\pm\delta}$(PLCCO), and show that compared to their hole doped counterparts, these electron-doped materials show a significant in-plane elastic scattering component. Since the copper deficiencies are intrinsic to PLCCO these could be the main source of the high in-plane scattering rate.

1. Introduction

An important test of the properties of the superconducting state is how non-magnetic and magnetic impurities affect it. There have been extensive theoretical [1,2,3] and experimental [4,5,6] studies of impurities in conventional superconductors. According to the pioneering work of Abrikosov and Gor'kov[1] and its extentions [2,3], although both magnetic and non-magnetic impurities are pair breakers in conventional superconductors, non-magnetic impurities have a weaker effect compared to magnetic impurities [4]. A magnetic perturbation in a conventional superconductor reduces the superconducting order parameter and leads to the appearance of quasiparticle excitations within the superconducting gap. Just the opposite is however expected to be true in the case of high-T_c cuprates, where magnetic impurities are predicted to have a weaker effect on superconducting properties [7,8]. In the high T_c cuprates however, these effects are difficult to discern from macroscopic measurements. For example, with the exception of Zn, resistivity and other bulk measurements indicate that both magnetic and non-magnetic impurities have similar effects on superconductivity [9,10,11,12]. Local probes like STM are therefore crucial to measure impurity effects on high T_c cuprates [4,7,13].Our approach in this paper is to obtain information on the quasiparticle scattering rate of PLCCO by analysing the lineshape of the STM spectra.

Superconductivity in electron doped materials arises within very narrow (10% to 18%) range of electron concentration. Our samples are optimally doped; 12% of trivalent Pr^{3+} is replaced by tetravalent Ce^{4+}. However, unlike their hole-doped counter parts mere doping is not sufficient to induce superconductivity in electron doped cuprates. They need to be post annealed. PLCCO samples used in this study were post annealed to 970°C under an Ar environment to achieve the superconducting transition temperature T_c = 24 K. Samples were cleaved in ultra high vacuum and inserted into the STM head held at 4K. From our previous studies [14,15], although the PLCCO surface lacks atomic scale features, it is possible to get reproducible differential conductance *(dI/dV(r,eV))* spectra from these samples. Figure 1 shows typical spectra obtained at different locations on PLCCO. Inhomogeneity, which is one of the hallmarks of high-T_c cuprates, is clearly visible in the figure as the spectra have different superconducting gaps. Our first observation is that compared to the prototypical hole-doped cuprate $Bi_2Sr_2CaCu_2O$ (Bi2212), the coherence peaks in this electron-doped sample are highly suppressed. Qualitatively, this indicates an increased scattering rate in PLCCO compared the Bi2212. However, in order to establish the magnitude and possible origins of this scattering we seek to obtain more quantitative information. Since there is no established microscopic theory for cuprate superconductors, we use a simple phenomenological picture to capture the spectral line shape [16,17]. This model captures the momentum space (k-space) variation in the superconducting gap [18,19] due to its d-wave symmetry.

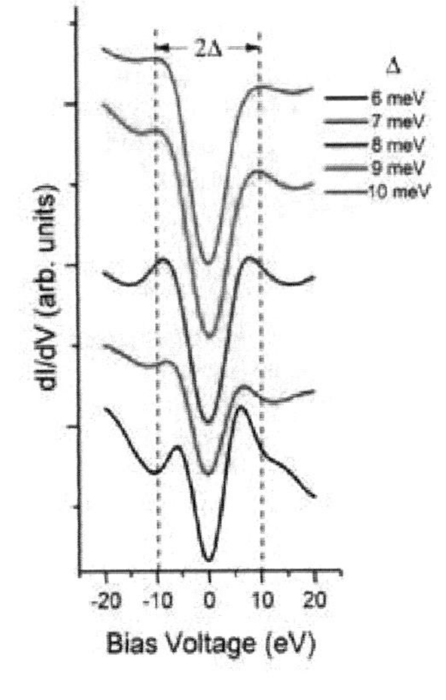

Figure 1: Typical STM spectrum taken with atomically resolved STM. Each spectrum is an average of five neighbouring spectra.

Modifying the BCS equations to account for d-wave pairing symmetry is straightforward and was done by K. Maki [20]. The tunnelling conductance is equivalent to the local density of states and can be written in terms of extended BCS theory for d-wave pairing as:

$$\frac{dI}{dV}(E,T) = \int_0^{\pi} \frac{df}{dE} d\theta \, Re\left(\frac{|E-i\Gamma|}{\sqrt{(E-i\Gamma)^2-\Delta(T)^2 Cos^2(2\theta)}}\right) \tag{1}$$

Where, "f" is Fermi Dirac distribution function at temperature T, $E=eV$ is the quasiparticle energy, Γ is quasiparticle scattering rate, and Δ is superconducting gap.

2. Energy Independent Scattering Rate

First we assume that the scattering rate has no energy dependence. By best fitting the integrated density of states for d-wave symmetry with the tunneling spectra we obtain the superconducting gap, Δ and the quasiparticle scattering rate, Γ. Figure 2 (a) shows the typical fit between theory and data. The histogram of Γ extracted from more than a thousand tunnelling spectra is shown in Figure 2(b). It is found that the mean value of the Γ is 2.8±0.76 meV. However, we find that the fits to our data with Γ held constant are not perfect. We therefore turn to similar studies on hole-doped cuprates. In $Bi_2Sr_2CaCu_2O_{8+\delta}$ (BSCCO2212), Alldredge et. al. reported that for an extensive doping range (8% to 22%) the constant scattering rate from near unitary scatterers plays only a subsidiary role. In contrast they find that using an energy dependent effective scattering rate ($\Gamma = \alpha E$), plays a key role at low temperatures [17]. Surprisingly, in our case we were not able to fit our spectra with the simple effective scattering rate, $\Gamma = \alpha E$. Instead, as shown in Figure 3 (a), the best fits were obtained with the combined scattering rate scattering rate $\Gamma = \Gamma_1 + \alpha E$.

This combined scattering rate is similar to that seen in angle resolved photoemission spectroscopy studies of underdoped Bi2212 used by Kaminski et al [21]. Similar to our observations, their studies indicate that the functional form of the scattering rate for underdoped and optimally doped samples has both energy independent and energy dependent parts with the

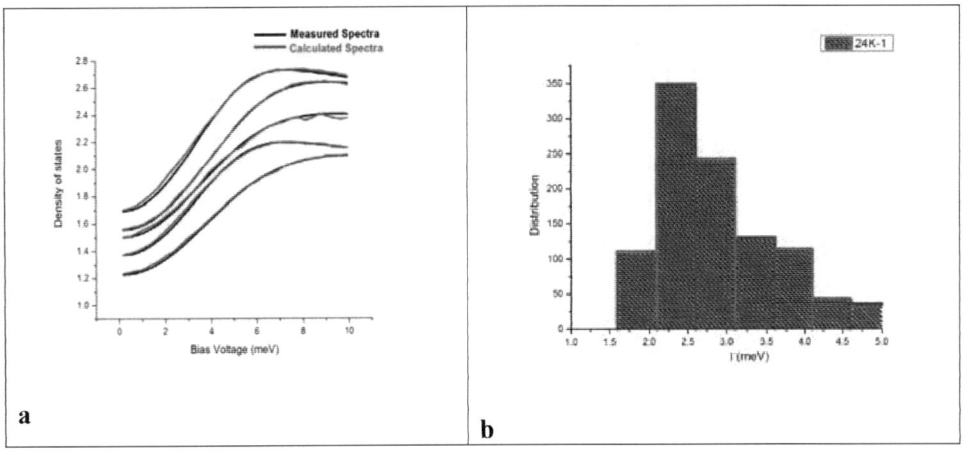

Figure 2.0: (a) Best fit with PLCCO spectra at 5 K using constant scattering rate Γ and **(b)** the histogram of Γ obtained from 1024 STM spectra

constant Γ_1 arising from the elastic contribution to scattering and the coefficient α representing the inelastic contribution. The histogram of inelastic scattering coefficient α is shown in Figure 3(b). The inelastic processes arise from exchange of dynamic spin fluctuations or phonons, whereas the elastic scattering can arise from impurities and disorder within the copper oxide planes (in-plane impurities) or from impurities in the neighbouring metal oxide planes (out-of-plane impurities). Scattering by out-of-plane impurities for example, has been claimed to be responsible for the electronic nanoscale inhomogeneity in the superconducting gap.[22,23] We find that unlike the hole doped cuprates, the in-plane constant elastic scattering (Γ_1) is much higher in PLCCO. Γ_1 has an average value of 1.5 meV, which is approximately 20% of the mean superconducting gap.

3. Sources of constant elastic scattering:

The high-temperature synthesis process of electron doped superconductors creates a Cu deficiency (~1%) in some of the as-grown electron doped cuprates[24, 25]. Kang et. al. suggested that these vacancies behave like the Zn impurities and locally destroy superconductivity [26]. While the post-annealing process repairs some of these vacancies, we expect that not all vacancies are repaired in the 24K PLCCO samples resulting in strong elastic scattering which is responsible for the large value of Γ_1.

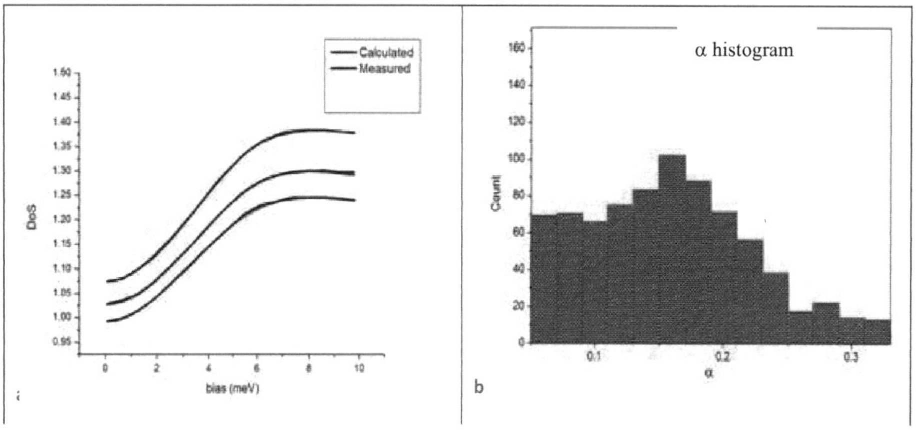

Figure 3.0: **(a)** Best fit with PLCCO spectra at 5K using $\Gamma = \Gamma_1 + \alpha E$. **(b)** α histogram obtained by fitting dI/dV spectra.

4. Conclusion:

We have found a very high constant in-plane elastic scattering rate in the electron doped superconductor PLCCO. Since this is not detected in the hole-doped counterparts, the constant elastic scattering is intrinsic to electron doped superconductors. Cu vacancies, which are naturally present in electron doped superconductors, act like non-magnetic impurities and can have a strong detrimental effect on superconductivity. Our experimental data indicate that most of the scattering in PLCCO occurs in the CuO_2 planes by non-magnetic unitary scatterers, which can be potentially attributed to these Cu vacancies.

Acknowledgements:

Work at Boston College is supported by U.S. NSF-CAREER-0645299 (V.M.) and DOE DESC0002554 and the work at King Fahd University of Petroleum and Minerals is supported by Deanship of Scientific Research (Project # JF100014). We thank Khalil A. Ziq at Department of Physics, King Fahd University of Petroleum and Minerals for valuable discussion.

References

[1] A. A. Abrikosov and L. P. Gor'kov, Zh. Eksp. Teor.Fiz.39, 1781 (1962), Sov.Phys. JETP 12, 1243(1961).
[2] L. Yu, Acta Phys. Sin. 21, 75 (1965); H. Shiba, Prog.Theor.Phys. 40, 435 (1968); A. I. Rusinov, Zh. Eksp.Teor.Fiz.56, 2047 (1969), Sov.Phys. JETP 29, 1101 (1969).
[3] J. Zittartz and E. Mu¨ller-Hartmann, Z. Phys. 232, 11 (1970); J. Zittartz, ibid. 237, 419 (1970).

[4] A. Yazdani, B. A. Jones, C. P. Lutz, M. F. Crommie, D. M. Eigler, "Probing the Local Effects of Magnetic Impurities on Superconductivity" Science 275, 1767 (1997).

[5] A. S. Edelstein, "Anomalous Gapless Superconductivity Due to Scattering from Localized Noninteracting Spins", Phys. Rev. Lett. 19, 1184 (1967)

[6] W. Bauriedl, P. Ziemann, and W. Buckel, "Electron-Tunneling Observation of Impurity Bands in Superconducting Manganese-Implanted Lead", Phys. Rev. Lett. 47, 1163(1981)

[7] E. W. Hudson, K. M. Lang, V. Madhavan, S. H. Pan, H. Eisaki, S. Uchida, J. C. Davis "Interplay of magnetism and high-T_c superconductivity at individual Ni impurity atoms in $Bi_2Sr_2CaCu_2O_{8+\delta}$" Nature411, 920 (2001)

[8] S. H. Pan, E. W. Hudson, K. M. Lang, H. Eisaki, S. Uchida& J. C. Davis "Imaging the effects of individual zincimpurity atoms on superconductivityin $Bi_2Sr_2CaCu_2O_{8+d}$", Nature, 403, 746 (2003)

[9] Maeda, A., Yabe, T., Takebayashi, S., Hase, M. &Uchinokura, K. "Substitution of 3d metals for Cu in $Bi_2(Sr_{0.6}Ca_{0.4})_3Cu_2O_y$" Phys. Rev. B 41, 4112 (1990).

[10] Ishida, K. et al. "Gapless superconductivity in Zn-doped $YBa_2Cu_3O_7$ studied by Cu NMR and NQR. Possibility of d-wave superconductivity in high-T_c oxides" Physica C 179, 29–38 (1991).

[11] D. A. Bonn, S. Kamal, Kuan Zhang, Ruixing Liang, D. J. Baar, E. Klein, and W. N. Hardy "Comparison of the influence of Ni and Zn impurities on the electromagnetic properties of $YBa_2Cu_3O_{6+x}$." Phys Rev. B 50, 4051 (1994).

[12] Fukuzumi, Y., Mizuhashi, K., Takenaka, K. & Uchida, S. "Universal superconductor-insulator transition and Tc depression in Zn-substituted high-Tccuprates in the underdoped regime." Phys. Rev.Lett. 76, 684 (1996).

[13] Ilija Zeljkovic, Zhijun Xu, Jinsheng Wen, Genda Gu, Robert S. Markiewicz, Jennifer E. Hoffman. "Imaging the Impact of Single Oxygen Atoms on Superconducting $Bi_{2+y}Sr_{2-y}CaCu_2O_{8+x}$", Science 337, 6092 (2012)

[14] F. C. Niestemski, S. Kunwar, S. Zhou, Shiliang Li, H. Ding, Ziqiang Wang, Pengcheng Dai & V. Madhavan."A distinct bosonic mode in an electron-doped high-transition temperature superconductor", Nature 450, 1058 (2007).

[15] Jun Zhao, F. C. Niestemski, Shankar Kunwar, Shiliang Li, P. Steffens,A. Hiess, H. J. Kang, Stephen D. Wilson, Ziqiang Wang, Pengcheng Dai, & V.Madhavan"Electron-Spin Excitation Coupling in an Electron Doped Copper OxideSuperconductor" Nature Physics 7, 719 (2011).

[16] Abhay N. Pasupathy,AakashPushp, Kenjiro K. Gomes,Colin V. Parker, Jinsheng Wen, ZhijunXu, GendaGu, Shimpei Ono, Yoichi Ando, Ali Yazdani "Electronic Origin of the InhomogeneousPairing Interaction in the High-TcSuperconductor $Bi_2Sr_2CaCu_2O_{8+\delta}$" Science 320, 196 (2008)

[17] J. W. Alldredge, Jinho Lee, K. McElroy, M. Wang, K. Fujita, Y. Kohsaka, C. Taylor, H. Eisaki, S. Uchida, P. J. Hirschfeld & J. C. Davis. "Evolution of the electronic excitation spectrum with strongly diminishing hole density in superconducting $Bi_2Sr_2CaCu_2O_{8+\delta}$" nature physics 4, 319 (2008)

[18] C. C. Tsuei, J. R. Kirtley, C. C. Chi, Lock See Yu-Jahnes, A. Gupta, T. Shaw, J. Z. Sun, and M. B. Ketchen . "Pairing Symmetry and Flux Quantization in a Tricrystal Superconducting Ring of $YBa_2Cu_3O_{7-\delta}$", Phys. Rev. Lett.73, 593 (1994).

[19] D. A. Wollman, D. J. Van Harlingen, W. C. Lee, D. M. Ginsberg, and A. J. Leggett. "Experimental determination of the superconducting pairing state in YBCO from the phase coherence of YBCO-Pb dc SQUIDs Phys. Rev. Lett.71, 2134 (1994).

[20] K. Maki, "Magnetotransport in spin-density waves", Phys. Rev. B 47, 11506 (1994).

Journal of Physics: Conference Series **449** (2013) 012008 doi:10.1088/1742-6596/449/1/012008

[21] A. Kaminski, H. M. Fretwell, M. R. Norman, M. Randeria, S. Rosenkranz, U. Chatterjee, J. C. Campuzano, J. Mesot, T. Sato, T. Takahashi, T. Terashima, M. Takano, K. Kadowaki, Z. Z. Li,10 and H. Raffy "Momentum anisotropy of the scattering rate in cuprate superconductors" Phys. Rev. B 71, 014517 (2005).
[22] M. M. Maska, Z. Sledz, K. Czajka, and M. Mierzejewski, "Inhomogeneity-Induced Enhancement of the Pairing Interaction in Cuprate Superconductors" Phys. Rev. Lett. 99, 147006 (2007)
[23] T. S. Nunner, B. M. Andersen, A. Melikyan, and P. J. Hirschfeld, "Dopant-Modulated Pair Interaction in Cuprate Superconductors" Phys. Rev. Lett.95, 177003 (2005).
[24] M. Matsuura, Pengcheng Dai, H. J. Kang, J. W. Lynn, D. N. Argyriou, K. Prokes, Y. Onose, and Y. Tokura, "Effect of a magnetic field on the long-range magnetic order in insulating Nd_2CuO_4 and nonsuperconducting and superconducting $Nd_{1.85}Ce_{0.15}CuO_4$" Phys. Rev. B 68, 144503 (2003)
[25] Shailing Li et. al. "Impact of oxygen annealing on the heat capacity and magnetic resonance of superconducting $Pr_{0.88}LaCe_{0.12}CuO_4$", PRB78, 014520 (2008)
[26] H. J. Kang, Pengcheng Dai, B. J. Campbell, P. J. Chupas, S. Rosenkranz, P. L. Lee, Q. Huang, S. L. Li, S. Komiya, and Y. Ando, "Microscopic annealing process and its impact on superconductivity in T′-structure electron-doped copper oxides", Nat. Mater. 6, 224 (2007).

Journal of Physics: Conference Series **449** (2013) 012009 doi:10.1088/1742-6596/449/1/012009

Electrostatic tuning of the superconductor to insulator transition of $YBa_2Cu_3O_{7-x}$ using ionic liquids

Xiang Leng, Javier Garcia-Barriocanal, J. Kinney, Boyi Yang, Yeonbae Lee, and A. M. Goldman

School of Physics and Astronomy, University of Minnesota, Minneapolis, Minnesota, USA

email: goldman@physics.umn.edu

Abstract. Ultrathin $YBa_2Cu_3O_{7-x}$ (YBCO) films were grown on $SrTiO_3$ (STO) substrates using a high-pressure oxygen sputtering system. The films were incorporated in a field effect transistor configuration to study the control of superconductivity by electrostatic charging. While devices using STO as both the substrate and gate dielectric have produced only small T_c shifts, a clear transition between superconducting and insulating behaviour was realized using an electronic double layer transistor employing the ionic liquid DEME-TFSI as a gate dielectric. Employing a finite size scaling analysis, curves of resistance vs. temperature were found to collapse into two branches over the temperature range from 6 K to 22K suggesting the existence of a quantum critical point. However the scaling failed at temperatures below this range, indicating the possible presence of an additional phase between the superconducting and insulating regimes. Further depletion of holes appears to result in the accumulation of electrons resulting in a change of the majority carriers from holes to electrons and the emergence of what appears to be very weak re-entrant superconductivity. By changing the polarity of the gate voltage, an underdoped film was tuned into the overdoped regime by accumulating additional holes. An unexpected two-step mechanism for electrostatic doping was revealed. Hall effect measurements exhibited anomalous features, which suggest the occurrence of an electronic phase transition near optimal doping.

1. Introduction

High temperature superconductors (HTSCs) with copper oxide planes are believed to be doped Mott insulators. Thus far many hole- [1. 2] and electron-doped [3, 4] HTSCs have been reported, with the doping process usually involving chemical substitution or oxygen stoichiometry modification. The modification of the electrical properties of such strongly correlated electron systems using an electric field as an external control parameter has been a long-standing goal in condensed matter physics [5, 6]. Field effect transistor concepts applied to HTSCs can provide a tool to control the superconducting condensate in a reversible manner, while keeping a fixed structure and avoiding altering the disorder associated with conventional approaches to doping. It can be a useful tool to study fundamental questions that still remain open regarding such systems.

Recently the development of electronic double layer transistors (EDLTs) that use ionic liquids (ILs) as gate dielectrics has been successfully employed to achieve levels of doping of the order of $10^{15}/cm^2$ [7–9]. Exploiting such large charge transfers, this approach has been used to tune the superconductor-insulator (SI) transition of $La_{2-x}Sr_xCuO_4$ (LSCO) [10] and of $YBa_2Cu_3O_{7-x}$ (YBCO) [11] as well as explore aspects of the phase diagram of YBCO. In the following we focus on work relating to YBCO films where we exploit the possibility of large transfers of charge facilitated through the use of EDLTs employing ILs. We will describe experiments in which carriers are depleted and display the transition to the insulating regime and beyond, and those in which they are accumulated so as to tune an underdoped sample into the overdoped regime, going across the top of the superconducting dome in the phase diagram. Studies of ***R(T)*** will be complimented by measurements of the Hall effect, which have been quite revealing in some instances.

Electronic Double Layer Transistors are configurations, which perform in a manner similar to conventional FETs. They contain a gate electrode, a semiconductor, an insulator or a metal into which charge can be accumulated or depleted, source and drain electrodes, and for purposes of characterization, electrodes for measuring longitudinal and transverse voltages relative to the current direction. In this instance the gate dielectric is an IL, although polymer electrolytes have also been employed. Ionic liquids are molten salts consisting of large ions such that their Coulomb interactions are sufficiently small to make them room temperature liquids. Upon applying a gate voltage, ions move to the surface of the semiconductor or insulator, forming an electric double layer (EDL) such that with the charge induced in the channel, acts as a capacitor of nano-scale thickness. This ultra-small effective thickness is what leads to the possibility of large charge transfer. A T_c shift of about 50 K has been observed in YBCO thin films by earlier workers using ILs [12]. However, no transition from superconductor to insulator was found. The reason is that the films studied were substantially thicker than the Thomas- Fermi screening length, which roughly determines the depth of the induced charge. Thus the gating process would affect the order of only one UC. In the case of films of thicknesses of 10 UCs or greater, which were studied, the proximity effect would dominate, and no transition to an insulating regime would be possible.

2. Experimental

Films were grown on (001)-oriented $SrTiO_3$ (STO) substrates using a high-pressure oxygen sputtering system employing a nominally stoichiometric YBCO target. Previous work by others, using the same technique, has shown that high pressure oxygen sputtering can provide high quality epitaxial ultrathin YBCO films [13, 14]. Amorphous Al_2O_3 films were deposited as masks, and the substrates were etched in buffered (10:1) HF solution and annealed in an O_2/O_3 mixture for 6 h at 750^{0}C. Atomic force microscopy (AFM) was used to make sure that the substrate surfaces were clean and were TiO_2 terminated. The oxygen pressure during deposition was 2.0 mbar and the substrate temperature was 900^{0}C. After deposition, films were cooled in an 800 mbar O_2 atmosphere and annealed at 500^{0}C for 30 min. Films were characterized *ex situ* by AFM and x-ray diffraction. Thickness was measured using x-ray reflectivity. A series of samples with thicknesses ranging from 5 to 10 UC was fabricated and measured using standard four-probe techniques. Films thinner than or equal to 6 UC were insulating whereas thicker ones were superconducting. This suggests that the first 5 to 6 UC are insulating and that a 7 UC film has a superconducting layer that is actually only 1 to 2 UC thick.

Ultrathin YBCO films are sensitive and react with most chemicals so we did not carry out any processing after film growth other than the deposition of Pt electrodes. The devices used were similar to the one shown in Fig. 1 of Ref. 8. Films were covered with the ionic liquid N, N-diethyl-N-(2-methoxyethyl)-N-methylammonium bis (trifluoromethyl sulphonyl)-imide (DEME-TFSI) [8] and were quickly cooled down to 240 K. The ionic liquid condenses into a rubberlike state at 240 K, a temperature at which most chemical reactions are suppressed. If the device is kept at room temperature for several hours with ionic liquid on it, we see an increase in resistance and a drop in T_c, possibly due to chemical reaction. However, after cooling down and being kept below 240 K, no changes are seen over many days. For measurements, the gate voltage was changed at 240 K where it was held for 1 h before cooling down. The gate voltage was kept constant during measurements. Gate voltages were alwayschanged incrementally and the response of the films when hole doped was found to be reversible albeit with some hysteresis. We were not able to confirm this behavior in the electron-doped regime, which suggests that the changes effected in that regime may be chemical rather than physical.

3. The superconductor-insulator transition of YBCO

Curves of Resistance vs. temperature, $R(T)$, at various gate voltages of a 7 UC thick YBCO film are shown in Fig. 1. The sample starts as a superconductor with an onset temperature of about 77 K. Here the onset temperature is taken to be the temperature at which the sheet resistance falls to 90% of its normal value. We realized an insulator with a gate voltage (V_G) of only 1.52 V. We noticed that the onset temperature was a nonlinear function of V_G. There is a V_G threshold of about 0.3 V below which almost no change can be seen. Then the onset temperature drops by about 30 K as V increases from 0.5 to 1.1 V, which is about 5 K/0.1 V. However, as V_G increases from 1.1 to 1.25 V, a change of 0.2 to 0.3 V can induce an onset temperature shift of about 5 K. This nonlinearity suggests that the gate voltage cannot be used as a tuning parameter for quantitative analysis.

As V_G increases to 1.37 V, the $R(T)$ curves flatten out at the lowest temperature and then undergo a small upturn at $V_G \sim 1.40$ V, although their resistances initially decrease as temperature decreases. Finally, $R(T)$ evolves to an insulating state as V_G increases further, as can be seen clearly in the inset of Fig. 1. Interestingly, we have found that the resistance in the insulating side up to $V_G \sim 1.48$ V is thermally activated with an Arrhenius type relation. Above this voltage the logarithm of sheet resistance is fit by a $T^{-1/3}$ dependence consistent with 2D Mott variable range hopping.

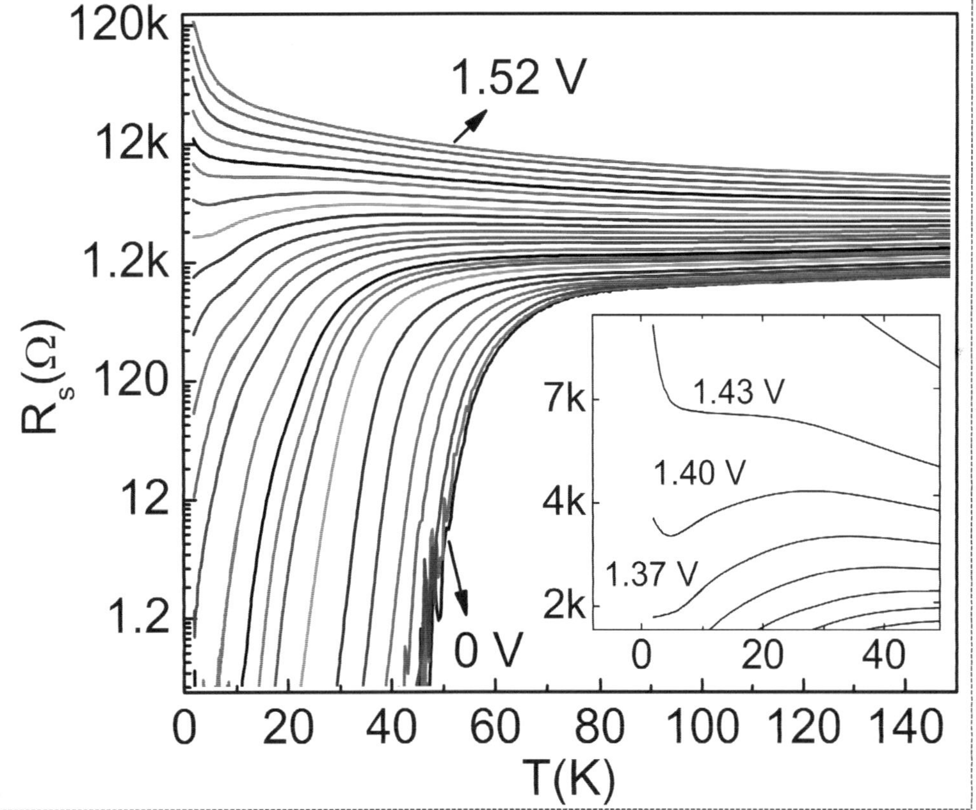

Figure 1. (color online). Logarithm of the sheet resistance vs. temperature with gate voltage varying from 0 to 1.52 V. From bottom to top: 0, 0.3, 0.5, 0.7, 0.8, 0.9, 1, 1.1, 1.12, 1.14, 1.17, 1.20, 1.25, 1.27, 1.30, 1.33, 1.35, 1.37, 1.40, 1.43, 1.44, 1.46, 1.48, 1.50, 1.52 (V). Inset: enlarged low temperature part near the transition between the superconducting and insulating regimes. (From Ref. 11.)

The low temperature upturn is similar to that observed in granular superconductors [15] and will dramatically affect any quantitative scaling analysis. We suggest that this is evidence of a mixed phase separating the superconducting and insulating regimes. Nevertheless, the thickness fluctuations in the single unit cell active layer could also contribute to this upturn. We also cannot rule out the possibility that we have not completely depleted the superconductivity of the underlayers.

For further quantitative analysis of the data, we require the induced carrier concentration. As discussed above, gate voltage is not simply related to the carrier concentration and the Hall resistance cannot be used because of the complicated electronic properties of YBCO. We assume that Drude behavior is found at high temperatures and take $1/R_s$ (200 K) to be proportional to the carrier concentration measured as carriers per in-plane Cu atom. Comparing this with the carriers per in-plane Cu atom derived from the phase diagram of bulk YBCO [16], we find that they match with a constant coefficient 0.165 as shown in Ref. 11. This is especially true in the transition regime, which is the focus of quantitative analysis. It is not true for carriers/Cu in excess of 0.10, necessitating reliance on the empirical relationship between T_c and carrier concentration to characterize the higher carrier concentration regime. This will be described in the next section.

Using measured $R(T)$ curves and calculated carrier concentrations, a phase diagram can be constructed and is plotted as Fig. 2. It is similar to the phase diagram of bulk high-T_c cuprates derived from chemical doping. Here it is measured continuously and perhaps more accurately, especially in the transition and insulating regimes. The data plotted here are in the linear regime of the I-V curves. There are nonlinear I-V curves deep in the insulating regime, which are denoted by white in the figure.

When a Quantum Critical Point (QCP) is approached, resistance isotherms as a function of the carrier concentration should cross at the critical resistance R_c and all resistance data should collapse onto a single scaling function. This was done in Ref. 10 in the case of LSCO, neglecting the lowest

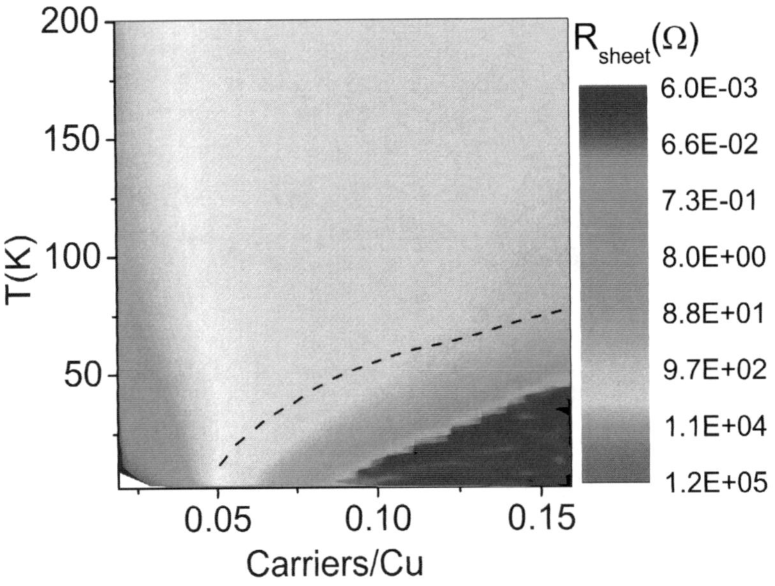

Figure 2. (color online). Plot of the $R(T)$ and calculated carrier concentration. This figure can be interpreted as a phase diagram. Different colors represent different sheet resistances. The dash line is the onset temperature. The white area in the bottom left-hand corner is a regime with nonlinear I-V curves. (From Ref. 11)

temperature isotherms. In the present work, the critical resistance per square was about 6.0 kΩ, which is very close to the quantum resistance given by $R_Q = h/4e^2 = 6.45$ kΩ considering that imprecise shadow masks were used to define the four-terminal configuration. It should be noted that in the work of Ref. 10 on LSCO the critical resistance was found to be the quantum resistance for electron pairs. The critical carrier concentration was x_c = 0.048 carriers/Cu, which is close to that derived from the bulk phase diagram (0.05). But this is not surprising given that the bulk phase diagram was used to calibrate the carrier concentration.

As shown in Ref 11, all the sheet resistance data from 6 to 22 K collapsed onto a single function $R_s = R_c f(|x-x_c|/T^{1/\nu z})$ with νz = 2.2. Assuming $z = 1$ [17], the value ν = 2.2 is close to that of the universality class of a metal- insulator transition in an anisotropic 2D system (7/3) [18] or that of the quantum percolation model (2.43) [19]. The value of νz = 2.2 for YBCO is different from νz = 1.5 found for LSCO [10]. This could be a consequence of the difference in the disorder or the structure of the two systems. Since the transition was initially traversed by the depletion of carriers, one can rule out the possibility that layers underneath the top layer have been rendered conducting upon the application of voltage. However, the scaling of the YBCO data breaks down below 6 K, which is a consequence of the non-monotonic behavior of $R(T)$ at low temperature at gate voltages in the transition region. These results suggest that the transition from superconductor to insulator is indirect and may involve an intermediate phase and possibly multiple quantum critical points.

4. Accessing the overdoped regime

The possibility of accumulating charge carriers by the field effect, involving solid dielectric, to tune an underdoped cuprate into the overdoped regime has hitherto remained elusive. For the 123 family of cuprate compounds, it has been shown that the CuO_2 planes are only indirectly affected by the electric field since the injected holes are mainly induced in the CuO_x chains [20]. Moreover, the systematic preparation of YBCO bulk samples by tuning the oxygen stoichiometry in the overdoped region is difficult to achieve and the highest level of over-doping is limited by the oxygen stoichiometric concentration, 0.194 holes/Cu for $YBa_2Cu_3O_7$ [21].

We have exploited the high local electric fields (10^9 V/m) induced by an ionic liquid at the surface of the sample to tune the concentration of holes in the superconducting condensate across the top of the superconducting dome. The experiment reveals that the electrostatic doping of YBCO involves a different doping mechanism, as well as an anomalous normal resistance behavior in the overdoped regime. Surprisingly, we also find that the Hall number measured at 180 K displays a maximum around the optimal doping level that suggests the occurrence of an electronic phase transition separating the underdoped and overdoped regimes.

We monitored the gating process of a 7 unit-cell YBCO film by measuring the sheet resistance as a function of temperature (see Fig. 3). The fresh sample, without any applied gate voltage, has a transition temperature $T_c \sim 42$ K. We determined T_c as the temperature corresponding to the crossing of the extrapolation of the fastest falling part of the $R(T)$ curve to zero resistance. After the application of a negative gate voltage, T_c increases and the normal state sheet resistance (the metallic region of the curve) drops. For higher negative gate voltages, V_G = -2.24 to -2.38 V, the rate of change of T_c and the normal resistance slows down and saturates. T_c reaches a maximum value of 67 K and the normal resistance reaches a minimum of 750Ω at 180 K. We also observed small fluctuations in both T_c and the normal resistance in this regime. As shown in Fig. 3(b), upon a further increase of the negative gate voltage, T_c decreases. In addition, the normal resistance increases with increasing negative gate voltage. For the highest negative gate voltages (- 2.46V $< V_G <$ -2.56 V) we note that the superconducting transition is not abrupt and a second transition is turned on at the lowest temperature part of the $R(T)$ curve. Next we reversed the polarity of the gating [see Fig. 3(c)], and we observed that after a threshold voltage of about +0.3 V, T_c starts to increase and the normal resistance starts to drop. With further increase in the positive gate voltage, as shown in Fig. 3(d), T_c drops and the normal resistance increases, and the sample returns to its initial state.

In order to obtain an independent variable to describe the process of accumulation or depletion of holes we inferred an effective hole doping "p" by using the universal empirical parabolic relation $T_c/T_{c,max}$ $1 = 1-82.6(p - 0.16)^2$ [22], which has been proved to be true for bulk high T_c superconductors including 123 compounds [23]. Using this calculated number of holes, the values of T_c and of the normal resistance at 180 K have been plotted as Figs. 2(b) and 2(c) in Ref. 24. Without going into detail we present the conclusions.

The electrostatic doping of YBCO appears to be a two-step process. First carriers are induced only on the CuO_x chains, and second, once a threshold concentration is reached, carriers can be induced on the CuO_2 planes. In the second step, the carriers could be indirectly doped on the CuO_2 planes through intra-cell charge transfer [20]. This two-step electrostatic doping process appears to be different from that of chemical doping, in which any changes of the hole doping under the superconducting dome will affect the normal resistance and the superconducting properties, simultaneously.

The normal resistance at 180 K exhibits a minimum at the optimal doping point. As stated before this is a surprising behavior since in the overdoped region of the general bulk phase diagram of cuprates, the normal state is a Fermi liquid and a lowering of the normal resistance would be expected as the doping level increases. We have no clear or definitive explanation for the observation of an increasing resistance with increasing doping in the overdoped regime.

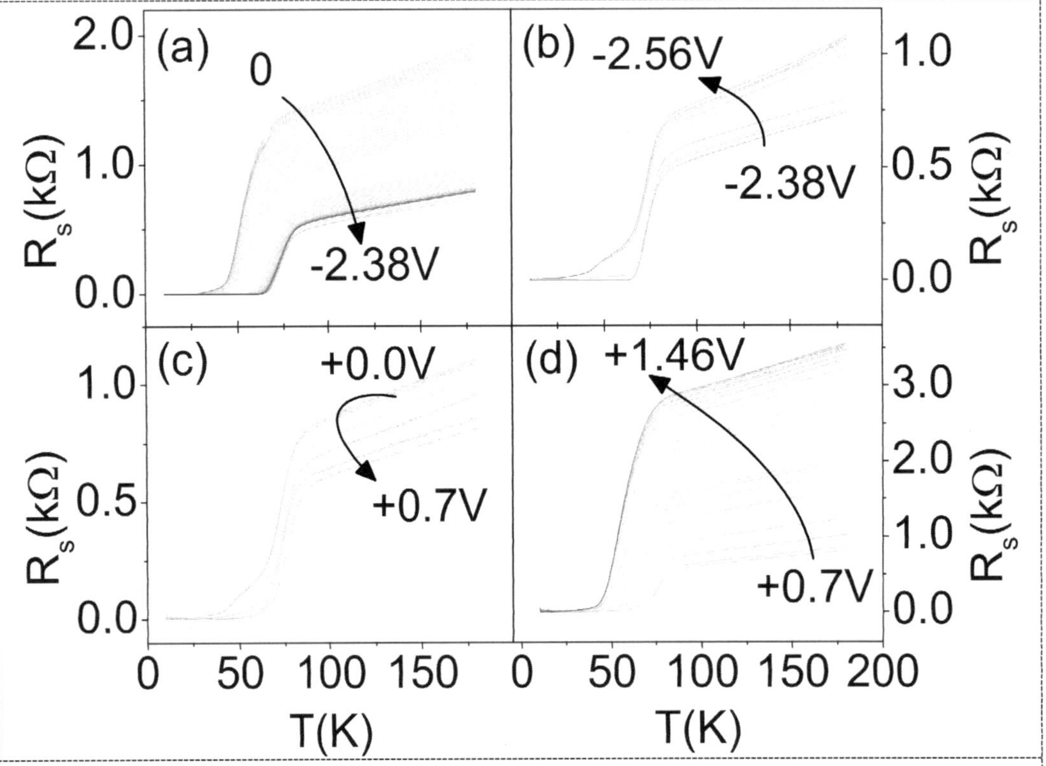

Figure 3. (color online). Temperature dependence of the sheet resistance at different gate voltages. (a) From 0 to -1.2 V, V_G was changed in -0.1 V increments and then from -1.2 to -2.38 V, the increment was -0.02 V. (b) From -2.38 to -2.56 V, V_G was changed in increments of -0.02 V. (c) Negative V_G was then removed and a positive V_G was added. From 0 to 0.7 V, V_G was changed in 0.1 V increments. (d) V_G = 0.7, 0.75, 0.80, 0.85, 0.90, 0.95, 1.0, 1.05, 1.07, 1.10, 1.13, 1.15, 1.17, 1.20 V, then the increment was changed to 0.02 V until V_G = 1.46 V. (From Ref. 24)

To obtain further insight into this behavior we measured the Hall resistance at 180 K at different gate voltages. Figure 4 shows the normalized Hall number (n_H) as a function of V_G (a) and effective hole doping (b). The Hall number is calculated using $n_H = 1/R_H e$, R_H being the Hall coefficient determined from the slope of the linear fit of the Hall resistance vs. magnetic field. Interestingly, we find that the Hall number shows a peak at a doping level of $\boldsymbol{p} \sim 0.15$ holes/Cu ($\boldsymbol{V_G}$ = -2.24 V) which does not coincide with the maximum and minimum of T_c and the normal resistance, respectively, (p = 0.16 holes/Cu or $\boldsymbol{V_G}$ = -2.38 V). Moreover, in between these levels of doping we observe fluctuations in both T_c and the normal resistance.

Experimentally, an anomalous dependence of the normal state Hall coefficient on the doping level has been reported in chemically doped YBCO [25], $Bi_2Sr_{1.51}La_{0.49}CuO_{6+\delta}$ (BSLCO) [26] and LSCO [27]. For YBCO the Hall conductivity at 125 K showed an anomalous behavior around the 60-K phase and in BSLCO and LSCO, the Hall number peaks around the optimal doping level at low temperature when superconductivity is suppressed by an ultra high magnetic field. These anomalous behaviors are commonly interpreted as evidence of an unusual electronic state or a sudden change in the Fermi surface, although chemical doping might also change the magnetic coupling and thus the Hall scattering rate. For electrostatic doping, we do not believe that an electric field would change the magnetic coupling. The $\boldsymbol{n_H}$ peak then suggests that an electronic phase transition is occurring, which is supported by the emergence of fluctuations in T_c and the normal state resistance at this same level of doping.

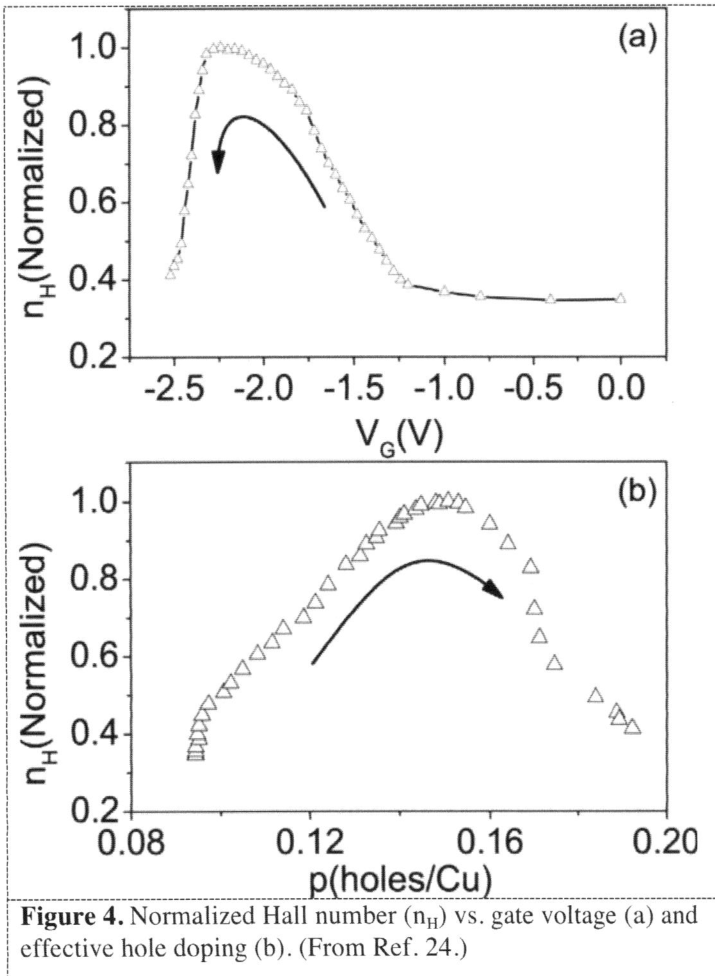

Figure 4. Normalized Hall number (n_H) vs. gate voltage (a) and effective hole doping (b). (From Ref. 24.)

The peak of the Hall number is found to be at $\boldsymbol{p} \sim 0.15$, to the left of the optimal doping point (p = 0.16). For a doping level of $\boldsymbol{p} \sim 0.15$ the value expected for T*, the pseudogap crossover temperature, roughly matches 180 K [28]. Quantum oscillation experiments [18, 29–31] showed that while there is a large hole-like Fermi surface in the overdoped regime, there are only small pockets in the underdoped regime. Angle Resolved Photoemission Spectroscopy also showed that the transition from the overdoped to the underdoped regime is accompanied by an electronic reconstruction of the Fermi surface [32–36]. Thus an electronic phase transition, which would reconstruct the Fermi surface is expected around the optimal doping regime and could be confirmed by our measurements.

5. The electron doped regime

The electron-doped side of 123 compounds such as YBCO cannot be reached by oxygen manipulation because of the limits of chemical stoichiometry. To our knowledge, the only material reported thus far is lanthanum-doped YBCO, with lanthanum substituted on both the yttrium and barium sites, $Y_{1-z}La_z(Ba_{1-x}La_x)_2Cu_3O_y$ (YLBLCO)[37]. These 123 compounds exhibit ambipolar transport on both the hole- and electron-doped regimes, but superconducting behaviour on the electron-doped side is not found. Extreme hole reduction and subsequent electron accumulation was demonstrated in YBCO films by an electrochemical technique by Nojima *et al.* [38]. After depleting the holes and driving the film insulating, additional electrons resulted in metallic *n*-type YBCO but with no sign of superconductivity. This work was done using ion-gels consisting of polyethylene oxide and an alkali metal perchlorate

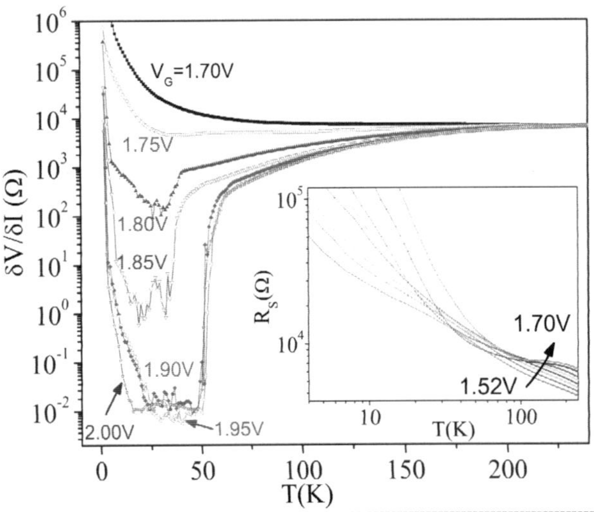

Figure 5. (color online) Differential resistance ***δV/δI*** vs. temperature at gate voltages from 1.70 V to 2.00 V. In this region the I-V curves are not linear at low temperatures (see Fig. 2) and the slope of the I-V curve at the lowest current limit is used to derive the differential resistance. Inset: Sheet resistance vs. temperature at gate voltage from 1.52 V to 1.70 V. From bottom to top, V_G = 1.52, 1.54, 1.56, 1.58, 1.60, 1.63, 1.67, 1.70 (V). In this region the I-V curves are linear. The lowest differential resistances for the two highest gate voltages are effectively in the noise, suggesting a range of zero or near zero resistance behavior.

We have found that after depleting the holes and driving a YBCO film insulating using the ionic liquid DEME-TFSI, further addition of electrons results in metallic behavior and indications of superconductivity. Hall effect measurements suggested that the majority carriers are electrons, implying that electron-doped YBCO may be a superconductor. However the superconductivity weakens with decreasing temperatures and the ground state is apparently an insulator. Because the samples did not exhibit reversible behaviour we cannot rule out the possibility that the observations are not due to electrochemical rather than electrostatic effects.

A nominally 7 unit-cell (UC) thick YBCO film has only 1-2 UCs that are active and without a gate voltage is an underdoped hole superconductor. Its superconductivity can be completely eliminated with a 1.52 V positive gate voltage due to the depletion of holes. In Fig. 5 we plot the differential resistance vs. temperature, at different gate voltages and it is apparent that superconductivity begins to be recovered when V_G =1.70 V. The onset of the transition can be as high as 60 K at the highest gate voltage (V_G =2.0 V). In preparing this plot we used the differential resistance measured in the limit of the lowest current in the regions where the *I* -*V* curves were nonlinear. When the *I* - *V* curves are linear, the differential resistance determined this way is the same as the sheet resistance measured at constant current. The regime in which local minima in the resistance vs. temperature curves developed will be shown to be dominantly electron doped through the interpretation of Hall effect data.

High temperature ***R(T)*** curves exhibit metallic behavior for V_G > 1.75 V but do not have a simple T^{β} power law dependence. We know that in the hole-doped cuprates such as YBCO and LSCO, the high temperature *R(T)* curves are linear [39] and in electron-doped cuprates such as $Nd_{2-x}Ce_xCuO_4$ (NCCO), the resistance shows a T^2 dependence[40]. Our data exhibit a totally different behavior and thus the mechanism of superconductivity might be different in an electrostatically charged system with carriers that are electrons. Another surprising feature is that the $\delta V/\delta I$ increases sharply as the temperature is lowered, resulting in local minima in the ***R(T)*** curves.

Since we are applying a positive gate voltage, which causes depletion of holes or accumulation of electrons in the system, the superconducting state at this high gate voltage would be expected to be electron doped. In Fig. 6 we plot the high temperature (T = 200 K) Hall resistance vs. magnetic field for different gate voltages. The slope of this curve can be used to determine the Hall coefficient (R_H). Although we cannot use the Hall coefficient to calculate the carrier concentration but we can still use it ($1/eR_H$) to estimate the relative changes of the carriers [41] and to determine their sign [42]. The upper panel shows positive Hall coefficients at low gate voltages which are consistent with the fact that the sample starts as a hole-doped superconductor. As the gate voltage increases the Hall coefficient increases thus $1/eR_H$ drops which confirms the depletion of holes. The curve for V_G = 1.70 V is very noisy and we cannot determine any slope with gate voltages between V_G =1.70V and 1.90 V.

The next measurable slope is found at V_G = 1.95 V and the Hall slope has changed sign, which strongly suggests that the carriers are electrons and that any superconductivity is electron-doped. The other feature is that the Hall coefficients in the lower panel are much larger than the ones in the upper panel. This suggests the carrier concentration on the electron-doped side is much smaller than the one on the hole doped side. It is long believed that the electron-doped superconductors have both electrons and holes as charge carriers [43]. The presence of holes on the electron-doped side can cause the measured net carrier concentration to be much smaller than the actual value of the electron concentration. The reason the regime of the crossover between holes and electrons is noisy is not understood.

The film reported here, before any gating, exhibited an onset temperature for superconductivity of 77 K. The differential resistance in the region of the local minimum dropped at about 50 K by several orders of magnitude from its normal state value and effectively into the noise of the measurement. Except for the reentrance to the insulating state with decreasing temperature, this regime of the phase diagram is very similar to what is found for the superconducting regime on the hole-doped side.

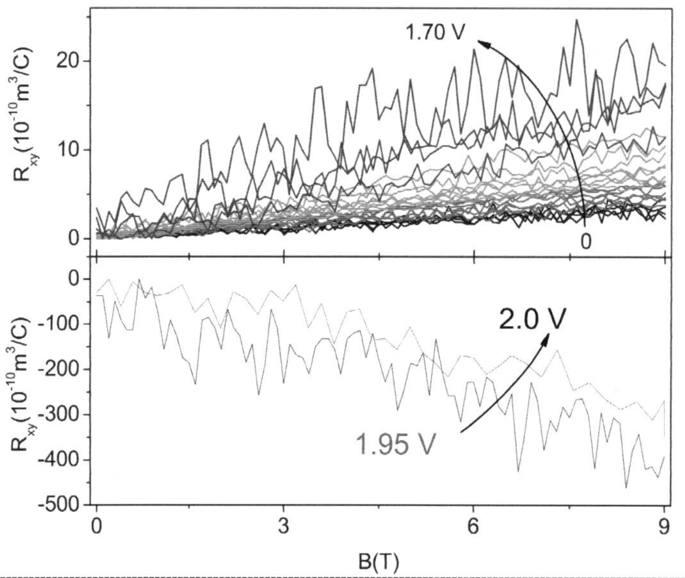

Figure 6. (color online) Hall resistance vs. magnetic field at different gate voltages. Upper panel: from bottom to top, V_G= 0, 0.3, 0.5, 0.7, 0.8, 0.9, 1, 1.1, 1.12, 1.14, 1.17, 1.20, 1.25, 1.27, 1.30, 1.33, 1.35, 1.37, 1.40, 1.43, 1.44, 1.46, 1.48, 1.50, 1.52, 1.54, 1.56, 1.58, 1.60, 1.63, 1.67, 1.70 (V). There is a gap in the data between gate voltages of 1.70 and 1.95V because of noise.

It is important to attempt to explain the reentrant behavior of the resistance as a function of temperature in the electron-doped regime. Along these lines we present a number of possible scenarios. First, local minima in ***R(T)*** curves are features usually seen in disordered or granular superconductors as an indication of a local superconducting state[44-46]. Such transitions are often referred to as being "quasi-reentrant" superconducting transitions since the resistance usually fails to fall to zero with decreasing temperature before beginning to increase with decreasing temperature. The film in the electron-doped regime may behave as an intrinsic granular system with superconducting islands connected by superconducting weak links and/or tunneling junctions. The resistance would then drop at the onset temperature due to the formation of superconducting islands. Before the establishment of superconducting coupling through weak links, the resistance would be controlled by quasiparticle tunneling. After the islands form and the temperature is decreased further, the quasiparticle tunneling paths are partially suppressed as the gap opens up and the resistance increases, resulting in a local minimum in the ***R(T)*** curve. In configurations in which the superconducting coupling dominates as the temperature is lowered, the superconducting islands will become coherently coupled and a global superconducting state will be achieved. However, a system in which the superconducting channels do not percolate will be an insulator even at $\boldsymbol{T} = 0$.

If the local minima are a consequence of the disorder in the film, the latter must be a very disordered system since the film is insulating at the lowest temperature. In the study by Nojima *et al.* of electrochemically induced electron-doped YBCO [38], the sample was quite thick (50 nm), and the metallic state also disappeared at low temperatures. It was suggested by these authors that localization effects are more serious on the electron-doped side. The film in the present work has only a 1 to 2 UC thick active layer. Thus it can be viewed as a 2-dimensional (2D) system. Such a 2D system might suffer more significant localization effects then a thicker film with interlayer coupling. Moreover, the charging process might also introduce disorder, as the layer of induced carriers would sample a random potential due to the ions of the electronic double layer. The presence of disorder may not be the explanation of the reentrant behavior as thin films prepared by high pressure oxygen sputtering are in general expected to be of high quality.

A second possibility is that electrostatically induced electron-doped reentrant superconductivity is mediated by spin fluctuations, but at low temperatures long-range antiferromagnetic (AF) order takes over. When this occurs, the spin fluctuations would then freeze out, suppressing the excitations that mediate the superconductivity. In the phase diagram of typical electron-doped superconductors, the insulating state protrudes into the superconducting regime at low temperature, which is different from what is found in hole-doped systems. This is found in the commonly studied electron-doped cuprate, NCCO, in which the AF state is very robust and persists to a very high doping level [4]. This type of reentrance would be distinct from that found in magnetic superconductors such as $ErRh_4B_4$ in that in this case the fluctuations associated with the magnetic order are intimately connected to the superconductivity, whereas they are not in the case of the latter [47].

It is possible that there are electrochemical reactions occurring in the ionic liquid based electrostatic charging process, which drives YBCO into the electron-doped regime. The fact that we could not demonstrate the electron charging process was reversible leaves this as a strong possibility. In the work of Nojima *et al.*, an electrochemical process was proposed [38], in which the reduction of oxygen atoms, first in the CuO_x chains and then in the CuO_2 planes, was the main mechanism for hole depletion and electron accumulation. In the present work the gate voltage was changed at 240 K and the device was always kept below this temperature during measurements. At these temperatures conventional electrochemical reactions are largely suppressed. On the other hand, the huge local electric fields in the electronic double layer (> 10^9 V/m) might also change the valence state of the Cu in the CuO_2 planes from Cu^{2+} to Cu^{1+}, resulting in electron doped metallic behavior and superconductivity. Although the superconductivity occurs on the CuO_2 planes, the interaction between the CuO_x chains and CuO_2 planes could be similar to what happens in the YLBLCO system[37]. The possible electrochemical origin of these observations would not detract from the conclusion that electron-doped YBCO has been realized which shows tendencies towards superconductivity.

Finally, one can rule out an artifact that might occur in a four-terminal measurement of resistance that yields zero resistance. This is the possibility that the voltage probes are not sampling a region of the sample where there is a nonzero current density. This might occur if the electrical transport were to occur through the IL rather than through the film. However all of the measurements were obtained at temperatures at which the ionic liquid DEME-TFSI is strongly insulating. As a consequence this is an unlikely occurrence.

6. Summary

A transition from superconducting to insulating behavior has been induced in several ultrathin YBCO films prepared by high pressure oxygen sputtering and modified through electrostatic charging using an EDLT device configuration. A striking feature of the data is the similarity between the phase diagram and the bulk phase diagram. This is surprising given that the active layer in the film is only one to two UCs and that there are high electric fields in the double layer. The transition between superconducting and insulating behavior is not direct and resembles the transition found in granular ultrathin films. It is not known that this is a general result. We have also accumulated holes, changing from the underdoped to the overdoped regime. The results reveal a two-step doping mechanism that may be different from conventional chemical doping. The normal state resistance increased with hole concentration in the overdoped regime, which is different from what is found in bulk systems. There was also a peak in the high temperature (180K) Hall number at $\boldsymbol{p} \sim 0.15$. This suggests that there is an electronic phase transition on the Fermi surface near the optimal doping point. Finally, we have electrostatically depleted the holes in a YBCO film and recovered indications of superconductivity when the majority of the carriers are electrons as determined from the sign of the Hall resistance. The superconductivity does not persist to the lowest temperatures studied. Thus the ground state in this regime would appear to be insulating. However we were not able to test reversibility in this regime, in contrast with the hole doped regime. Thus it is an open question as to whether the apparent electron doped regime of YBCO realized in these experiments is a consequence of a physical process involving electrostatic charging or is due to chemical reactions at the interface between the ionic liquid and the YBCO film, which is an alternative scenario

Acknowledgements
The authors would like to thank S. Bose for his assistance with sample preparation and C. Geppert for his help with the measurements. This work was supported by the National Science Foundation under grants NSF/DMR-0709584 and 0854752. Part of this work was carried out at the University of Minnesota Characterization Facility, a member of the NSF-funded Materials Research Facilities Network via the MRSEC program, and the Nanofabrication Center, which receives partial support from the NSF through the NNIN program. JGB thanks the Spanish Ministry of Education for the financial support through the National Program of Mobility of Human Resources (2008-2011).

References
[1] J. G. Bednorz and K. A. Muller 1986 *Zietschrift fur Physik B Condensed Matter* **64** 189
[2] M. K. Wu *et al.* 1987 *Phys. Rev. Lett.* **58** 908
[3] Y. Tokura H. Takagi and S. Uchida 1989 *Nature* **337** 345
[4] N. P. Armitge, P. Fournier, and R. L. Greene 2010 *Rev. Mod. Phys.* **82** 2421
[5] C. H. Ahn *et al.* 2006 *Rev. Mod. Phys.* **78** 1185
[6] C. H. Ahn *et al.* 1999 *Science* **284** 1152
[7] J. T. Ye *et al.* 2009 *Nature Mater.* **9** 2009
[8] Y. Lee *et al.* 2011 *Phys. Rev. Lett.* **106** 136809
[9] H. Shimotani *et al.* 2007 *Appl. Phys. Lett.* **91** 082106
[10] A. T. Bollinger *et al.* 2011 *Nature (London)* **472**
[11] X. Leng *et al.* 2011 *Phys. Rev. Lett.* **107** 027001
[12] A. S. Dhoot *et al.* 2010 *Adv. Mater.* **22** 2529
[13] M. Varela *et al.* 2001 *Phys. Rev. Lett.* **86**,5156
[14] M. Varela *et al. 1999 Phys. Rev. Lett.* **83** 3936
[15] H. M. Jaeger *et al.* 1989 *Phys. Rev. B* **40** 182
[16] N. Doiron-Leyraud *et al.* 2007 *Nature (London)* **447** 565
[17] I. F. Herbut 2001 *Phys. Rev. Lett.* **87** 137004
[18] M. Ruhlander and C. M. Soukoulis 2001 *Phys. Rev. B* **63** 085103
[19] D.-H. Lee, Z. Wang, and S. Kivelson 1993 *Phys. Rev. Lett.* **70** 4130
[20] M. Salluzzo *et al.* 2008 *Phys. Rev. Lett.* **100** 056810
[21] R. Liang, D. A. Bonn, and W. N. Hardy 2006 *Phys. Rev. B* **73** 180505
[22] M.R. Presland *et al.* 1991 *Physica (Amsterdam)* **176C** 95
[23] J. L. Tallon *et al.* 1995 *Phys. Rev. B* **51** 12911
[24] X. Leng *et al.* 2012 *Phys. Rev. Lett.* **108** 067004
[25] K. Segawa and Y. Ando 2001 *Phys. Rev. Lett.* **86** 4907
[26] F. F. Balakirev *et al.* 2003 *Nature (London)* **424** 912
[27] F. F. Balakirev *et al.* 2009 *Phys. Rev. Lett.* **102** 017004
[28] J. L. Tallon *et al.* 1999 *Phys. Status Solidi B* **215** 531
[29] C. Jaudet et al. 2008 *Phys. Rev. Lett.* **100** 187005
[30] B. Vignolle *et al.* 2008 *Nature (London)* **455** 952
[31] D. LeBoeuf *et al.* 2007 *Nature (London)* **450** 533
[32] M. R. Norman *et al.* 1998 *Nature (London)* **392** 157
[33] K. M. Shen *et al.* 2005 *Science* **307** 901
[34] M.Plate *et al.* 2005 *Phys.Rev.Lett.* **95** 077001

[35] M. A. Hossain *et al.* 2008 *Nature Phys.* **4** 527
[36] M. R. Norman 2010 *Physics* **3** 86
[37] K. Segawa *et al.* 2010 *Nat. Phys.* **6** 579
[38] T. Nojima *et al.* 2011 *Phys. Rev. B* **84** 020502
[39] M. Gurvitch *et al.* 1988 *Physica C* **153-155** 1369
[40] C. Tsuei, A. Gupta, and G. Koren 1989 *Physica* C **161** 415
[41] Y. Ando *et al.* 2000 *Phys. Rev. B* **61** R14956
[42] T. D. Stanescu and P. Phillips 2004 *Phys. Rev. B* **69**, 245104
[43] W. Jiang *et al.* 1994 *Phys. Rev. Lett.* **73** 1291
[44] B. G. Orr, H. M. Jaeger, and A. M. Goldman 1985 *Phys. Rev. B* **32** 7586
[45] M. Kunchir *et al.* 1987 *Phys. Rev. B* **36**, 4062
[46] A. Gerber *et al.* 1990 *Phys. Rev. Lett.* **65** 3201
[47] W. A. Fertig *et al.* 1977 *Phys. Rev. Lett.* **38** 987

High Field determination of superconducting fluctuations in high-T$_c$ cuprates

F. Rullier-Albenque[1], H. Alloul[2], D. Colson[1] and A. Forget [1]

[1] Service de Physique de l'Etat Condensé, Orme des Merisiers, CEA Saclay (CNRS URA 2464), 91191 Gif sur Yvette cedex, France

[2] Laboratoire de Physique des Solides, UMR CNRS 8502, Université Paris Sud, 91405 Orsay, France

E-mail: florence.albenque-rullier@cea.fr

Abstract. Large pulsed magnetic fields up to 60 Tesla are used to suppress the contribution of superconducting fluctuations (SCF) to the ab-plane conductivity above T_c in a series of $YBa_2Cu_3O_{6+x}$ single crystals. The fluctuation conductivity is found to vanish nearly exponentially with temperature, allowing us to determine precisely the field $H_c'(T)$ and the temperature T_c' above which the SCFs are fully suppressed. T_c' is always found much smaller than the pseudogap temperature A careful investigation near optimal doping shows that T_c' is higher than the pseudogap $T^\star$, which indicates that pseudogap cannot be assigned to preformed pairs. For nearly optimally doped samples, the fluctuation conductivity can be accounted for by gaussian fluctuations following the Ginzburg-Landau scheme. A phase fluctuation contribution might be invoked for the most underdoped samples in a T range which increases when controlled disorder is introduced by electron irradiation. Quantitative analysis of the fluctuating magnetoconductance allows us to determine the critical field $H_{c2}(0)$ which is found to be quite similar to $H_c'(0)$ and to increase with hole doping. Studies of the incidence of disorder on both T_c' and $T^\star$ enable us to propose a three dimensional phase diagram including a disorder axis, which allows to explain most observations done in other cuprate families.

1. Introduction

One of the most puzzling feature of the high-T_c cuprates is the existence of the so-called pseudogap phase in the underdoped region of their phase diagram. After the first evidence of an anomalous drop of the spin susceptibility detected by NMR experiments in underdoped YBCO well above T_c [1], a lot of unusual properties have been observed in the pseudogap phase [2]. However its exact relationship with superconductivity is still highly debated. In fact there does not exist up to now a unique representation for the pseudogap line $T^\star$ as illustrated in Fig.1. Either T^* is found to merge with the superconducting dome in the overdoped part of the phase diagram, or to cross it near optimal doping. In the first case, it has been proposed that the pseudogap could be ascribed to the formation of superconducting pairs with strong phase fluctuations [3]. This scenario has been supported by the observation of a large Nernst effect and of diamagnetism above T_c, which delineates another line T_ν below which strong superconducting fluctuations and/or vortices persist in the normal state [4]. In the second approach, the pseudogap and the superconducting phases arise from different, even competing, underlying mechanisms and are associated with different energy scales [5].

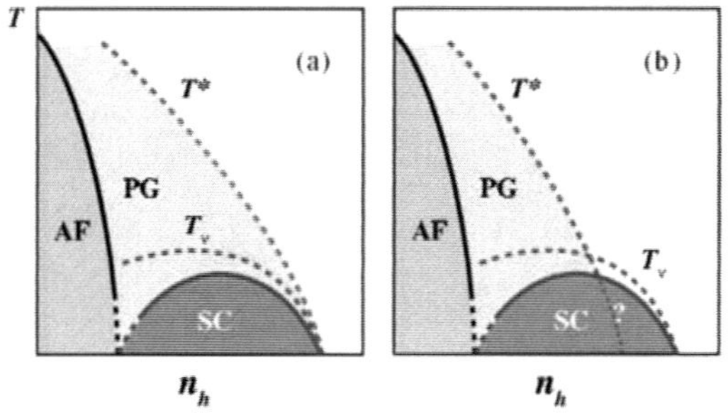

Figure 1. Different scenarios for the phase diagram of the high-T_c cuprates. While in (a) T^* merges with T_c in the strongly overdoped regime, in (b) $T^\star$ intersects the superconducting dome near optimal doping. T_ν represents the onset of the Nernst signal. From ref.[8]

While superconducting fluctuations (SCF) are expected to be especially large in these anisotropic compounds with low superfluid density, there is not a clear consensus about the temperature range above T_c in which SCFs survive. In this paper, we will present our results on superconducting fluctuations for a series of $YBa_2Cu_3O_{6+x}$ single crystals from underdoping to slight overdoping. We have used an original method based on the measurements of the magnetoresistance in high pulsed magnetic fields. We have initiated this method in ref.[6] in underdoped compounds, and have done complementary measurements for various O contents [7]. In this later paper we gave an extensive report on the data and their analysis. Here we present a simplified comprehensive summary of our main results and discuss them in the context of recent experimental results reported by others at this conference.

The principle of our study is to use high magnetic fields to determine the normal state resistivity and to extract with high accuracy the superconducting fluctuations (SCF) contributions to the conductivity and their dependences as a function of temperature and magnetic field (section 2). We are thus able to determine the threshold values of the magnetic field H_c' and temperature T_c' above which the normal state is completely restored. In the same set of transport data, we could compare the values of T_c' and of the pseudogap temperature $T^\star$ as a function of doping [8] (sec.3). We have added in the present paper another comparison of these two temperature scales by using previous NMR data to determine $T^\star$ [1]. We will also show how our results can be analysed in the framework of the Ginzburg-Landau approach, making it possible to extract microscopic parameters of the superconducting state such as the zero-temperature coherence length (sec.4-5). The effect of disorder introduced by electron irradiation at low temperature will be also presented (sec.6).

2. Experimental

Details on the experimental conditions concerning the different single crystals and the high-field experiments as well as the method used to extract the SCF contribution to the conductivity are given in ref.[7]. Four different single crystals of $YBa_2Cu_3O_{6+x}$ have been studied. They are labelled with respect to their critical temperatures measured at the mid-point of the resistive transition: two underdoped samples UD57 and UD85, an optimally doped sample OPT93.6 and a slightly overdoped one OD92.5, corresponding to oxygen contents of approximately 6.54, 6.8, 6.91 and 6.95 respectively. Some of these samples have been irradiated by electrons at low T, which allows us to introduce a well controlled concentration of defects in the CuO_2 planes [9].

The transverse MR of the different samples have been measured in a pulse field magnet up to 60T at the LNCMI in Toulouse. An example of the transverse MR curves measured on the OPT93.6 sample is illustrated in fig.2 for T ranging from above T_c to 150K. In the normal state well above T_c, it was shown that the transverse MR increases as H^2 both in optimally doped and in underdoped YBCO [10]. This is indeed what is found also here for H up to 60T and for $T \gtrsim 140$K (see inset of fig.2). At lower T, some downward departure from this H^2 behavior is observed for low values of H which we attribute to the destruction of SCFs by the magnetic field. The normal state behavior is only restored above a threshold field $H_c'(T)$ which increases with decreasing temperatures. In ref.[7], we have shown that the magnetoresistance coefficients measured at low T above $H_c'(T)$ are in continuity with those measured at higher temperatures and low field, which is a strong indication that the effect of the field is to restore the normal state.

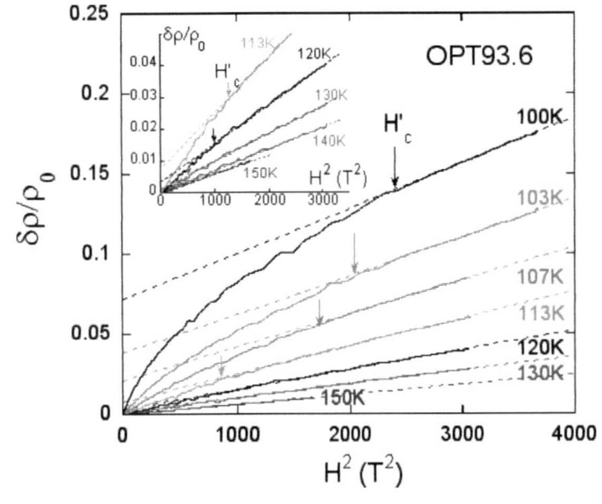

Figure 2. Field variation of the resistivity increase normalized to its zero-field value $\delta\rho/\rho_0$ plotted versus H^2 for decreasing temperatures down to $T \simeq T_c$ in the optimally doped sample OPT93.6. The inset shows an enlargement of the curves for the highest temperatures. (from ref.[7]).

This experimental approach allows us to single out the normal state properties and determine the SCF contributions to the transport. In particular, the extrapolation down to $H = 0$ of the H^2 normal state MR above $H_c'(T)$ gives us the value of the normal state resistivity $\rho_n(T)$. The way to extract the fluctuating conductivity and its dependence with temperature $\Delta\sigma_{SF}(T, 0)$ and magnetic field $\Delta\sigma_{SF}(T, H)$ is explained in details in ref.[7].

3. Onset of superconducting fluctuations and pseudogap

The T dependences of the zero-field superconducting conductivities $\Delta\sigma_{SF}(T, 0)$ are reported in Fig.3 for the OPT93.6 and UD57 samples and compared to the off-diagonal Peltier conductivity deduced from the Nernst measurements on the same samples [11]. We observe that T_c' is always found larger that the onset of Nernst signal T_ν. This might come from the difficulty to choose a good criterion to determine T_ν as the minimum in $\alpha_x y/B$ may hide the real onset of SCFs. On the contrary, in Fig.4 where $\Delta\sigma_{SF}(T, 0)$ are plotted in a semi-logarithmic scale, one can note that this quantity vanishes very fast. This allows us to define a precise criterion to determine the onset temperature T_c', corresponding here to $\Delta\sigma_{SF}(T, 0)$=1 x $10^3(\Omega m)^{-1}$. Of course decreasing or improving the sensitivity for SCF detection will result in decrease or increase of T_c', which might explain the different temperature ranges of SCFs derived from different experimental probes. If we were able to improve our sensitivity by an order of magnitude, the values of T_c' could only be increased by $\sim$ 15K and would correspond to an extremely low SCF contributions to the conductivity, about four orders of magnitude lower than at measured at T_c.

In the case of the optimally doped compound, our value of T_c' is in very good agreement with the onset temperature determined by magnetic susceptibility [12]. However these values are found larger than those determined by microwave ac conductivity measurements [13]. This can be likely ascribed to the fact that a field of 16T is assumed to be sufficient to suppress all superconductivity above the zero-field T_c in this latter work, which is clearly in contradiction with our results. On the other hand, recent terahertz conductivity measurements in LSCO show that the signatures of the fluctuations only persist in a very narrow range, at most 16K above

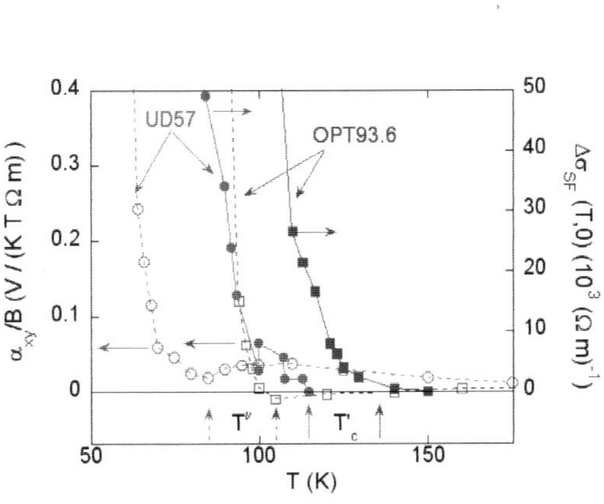

Figure 3. Comparison of the onset temperature for SCFs extracted from Nernst measurements (open symbols) and from $\Delta\sigma_{SF}(T,0)$ (full symbols) for the YBCO OPT93.6 (squares) and UD57 (circles) samples (from ref.[7]). Lines are guides for the eyes.

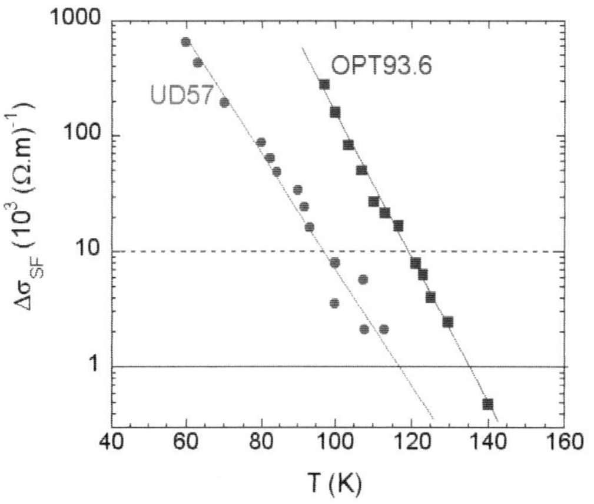

Figure 4. $\Delta\sigma_{SF}(T,0)$ for the OPT93.5 and UD57 samples in a semi-logarithmic scale. T_c' is defined as the temperature above which $\Delta\sigma_{SF}(T,0)$ is lower than 1 x $10^3(\Omega m)^{-1}$ (full line). For a lower sensitivity ($\Delta\sigma_{SF}(T,0) =$ 10 x $10^3(\Omega m)^{-1}$- dashed line) the values of T_c' would be smaller by 15-20K.

T_c [14]. This has to be contrasted with determinations from Nernst measurements [4] or high field magnetoresistance measurements comparable to ours [15] which give much larger onset temperatures. This clearly indicates that these different types of experiments do not detect the SCFs with a similar sensitivity as ours, or that they probe different types of SCFs (namely phase versus amplitude).

Independently of these remarks, all the recent experimental data now point *to an onset temperature for superconducting fluctuations well below the pseudogap temperature.* One can also observe in Fig.5 that T_c' is only slightly dependent on hole doping, increasing from $\backsim$ 120K to $\backsim$ 140K from the UD57 sample to the optimally doped one OPT93.6. This is very similar to what has been found from Nernst or magnetization experiments in Bi2212 [16]. However this strongly contrasts with the pseudogap temperature $T^\star$ which decreases with increasing doping.

In order to compare the extension of the SCFs with respect to the opening of the pseudogap, the drop of the electronic susceptibility as measured by the Y NMR Knight shift [17] is reported in Fig.6 together with the fluctuation conductivity for the UD57 sample. These data clearly evidence that the electronic states lost by the opening of the pseudogap at $T^\star$ are not redistributed into the formation of preformed pairs in a precursor superconducting state as nearly half of the susceptibility has already been suppressed at T_c'. This strongly indicates that these two states are not related. The situation is more delicate for the optimally doped sample for which $T^\star$ becomes comparable to T_c'. In this case, a careful examination of the resistivity data allowed us to determine both the onset of SCFs and the pseudogap temperature with the same experimental sensitivity criterion [8]. The fact that the T_c' line crosses the pseudogap line near optimal doping, as reported in Fig.5, unambiguously proves that the pseudogap phase cannot be a precursor state for superconductivity.

4. Quantitative analysis of the paraconductivity in the Ginzburg-Landau approach
The variations of $\Delta\sigma_{SF}(T)$ are reported versus $\epsilon = \ln(T/T_c)$ in fig.7 for the four hole dopings studied. Except for the UD57 sample, it is striking to see that the experimental

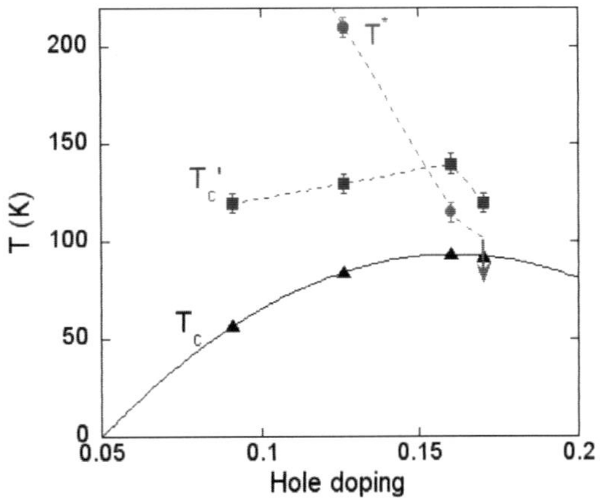

Figure 5. The values of T_c' (■) and $T^\star$ (●) are plotted versus the hole doping for the four samples studied (from ref.[7]). The solid line indicates the superconducting dome. Contrary to T_c' that is rather insensitive to hole doping, $T^\star$ is found to decrease with increasing doping and intersects the T_c' line near optimal doping [8].

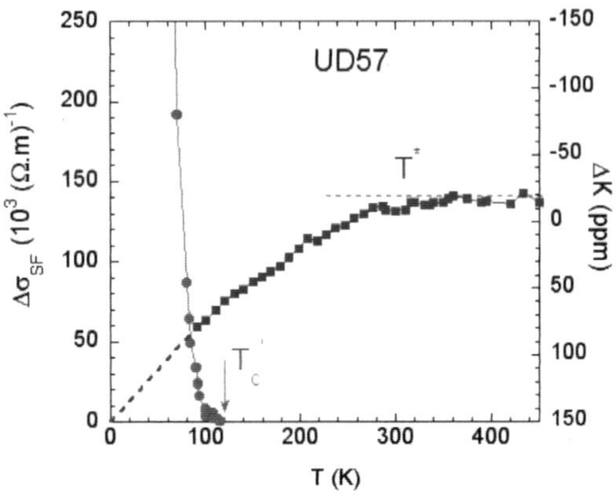

Figure 6. The fluctuation conductivity (●, left scale) is compared to the Y NMR Knight shift [1], [17] (■, right scale) for the UD57 sample. It is remarkable to see that nearly half of the susceptibility has already been lost at the onset of SCFs.

data collapse on a single curve. For $\epsilon \lesssim 0.1$ these results can be well accounted for by gaussian fluctuations within the Ginzburg-Landau (GL) theory [18]. In this approach the excess

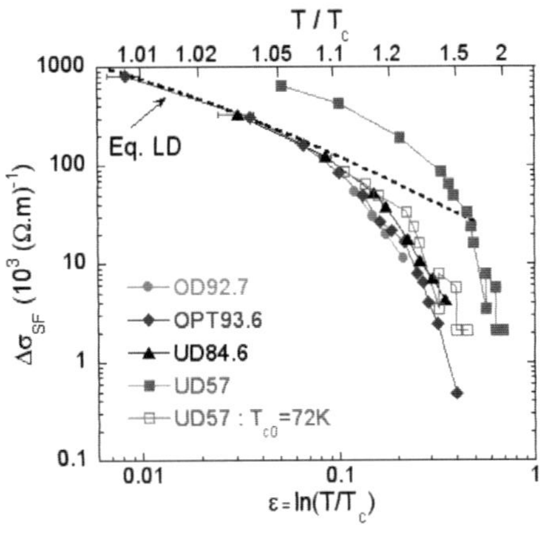

Figure 7. Superconducting fluctuation conductivity $\Delta\sigma_{SF}$ for the four pure samples studied plotted versus $\epsilon = ln(T/T_c)$ (from ref.[7]). Values of T_c have been taken here at the midpoint of the resistive transition, and error bars for ϵ using the onset and offset values of T_c are indicated. The dashed line represents the expression of Eq.(2) with $s = 11.7$Å, and $\xi_c(0) \simeq 0.9$Å. The data for the most underdoped sample can be matched with the other ones if one takes $T_{c0} = 72$K for the actual T_c instead of 57.1K. Full lines are guides to the eye.

fluctuating conductivity, called here paraconductivity, is related to the temperature dependence of $\xi(T)$, the superconducting correlation length of the short-lived Cooper pairs, which is expected to diverge with decreasing temperature as:

$$\xi(T) = \xi(0)/\sqrt{\epsilon}, \tag{1}$$

where $\xi(0)$ is the zero-temperature coherence length and $\epsilon = \ln(T/T_c) \simeq (T - T_c)/T_c$ for $T \gtrsim T_c$. More generally, the temperature dependence of $\Delta\sigma_{SF}(T)$ is given by the Lawrence-Doniach (LD) expression which takes into account the layered structure of the high-T_c cuprates [19]:

$$\Delta\sigma^{LD}(T) = \frac{e^2}{16\hbar s}\frac{1}{\epsilon\sqrt{1+2\alpha}}, \tag{2}$$

where the coupling parameter $\alpha = 2(\xi_c(T)/s)^2$ with $\xi_c(T) = \xi_c(0)/\sqrt{\epsilon}$. Sufficiently far from T_c, one expects $\xi_c(T) \ll s$ and Eq.2 reduces to the well-known 2D Aslamazov-Larkin expression:

$$\Delta\sigma^{AL}(T) = \frac{e^2}{16\hbar s}\epsilon^{-1} = \frac{e^2}{16\hbar s}\frac{\xi^2(T)}{\xi^2(0)}. \tag{3}$$

The only parameters in this expression are the value of the interlayer distance s and the value taken for T_c which can have a huge incidence on the shape of the curve especially for $(T-T_c)/T_c < 0.01$. It can be seen in Fig.7 that the data are reasonably fitted by the LD expression (Eq.2) in the small temperature range $0.03 \leq \epsilon \leq 0.1$ if one takes $\xi_c(0) \simeq 0.9$Å. We have assumed here, as usually done, that the CuO_2 bilayer constitutes the basic two-dimensional unit, and s is then taken as the unit-cell size in the c direction: $s = 11.7$Å.

For the UD57 sample, $\Delta\sigma_{SF}(T)$ is found to be much larger (about a factor four at $\epsilon = 0.05$) than for the other doping contents. Quite surprisingly, it is possible to recover a good matching with these latter data by assuming a effective mean field temperature T_{c0} different from the actual T_c. This is illustrated by the empty symbols in fig.7 using $T_{c0} = 72$K. This points to an additional origin of SCFs below T_{c0} which might be ascribed to phase fluctuations of the order parameter. This is discussed in more details in ref.[7].

For all the samples, one can see in fig.7 that $\Delta\sigma_{SF}(T)$ vanishes very rapidly for $\epsilon \gtrsim 0.1$. This behaviour which has been noticed previously in many studies is particularly well defined here given the method used to extract the fluctuating conductivity. An extension of the AL theory including short wave length fluctuations [20] has been invoked to explain the steeper decrease of $\Delta\sigma_{SF}(T)$. This has also been treated by inserting a cutoff phenomenologically, implying that the density of fluctuating pairs vanishes at T_c' as detailed in ref.[7].

5. Field variation of the SCF conductivity: H_c' and upper critical field H_{c2}

From the data reported in Fig.2, we can also extract the magnetic field $H_c'(T)$ above which the MR recovers a H^2 dependence, which we take as the sign that the normal state is completely restored. As T decreases, it becomes difficult to ascertain that the normal state is fully reached when $H_c'(T)$ becomes comparable to the highest available field. This makes it difficult to precisely deduce values of $H_c'(T)$ larger than 45T.

The evolution of $H_c'(T)$ are plotted in fig.8 for the four samples. One can see that $H_c'(T)$ increases rapidly with decreasing T and displays a linear variation near T_c'. We have tentatively tried to estimate a low T extrapolation of H_c' using a parabolic T variation as applied for the critical field of classical superconductors:

$$H_c'(T) = H_c'(0)[1-(T/T_c')^2]. \tag{4}$$

The fitting curves are displayed as dashed lines in Fig.8 and show that $H_c'(0)$ increases with hole doping and reaches a value as high as ~ 150 Tesla at optimal doping.

A precise analysis of the fluctuation magnetoconductivity $\Delta\sigma_H(T,H)$ is a valuable tool to extract different microscopic parameters of high-T_c cuprates, such as the value of $H_{c2}(0)$ not directly accessible from experiments. $\Delta\sigma_H(T,H)$ can be written out as:

$$\begin{aligned}\Delta\sigma_H(T,H) &= \Delta\sigma(T,H) - \Delta\sigma_n(T,H)\\ &= \Delta\sigma_{SF}(T,H) - \Delta\sigma_{SF}(T,0).\end{aligned} \tag{5}$$

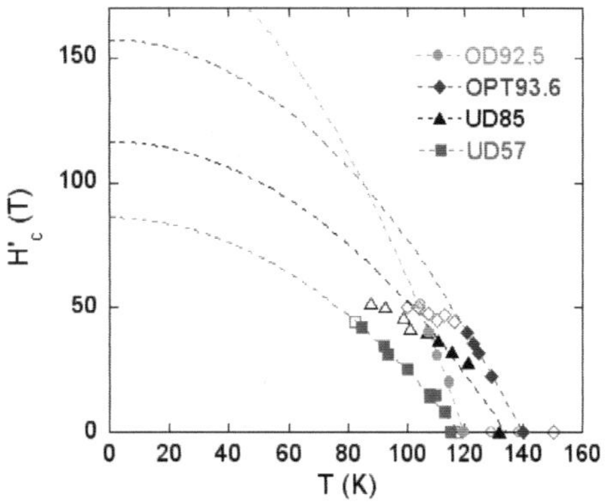

Figure 8. The field H_c', at which the SC fluctuations disappear and the normal state is fully restored, is plotted versus T for the four pure samples studied. Dashed lines represent the fitting curves to Eq.4 using data with closed symbols. When $H_c'(T) \gtrsim 40T$ (empty symbols), the data are somewhat underestimated as the maximum applied field is not sufficient to restore the normal state.(from ref.[7]).

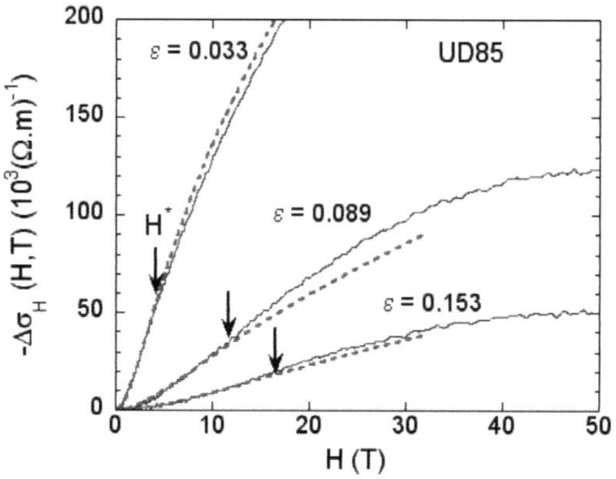

Figure 9. Evolution of the fluctuation magnetoconductivity $-\Delta\sigma_H(T,H) = \Delta\sigma_{SF}(T,0) - \Delta\sigma_{SF}(T,H)$ as a function of H for the UD85 sample at different temperatures: 87.5, 92.4 and 98.6K. The dotted lines represent the computed results from the ALO contribution with $H_{c2}(0) = 125(5)$T. They deviate from the data beyond the $H^\star$ field values shown by arrows [7].

It has been very often assumed that the second term of the first equation can be neglected as being only weakly dependent on magnetic fields. However, our study clearly shows that this is not the case (see for instance the data displayed in fig.2). Thus our method provides here a correct determination of $\Delta\sigma_H(T,H)$.

In the GL approach, the evolution of the fluctuation magnetoconductivity with H comes from the pair-breaking effect which leads to a T_c suppression. In the case of interest, the major contribution results from the AL process, and more particularly from the interaction of the field with the carrier orbital (ALO) degrees of freedom. The detailed analysis and discussion of the fluctuation magnetoconductivity are reported in ref.[7]. The important point here is that the analysis of the fluctuation magnetoconductivity can give a direct determination of the coherence length, and then of $H_{c2}(0)$, as the same coherence length governs the fluctuating and the superconducting regimes. Thus a mirror field $H^\star(T)$ of the upper critical field $H_{c2}(T)$ can be determined above T_c in the GL approach [18]. Fig.9 shows the data for the UD85 sample together with the fits using the ALO expression with $H_{c2}(0) = 125(5)$T being the only adjustable parameter. We checked that, as predicted by the theory, the fits are valid as long as $H \lesssim H^\star(T) = \epsilon H_{c2}(0)$.

Let us mention here that the analysis of the magnetoconductivity in terms of the ALO expression has to be restricted to the temperature range where the fluctuating conductivity can be also well described within the GL framework. For instance for the UD-85 sample, it is not possible to fit the $\Delta\sigma_H(H,T)$ curves with the same value of H_{c2} for $T > 99$K ($\epsilon > 0.16$) above which $\Delta\sigma_{SF}(T)$ starts to deviate significantly from the LD expression (see fig.7). This would give values of H_{c2} correspondingly smaller as the temperature is higher. Moreover, in the case of the UD57 sample, the fits of the $\Delta\sigma_H(H,T)$ curves can only be performed for temperatures above the mean field temperature T_{c0} using the value of $T_{c0} = 72$K in the ALO expression. It can be seen in table 1 that the value of H_{c2} extracted from the low-field part of the magnetoconductivity data matches very well that of $H_c'(0)$ obtained in a completely different way. Let us emphasize here that the comparison can only be indicative as the use of Eq.4 is not granted. So the value of $H_c'(0)$ could as well be a lower bound of the field above which the SCFs are suppressed. However, the fact that both fields are comparable for all the doping contents investigated highlights the consistency of our data analysis.

Table 1. Values of $H_{c2}(0)$ extracted from the fluctuation magnetoconductivity. They are comparable to the extrapolated values of the onset field of SCF $H_c'(0)$.(from ref.[7])

Sample	UD57	UD85	OPT93.6	OD92.5
$H_{c2}(0)$(T)	90(10)	125(5)	180(10)	200(10)
$H_c'(0)$(T)	86(10)	115(5)	155(10)	207(10)

The important result here is to show that the superconducting gap which is directly related to $H_{c2}(0)$ increases smoothly with increasing hole doping from the underdoped to the overdoped regime, contrary to the pseudogap which decreases. This is a strong indication that the gap determined here can thus be assimilated to the "small" gap detected recently by different techniques, while the pseudogap would be rather connected with the "large" gap [5].

One can point out that rather different values of H_{c2} have been reported in ref.[21] from the analysis of the T dependence of the magnetoconductivity at 1T in untwinned YBCO crystals. Even if this study was performed on different single crystals, one can conjecture that the very few data points used in that case to fit the fluctuation magnetoconductivity in a temperature range where the validity of the GL approach was not really checked, might be the source of large errors in the determination of H_{c2}.

6. Influence of disorder

It is now well admitted that the properties of cuprates are strongly dependent on disorder. We have studied for long the effect of the introduction of controlled disorder by electron irradiation and the way it affects the transport properties [22]. In particular, we have shown that similar upturns of the low-T resistivity are found for controlled disorder in YBCO and in some "pure" low-T_c cuprates, which indicates the existence of intrinsic disorder in those families [23].

We have also carried out magnetoresistance measurements in some OPT93.6 and UD57 samples irradiated by electrons. When T_c is decreased by disorder, we find that both T_c' and $H_c'(0)$ are also affected. The reduction in T_c' nearly follows that in T_c for the underdoped sample while it is slightly larger for the OPT sample. Consequently, when T_c is decreased by disorder, the relative range of SCFs with respect to the value of T_c expands considerably. For instance, in ref.[7], we still detect $T_c' \sim 60$K in an UD57 irradiated sample with $T_c = 4.5$K.

These results allow us to draw important conclusions on the cuprate phase diagram. Indeed, contrary to T_c, T_c' or H_c', the pseudogap temperature $T^\star$ has been found very early to be quite robust to disorder [24]. This is another indirect evidence that the pseudogap phase is not related to superconductivity. We want also to emphasize here that specific effects induced by disorder are probably at the origin of many confusions in the study of high-T_c cuprates. This leads us to propose in fig.10 a 3D phase diagram where the effect of disorder has been introduced as a third axis.

There, in the pure systems, the occurrence of SCFs and the difficulty to separate the SC gap from the pseudogap in zero-field experiments justifies that the T_c' line could often be taken as a continuation of the $T^\star$ line.

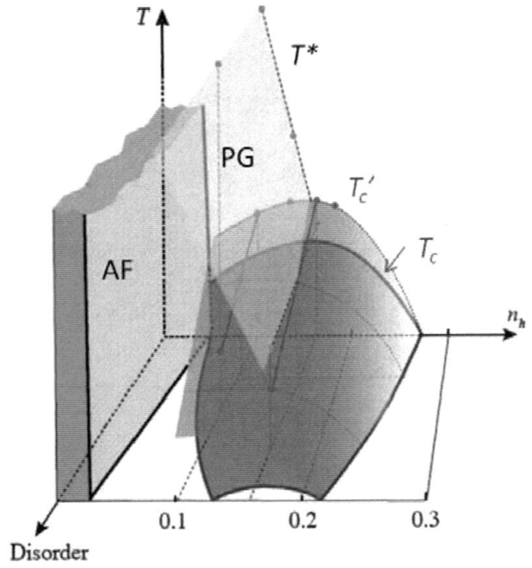

Figure 10. Phase diagram constructed on the data points obtained here, showing the evolution of T_c' the onset of SCF, with doping and disorder (from ref.[7]). The fact that the pseudogap and the SCF surfaces intersect each other near optimum doping in the clean limit is apparent. These surfaces have been limited to experimental ranges where they have been determined experimentally.

It can also be seen in this figure that the respective evolutions with disorder of the SC dome and of the amplitude of the SCF range explains the phase diagram often shown in a low-T_c cuprate such as Bi-2201 and sketched in fig.1(a). Finally, for intermediate disorder, the enhanced fluctuation regime with respect to T_c observed in the Nernst measurements for the $La_{2-x}Sr_xCuO_4$ can be reproduced as well [4].

7. Conclusion

We have presented here a condensed report of our quantitative study of the superconducting fluctuation conductivity in YBCO that was extensively detailed in previous publications [6]-[8]. Our data give important determinations of some thermodynamic properties of the SC state of high-T_c cuprates which are not accessible otherwise, as flux flow dominates near T_c in the vortex liquid phase and the highest fields available so far are not sufficient to overcome H_{c2} and then to reach the normal state at $T = 0$.

The consistency of our data analyses, which establish that the SCFs do match quantitatively the expectations from the 2D GL approach, strongly justifies the method used to determine the superconducting fluctuation conductivity from the deviations of the magnetoresistance from the H^2 normal state observed well above T_c.In a metallic state, deviations of the MR from an H^2 behavior could of course be expected in case of Fermi surface reconstruction. A two-carrier model has been indeed proposed to explain some MR data in $YBa_2Cu_4O_8$ [25] in which quantum oscillations could be detected at low T. In that approach the contributions of the SCFs has been completely ignored above T_c, though they should undoubtedly be present.

For our three higher oxygen content samples around optimal doping, Fermi Surface reconstructions have never been observed so far. As for our underdoped sample, it has a lower doping and lower T_c than the ortho-II ordered $YBCO_{6.5}$ sample in which Fermi surface reconstruction has been detected at the highest T of about 60K [26]. Our UD sample being twinned and thus with poor oxygen order, a reconstruction, if any, should then occur at even lower T. So, if such reconstruction effects do occur in this very sample, their influence should not extend in the high temperature range of our measurements (from 70K up to 130K). The oxygen disorder could play a role in reducing the T_c of that sample. As in our lower T_c samples in which disorder has been introduced on purpose, this could justify that, in presence of disorder, we need to introduce a mean field $_Tc$ value higher than the zero field T_c to interpret the data in a GL approach [7].

Finally, contrary to what was claimed by others in this conference [27], our data unambiguously show that the superconducting fluctuations vanish abruptly with increasing temperature, allowing us to define an onset temperature T_c' and an onset magnetic field H_c' above which the SCF contribution to conductivity becomes unmeasurable. The comparison between the huge drop of the electronic susceptibility determined by NMR from $T^\star$ and the emergence of superconducting fluctuations at T_c' well below $T^\star$ clearly indicates that these two temperature scales are not connected with each other.

We therefore have evidenced that the fluctuation conductivity can be very well accounted for in terms of the first series expansion of gaussian fluctuations in a limited temperature range, but require extension of the theory to explain the sharp drop at higher T. Moreover, the analysis of the fluctuating magnetoconductivity in this temperature range allows us to demonstrate that the pairing energy and SC gap both increase with doping, confirming then that the pseudogap has to be assigned to an independent magnetic order or crossover due to the magnetic correlations.

Acknowledgments
We acknowledge collaboration with C. Proust, B. Vignolle and G. Rikken at the LNCMI. This work has been performed within the "Triangle de la Physique" and was supported by ANR grants "OXYFONDA" NT05-4 41913 and "PNICTIDES". The experiments at LNCMI-Toulouse were funded by the FP7 I3 EuroMagNET.

[1] Alloul H Ohno T and Mendels P 1989 *Phys. Rev. Lett.* **63**, 1700
[2] Timusk T and Statt B 1999 *Rep. Prog. Phys.* **62** 61
[3] Emery VJ and Kivelson SA 1995 *Nature* **374** 434
[4] Wang Y Li L and Ong NP 2006 *Phys. Rev.* B **73** 024510
[5] Hüfner S Hossain MA Damascelli A and Sawatzky GA 2008 *Rep. Prog. Phys.* **71** 062501
[6] Rullier-Albenque F Alloul H Proust C Colson D and Forget A 2007 *Phys. Rev. Lett.* **99** 027003
[7] Rullier-Albenque F Alloul H Rikken G 2011 *Phys. Rev.* B **84** 014522
[8] Alloul H Rullier-Albenque F Vignolle B Colson D and Forget A 2010 *Europhys. Lett.* **91** 37005
[9] Legris A Rullier-Albenque F.Radeva E and Lejay P 1993 *J. Phys.* I France **3** 1605
[10] Harris JM Yan YF Matl P Ong NP Anderson PW Kimura T and Kitazawa K 1995 *Phys. Rev. Lett.* **75**, 1391
[11] Rullier-Albenque F Tourbot R Alloul H Lejay P Colson D and Forget A 2006 *Phys. Rev. Lett.* **96** 06700
[12] Li L Wang Y Komiya S Ono S Ando Y Gu GD and Ong NP 2010 *Phys. Rev. B***81** 054510
[13] Grbić MS Požek M Paar D Hinkov V Raichle M Haug D Keimer B Barišić N and Dulčić A 2011 *Phys. Rev. B 83* 144508.
[14] Bilbro LS Valdés Aguilar R Logvenov G Pelleg O Božović I and Armitage NP 2011 *Nature Physics* **7** 298
[15] Rourke PMC Mouzopoulou I Xu X Panagopoulos C Wang Y Vignolle B Proust C Kurganova V Zeitler U Tanabe Y Adachi T Koike and Hussy NE 2011 *Nature Physics* **7** 455
[16] Wang Y Li L Naughton MJ Gu GD Uchida S and Ong NP 2005 *Phys. Rev. Lett.* **95** 247002
[17] Alloul H Mahajan A Casalta H Klein O (1993) *Phys. Rev. Lett* **70** 1171
[18] For a review see Larkin A and Varlamov AA 2005 *Theory of fluctuations in superconductors* (Oxford University Press, Oxford.
[19] Lawrence WE and S. Doniach 1971 Proceedings 12th International Conference on Low Temperature Physics, Kyoto 1970 (E. Kanda, Keigaku, Tokyo) 361
[20] Reggiani l Vaglio R Varlamov A.A (1991) *Phys. Rev. B* **44** 9541
[21] Ando Y Segawa K (2002) *Phys. Rev. Lett.* **88** 167005
[22] Rullier-Albenque F Alloul H Tourbot R 2003 *Phys. Rev. Lett.* **91** 047001
[23] Rullier-Albenque F Alloul H Balakirev F Proust C. 2008 *Europhys. Lett.* **81** 37008
[24] Alloul H Mendels P Casalta H Marucco JF Arabski J (1991) *Phys. Rev. Lett.* **67** 3140
[25] Rourke PMC. Bangura AF. Proust C Levallois J Doiron-Leyraud N LeBoeuf D Taillefer L Adachi S Sutherland SL and Hussey NE 2010 *Phys. Rev. B* **82**, 020514(R)
[26] Doiron-Leyraud N Proust C LeBoeuf D Levallois J Bonnemaison JB Liang R Bonn DA Hardy WN Taillefer L 2007 *Nature* **447** 565
[27] Taillefer L, Chang J et al. talks at this M2S conference

Pseusogap in Cuprates by Electronic Raman Scattering

A. Sacuto, S. Benhabib, Y. Gallais, S. Blanc, M. Cazayous, M.-A. Méasson

Laboratoire Matériaux et Phénomènes Quantiques (UMR 7162 CNRS), Université Paris Diderot-Paris 7, Bat. Condorcet, 75205 Paris Cedex 13, France

J. S. Wen, Z. J. Xu and G. D. Gu

Matter Physics and Materials Science, Brookhaven National Laboratory (BNL), Upton, NY 11973, USA

E-mail: alain.sacuto@univ-paris-diderot.fr

Abstract. We present Raman experiments on underdoped and overdoped $Bi_2Sr_2CaCu_2O_{8+\delta}$ (Bi-2212) single crystals. We reveal the pseudogap in the electronic Raman spectra in the B_{1g} and B_{2g} geometries. In these geometries we probe respectively, the antinodal (AN) and nodal (N) regions corresponding to the principal axes and the diagonal of the Brillouin zone. The pseudogap appears in underdoped regime and manifests itself in the B_{1g} spectra by a strong depletion of the low energy electronic continuum as the temperature decreases. We define a temperature T^* below which the depletion appears and the pseudogap energy, ω_{PG} the energy at which the depeletion closes.

The pseudogap is also present in the B_{2g} spectra but the depletion opens at higher energy than in the B_{1g} spectra. We observe the creation of new electronic states inside the depletion as we enter the superconducting phase. This leads us to conclude (as proposed by S. Sakai et al. [1]) that the pseudogap has a different structure than the superconducting gap and competes with it. We show that the nodal quasiparticle dynamic is very robust and almost insensitive to the pseudogap phase contrary to the antinodal quasiparticle dynamic. We finally reveal, in contrast to what it is usually admitted,an increase of the nodal quasiparticle spectral weight with underdoping. We interpret this result as the consequence of a possible Fermi surface disturbances in the doping range $p = 0.1 - 0.2$.

1. Introduction

The pseudogap phase has been first revealed by nuclear magnetic resonance (NMR) [2, 3] and then it has been extensively studied by transport [4, 5, 6, 7, 8], angle resolved photoemission (ARPES) [9], optics [10, 11, 12] and scanning tunneling spectroscopy (STS) [13]. In comparison, Raman scattering investigations of the pseudogap remain relatively scarce. One reason is that the pseudogap was first detected in the B_{2g} geometry (nodal region) [14, 15, 16] which introduced confusion with respect to ARPES data where the pseudogap sets in the antinodal region.

In fact, Electronic Raman scattering investigations in the normal state have also revealed signatures of the pseudogap in B_{1g} geometry (antinodal region) [17, 18] however its effect has not been clearly quantified and well understood. For example the doping evolution of the pseudogap energy ω_{PG} has not been studied in details and the doping level where the pseudogap disappears has not been determined.

Here, we present Raman experiments on underdoped and overdoped $Bi_2Sr_2CaCuO_{2+\delta}$ (Bi-2212) single crystals. In B_{1g} geometry the pseudogap manifests itself by a strong depletion in the low energy Raman response function which develops as the temperature decreases. The pseudogap effect in B_{2g} geometry is more subtle. We detect a depletion in the Raman response function but at finite energy. At low energy the Raman response function exhibits a linear slope which increases with cooling as expected for a conventional metal [15]. Interestingly, we reveal that supplementary electronic states are created inside the energy range of the depletion as we enter in the superconducting state. We interpret our experimental findings as the unforseen $s-$wave anisotropic character of the pseudogap. This leads us to conclude that the pseudogap and the superconducting gap have different symmetries and most likely compete with each other [1].

Finally we focus on the low energy Raman responses and extract the $\zeta = \Gamma/(Z^*)^2$ ratio where Γ is the static scattering rate and Z^*, the renormalized quasiparticle spectral weight [19].

In the antinodal region, ζ is temperature independent inside the pseudogap phase while it increases with temperature outside the pseudogap phase as expected for a metallic-like behavior for which the scattering rate has to increase with temperature.

In contrast in the nodal region ζ exhibits a metallic-like temperature dependence both inside and outside the pseudogap phase. The equation of ζ depends both on Z^* and Γ which are difficult to disentangle in a first approach. However, a simple modeling of the Raman response function in Drude like conductivity allows us to follow separately the doping and temperature dependences of these two physical quantities in both the nodal and antinodal regions. We show that Z^*_N exhibits an unexpected increase with underdoping while Z^*_{AN} decreases. Simultaneously the static scattering rate Γ_{AN} strongly increases while the Γ_N exhibits only small changes. We interpret this result as a consequence of a Fermi surface disturbances between 0.11 and 0.22 in which well defined quasiparticles are preserved and even enhanced in the nodal region while they are destroyed in the antinodal region as the doping level is reduced. We find that Γ_N exhibits a quasi linear temperature dependence in and out of the pseudogap phase.

2. Experimental Results

The Bi-2212 single crystals have been grown by floating zone method. Doping is achieved by changing the oxygen content only. The detailed procedures of the crystal growth are described elsewhere [20, 21]. Raman experiments have been carried out using a triple grating spectrometer (JY-T64000) equipped with a nitrogen cooled CCD detector. We used the 532 nm excitation line from a diode pumped solid state laser (DPSS) and the 514.52 nm line from a Ar^+,Kr^+ laser. Temperature dependent measurements have been carried out in a ARS closed cycle cryostat. All measurements have been corrected for the Bose factor and the instrumental spectral response. The B_{1g} and B_{2g} geometries have been obtained using crossed light polarizations at 45° from the $Cu-O$ bond directions and along them respectively.

Special care has been devoted to make reliable quantitative comparisons between the Raman intensities of distinct crystals with different doping levels measured in the same geometry, and between measurements in distinct geometries for crystals with the same doping level. We have performed all the measurements during the same run and the crystals with various doping levels have been mounted on the same sample holder in order to keep the same optical configuration. The optical constant, deduced from spectroscopic ellipsometry measurements on Bi-2212 single crystals, have also been taken into account in order to have comparable Raman intensities of crystals with distinct doping levels.

In Figure 1 are displayed the temperature dependence of the B_{1g} and B_{2g} Raman response functions for two distinct Bi-2212 single crystals. The first one is an underdoped (UD) crystal

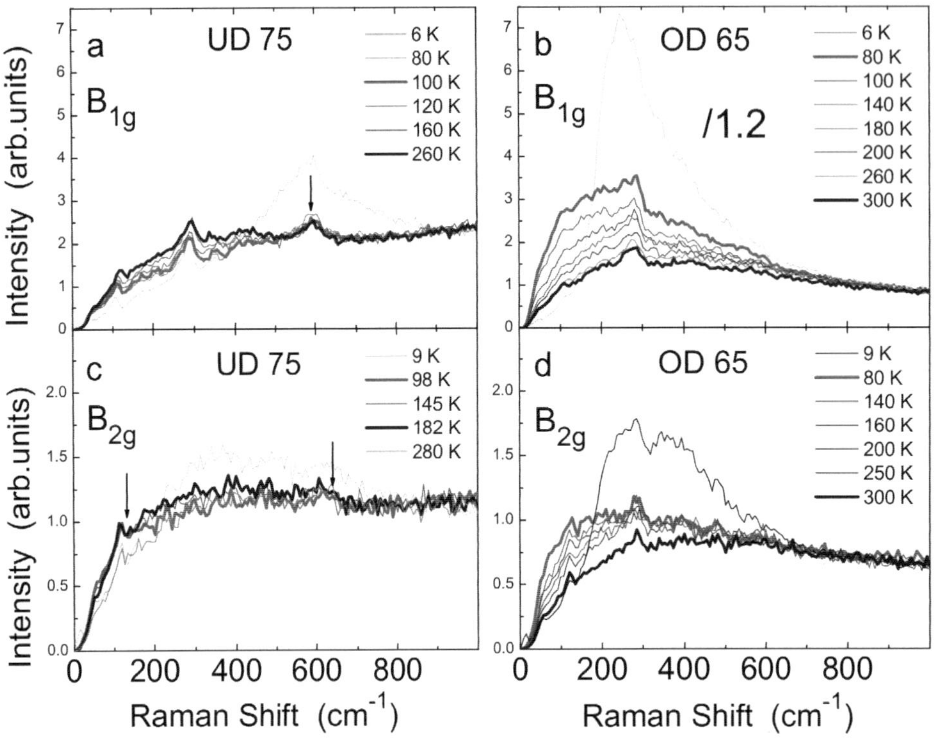

Figure 1. B_{1g} and B_{2g} Raman response functions of underdoped and overdoped Bi-2212 single crystals. Spectra have been measured with the 532 nm laser line. The UD 75 and OD 65 single crystals have been measured after two successive exfoliations. In pannel b, the Raman intensities have been divided by 1.2

$p = 0.11$ with a $T_c = 75$ K and the second one is overdoped (OD) $p = 0.22$ with $T_c = 65$ K. The Raman response functions in the superconducting state (black thin curves) exhibit a pair breaking peak. This peak is much more pronounced in the B_{1g} than in B_{2g} geometry (for the same doping level) as expected for a d-wave superconducting gap [22]. We can also notice that the B_{1g} superconducting peak increases in intensity while its energy decreases with doping. This has already been pointed out in our previous works [23, 24] and we have assigned the integrated area under the B_{1g} pair breaking peak to the density of Cooper pairs in the antinodal region. In the normal state (Fig.1-a) the B_{1g} Raman response function of the UD 75 crystal exhibits a strong depletion of the low frequency electronic continnum as the temperature decreases. The thick curves underline the low energy depletion of the electronic background in the Raman spectra. It extends from the lowest energy to approximatively 600 cm^{-1} (see arrow). We define the end of this depletion in the spectra as the pseudogap energy ω_{PG}. Such a low energy depletion has been qualitatively reproduced by recent cluster dynamical mean field methods [1, 25].

The non trivial fact is that a depletion is also detected in the B_{2g} geometry where it opens at finite energy, above 150 cm^{-1}, and closes around 600 cm^{-1} (see arrows in Fig.1-c and 2-a). This has also been reported in previous works [14, 15, 16]. Below 150 cm^{-1}, the B_{2g} Raman response is not altered by the depletion and the slope increases as the temperature decreases. This is expected for a metal since the low energy slope of the Raman response is proportional to the quasiparticle lifetime [22] . In sharp contrast, no trace of depletion is detected in the Raman spectra of the OD 65 crystal in both B_{1g} and B_{2g} geometries (see figs 1-b,d) and the low energy slopes of the Raman response functions increase with cooling as expected for a metal [22, 26].

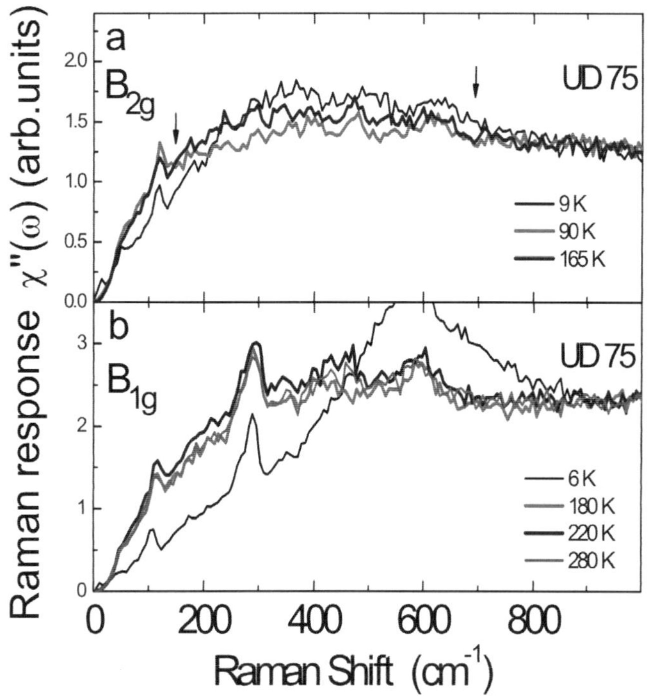

Figure 2. Selected (a) B_{2g} and (b) B_{1g} Raman spectra of UD 75 compound.

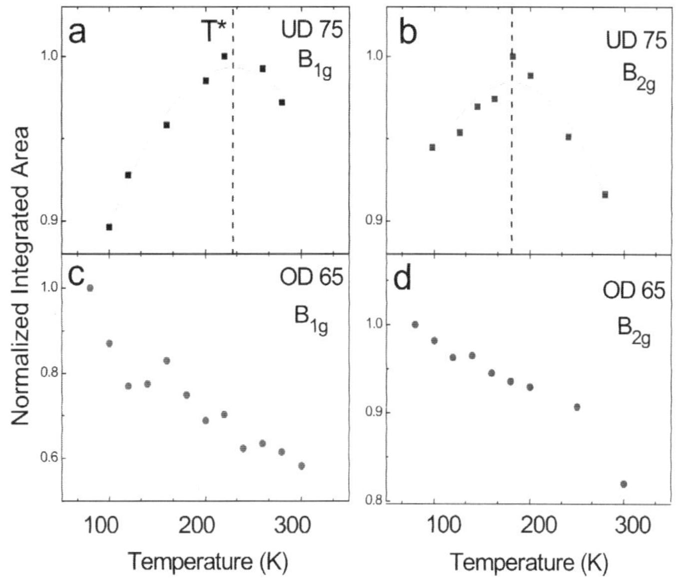

Figure 3. Temperature dependence of the normalized integrated area of the Raman response functions shown in fig.1. Normalization has been achieved by dividing the areas by their maximum value. The vertical dashed line indicates T^* and the thin line corresponds to a polynomial fit.

We can estimate the pseudogap temperature T^* by studying the temperature dependence of the normalized integrated area of the B_{1g} Raman response (up to 800 cm^{-1}). The integrated area are plotted in Fig. 3. In fig. 3-a, (ud 75), the positive slope corresponds to a filling of the depletion as the temperature increases. The electronic background is then progressively restored and we can define the temperature T^* for which the sign of the slope changes. It is approximatively 230 K (see dashed line). Beyond, T^* the integrated electronic continuum starts to decrease. The slope of the integrated area as a function of temperature is negative. This is a consequence of the decrease of the low energy slope of the Raman response with temperature. This is can be seen in fig.2-b where the low energy slope of the B_{1g} Raman response decreases between 220 and 280 K. This corresponds to a metallic behavior [15, 22, 26].

The B_{2g} integrated area of UD 75 exhibits a similar temperature behaviour as shown in fig. 3-b. However, the extracted T^* value is smaller than the B_{1g} one. A possible reason for this behaviour is that the B_{2g} depletion opens at finite energy while the low electronic continuum still exhibit a metallic behaviour. The low energy continuum thus decreases as temperature raises playing against the filling of the depletion. This effect may reduce the extracted T^* in the nodal region with respect to the antinodal one.(see ref. [1]). In the following we will assume that the value of T^* directly related to the pseudogap temperature is the B_{1g} one.

In the overdoped case (see fig.3-c,d) the temperature dependence of the integrated area in B_{1g} and B_{2g} geometries exhibit negative slopes down to the lowest temperatures. This corresponds to a continuous intensity decrease of the electronic background with temperature as expected for a metal[22]. Our experimental findings indicate that the pseudogap phase only sets in below $p = 0.22$.

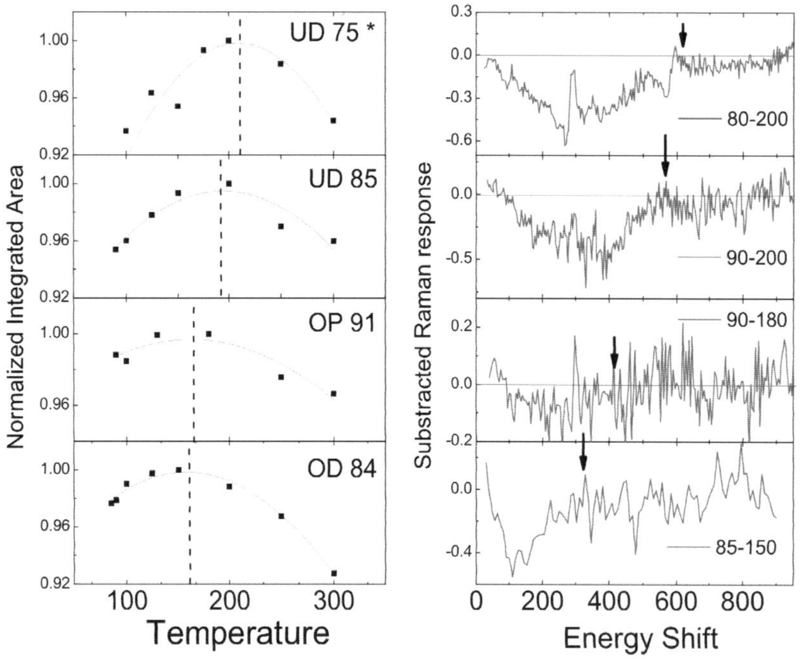

Figure 4. left panel : Normalized integrated area for the B_{1g} Raman spectra of Bi-2212 crystals. Normalization has been made from the maximum. The vertical dashed line indicates T^* and the thin line corresponds to a polynomial fit; right panel: Subtracted Raman responses measured close to T^* and just above T_c. Measurements have been performed with the 514.52 nm laser line. The UD 75 single crystal has been measured without exfoliation.

In order to map T^* versus doping, we have performed Raman measurements on Bi-2212 crystals with various doping levels. This is shown in the left panel of fig.4 where the integrated area of the B_{1g} Raman response function up to 800 cm^{-1} is plotted for various dopin levels. The dash lines correspond to the temperature from which the slope of the integrated area curve changes of sign. We observe that T^* increases with underdoping. We have also defined the pseudogap energy by subtracting the Raman response measured close to T^* from the one just above T_c. The subtracted spectra are shown in the right panel of fig.4. The pseudogap energy ω_{PG} corresponds to the energy for where the depletion closes. We see that ω_{PG} also increases with underdoping.

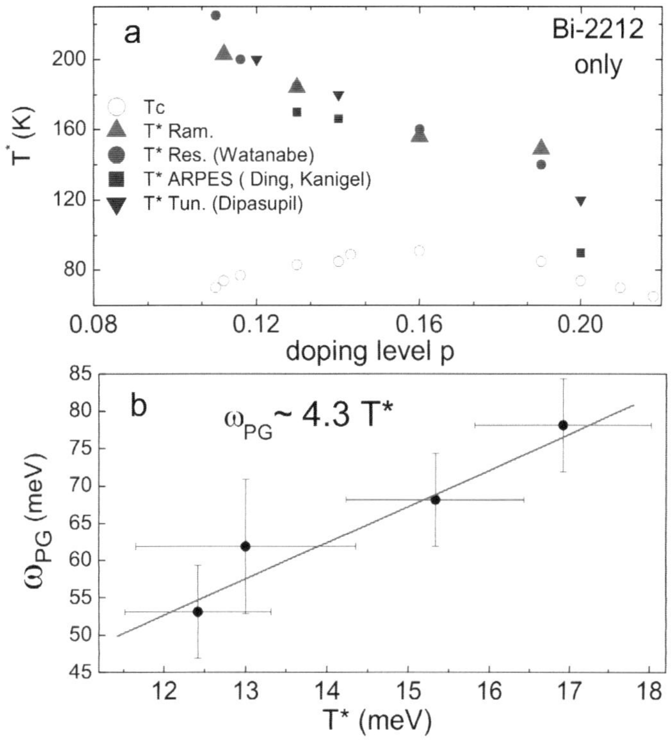

Figure 5. (a) Doping evolution of the pseudogap temperature T^* obtained from various techniques on Bi-2212 compound.(b)ω_{PG} versus T^*

We have also plotted the pseudogap energy obtained from distinct probes on the same Bi-2212 system: resistivity [27], ARPES [28, 29] tunneling [30] and our Raman data in B_{1g} geometry. We can see in fig. 5a that all the measurements are consistent each other. We observe that T^* exhibits the same doping evolution and the same order of magnitude. Typically T^* increases from 100 K to 230 K between $p = 0.1$ and 0.2. This makes us confident about our procedure to define T^*. We have also plotted in fig.5-b ω_{PG} versus T^* and we find $\omega_{PG} \approx 4T^*$ as it was observed from tunneling [31].

3. Two Distinct Structures for the Pseudo gap and the Supercoducting gap

We now discuss the implications of our findings on the pseudogap structure and its symmetry with respect to the superconducting gap.

The starting point is the unusual depletion of the B_{2g} Raman response which sets in an intermediate energy range. This has recently suggested a new interpretation for the pseudogap structure [1] deduced from cellular dynamical mean field theory (CDMFT) [32]. The pseudogap would be distinct from the $d-$wave symmetry [1]. Indeed the opening of a $d-$wave pseudogap around the antinodes (as sketches in fig.6-a) should generate a continuous loss of electron-hole pair excitations in a large energy range from very low to high energy even in nodal region. Inside the k-window fixed by the angular extension of the B_{2g} Raman vertex, very low lying energy excitations are allowed but there is no simple way to get a depletion for intermediate energy range only. In contrast an s-wave anisotropic pseudogap which sets in away from the Fermi surface in the nodal region, like the one depicted in fig. 6-b, leads naturally to a depletion at finite energy. This is because low energy excitations are allowed in the nodal region while the intermediate energy ones are forbidden until the pseudogap energy is exceeded (see fig.4-b). Our view is consistent with recent ARPES data where particle-hole asymmetry has been found in the nodal region. It was obtained on states thermally occupied above the Fermi level [33]

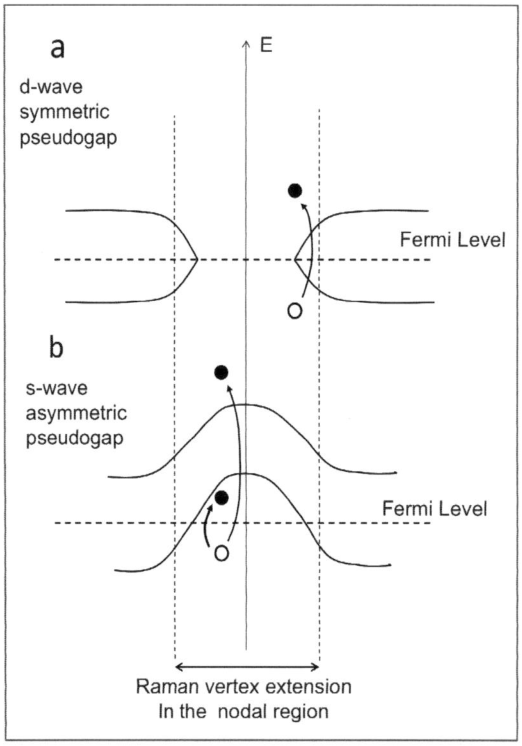

Figure 6. (a) d-wave symmetric and (b) s-wave asymmetric shapes of the pseudogap in the nodal region. electron-hole pair excitations are sketched for the both scenarios. The band structure below the Fermi level looks similarly between d and s-wave pseudogaps but are different above the Fermi level.

The second crucial point that our data reveal is the presence of additional electronic states generated inside the pseudogap when we enter the superconducting state. In the B_{2g} geometry,(see fig. 2-a) the superconducting peak emerges inside the energy range $150-600\ cm^{-1}$ where the depletion sets in. This is also observed in the B_{1g} geometry (see fig. 2-b). The low energy edge of the superconducting peak is located near $400\ cm^{-1}$ inside the energy range of the depletion. Creation of new states within the pseudogap is in agreement with recent advances in dynamical mean field theory which propose a competition between the superconducting gap and the pseudogap[1, 34, 35].

These two experimental facts make very improbable a performed pairs scenario for the pseudogap phase. The first reason is that we should have the same symmetry for both the pseudogap and the superconducting gap. This is not the case here. The second reason is that we should expect an enhancement of the depletion in the superconducting state if this last one results from preformed pairs. This is clearly not the case since the pair breaking peak emerges inside the depletion. Consequently, our experimental findings show that the pseudogap and superconducting gap are distinct in origin. This answers a long standing question introduced by Norman in 2005[36]: "The pseudogap: friend or foe of high Tc?". It is a foe. Our findings bring additional experimental evidence to recent studies in ARPES [37, 38], STS [39],optics [40] and transport [41, 42] all showing that the pseudogap and superconducting phases are in competition.

4. Nodal versus Antinodal Low Energy Quasiparticle Dynamics

Here we focus on the very low energy side of the B_{1g} and B_{2g} Raman response functions in and out of the pseudogap phase.

The inverse of the low energy slope of the electronic Raman response function in the normal state is given by $\zeta(T,\phi) = \frac{1}{(\gamma(\phi))^2 N_F(\phi)} \cdot \frac{\Gamma(T,\phi)}{(Z^*)^2(T,\phi)}$ according to [43]. Γ and Z^* are respectively the static scattering rate and the renormalized spectral weight of the quasiparticles. N_F is defined as the density of states at the Fermi level above T* and γ the Raman vertex. The angle ϕ defines a direction in the Brillouin zone with respect to the principal axis. ϕ is equal to 0 and $\pi/4$ for the AN and N directions respectively.

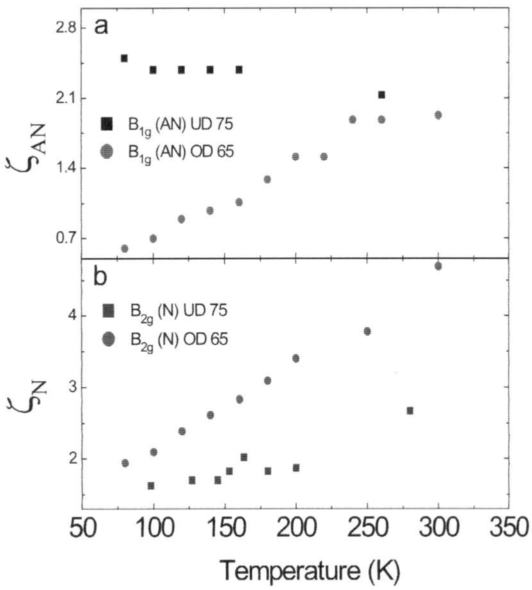

Figure 7. Temperature dependence of (a) ζ_{AN} and (b) ζ_N for the UD 75 and OD 65 compounds.ζ_{AN} has been extracted from the Raman spectra in fig.1.

Fig. 7-a shows the temperature dependences of ζ_{AN} for the UD 65 and OD 65 crystals. ζ_{AN} exhibits different temperature dependences depending on the doping level. ζ_{AN} is almost temperature independent for UD 75 while it exhibits a quasi linear temperature dependence for OD 65.

In contrast, ζ_N for both the OD 65 and UD 75 increases continuously upon heating (see Fig. 7-b). We can also note that the temperature dependence of ζ_N is stronger for the OD 65 than for the UD 75 compound. Even above T^* (230 K), it subsists a factor of about 1.6 between the ζ_N values for the two doping levels (see fig.7-b). What does it means ? Does the nodal static scattering rate is larger for OD 65 than for UD 75 or/and the nodal quasiparticle spectral weight is smaller in the OD case than in the UD one ?

A decrease of Z_N^* with doping could be considered as paradoxical and unexpected since an increase of the doping level should a priori enhance the quasiparticle spectral weight. In fact, the quasiparticles dynamic at the nodes and antinodes have to be considered separately. This dichotomy is clearly seen at 300 K (above T^*) where the electronic continuum intensities in B_{1g} and B_{2g} exhibit opposite doping depedences [23]: the intensity of the B_{2g} low energy continuum decreases while the B_{1g} one increases with doping(see fig.8-a and 8-b).

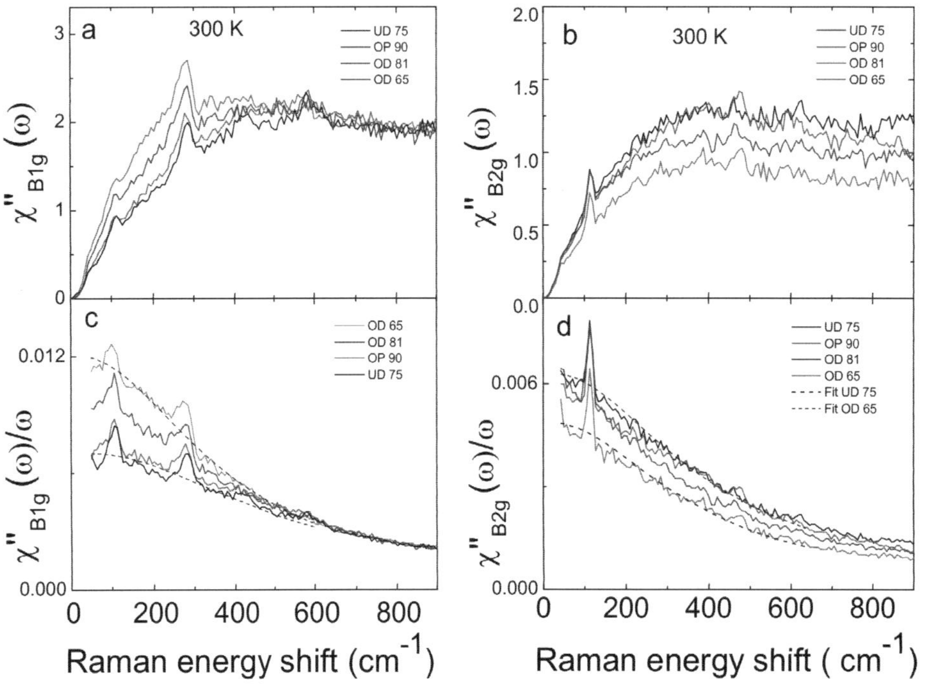

Figure 8. (a) and (b) B_{1g} and B_{2g} Raman spectra measured at 300 K for various doping levels; (c) and (d) Raman spectra (a,b) divided by ω. The dash lines correspond to the Drude fits for selected spectra.

In order to emphasize this, we have reported in figs. 8-a and -b the Raman responses $\chi''(\omega)$ divided by the frequency, ω. According to the Shrastry-Shraiman relation [44] it is related to the real part of the optical conductivity $\chi''(\omega)/\omega$=Re$\sigma(\omega)$. We then observe (see Fig. 8-c and d) a "Drude" peak which grows up in the B_{1g} Raman like conductivity spectra whereas it decreases in intensity as the doping level is increased in the B_{2g} geometry.

This behaviour is in contradiction with recent cluster dynamical mean field calculations where the Drude peak of the calculated B_{2g} Raman spectra is seen to grow up and sharpen as the doping level is increased [25]. We can notice that the ab plane optical conductivity measurements also exhibit an enhancement of the a Drude peak with doping but optical conductivity is not a k-selective probe [45]. We do not yet understand this discrepancy but we suspect that an ingredient is missing such as a possible reconstruction of the Fermi surface with doping [41, 46].

We have fitted the $\chi''(\omega)/\omega$ curves with a standard Drude expression assumed to hold at sufficiently low energy [43]: $\Gamma(0)(Z^*)^2/(\Gamma^2(0)+\omega^2)$. $\Gamma(0)$ is the static scattering rate. This hypothesis makes sense since the scattering rate measured from optical measurements on the same compounds is rougthly constant up to 550 cm^{-1} [45]. The Drude expression (just above) is particulary useful here because it allows us to disentangle between Γ and Z^*.

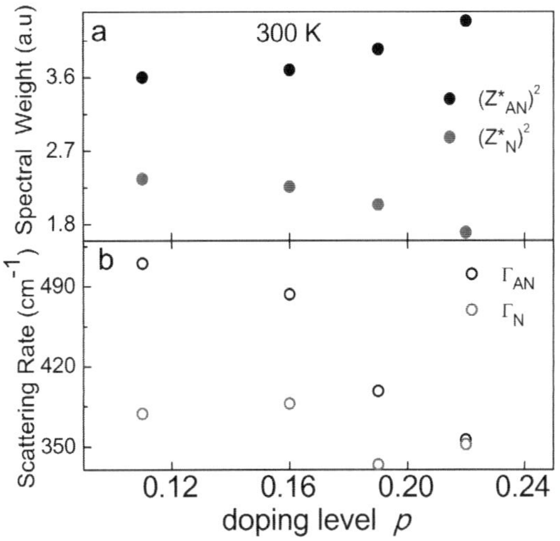

Figure 9. Doping dependence of (a) the quasiparticle spectral weights and (b) the scattering rates in the nodal and antinodal regions. Data have been extracted from the Drude fits of the Raman like conductivity spectra of fig.8.

The doping dependence of $(Z^*)^2$ and Γ extracted from the Drude fits at 300 K in B_{1g} and B_{2g} geometries are reported in Fig. 9-a and b. In fig. 9-a, we observe a decrease of $(Z_N^*)^2$ of roughtly 40% with doping while $(Z_{AN}^*)^2$ increases. This unexpected decrease of $(Z_N^*)^2$ with doping is also detected from the Drude fits performed on the B_{2g} Raman like conductivity spectra measured at different temperatures for the UD 75 and OD 65 compounds (see fig. 10-a).

In Fig. 9-b, Γ_{AN} exhibits a clear decrease of about 45% upon doping: the antinodal quasiparticles are better defined at high doping level than at low one. On the contrary, the nodal scattering rate Γ_N, is barely doping dependent and shows only a small decrease of 7%. Such variations of the static scattering rate are consistent with previous Raman investigations [15, 26].

From this analysis, it clearly appears that the main contribution responsible for the large difference between ζ_N between OD 65 and UD 75 (at 300 K) comes from a decrease of $(Z_N^*)^2$ rather than a change in Γ_N with doping. Notice here that the doping evolutions of the Raman vertex and the density of states N_F are unable to explain the increase of ζ_N with underdoping. The first one (which is proportional in a first approximation to t', the second nearest-neighbor intraplane hopping integral, is expected to decrease with underdoping [47] while N_F around the nodes is expected to be almost constant in this doping range $(0.11 - 0.22)$ [47].

We would like to mention here that the Z_N^* increase with underdoping is perfectly consistent with our previous investigations in the superconducting state. We have shown that the nodal slope of the B_{2g} superconducting Raman response $\frac{N_F(Z_N^*)^2}{v_\Delta}$ is almost constant [23, 48]. The most likely scenario for this is that v_Δ, the nodal slope of the superconducting gap increases with underdoping [24, 49]. This implies that Z_N^* increases with underdoping.

Such a strange doping evolution of Z_N^* is not yet elucidated. However we believe that Fermi surface disturbances could induce this kind of behaviour.

By applying the Drude fit procedure on the Raman like conductivity spectra, we can also extract the temperature dependence of both $(Z_N^*)^2$ and Γ_N. This is shown in fig. 10. Surprisingly

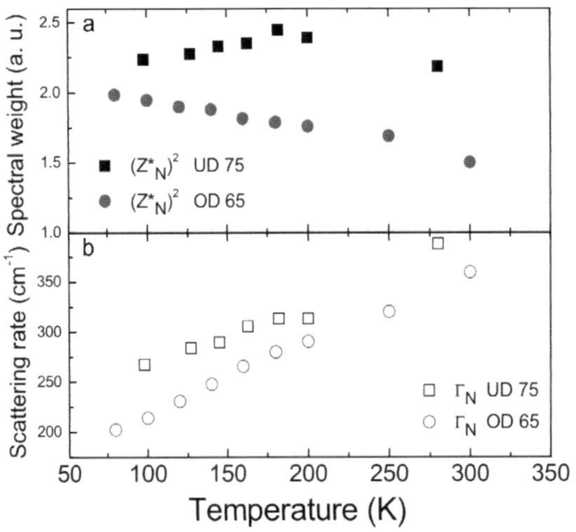

Figure 10. Temperature dependence of (a) the quasiparticles spectral weight and (b) the scattering rates for the UD 75 and OD 65 crystals.

we find that Z_N^* exhibits a weak temperature dependence for UD 75 (7%) and a larger one of about 30% for OD 65. We do not have at present an explanation for this behaviour. We also observe as already mentioned that the $(Z_N^*)^2$ values (in the temperature range considered) are larger for the UD 75 than for the OD 65 compounds. In fig. 10-b, the static scattering rate, Γ_N, shows a quasi linear temperature dependence for both the UD 75 and OD 65 compounds. It is quite similar to the temperature dependence of the scattering rate obtained from earlier Raman experiments [15]. The temperature dependence of Γ_N is quite robust in the doping range $p = 0.11$ and 0.22. We have to investigate Raman measurements at higher doping level to detect the T^2 behaviour seen from quantum oscillations measurements on strongly overdoped compounds (OD 15K) [7].

5. Conclusion

The pseudogap manifests itself in different manners in the B_{1g} and B_{2g} Raman response functions of underdoped Bi-2212 single crystals. In the B_{1g} spectra, it corresponds to a strong depletion of the low energy electronic continuum as the temperature decreases. This has permitted us to identify both the pseudogap temperature and the pseudogap energy as a function of doping level. The effect of the pseudogap in the B_{2g} Raman spectra is more subtle. The depletion of the electronic continuum settles at a higer energy range than in the B_{1g} geometry. Moreover, superconductivity brings supplementary electronic states inside the energy range where the depletion develops. This is in contradiction with a preformed pairs scenario and lead us to suggest different structures for the pseudogap and the superconducting gap [1]: a d-wave superconducting gap and an s-wave anisotropic pseudogap. We show that the antinodal and nodal quasiparticle dynamics are distinct. The nodal scattering rate exhibits a quasi linear temperature dependence both inside and outside the pseudogap phase while the antinodal scattering rate is strongly altered and becomes non temperature dependent inside the pseudogap phase. This reveals the robutness of the nodal quasiparticles with doping range $p = 0.1 - 0.2$. Finally, we find that the nodal quasiparticles spectral weight increases with underdoping while the antinodal one decreases. We believe these observations deserve to be more deeply investigated.

5.1. *Acknowledgments*
We thanks S. Sakai, M. Civelli, A. Georges, A. Millis, G. Orso and R. Lobo for very fruitful discussions.

6. References

[1] Sakai S *et al.* 2012 arXiv:1207.5070 submitted
[2] Warren W W *et al.* 1989 *Phys. Rev. Lett.* **62** 1193
[3] Alloul H *et al.* 1989 *Phys. Rev. Lett.* **63** 1700
[4] Ando Y *et al* 2004 *Phys. Rev. Lett.* **93** 267001
[5] Loram J W *et al* 1994 *J. Supercond.* **7**, 243
[6] Tallon J L *et al* 2001 *Physica C* **349**, 53
[7] Hussey N E *et al* 2008 *J. Phys. : Condens. Matter* **20** 123201
[8] Taillefer L , 2010 *Annual Review of Condensed Matter Physics* **1** 51
[9] Damascelli A *et al* 2003 *Rev. Modern. Phys.* **75** 473
[10] Homes C C *et al* 1993 *Phys. Rev. Lett.* **71** 1645
[11] Timusk T and Statt B 1999 *Rep. Prog. Phys.* **62** 122
[12] Lobo R 2011 *High-temperature superconductors* ed. Qiu X G *Woohead Publishing Oxford Cambridge, Philadelphia* 103
[13] Fischer O *et al* 2007 *Rev. Modern. Phys.* **79** 0610672
[14] Nemetschek R *et al.* 1997 *Phys. Rev. Lett.* **78**, 4837
[15] Opel M *et al.* 2000 *Phys. Rev. B* **61**, 9752
[16] Gallais Y *et al.* 2005 *Phys. Rev. B* **71**, 012506
[17] Blumberg G *et al. Science* **278** 1427
[18] Guyard W *et al.* 2008 *Phys. Rev. B* **77** 024524
[19] Z^* is equivalent to $Z\Lambda$ defined in ref. [48].
[20] Wen J S *et al.* 2008 *J. of Crystal Growth.* **310** 1401
[21] Mihaly L *et al.* 1993 *Rev. Sci. Instrum.* **64** 2397
[22] Devereaux T P and Hackl R 2007 *Rev. Mod. Phys.* **79** 175
[23] Blanc S *et al.* 2009 *Phys. Rev. B* **80** 140502R
[24] Sacuto A *et al* 2011 *C. R. de Physique* **12** 480
[25] Lin N *et al.* 2010 *Phys. Rev. B* **82** 045104 ; Lin N *et al.* arXiv: 1206.0023v1
[26] Venturini F *et al.* 2002 *Phys. Rev. Lett.* **89** 107003
[27] Watanabe T *et al* 2007 *Phys. Rev. Lett.* **79** 2113
[28] Ding H *et al* 1996 *Nature* **382** 51
[29] Kanigel A *et al. Nat. Phys.* **2** 447
[30] Dipasupil I *et al.* 2002 *J. Phys. Soc. Jpn* **71** 1535
[31] Kugler M *et al.* 2001 *Phys. Rev. Lett.* **86** 4911
[32] Kotliar G *et al.* 2006 *Rev. Mod. Phys.* **78** 865; Maier T *et al.* 2005 *Rev. Mod. Phys.* **78** 1027; Tremblay A -M S *et al.* 2006 *Low Temp. Phys.* **32** 424
[33] Johnson P. (private communication); Yang H B *et al.* 2008, *Nature* **456** 77
[34] Gull E *et al.* arXiv:1207.2490v1
[35] Sordi G *et al.* 2012 *Sci. Rep.* **2** 547
[36] Norman M R *et al.* 2005 *Adv. Phys.* **54** 712
[37] Kondo T *et al.* 2009 *Nature* **457** 296
[38] He R H *et al.* 2011 *Science* **331** 1579
[39] Kohsaka Y *et al.* 2007 *Science* **315** 1380; Kohsaka Y *et al.* 2008 *Nature* **454** 1072
[40] Bernhard C *et al.* 2008 *J. Phys. Chem. Solids* **69**, 3064
[41] leBoeuf D *et al.* 2007 *Nature* **450** 533
[42] Rullier-Albenque F *et al.* 2012 *EPJ Web of Conferences* **23** 00016
[43] Devereaux T P 2003 *Phys. Rev. B* **68** 094503
[44] Shastry B S and Shraiman B I 1990 *Phys. Rev. Lett.* **65** 1068; Shastry B S and Shraiman B I 1991 *Int. Journ. Mod. Phys. B* **5** 365
[45] Hwang J *et al.* 2007 *J. Phys. Cond. Matter* **19** 125208
[46] Norman M R 2010 *Physics* **3** 86
[47] Kordyuk A A *et al.* 2003 *Phys. Rev. B* **67** 064504
[48] Le Tacon M *et al.* 2006 *Nature Phys.* **2** 537
[49] Blanc S *et al.* 2010 *Phys. Rev. B* **82** 144516

Journal of Physics: Conference Series **449** (2013) 012012 doi:10.1088/1742-6596/449/1/012012

Evidence for Intra-Unit-Cell magnetic order in the pseudo-gap state of high-T_c cuprates

Y. Sidis and P. Bourges

Laboratoire Léon Brillouin, CEA-CNRS, CE-Saclay, 91191 Gif sur Yvette, France

E-mail: yvan.sidis@cea.fr

Abstract.

The existence of the mysterious pseudo-gap state in the phase diagram of copper oxide superconductors and its interplay with unconventional *d-wave* superconductivity has been a long standing issue for more than a decade. There is now a growing number of experimental indications that the pseudo-gap phase actually corresponds to a symmetry breaking state. In his theory for cuprates, C. M. Varma proposes that the pseudo-gap is a new state of matter associated with the spontaneous appearance of circulating current loops within CuO_2 unit cell. This intra-unit-cell order breaks time reversal symmetry, but preserves lattice translation invariance. Polarized elastic neutron scattering measurements provide evidence for an intra-unit-cell magnetic order inside the pseudo-gap state. This order could be produced by the orbital-like magnetic moments induced by the circulating current loops. The magnetic order displays the same characteristic features in $HgBa_2CuO_{4+\delta}$, $YBa_2Cu_3O_{6+x}$ and $Bi_2Sr_2CaCu_2O_{8+\delta}$ demonstrating that this genuine phase is ubiquitous of the pseudo-gap of high temperature copper oxide materials. We review the main properties characterizing this intra-unit-cell magnetic order and discuss its interplay or competition with other spin and charge instabilities.

1. Pseudo-gap state and circulating current loop theory

In the phase diagram of high temperature superconducting copper oxides, various anomalous electronic properties have been reported in addition to the occurrence of unconventional *d-wave* superconductivity. In normal state, these materials exhibit non-Fermi liquid properties and enter a mysterious pseudo-gap regime, characterized by the loss of density of states on certain portions of the Fermi surface below a temperature $T^\star$. A large variety of theoretical models has been designed to account for these exotic electronic properties and to shed light on their interplay with non conventional superconductivity. Among them, the pseudo-gap theory of C. M. Varma [1, 2, 3] proposes that the pseudo-gap state is a long range ordered state and that fluctuations of its order parameter control not only the non Fermi liquid properties, but also are likely to mediate the *d-wave* superconducting pairing [4]. Derived from the 3-band Hubbard model, this mean-field theory postulated that strong electronic interaction involving the Coulomb repulsion between copper and neighboring planar oxygens give rise to the spontaneous appearance of circulating current (CC) loops within CuO_2 unit cell. The CC-loops appear by pairs, turning clockwise and anti-clockwise. This Intra-Unit-Cell (IUC) order breaks time reversal symmetry, but preserves lattice translation invariance. Beyond the mean-field theory proposed by C. M. Varma, the existence of a CC-loop order is challenged by other theoretical studies. The numerical analysis carried by Thomale and Greiter [5, 6] does not confirm their existence. The quantum

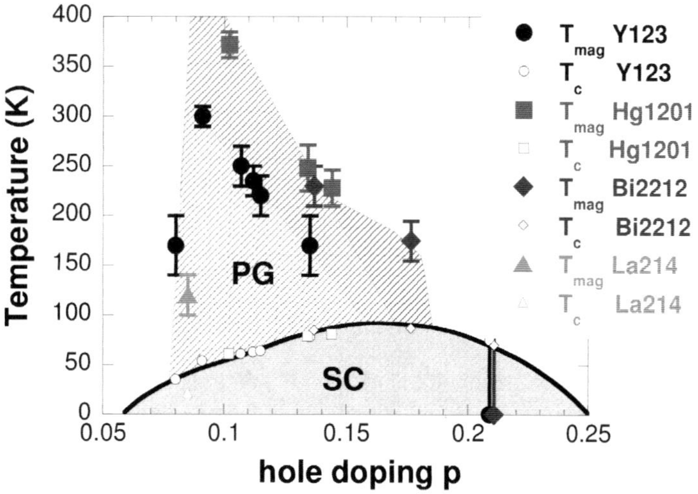

Figure 1. (color online) Variation of the temperature T_{mag} at which the IUC magnetic order appears as a function the hole doping level for four cuprate families: $YBa_2Cu_3O_{6+x}$ (Y123 - ●)) [11, 13, 14], $HgBa_2CuO_{4+\delta}$ (Hg1201- ■) [15, 16], $Bi_2Sr_2CaCuO_{8+\delta}$ (Bi2212 - ♦) [17], $La_{2-x}Sr_xCuO_4$ (La214 - ▲) [18]. The open symbols stand for the value of the superconducting critical temperature T_c for each sample. Note that the value of T_c is a function not only of p but also of disorder, possibly leading to systematic differences in carrier concentration estimates. p for Hg1201 is estimated using the $T_c(p)$ relationship for Y123 [20].

variational Monte Carlo simulation of Weber and co-workers [7] questions the stability of the CC-loops suggested by Varma's model, but suggests that a CC-loop order could become much more stable in a 5-band Hubbard model, where apical oxygen orbitals are also included. In addition, the existence of CC-loops in ladder copper oxide materials is not fully excluded [8, 9, 10]. Beyond pure theoretical considerations, several experimental observations provide strong encouragement for models based CC-loop order in copper oxide materials.

2. Evidence for an Intra-Unit-Cell magnetic order

Polarized neutron scattering studies carried in bilayer $YBa_2Cu_3O_{6+x}$ (Y123) [11, 12, 13, 14] and monolayer $HgBa_2CuO_{4+\delta}$ (Hg1201) [15, 16] have reported experimental evidence of a long range 3D magnetic order hidden in the pseudo-gap state. The existence of such a long range ordered magnetic state has been confirmed recently in bilayer $Bi_2Sr_2CaCuO_{8+\delta}$ (Bi2212) [17]. In monolayer $La_{2-x}Sr_xCuO_4$ (La214), a similar magnetic order has also been observed [18], but in this system it remains 2D and short ranged. Figure 1 shows the variation of the temperature T_{mag} at which the magnetic order settles in as a function of the hole doping level (p).

In all the three compounds with a superconducting critical temperature T_c as high as ~90 K at optimal doping, the magnetic order develops concomitantly with the pseudo-gap state, at least for a hole doping level larger than p=0.09. In Hg1201 and Y123, the magnetic ordering temperature T_{mag} matches the pseudo-gap temperature $T^\star$ determined by resistivity measurements [11, 15, 19]. In Bi2212 [17], the hole doping dependence and the magnitude of T_{mag} and $T^\star$ determined by various techniques [21, 22, 23, 24] are quite consistent. The fact that $T^\star$ does not decrease linearly with increasing hole doping and likely diplays a shoulder around optimal doping in Bi2212 [25] is well reproduced by the hole doping dependence of T_{mag} (Fig. 1). In addition, the comparison of the polarized neutron scattering measurements and tunneling spectroscopy measurements in Bi2212 [24] suggest that the variation of the ordered magnetic moment upon increasing hole doping seems to reproduce the one of the pseudo-gap energy Δ_{PG}.

This novel 3D magnetic order does not seem to depend strongly on the structural properties of the materials. In contrast with Y123 or Hg1201, Bi2212 has a body centered structure. The magnetic order seems to remain 3D independently from the nature of the stacking of Cu sites along the **c** axis. Furthermore, Hg1201 is a tetragonal system, Y123 is orthorhombic along the Cu-O bonds and Bi2212 is orthorhombic along diagonals. The influence of orthorhombicity on details of the magnetic structure is a relevant issue, but dedicated neutron scattering studies are still under progress.

Whatever the system where it has been observed, this novel magnetic state preserves the lattice translation invariant, but, at variance with ferromagnets, does not give rise to a uniform magnetization [19]. These observations imply the existence of an IUC (antiferro-)magnetic order. The overall symmetry of this order is consistent with the so-called CC-θ_{II} phase proposed by C. M. Varma [3]. Polarized neutron scattering studies may have detected the distribution of static magnetic fields generated by the CC-loops. The possible detection of these magnetic fields by local probes such as NMR, NQR or μSR is still debated [26, 27, 28, 29, 30], but should help to get a deeper understanding of the intrinsic nature of the IUC magnetic order.

Beyond polarized neutron scattering experiments, other measurements carried out in Y123 provide direct or indirect indications that a symmetry breaking state develops below the pseudo-gap temperature. Indeed, resonant ultrasound spectroscopy measurements have recently reported the first thermodynamic evidence of a true phase transition taking place when entering the pseudo-gap [31]. In addition, the observation of anomalies in the second derivative of the magnetization below T_{χ} [32] can be understood in term of a coupling of the CC-loop order parameter to the uniform susceptibility [33]. Furthermore, the spontaneous appearance of an a-b anisotropy in the Nernst coefficient at T_{ν} [34] could be interpreted as a violation of the Onsager Reciprocity principle in the CC-loop state which breaks time reversal symmetry [35]. Finally, a polar Kerr effect has been observed within the pseudo-gap state below T_K [36]. This effect has been initially interpreted as a signature of a time reversal symmetry broken state [36], but actually ascribed to a gyrotopic order [37]. As shown in Fig.2, the hole doping dependencies of the different temperatures T_{mag}, T_{χ}, T_K and T_{ν} display a certain similarity, suggesting that the related phenomena could share a common origin.

While the appearance of an IUC magnetic order can be ascribed to CC-loops, to date there is no direct observation of these CC-loops in superconducting cuprates. Alternatively, a resonant X-ray diffraction study in non superconducting CuO has claimed the first direct observation of CC-loops in CuO_2 unit cell, i.e the basic building block of superconducting copper oxide materials [38], but has been challenged theoretically [39]. Nevertheless, such a kind of experiments in the superconducting cuprates would be very illuminating in resolving the question of whether a CC-loop description is appropriate for the observed neutron scattering signal in the pseudo-gap phase [39].

3. Time reversal symmetry breaking

Since the discovery of the pseudo-gap regime, by pioneering NMR measurements in the early 90's [40], the most accurate information concerning the electronic properties of this phase have been provided by tunneling spectroscopy [41] and angle resolved photo-emission (ARPES) [42] in bilayer Bi2212. In the pseudo-gap state, left-circularly polarized photon give a different photocurrent from right-circularly photons in ARPES measurements [43]. This circular dichroism would imply that time reversal symmetry is spontaneously broken in the pseudo-gap state, a phenomenon in good agreement with a CC-loop order [44].

The recent observation of an IUC magnetic order in Bi2212 further confirms that time reversal symmetry is broken in the pseudo-gap state. The temperature dependencies of polarized neutron

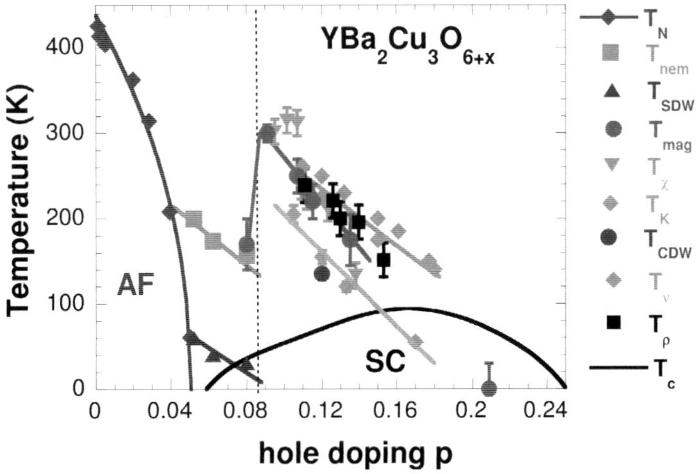

Figure 2. (color online) $YBa_2Cu_3O_{6+x}$ phase diagram as a function of the hole doping p. In addition to the antiferromagnetic (AF) state and the d-wave superconducting (SC) state, several phases can be detected. In the lighly doped regime ($0.05 < p < 0.085$), incommensurate spin fluctuations develop a quasi-1D anisotropy spontaneously below T_{nem} [59]. The C_4 rotational symmetry is first broken, but the translation invariance is preserved. At lower temperature, below T_{SDW}, the spin correlations freeze and a spin density (SDW) state settles in [59]. In the underdoped regime ($p > 0.085$), the pseudo-gap state develops below $T^\star$ that can be determined by the departure of the electrical resistivity from a linear T-dependence at T_ρ [34]. The IUC-magnetic order appears at $T_{mag} \sim T^\star$ [14]. Around that temperature an anomaly can be observed in the second derivative of the magnetization at T_χ [32] and the Nernst coefficient exhibits a net a-b anisotropy at T_ν [34]. Below T_{mag}, a Kerr effect is observed at T_K [36]. Around the specific hole doping $p \sim 1/8$, a charge density wave (CDW) instability develops below a temperature T_{CDW} [68], close to T_K.

measurement in samples UD-85 and OD-70 [17] and of the dichroic effects in samples UD-85 and OD-65 [43] match with each other (Fig. 3). Keeping in mind that ARPES measurements are performed on thin films and polarized neutron measurement on large single crystals, the onset of time reversal symmetry breaking found by both types of measurements for samples UD-85 is in a good agreement (Fig. 3). Likewise, the observations of dichroic effect at antinodal wave vectors and of a magnetic signal on Bragg reflections (1,0,L) indicate that the phase responsible for both effects possess the same symmetry as the CC-θ_{II} phase proposed in the CC-loop theory for the pseudo-gap [44].

It is worth pointing out that the reported dichroic effect in ARPES and its interpretation are still controversial and the subject of a long running debate [45, 46, 47, 48, 49], since it has been argued that structural effects could also account for the dichroic signal reported in ARPES. Whenever correct for the analysis of ARPES measurements, these criticisms do not hold for polarized neutron diffraction measurements, since structural effects cannot produce a spin flip signal varying with the neutron spin polarization as observed experimentally.

Beyond circularly polarized ARPES measurements and polarized neutron diffraction, the observations of a polar Kerr effect at a temperature T_K in bilayer Y123 [36] and in monolayer $Pb_{0.55}Bi_{1.5}Sr_{1.6}La_{0.4}Cu_{6+\delta}$ (Pb-Bi2201) [50] have been first presented as a direct evidence for time reversal symmetry breaking within the pseudo-gap state. This interpretation is now challenged by the fact the sign of the rotation of the polarization angle is the same on opposite surfaces of the sample, while in the usual Kerr effect the rotation angle must reverse. While T_K

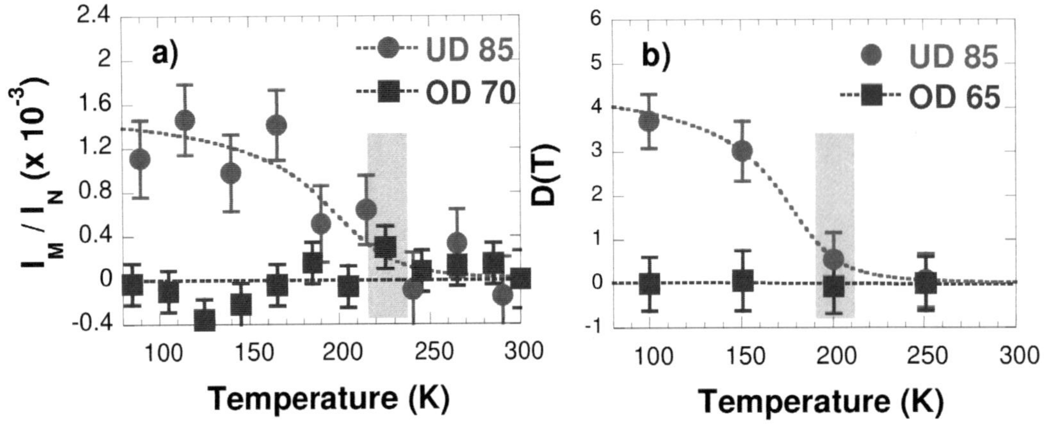

Figure 3. (color online) Comparison between polarized neutron scattering measurements performed on Bi2212 single crystals [17] and circular dischroism obtained from ARPES measurements on Bi2212 thin films [43]. a) The full magnetic intensity (I_M) normalized by the nuclear intensity (I_N) measured at the Bragg reflection (1,0,1). This normalization procedure is useful for comparing different samples. (I_M) is about 3 orders of magnitude smaller than (I_N). Raw polarized neutron scattering data and a detailed description of the method used to extract such a weak magnetic intensity have been recently given in [17, 19]. b) Circular dischroism D(T) at the anti-nodal wave vectors, reproduced from [43]. The data associated with underdoped samples (UD- T_c) are shown with symbols ● and those associated with overdoped samples (OD- T_c) with symbols ■.

coincides with $T^\star$ in Pb-Bi2201, it appears deep inside the pseudo-gap state in Y123 (Fig. 2). Starting from a CC-loop order breaking time reversal symmetry at $T^\star$, distinct theoretical models propose different routes to account for the observed polar Kerr once additional chiral properties are considered [51, 52].

All experimental reports available in monolayer systems (Hg1201, Pb-Bi2201) and bilayer systems (Y123, Bi2212) provide compelling evidence that the pseudo-gap phase is actually an ordered state, whose order parameter and broken symmetries remains to be clearly identified.

4. Static and dynamical magnetic properties

Neutron scattering technique can probe simultaneously the nuclear and magnetic correlations. Spin density wave (SDW) -like instabilities break translation invariance and produces a magnetic response at wave vectors distinct from the nuclear Bragg reflections. In cuprates, static or fluctuating SDWs are usually located around the antiferromagnetic (AF) wave vector and characterized by the wave vectors $\mathbf{q}_{SDW}=\mathbf{q}_{AF}\pm(\delta,0)$ (and/or $\pm(0,\delta)$) [53], where $\mathbf{q}_{AF}$=(0.5,0.5) stands for the planar AF wave vector, given in reduced lattice units (tetragonal notation) (Fig. 4.a). At variance, the IUC magnetic order implies the existence of staggered magnetic moments within the unit cell and corresponds to a q=0 magnetic instability. Since the IUC magnetic order preserves the lattice translation invariance, the weak magnetic signal that it produces in a neutron scattering experiment is superimposed with the much stronger nuclear scattering. The use of spin polarized neutron scattering technique is then essential to disentangle the nuclear and magnetic neutron scattered intensities. A full polarization analysis allows a determination of the orientation of the ordered magnetic moments.

The search for a long range 3D magnetic order in the pseudo-gap phase has been performed on Bragg reflections (1,0,L)-(0,1,L) with integer L values in monolayer Hg1201 [15, 16] and in bilayers Y123 [11, 12, 13, 14] and Bi2212 [17]. In the case of monolayer La214 [18], the IUC order is quasi-2D and can be observed for any L value. In Ref. [17], the magnetic intensities along the **c** axis have been compared for these 4 cuprates families. Interestingly, these intensities exhibit for all systems a similar decay with $\frac{2\pi}{c}$L (the component of the wave vector perpendicular to the CuO_2 plane), suggesting a common magnetic origin. For all cuprate families, the polarization analysis demonstrates that the magnetic moments cannot be strictly perpendicular to the CuO_2 planes. A conservative estimate of 45 ± 20° can be given for the tilt angle of these moments with respect to the **c** axis. While Polar Kerr effect is usually associated with the appearance of ferromagnetic moments, it can also be mediated by magneto-electric coupling for an antiferromagnetic order. The tilt of magnetic moments could be an important feature which lowers the symmetry of the IUC magnetic structure and allows the magnetic structure to be active to produce a Kerr effet [54].

The origin of the static magnetic order reported by polarized neutron diffraction is still an open issue. The IUC-cell order could be induced by CC-loops in the CC-θ_{II} phase. This phase is characterized by two CC-loops per CuO_2 unit cell and the loop pattern is fourfold degenerated. However, the CC-loops are confined with the CuO_2 planes, generating orbital magnetic moments perpendicular to the planes. This is at variance with the experimental observation. Recently, two Ising magnetic mode has been discovered in the pseudo-gap state of Hg1201 [55, 56]. These new modes could be ascribed to magnetic excitations associated with a change of configuration of the CC-loop pattern. The two excitations corresponding to a $\pm 90^o$ rotation of a loop pattern are indeed active in the spin flip channel in inelastic polarized neutron scattering. The quantitative analysis of the characteristic mode energies and their weak dispersion leads to the conclusion that the ground state cannot reduce to one of the four states of the CC-θ_{II} phase, but should be made of their quantum superposition [57, 58]. This produces a quantum interference phenomenon in neutron diffraction measurement: the tilt angle of the magnetic moment reported in polarized neutron scattering measurements would then highlight the degree of admixture of the four orthogonal states of the CC-θ_{II} phase.

Alternatively, it has been proposed that CC-loops could be delocalized on the CuO_6 octahedra for a monolayer like Hg1201 [7] or on CuO_5 pyramids in bilayer systems like Y123 or Bi2212 [30]. This model should produce a magnetic signal on Bragg reflection (00L), but no signal has been observed at Bragg position (002) neither in Y123 [11] nor in Hg1201 [15]. Furthermore, this model fails to account for the polarization analysis carried out on the Bragg reflection (100) [19].

5. Competing spin instabilities

In La214 [18] for a hole doping level p=0.085, the quasi-2D and short range IUC magnetic order appears at a rather low temperature $T_{mag} \sim$120 K. Below this temperature, the dynamical spin response at the wave vector $\mathbf{q}_{SDW}$ exhibits two marked modifications: (i) an enhancement of the low energy spin fluctuations intensity and (ii) an order parameter-like increase of the incommensurability parameter δ. Both effects indicate a deep interplay between the q=0 magnetic order and the incommensurate spin fluctuations. In lightly doped Y123 [14] ($p \sim 0.08$), the 3D long range IUC order develops at $T_{mag} \sim$170 K (Fig. 1), i.e very close to the temperature $T_{nem} \sim$ 150 K (Fig.2). At this characteristic temperature, the low energy incommensurate spin fluctuations start exhibiting a net a-b anisotropy [59]. T_{nem} is usually interpreted as the onset temperature at which the C_4 rotational symmetry is spontaneously broken by strong electronic correlations pushing the system into an electronic nematic state. In lightly doped Y123, the orthorhombic lattice distortion serves as weak orientational field and is essential for the observation of the a-b anisotropy. Since this state also preserves the translation invariance, it corresponds to another competing q=0 electronic instability. At much lower temperature, spin correlations freeze at T_{SDW} (Fig. 2), giving rise to a spin density wave (SDW) state that

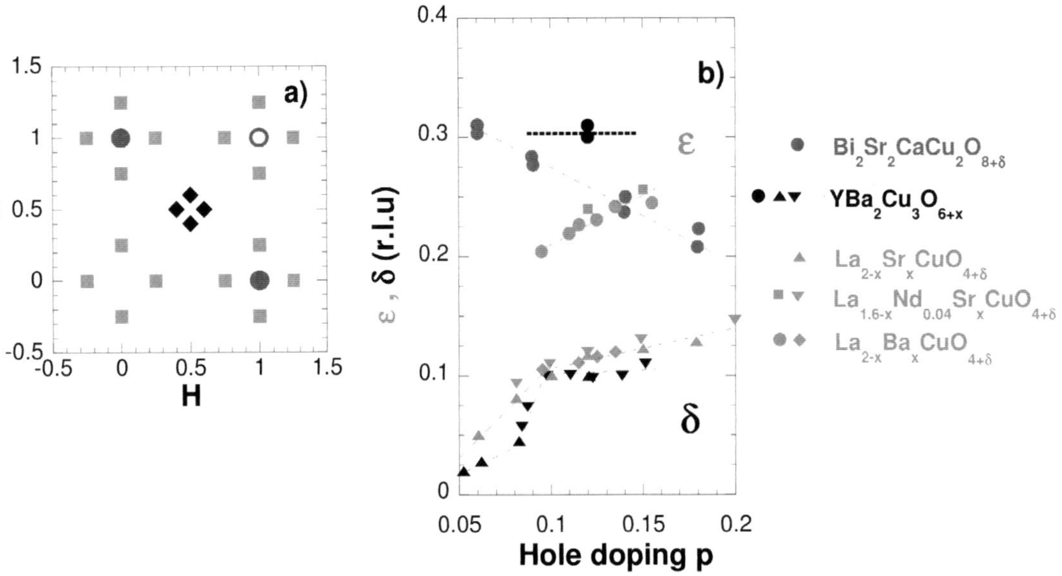

Figure 4. (color online) a) Location in the reciprocal space of the different kinds of spin, charge and orbital-like magnetic signals. The IUC-magnetic order is observed at the planar waves vector **q** =(1,0) and (0,1) (●), but should be observable at **q** =(1,1) (○). A fluctuating or static charge density wave (CDW) is characterized by the planar wave vectors $\mathbf{q}_{CDW}=\pm(\epsilon,0)$ (and/or $\pm(0,\epsilon)$ (■). A fluctuating or static spin density wave (SDW) is associated with the planar wave vectors $\mathbf{q}_{SDW}=\mathbf{q}_{AF}\pm(\delta,0)$ (and/or $\pm(0,\delta)$) (♦), where $\mathbf{q}_{AF}$=(0.5,0.5). b) Hole doping dependencies of the incommensurability parameters ϵ and δ for 3 cuprate families: (i) $La_{2-x}Sr_xCuO_4$ [72], $La_{1.6-x}Nd_{0.4}Sr_xCuO_4$ [73, 74, 75, 53, 76, 77], $La_{2-x}Ba_xCuO_4$ [78] (Light green symbols) ,(ii) $YBa_2Cu_3O_{6+x}$ [59, 60, 68, 69](black symbols), (iii) $Bi_2Sr_2CaCu_2O_{8+\delta}$ [64] (red symbols). The parameter δ is measured using neutron scattering technique only, while ϵ can be determined using either by X-ray scattering or neutron scattering technique. For $Bi_2Sr_2caCu_2O_{8+\delta}$, ϵ is estimated from STM measurements.

can further coexist with the superconducting phase [59]. In underdoped Y123 [14] (p > 0.09), the 3D IUC magnetic order is well developed and settles in at $T_{mag} \simeq T^{\star}$. In parallel, low energy spin fluctuations around the AF wave vector are gapped. Through the substitution of a few percents of non magnetic Zn impurities (known to strongly weaken the superducting order), neither T_{mag} [14] nor $T^{\star}$ [61] are affected. On the contrary, the average magnetic moment associated with the IUC magnetic order drops down. This indicates that the IUC order is likely to be destroyed around Zn impurities, but remains unaffected far away. Simultaneously, quasi-1D incommensurate spin fluctuations are restored at low energy, suggesting that a competing SDW order could develop where the IUC magnetic order is suppressed. All these studies suggest a competition between the IUC magnetic state and an incipient SDW state.

6. Competing charge instabilities

The existence of an IUC order associated with the pseudo-gap state in Bi2212 system is not only supported by polarized neutron scattering measurements. In Bi2212, the analysis of STM images at Δ_{PG}, the energy usually associated with the pseudo-gap, points towards an IUC order in the pseudo-gap state [62]. At Δ_{PG}, the unbalance of the electronic densities measured at planar wave vector $\mathbf{q}_x=(1,0)$ and $\mathbf{q}_y=(0,1)$ suggests that C_4 rotational symmetry is broken in the pseudo-gap state. The analysis of spectroscopic imaging STM is consistent with an IUC

Journal of Physics: Conference Series **449** (2013) 012012 doi:10.1088/1742-6596/449/1/012012

electronic nematic order, yielding distinct electronic density on O sites along **a** and **b** directions [62]. In principle, the IUC electronic nematicity is different from the IUC loop order. However, the mean-field analysis of the different IUC-ordering possibilities in the 3-band Emery model indicates that the electronic nematic order and the CC-loop order could actually coexist [63].

The IUC electronic nematicity inferred from spectroscopic imaging STM studies corresponds to a long range order which coexists with a (short range) charge density wave (CDW) order [62]. This order break the C_4 rotational symmetry and the lattice translation invariance. The CDW correlations are characterized by the planar wave vectors $\mathbf{q}_{CDW}=\pm(\epsilon,0)$ (and/or $\pm(0,\epsilon)$) (Fig. 4.a). According to STM measurements in Bi2212, the incommensurability parameter ϵ is typically of the order of 0.25 r.l.u and decreases linearly with increasing hole doping [64] (Fig. 4.b). Furthermore, the fluctuating CDW order develops below the onset of the pseudo-gap state and can be viewed as a consequence of the pseudo-gap rather than its cause [65]. Likewise, a local CDW order has been recently associated with topological defects in the electronic nematic state [66].

In Y123, the observation of an a-b anisotropy in the Nernst coefficient [34] close to $T^\star$ (Fig 2) is often interpreted as an indication that the C_4 rotation symmetry is spontaneously broken within the pseudo-gap state and associated with an electronic nematic state [67]. At lower temperature, hard X-ray [68] and soft X-ray resonant scattering [69] studies reported evidence for a charge modulation along wave vectors $\mathbf{q}_{CDW}=\pm(\epsilon,0)$ and $\pm(0,\epsilon)$, with $\epsilon\sim 0.3$ r.l.u (Fig.4.b). Considering X-ray measurements, one cannot yet unambiguously distinguish between an equal distribution of domains of two uniaxial CDWs with mutually perpendicular propagating wave vectors or a single CDW with biaxial charge modulation. However recent sound velocity measurements strongly support a single CDW with biaxial charge modulation [71], but is at variance with C_4 rotation symmetry breaking inferred from Nernst coefficient. The intensity of the CDW signal grows on cooling down to T_c below which it is suppressed. Under an external magnetic field, the CDW is strengthened while superconductivity is depressed. NMR measurements [70] further indicate that a static CDW shows up only under magnetic field, but should be absent at zero field. As a consequence the CDW reported in X-ray measurements should be still fluctuating. In Y123, the evidence of a CDW instability is well documented in a narrow hole doping range around $p\sim 1/8$ ($0.09 < p < 0.13$) (Fig. 4.b, solid line). The CDW instability develops at T_{CDW} well below the pseudo-gap temperature $T^\star$ or the temperature associated with the IUC magnetic order T_{mag} (Fig. 2).

In La214, there is no direct observation of a CDW order neither by neutron scattering nor by X-ray scattering. At variance, once the system is co-doped with Nd ($La_{1.6}Nd_{0.04}Sr_xCuO_4$) [73, 74, 75, 76, 77] or Sr replaced by Ba ($La_{2-x}Ba_xCuO_4$) [78], a CDW order can be observed below the LTO-LTT structural transition and followed by a SDW order. In that specific cuprate family, ϵ is about twice δ, suggesting a strong interplay between charge and spin modulations. These combined charge and spin modulations are usually interpreted within stripe models where doped hole segregate into charge stripes separating hole poor AF domains ([53] and references therein). The stripe order develops around the characteristic hole doping $p\sim 1/8$, where stripes could be pinned down onto the lattice. In La214 (p=0.085) [18], where there is neither indication for a CDW order nor for a SDW order, the IUC magnetic order remains quasi-2D and at short range. The planar magnetic correlation lengths are consistent with a picture where CC-loops could be confined within 2-leg copper ladders. In this scenario, the CC-Loop order would be frustrated by the competing stripe instability, which seems much stronger in that cuprate family than in any other superconducting cuprate materials.

More theoretical and experimental work is needed to shed light on the interplay between the pseudo-gap phenomenon, the IUC orders (i.e q=0 instabilities) and the spin and/or charge density wave instabilities. To this respect, one can notice that for decades the role of oxygen in cuprates has been overlooked, whereas an IUC magnetic order could be bound to CC-loops

[1, 2, 3] or orbital oxygen moments [79]. In addition, the observed charge modulations could be produced by a valence bond state or quadrupolar density wave state [80], related to a non uniform electronic densities on oxygen sites. Thus, the internal degrees of freedom of CuO_2 plaquettes could become relevant to shed light on the physics of cuprates.

7. Superconducting state

While polarized neutron scattering measurements show the existence of an IUC magnetic order in the normal state below $T^\star$, the evolution of this order upon entering the superconducting state is still an open and interesting issue. Indeed time reversal symmetry is violated in this state, whereas it is preserved in a pure *d-wave* superconducting state [81]. However the search for indications of the competition or coexistence between these distinct electronic states using polarized neutron scattering technique has been limited by a technical problem. To date, the elastic polarized neutron scattering measurements have been performed using a standard longitudinal polarization set-up [19], for which the neutron polarization is maintained on the sample using a tiny magnetic guide field ($\sim$ 10 Gauss). Since a depolarization of the neutron beam can take place when cooling down below T_c, most of polarized neutron scattering studies have been restricted to the normal state. In early studies carried out in Y123 [11], a few data in the superconducting state were reported, but no definitive conclusion can be drawn from these data, since the issue of depolarization of the neutron beam was not addressed. To overcome such a technical problem, measurements in the superconducting state require to place the sample in a zero magnetic field chamber, already available in a more sophisticate spherical polarization set-up (such Cryo-PAD or Mu-PAD). Using this device, elastic polarized neutron measurements will be safely extended into the superconducting state in a close future.

8. Conclusion

Polarized elastic neutron scattering measurements indicate that an IUC magnetic order develops in the pseudo-gap state. This novel magnetic order displays the same characteristic features in monolayer Hg1201 and bilayer Y123 and Bi2212, demonstrating that this genuine phase is ubiquitous in the pseudo-gap of high temperature copper oxide materials. The polarized neutron scattering and the circularly polarized photoemission out over various cuprate families all point towards a breaking of time reversal symmetry up on entering the pseudo-gap state. The intrinsic nature of the IUC magnetic order remains to be understood, but, to date, the few information we have concerning the magnetic pattern within the CuO_2 unit cell are consistent with the CC-loop model (phase CC-θ_{II}) proposed by C. M. Varma [1, ?, 3]. Nevertheless, the stability of the CC-loop state is still an open theoretical issue, in addition to the ability of such a q=0 instability to produce a gap in the charge excitation spectrum. New theoretical and experimental developments are needed to shed light on the interplay between the observed IUC magnetic order and other electronic states, such as an IUC electronic nematicity, incipient CDW/SDW orders and *d-wave* superconductivity.

References

[1] Varma C M 1997 *Phys. Rev. B* **55** 14554
[2] Varma C M 1999 *Phys. Rev. Lett.* **83** 3538
[3] Varma C M 2006 *Phys. Rev. B* **73** 155113
[4] Aji V, Shekhter A and Varma C M 2010 *Phys. Rev. B* **81** 064515
[5] Thomale R and Greiter M 2007 *Phys. Rev. Lett* **99** 027005
[6] Thomale R and Greiter M 2008 *Phys. Rev. B* **77** 094511
[7] Weber C, Lauchli A, Mila F and Giamarchi T 2009 *Phys. Rev. Lett.* **102** 017005
[8] Chudzinski P, Gabay M and Giamarchi T 2008 *Phys. Rev. B* **78** 075124
[9] Chudzinski P, Gabay M and Giamarchi T 2010 *Phys. Rev. B* **81** 165402
[10] Nishimoto S, Jeckelmann E and Scalapino D J 2009 *Phys. Rev. B* **79** 205115

[11] Fauqué B, Sidis Y, Hinkov V, Pailhès S, Lin C T, Chaud X and Bourges P 2006 *Phys. Rev. Lett.* **96** 197001
[12] Sidis Y, Fauqué B, Aji V and Bourges P 2007 *Physica B* **397** 1
[13] Mook H A, Sidis Y, Fauqué B, Balédent V and Bourges P 2008 *Phys. Rev. B* **78** 020506
[14] Balédent V, Haug D, Sidis Y, Hinkov V, Lin C T and Bourges P 2011 *Phys. Rev. B* **83** 104504
[15] Li Y, Balédent V, Barisic N, Cho Y, Fauqué B, Sidis Y, Yu G, Zhao X, Bourges P and Greven M 2008 *Nature* **445** 372
[16] Li Y, Balédent V, Barisic N, Cho Y C, Sidis Y, Yu G, Zhao X, Bourges P and Greven M 2011 *Phys. Rev. B* **84** 224508
[17] De Almeida-Didry S, Sidis Y, Balédent V, Giovannelli F, Monot-Laffez I and Bourges P 2012 *Phys. Rev. B* **86** 020504
[18] Balédent V, Fauqué B, Sidis Y, Christensen N B, Pailhès S, Conder K, Pomjakushina E, Mesot J and Bourges P 2010 *Phys. Rev. Lett.* **105** 027004
[19] Bourges P and Sidis Y 2011 *C. R. Physique* **12** 461
[20] Liang R, Bonn D A, and Hardy W N 2006 *Phys. Rev. B* **73** 180505
[21] Raffy H, Toma V, Murrills C and Li Z, 2007 *Physica C* **460-462** 851 ; Konstantinovic Z, Li Z and Raffy H 1999 *Physica C* **259-261** 567
[22] Ding H, Yokoya T, Campuzano J C, Takahashi T, Randeria M, Norman M R, Mochiku T, Kadowaki K and Giapintzakis J 1996 *Nature* **382** 51
[23] Ishida K, Yoshida K, Mito T, Tokunaga Y, Asayama K, Nakayama Y, Shimoyama J and Kishio K 1998 *Phys. Rev. B* **58** R5960
[24] Dipasupil R M, Oda M, Momono M. and Ido M 2002 *J. Phys. Soc. Jpn* **71** 1535
[25] Vishik I M, Hashimoto M, He R H, Lee W S, Schmitt F, Lu D H, Moore R G, Zhang C, Meevasana W, Sasagawa T, Uchida S, Fujita K, Ishida S, Ishikado M, Yoshida Y, Eisaki H, Hussain Z, Devereaux T P and Shen Z X 2012 *PNAS* **109** 18332
[26] Sonier J E, Brewer J H, Kiefl R F, Miller R I, Morris G D, Stronach C E, Gardner J S, Dunsiger S R, Bonn D A, Hardy W N, Liang R and Heffner R H 2001 *Science* **292** 1692
[27] Sonier J E, Brewer J H, Kiefl R F, Heffner R H, Poon K F, Stubbs S L, Morris G D, Miller R I, Hardy W N, Liang R, Bonn D A Gardner J S, Stronach C E and Curro N J 2002 *Phys. Rev. B* **66** 134501
[28] MacDougall G J, Aczel A A, Carlo J P, Ito T, Rodriguez J, Russo P L, Uemura Y J, Wakimoto S and Luke G M 2008 *Phys. Rev. Lett.* **101** 017001
[29] Strässle S, Graneli B, Mali M, Roos J and Keller H 2011 *Phys. Rev. Lett.* **106** 097003
[30] Lederer S and Kivelson S A 2012 *Phys. Rev. B* **85** 155130
[31] Shekhter A, Ramshaw B J, Liang R, Hardy W N, Bonn D A, Balakirev F F, McDonald R D, Betts J B, Riggs S C and Migliori A 2013 *Nature* to appear; arXiv:1208.5810
[32] Leridon B, Monod P and Colson D 2009 *Europhys. Lett.* **87** 17011
[33] Gronsleth M S, Nilssen T B, Dahl E K, Stiansen E B, Varma C M, and Sudbo A 2009 *Phys. Rev. B* **79** 094506
[34] Daou R, Chang J, LeBoeuf D, Cyr-Choiniére O, Laliberté F, Doiron-Leyraud N, Ramshaw B J, Liang R, Bonn D A, Hardy W N and L Taillefer 2010 *Nature* **463** 520
[35] Varma C M , Yakovenko V M and Kapitulnik A arXiv:1007.1215
[36] Xia J, Schemm E, Deutscher G, Kivelson S. A, Bonn D A, Hardy W. N, Liang R, Siemons W, Koster G, Fejer M M and Kapitulnik A 2008 *Phys. Rev. Lett.* **100** 127002
[37] Hosur P, Kapitulnik A, Kivelson S A, Orenstein J and Raghu S 2013 *Phys. Rev. B* **87** 115116
[38] Scagnoli V, Staub U, Bodenthin Y, De Souza R A, Garcia-Fernández M, Garganurakis M, Boothroyd A T, Prabhakaran D and Lovesey S W 2011 *Science* **332** 696
[39] Di Matteo S and Norman M R 2012 *Phys. Rev. B* **85** 235143
[40] Alloul H, Ohno T and Mendels P 1989 *Phys. Rev. Lett.* **63** 1700
[41] Fisher S, Kugler M, Maggio-Aprile I, Berthod C and Renner C 2007 *Rev. Mod. Phys.* **79** 353
[42] Damascelli A, Hussain Z and Shen Z X 2003 *Rev. Mod. Phys.* **75** 473
[43] Kaminski A, Rosenkranz S, Fretwell H M, Campuzzano J C, Li Z, Raffy H, Cullen W G, You H, Olson S G, Varma C M and Höchst H 2002 *Nature* **416** 610
[44] Simon M E and Varma C M 2002 *Phys. Rev. Lett.* **89** 274003
[45] Borisenko S V, Kordyuk A A, Kim T K, Nenkov K A, Fink J, Grazioli C, Turchini S and Berger H 2004 *Phys. Rev. Lett.* **92** 207001
[46] Di Matteo S and Norman M R 2007 *Phys. Rev B* **76** 014510
[47] Arpiainen V, Bansil A and Lindroos M 2009 *Phys. Rev. Lett.* **103** 067005
[48] Norman M R, Kaminski A, Rosenkranz S and Campuzano J C 2010 *Phys. Rev. Lett.* **105** 189701
[49] Lindros M, Arpiainen V and Bansil A 2010 *Phys. Rev. Lett.* **105** 189702
[50] He R H, Hashimoto M, Karapetyan H, Korolek J. D, Hinton J P, Testaud J P, Nathan V, Hyao H, Takanaka

K, Meevasana W, Moore R G, Lu D H, Mo S K, Ishikado M, Eisaki H, Hussain Z, Deveraux T P, Kivelson S A, Orenstein J, Kapitulnik A and Shen Z X 2011 *Science* **231** 1579

[51] Aji V, He Y and Varma C M arXiv:1211.1391

[52] Pershoguba S S, Kechedzhi K and Yakovenko V M arXiv:1303.2982

[53] Tranquada J M 2007 *Treatrise of high temperature superconductivity,* J R Schrieffer, (Springer-verlag, New York) pp 257-298

[54] Orenstein J 2011 *Phys. Rev. Lett.* **107** 067002

[55] Li Y, Balédent V, Yu G, Barisić N, Hradil K, Mole R A, Sidis Y, Steffens P, Zhao X, Bourges P and Greven M 2010 *Nature* **468** 283

[56] Li Y, Yu G, Balédent V, Li Yangmu, Barisić N, Zhao X, Hradil K, Mole R. A, Sidis Y, Steffens P, Bourges P and Greven M 2012 *Nat. Phys. 2012* **8** 407

[57] Varma C M 2010 *Nature* **468** 184

[58] He Yan and Varma C M 2011 *Phys. Rev. Lett.* **106** 147001; 2012 Phys. Rev. B **85** 155102

[59] Haug D, Hinkov V, Sidis Y, Bourges P, Christensen N B, Ivanov A, Keller T, Lin C T and Keimer B 2010 *New J. Phys.* **12** 105006

[60] Dai P, Mook H A, Hunt R D and Dogan F 2001 *Phys. Rev. B* **63** 054525

[61] Alloul H, Bobroff J, Gabay M and Hirschfeld P J 2009 *Rev. Mod. Phys.* **81** 45

[62] Lawler M J, Fujita K, Lee J, Schmidt A R, Kohsaka Y, Kim C K, Eisaki H, Uchida S, Davis J C, Sethna J. P and Kim E A 2010 *Nature* **466** 347

[63] Fischer M H and Kim E A 2011 *Phys. Rev. B* **84** 144502

[64] Kohsaka Y, Taylor C, Wahl P, Schmidt A, Lee Jhinhwan, Fujita K, Alldredge J W, McElroy K, Lee Jinho, Eisaki H, Uchida S, Lee D H and Davis J C 2008 *Nature* **454** 1072

[65] Parker C V, Aynajian P, da Silva Neto E H, Pushp A, Ono S, Wen J, Xu Z, Gu G and Yazdani A 2010 *Nature* **468** 677

[66] Mesaros A, Fujita K, Eisaki H, Uchida S, Davis J. C, Sachdev S, Zaanen J, Lawler M and Kim E A, 2012 *Science* **333** 426

[67] Hackl A and Vojta M (2009) *Phys. Rev. B (R)* **80** 220514 (2009)

[68] Chang J, Blackburn E, Holmes A T, Christensen N B, Larsen J, Mesot J, Liang R, Bonn D A, Hardy W N, Watenphul A, Zimmermann M V, Forgan E M and Hayden S M 2012 *Nat. Phys.* Physics **8**, 871-876

[69] Ghiringhelli G, Le Tacon M, Minola M, Blanco-Canosa S, Mazzoli C, Brookes N B, De Luca G M, Frano A, Hawthorn D G, He F, Loew T, Moretti Sala M, Peets D C, Salluzzo M, Schierle E, Sutarto R, Sawatzky G A, Weschke E, Keimer B and Braicovich L 2012 *Science* **337**, 821

[70] Wu T, Mayaffre H, Krämer S, Horvatic M, Berthier C, Hardy W N, Liang R, Bonn D. A and Julien M H 2011 *Nature* **477** 191

[71] LeBoeuf D, Krämer S, Hardy W N, Liang R, Bonn D A and Proust C 2013 *Nat. Phys.* **9**, 79

[72] Yamada K, Lee C H, Kurahashi K , Wada J, Wakimoto S, Ueki S, Kimura Y, Endoh Y, Hosoya S, Shirane G, Birgeneau R J, Greven M, Kastner M A and Kim Y J 1998 *Phys. Rev. B* **57** 6165

[73] Tranquada J M, Sternlieb B J, Axe J D, Nakamura Y and Uchida S 1995 *Nature* **375** 561

[74] Tranquada J M, Axe J D, Ichikawa N, Nakamura Y, Uchida S, and Nachumi B 1996 *Phys. Rev. B* **54** 7489

[75] Tranquada J M, Axe J D, Ichikawa N, Moodenbaugh A R, Nakamura Y and Uchida S 1997 *Phys. Rev. Lett.* **78** 338

[76] Niemöller T, Ichikawa N, Frello T, Hünnefeld H, Andersen N H, Uchida S, Schneider J R and Tranquada J M 1999 *Eur. Phys. J. B* **12** 509

[77] Zimmermann M V, Vigliante A, Niemöller T, Ichikawa N, Frello T, Madsen J, Wochner P, Uchida S, Andersen N H, Tranquada J M, Gibbs D and Schneider J R 1998 *Europhys. Lett.* **41** 629

[78] Hücker M, Zimmermann M V, Gu G D , Z J Xu, Wen J S, Xu Guangyong, Kang H J, Zheludev A and Tranquada J M 2011 *Phys. Rev. B* **83** 104506

[79] Moskvin A S 2012 *JETP Lett.* **96** 385

[80] Efetov K B, Meier H and Pépin C arXiv:1210.3276

[81] B. Leridon K, Ng T K and Varma C M 2007 *Phys. Rev. Lett.* **99** 027002

Evidence of Absence of Loop-Current Order in $La_{2-x}Sr_xCuO_4$ from μSR Measurements

J E Sonier[1,2], W Huang[3], V Pacradouni[4], M P Kennett[1] and S Komiya[5]

[1] Department of Physics, Simon Fraser University, Burnaby, Bristish Columbia V5A 1S6, Canada
[2] Canadian Institute for Advanced Research, Toronto, Ontario, M5G 1Z8, Canada
[3] Department of Physics and Astronomy, McMaster University, Hamilton, Ontario L8S 4M1, Canada
[4] Powertech Labs Inc., Surrey, British Columbia V3W 7R7, Canada
[5] Central Research Institute of Electric Power Industry, Yokosuka, Kanagawa 240-0196, Japan

E-mail: jsonier@sfu.ca

Abstract. We discuss the results of a recent high precision search for orbital-like magnetic order in the pseudogap regime of $La_{2-x}Sr_xCuO_4$ single crystals using zero-field muon spin relaxation (ZF-μSR). No evidence of proposed loop-current ordered states or any other form of magnetic order is found. We explain why recent experiments refute the argument that this null result is due to the positive muon μ^+ locally destroying the loop-current order.

1. Introduction

Polarized neutron diffraction has revealed two dimensional (2D), short range magnetic order in $La_{1.915}Sr_{0.085}CuO_4$ well below the onset of the pseudogap temperature T^* [1]. The magnetic ordering bears some resemblance to an unusual 3D long range magnetic order that onsets near T^* in $YBa_2Cu_3O_{7-\delta}$ and $HgBa_2CuO_{4+\delta}$ [2, 3, 4, 5, 6]. The main features of the magnetic order lend support to a proposed circulating-current model of the pseudogap state [7]. This connection assumes the magnetic order exists throughout the sample. As a volume-integrated technique, neutron diffraction does not provide information on the magnetic volume fraction, and hence local magnetic probes such as nuclear magnetic resonance (NMR) or muon spin relaxation (μSR) are needed. Zero-field (ZF) μSR measurements on a $YBa_2Cu_3O_{6.6}$ single crystal investigated by polarized neutron diffraction indicate that the unusual magnetic order is present in only about 3 % of the sample [8]. On the other hand, such magnetic order is not observed by ZF-μSR in higher-quality $YBa_2Cu_3O_{7-\delta}$ crystals. While ZF-μSR studies of $La_{2-x}Sr_xCuO_4$ also show no evidence for this kind of magnetic order [9], the possibility of a slight magnetic volume fraction was not considered. Moreover, the contribution of the nuclear moments of $La_{2-x}Sr_xCuO_4$ to the ZF-μSR signal was not established in these earlier studies. In the present paper we summarize recent ZF-μSR measurements of $La_{2-x}Sr_xCuO_4$ that address these issues [10].

2. Results and Discussion

Figure 1 shows the ZF-μSR time spectrum for $La_{1.85}Sr_{0.15}CuO_4$ just above T^* along with simulations of the ZF-μSR signal expected in the Θ_{II} state of the circulating current theory

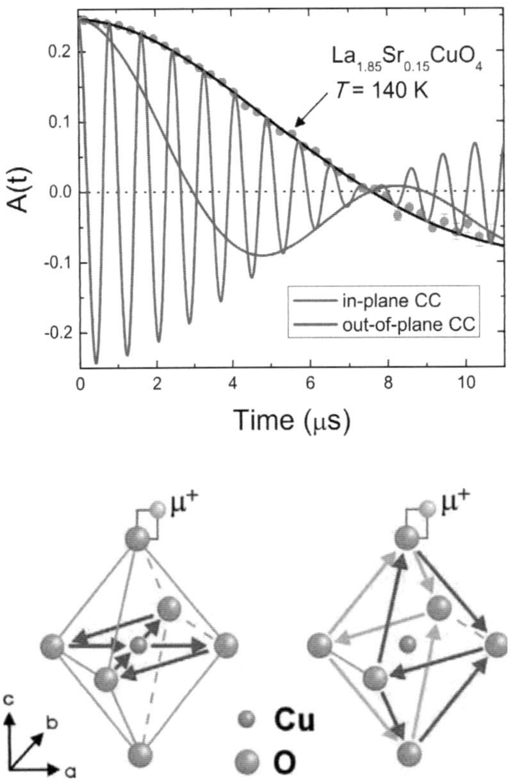

Figure 1. ZF-μSR asymmetry spectrum of $La_{1.85}Sr_{0.15}CuO_4$ at $T = 140$ K (green circles) and simulations of the ZF-μSR signal for the proposed Θ_{II} phase. The red curve corresponds to the original proposal of circulating currents in the CuO_2 layers as depicted in the lower left figure. The grey curve corresponds to a similar pattern of circulating currents that flow out and back into the CuO_2 layers through the apical oxygen atoms (as shown in the lower right figure). Both simulated ZF-μSR signals are damped by the nuclear contribution that is the source of the relaxation of the spectrum for $La_{1.85}Sr_{0.15}CuO_4$ at $T = 140$ K. The rectangle connecting the μ^+ to the apical oxygen in the bottom figures emphasizes that the location of the muon is in the a-c plane at (0.120a, 0, 0.219c).

[7, 11]. The simulations assume the maximum ordered magnetic moment of 0.02 μ_B determined from the spin-polarized neutron experiments on $La_{1.915}Sr_{0.085}CuO_4$ (ref. [1]), that the loop-current order resides in the full sample volume, and that the μ^+ stops at the (0.120a, 0, 0.219c) site in the crystal lattice. Recently, this muon site in $La_{2-x}Sr_xCuO_4$ was accurately determined by calculating the contributions to the ZF-μSR signal of the dipolar magnetic and quadrupolar electrostatic interactions of the μ^+ with the host nuclei [10]. The nuclei are the source of the relaxation of the ZF-μSR asymmetry spectrum shown in figure 1. Simulated ZF-μSR signals of the Θ_{II} phase are shown in figure 1 for the original proposal of orbital currents flowing in the CuO_2 plane, and for currents flowing out and back into the CuO_2 layers through the apical

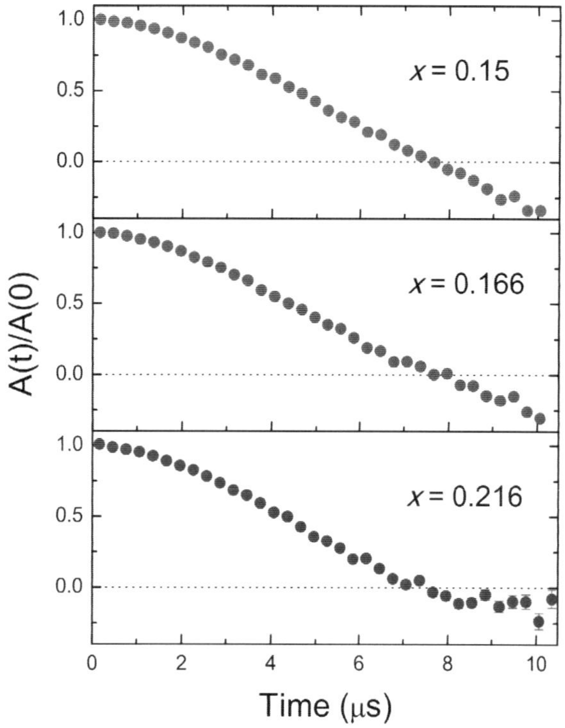

Figure 2. ZF-μSR asymmetry spectra of $La_{2-x}Sr_xCuO_4$ at T = 2.5 K, normalized with respect to the signal amplitude at time $t = 0$. This temperature is below T^* for all three hole concentrations [13].

oxygen atoms. The latter scenario has been proposed to explain the orientation of the magnetic moments observed in the neutron experiments [12]. Disorder and/or slow fluctuations of the circulating-current pattern will increase the damping of the ZF-μSR signal. Regardless, we find that the ZF-μSR spectra at low temperatures do not deviate significantly from that above T^* (see figure 2). Furthermore, the early-time spectra (figure 3) do not show a damped small-amplitude oscillating component, indicative of a minority volume fraction containing magnetic order. Hence these measurements show no evidence of loop-current order.

In a metal the single positive elementary charge of the μ^+ attracts a screening cloud of conduction electrons [15]. It has been argued that the screening of the μ^+ in cuprates locally reduces the charge density of the nearest CuO_2 layers to almost that of the lightly doped insulating state, which is below the range of charge density where loop-current order is proposed to exist [16]. These calculations show that the μ^+ reduces the charge density by $\sim$10 % in the nearest metallic layers, and $\sim$5 % in the next several layers, such that the field at the muon site is diminished by two to three orders of magnitude. More recently, the effect of the μ^+ charge on the local spin dynamics has also received theoretical attention [17]. The local electronic charge density distribution in cuprate superconductors must certainly be modified in response to the presence of the μ^+, but the degree of the disturbance has never been verified. Quite recently it has been demonstrated that the normal-state μ^+-Knight shift in x=0.15 and x=0.24

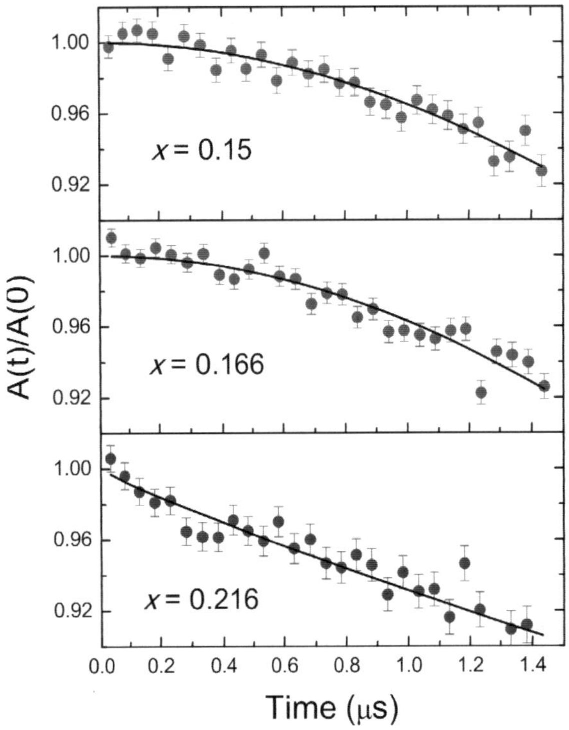

Figure 3. Early-time ZF-μSR asymmetry spectra of $La_{2-x}Sr_xCuO_4$ at $T = 2.5$ K. Note that the difference in the functional form of the signal for $x = 0.24$ may be attributed to the presence of Curie-like paramagnetism, as observed in the bulk magnetic susceptibility [14].

samples scales linearly with the bulk magnetic susceptibility $\chi_{\rm bulk}$ measured using a SQUID magnetometer [18]. The μ^+-Knight shift is a measure of the local magnetic susceptibility, and hence linear scaling with $\chi_{\rm bulk}$ is unambiguous proof that the μ^+ does not strongly perturb its local environment in $La_{2-x}Sr_xCuO_4$.

The theoretical calculations of refs. [16, 17] treat the μ^+ as an isolated positive charge impurity. On the other hand, potential energy calculations and experiments indicate that in oxides the μ^+ bonds to an oxygen atom to form the analogue of a hydroxyl bond. The μ^+ site identification in ref. [10] is in fact compatible with the μ^+ being bound to an apical oxygen atom in $La_{2-x}Sr_xCuO_4$. In $YBa_2Cu_3O_{7-\delta}$ the μ^+ apparently has two sites, bonding exclusively to an apical oxygen atom in the undoped $\delta = 1$ compound, but bonding exclusively to a CuO chain oxygen atom when $\delta = 0$ [19]. Forming a regular covalent O-μ bond requires the μ^+ to grab a lone pair of electrons from the outer shell of the bound oxygen atom. In this case the positive elementary charge added by the μ^+ is effectively transferred away from the muon to another bound atom. Density functional theory calculations are needed to verify this picture for cuprates, but the absence of experimental evidence of strong muon perturbative effects may be related to the μ^+ being bound.

2.1. Acknowledgments
This work was supported in part by the Natural Sciences and Engineering Research Council of Canada, and the Canadian Institute for Advanced Research. We thank Paul Percival for enlightening discussions.

3. References

[1] Balédent V, Fauqué B, Sidis Y, Christensen N B, Pailhès S, Conder K, Pomjakushina E, Mesot J and Bourges P 2010 *Phys. Rev. Lett.* **105** 027004
[2] Fauqué B, Sidis Y, Hinkov V, Pailhès S, Lin C T, Chaud X and Bourges P 2006 *Phys. Rev. Lett.* **96** 197001
[3] Mook H A, Sidis Y, Fauqué B, Balédent V and Bourges P 2008 *Phys. Rev. B* **78** 020506(R)
[4] Li Y, Balédent V, Barišić N, Cho Y, Fauqué B, Sidis Y, Yu G, Zhao X, Bourges P and Greven M 2008 *Nature* **455** 372
[5] Balédent V, Haug D, Sidis Y, Hinkov V, Lin C T and Bourges P 2011 *Phys. Rev. B* **83** 104504
[6] Li Y, Balédent V, Barišić N, Cho Y C, Sidis Y, Yu G, Zhao X, Bourges P and Greven M 2011 *Phys. Rev. B* **84** 224508
[7] Varma C M 2006 *Phys. Rev. B* **73** 155113
[8] Sonier J.E. *et al.* 2009 *Phys. Rev. Lett.* **103** 167002
[9] MacDougall G J *et al.* 2008 *Phys. Rev. Lett.* **101** 017001
[10] Huang W, Pacradouni V, Kennett M P, Komiya S and Sonier J E 2012 *Phys. Rev. B* **85** 104527
[11] Simon M E and Varma C M 2002 *Phys. Rev. Lett.* **89** 247003
[12] Weber C, Läuchli A, Mila F and Giamarchi T 2009 *Phys. Rev. Lett.* **102** 017005
[13] Hashimoto M, Yoshida T, Tanaka K, Fujimori A, Okusawa M, Wakimoto S, Yamada K, Kakeshita T, Eisaki H and Uchida S 2007 *Phys. Rev. B* **75** 140503(R)
[14] Nakano T, Oda M, Manabe C, Momono N, Miura Y and Ido M 1994 Phys. Rev. B **49** 16000
[15] Schenck A 1985 *Muon Spin Rotation Spectroscopy: Principles and Applications in Solid State Physics* (Adam Hilger Ltd, Bristol and Boston)
[16] Shekhter A, Shu L, Aji V, MacLaughlin D E and Varma C M 2008 *Phys. Rev. Lett.* **101** 227004
[17] Dang H T, Gull E and Millis A J 2010 *Phys. Rev. B* **81** 235124
[18] Kaiser C V, Huang W, Komiya S, Hussey N E, Adachi T, Tanabe Y, Koike Y and Sonier J E 2012 *Phys. Rev. B* **86** 054522
[19] Pinpank M 1999 *Physica C* **317-318** 299

Journal of Physics: Conference Series **449** (2013) 012014 doi:10.1088/1742-6596/449/1/012014

The Rise of T_c: A Promising Paradigm via Interfacial Mechanism

C W Chu[1,2,3], B Lv[1], L Z Deng[1], B Lorenz[1], B Jawdat[1], M Gooch[1], K Shrestha[1], K Zhao[1], X Y Zhu[1], Y Y Xue[1] and F Y Wei[1]

[1] Department of Physics and Texas Center for Superconductivity, University of Houston, Houston TX 77204-5002, USA

[2] Lawrence Berkeley National Laboratory, Berkeley CA 94720, USA

[3] Taiwan Comprehensive University System, Tainan, Taiwan

E-mail: cwchu@uh.edu

A tribute to those with courage in the search for higher T_c in the last 25 years:

Ralph Waldo Emerson once said "Whatever course you decide upon, there is always someone to tell you that you are wrong. There are always difficulties arising that tempt you to believe that your critics are right. To map out a course of action and to follow it to the end requires courage."

Abstract. Following the seminal detection of superconductivity at 35 K in the first cuprate high temperature superconductor Ba-doped ternary LCO (La214) by Mueller and Bednorz in 1986, the discovery of the first liquid nitrogen superconductor quaternary YBCO (Y123) with a transition temperature T_c of 93 K by Chu, Wu, and colleagues in 1987 has ushered in the modern era of high temperature superconductivity (HTS). In the ensuing 25 years, great progress has been made in all areas of HTS research and development, from materials to physics to devices. However, the T_c remains at 134 K in quaternary HBCCO (Hg1223) at ambient pressure achieved in 1993 by Schilling et al. and at 164 K in the same compound under high pressure in the same year by Gao et al. A holistic multidisciplinary enlightened empirical approach has been proposed to search for superconductors with higher T_c, which includes the discovery of new compounds empirically and the realization of novel mechanisms proposed by theorists. Recently, we have observed nonbulk superconductivity in single crystalline rare-earth doped Ca122 samples with an onset T_c of 49 K, higher than the T_c of any known compounds formed of the constituent elements at ambient or under high pressures. Subsequent systematic structural, electrical, magnetic, and calorimetric investigations suggest that the observation may be attributed to an interface-induced T_c, thus suggesting a possible new paradigm for higher T_c. In this presentation, our results will be presented and discussed, following a brief recollection of a few events in our lab in the continuing search for higher T_c, even though a number of review articles have appeared.

1. Introduction

Superconductivity is intellectually challenging and technologically promising. It provides the never-ending lure to scientists and technologists and the never-ending inspiration to those in the frontiers of science from quarks to Majorana fermions. As a result, superconductivity is considered a subject of pursuit only 101 years young. One of the major driving forces for its vitality is the continued quest for superconductivity at higher T_c.

The T_c-rise with time is shown in figure 1. T_c was 4 K in Hg when superconductivity was first discovered by Kamerlingh Onnes at Leiden in 1911 after liquefying gas helium [1], rose steadily but only slowly to 23 K in Nb_3Ge thin film by Gavaler et al. of Westinghouse and Testardi et al. of Bell Labs in 1973 [2], after improving the stoichiometry of the compound, and stayed there until 1986. However, it jumped to 35 K in the mixed-phase La-Ba-Cu-O samples when Mueller and Bednorz (MB) at Zurich decided to dive into the new path of perovskite oxides for a higher T_c in 1986 [3]. The genie of this seminal discovery was formally let out of the bottle on December 4 at the 1986 Fall MRS meeting in Boston when I announced our reproduction at Houston of the Zurich result at the end of my formal presentation on the oxide BPBO superconductor and, during the ensuing Q&A, Kitazawa of Tokyo immediately revealed their identification of the superconducting phase as La214 in MB's samples [4]. The world caught fire afterward, busily working around the clock on the new La214 superconductor, in spite of the initial skepticism expressed by many except for a lucky few, including my group.

In an attempt to understand the cause of the observation and to explore the possibility of further raising the T_c, we decided to determine the effect of the inter-atomic distance on the 35 K superconductivity in the La-Ba-Cu-O samples by the application of pressure. To our great surprise, the T_c increases rapidly at a rate about ten times that observed in previously known superconductors to a new record first at 40.2 K [5], and then at 52.5 K [6], exceeding the T_c-ceiling of ~ 30s K predicted by theory [7] and thus raising the prospect that higher T_c is possible through high pressure and/or chemical replacement. It should be noted that unstable resistive and ac magnetic anomalies were also detected above 77 K before the MRS meeting in some mixed-phase samples of La-Ba-Cu-O of ours [8], suggestive of superconductivity above the liquid nitrogen temperature for the first time. I shared

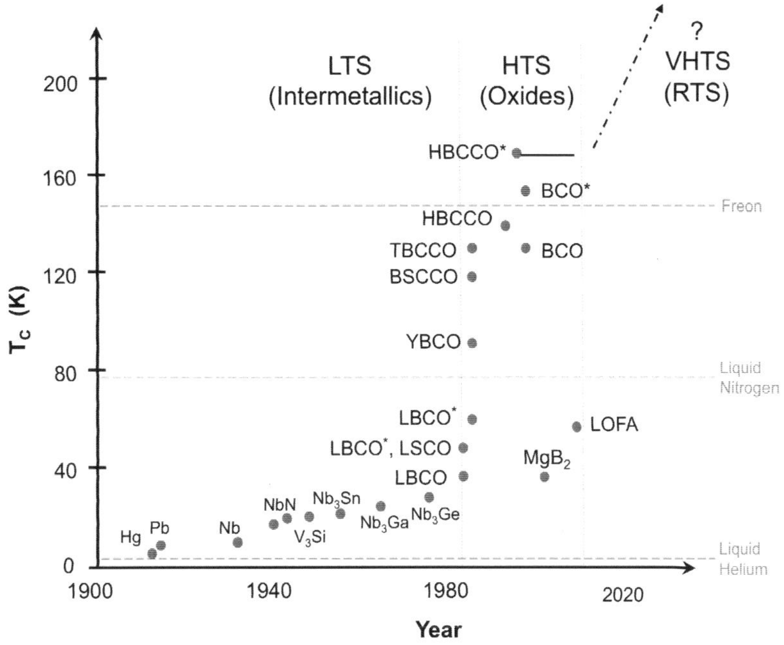

Figure 1. T_c-rise with time (* — under pressure).

the preliminary data at the MRS meeting at Boston with my former student M. K. Wu and invited him to join the search.

Like many in the field who believe that the best way to study the physics of materials is to use single-crystal samples, we tried but failed to grow single crystals of La214 following Kitazawa's phase identification. Unfortunately, two-thirds of our Pt-crucibles for crystal growth were destroyed. However, during this period of time, we found that as the polycrystalline samples were made purer with La214, the appearance of the high temperature resistive anomalies became less frequent. We decided to skip the study of single-crystalline La214 and tried to achieve higher T_c directly by stabilizing the resistive and magnetic anomalies detected sporadically by us above 77 K via chemical replacement and pressure application [9]. Replacements of Ba by the smaller isovalent Ca and Sr and of La by the smaller non-magnetic rare-earth (R) elements, Y, Yb, and Lu (thinking to avoid the magnetic ones, based on the conventional wisdom but finding out later that this was unnecessary as shown below), were proposed and attempted. Substitution of Ba by Sr indeed leads to a T_c increase but not by Ca. A resistive drop in the Y-Ba-Cu-O above 77 K was observed on January 28, 1987, and the superconducting state in these samples was firmly established by detecting the Meissner effect in these samples the very next day. Stable superconductivity with a T_c of 93 K above the liquid nitrogen temperature of 77 K was finally achieved in YBCO [10]. A subsequent high pressure test demonstrated that the newly discovered 93 K T_c YBCO belongs to a new material family different from that of La214 [11]. The crystal structure of YBCO was quickly determined in collaboration with Hazen et al. of Geophysical Laboratories at Washington DC to be $YBa_2Cu_3O_7$ (Y123) [12]. To probe the role of Y^{3+} in YBCO, a standard technique was employed by examining the effect of a small replacement of the Y^{3+} by a magnetic R^{3+} element such as Gd. We found the T_c was hardly affected and concluded that the role of Y in YBCO is to stabilize the crystal structure and not to participate in the superconducting process. We therefore replaced the Y in YBCO completely by R and obtained the new superconducting family of RBCO ($RBa_2Cu_3O_7$ or R123) for all Rs except Ce and Pr, with a T_c up to 98 K [13]. Extending the replacement of Y with the trivalent ions beyond the R^{3+}s, two more superconducting families BSCCO [$Bi_2Sr_2Ca_{n-1}Cu_nO_{2n+4}$ for n = 1, 2, 3, … or Bi22(n-1)n] with a T_c up to 105 K for n = 3 and TBCCO [$Tl_2Ba_2Ca_{n-1}Cu_nO_{2n+4}$ for n = 1, 2, 3, … or Tl22(n-1)n] with a T_c up to 125 K for n = 3 were discovered by Maeda et al. in Tsukuba [14] and Sheng and Hermann in Arkansas [15], respectively, in 1988. Five years later in 1993, Schilling et al. [16] in Zurich replaced the trivalent R with Hg and raised the T_c to the current record of 134 K in HBCCO [$HgBa_2Ca_{n-1}Cu_nO_{2n+3}$ for n = 1, 2, 3, …] for n = 3 at ambient, which was subsequently raised by the application of pressure at ~ 30 GPa to 164 K by Gao et al. of Houston and Mao et al. at Geophysical Laboratories at Washington DC [17].

In the last 25 years, a large number of cuprate HTSs has been discovered. They belong to the above families and their derivatives. These cuprate HTSs display a layered structure and can be represented by a generic formula, $A_mE_2R_{n-1}Cu_nO_{2n+m+2}$, with m = 1 or 2; n = 1, 2, 3, …; A = Bi, Tl, Pb, or Cu; E = Ca, Sr, or Ba; and R = Ca or a rare-earth element. The generic formula can be rewritten as $A_mE_2R_{n-1}Cu_nO_{2n+m+2} = [(EO)(AO)_m(EO)] + [(CuO_2)R_{n-1}(CuO_2)_{n-1}]$ that consists of two main alternately stacked blocks, namely the active block $[(CuO_2)R_{n-1}(CuO_2)_{n-1}]$, where supercurrent flows along the CuO_2-planes, and the charge-reservoir block $[(EO)(AO)_m(EO)]$, where doping takes place to introduce charge-carriers without defects to the CuO_2-planes. The active block comprises n (CuO_2)-layers interleaved by (n-1) R-layers, while the charge-reservoir block consists of m (AO)-layers bracketed by 2 (EO)-layers, as shown in figure 2. Almost every cuprate HTS can be designated as Am2(n-1)n with the (AO)-layer or 0m2(n-1)n without the (AO)-layer. Their parents are an antiferromagnetic Mott-insulator and made superconducting through doping. T_c increases with n up to 3 and decreases beyond due to the inhomogeneous doping and proximity effect for n > 3. In spite of the large number and their higher T_c, YBCO remains to be the choice HTS for science and application due to its higher J_c in high field, higher chemical stability and physical robustness, ease in epitaxial film growth and lower cost. As a result, it has been included in the 2000 White House Millennium Time Capsule "to honor the past

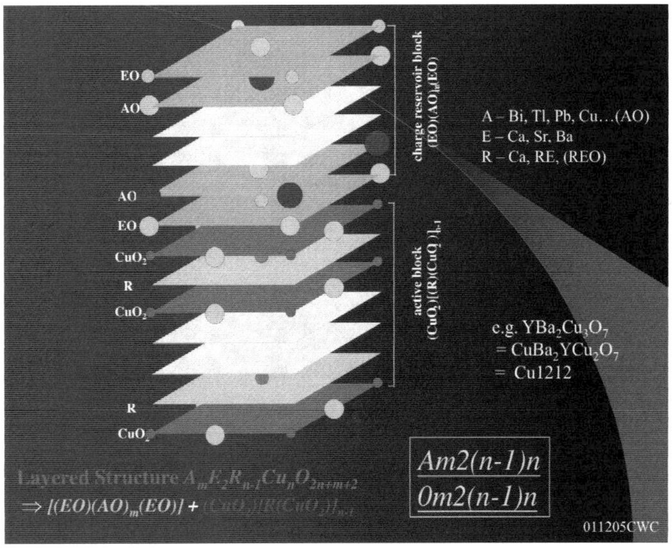

Figure 2. The schematic crystal structure and generic formula of the cuprate layered HTSs.

and imagine the future" as a tribute to the ingenuity of the Americans in the 20th century. Hope the full potential of YBCO can be realized long before our grandchildren open the time capsule in 2100.

In 2008, the layered Fe-based superconductor F-doped LaFeAsO (1111) with a T_c ~ 26 K was discovered by Hosono et al. at Tokyo [18], leading to the second explosion of superconductivity research activities in the hope of raising the T_c further because of the existence of a large number of compounds with similar structure and the unraveling of the role of magnetism due to the existence of a large amount of the magnetic Fe in them. Indeed, a new class of Fe-based pnictide and chalcogenide superconductors has been found with a maximum T_c of 56 K in the ensuing years. They all possess the FeAs or FeSe-layers and may be grouped into the following families as shown in figure 3: the RFeAsO (1111) with R = rare-earth elements and a maximum T_c of 56 K [19]; the AFe_2As_2, $A'Fe_2As_2$, and AFe_2Se_2 (122) with A = alkaline and A' = alkaline-earth elements and a maximum T_c of 38 K [20] in general and 49 K in special cases [21] (see below); AFeAs (111) with A = Li and Na and a maximum T_c of 25 K [22]; and FeSe (11) with a maximum T_c of 10 K [23] and 46 K [24] in bulk and single-layer form, respectively (see below).

2. Search for superconductors of higher T_c

Indeed, in the last 25 years, T_c has undergone a dramatic rise to the current record of 134 K at ambient [16] and 164 K under pressure [17]. Unfortunately, it has stagnated at these values since 1993. However, the intrinsic scientific lure and potential technological implication of superconductivity at a higher T_c have kept its search unabated. The ultimate goal is to discover a superconductor with a higher T_c, preferably at room temperature. It should be pointed out that the term of room temperature superconductor (RTS) can be a relative one. A superconductor is an RTS if its T_c is equal to the ambient temperature of the environment where a superconductor is tested and used, i.e. it can be the liquid nitrogen boiling point, the liquid natural gas boiling point, or 300 K around which we live. Therefore, there are two ways to achieve RTS: one is to increase the T_c and the other is to lower the ambient temperature. For practitioners in the HTS-field, we would like to raise the T_c to 300 K. Three questions immediately follow: the first is whether it is possible to achieve 300 K superconductivity. My answer is a simple "yes," because we have learned the following over the past 25 years: there is no evidence, experimental or theoretical, to show that room temperature superconductivity is impossible; whatever physics law doesn't say won't happen will happen; and history repeats itself always. The second is whether the effort to search for RTS is worth it in view of the potential challenge to use RTS for power or large current applications because of its possible small current carrying capability due to the small flux pinning force associated with its short coherence length $\xi \propto 1/T_c$ as a result of the high T_c. This is particularly true for a highly anisotropic superconductor, which has generally been considered to favor high T_c. However, a reduction in the anisotropy of an RTS can provide relief to such an impasse to a large extent. In addition, small-current applications of HTS can have equal, if not more, impacts on lives in this increasingly electricity-dependent society. The third question is how to find it. We believe a holistic multidisciplinary enlightened empirical approach, where knowledge and experience in different fields, insight, imagination, courage, and even luck play important roles, can be most fruitful [25].

In the general approach mentioned above, two paths have been adopted to raise T_c: to realize novel mechanisms inspired by theoretical models and to discover new compounds guided by experience and insight empirically. Among the many theoretical mechanisms proposed for higher T_c, interfacial mechanism in different forms has been the one most explored theoretically and experimentally since 1964 when Little and Ginzburg [26] first proposed it. However, an analytic study was not made until 1973 when Allender, Bray, and Bardeen (ABB) [27] considered a model system that consists of an interface between a thin metal-layer and a semiconductor. Their realistic estimations showed that a substantial T_c-enhancement is possible via the exchange of excitons with a higher characteristic energy should the stringent requirements on the interfaces and materials be met, although questions have also been raised [28]. As was pointed out by ABB, the realization of interfacial mechanism might not only lead to a substantial T_c-enhancement but would also be a notable achievement by itself. Numerous experiments inspired by the general concept of interface-induced T_c-enhancement have subsequently been carried out on samples with heterostructures that consist of insulating, semiconducting, metallic, and superconducting bilayers, trilayers, or superlattices [29]. Some have reported the observation of interfacial superconductivity with enhanced T_c. However, evidence for its existence remains elusive due to the technical difficulties in direct diagnoses. Interfacial superconductivity has often been inferred from the higher T_c detected in the artificially made layered samples as compared to that in the starting materials of the layers under ambient conditions without detailed information about the superconducting state. Even the T_c reported in these heterostructural samples appears to be often achievable in the original materials through either doping or straining.

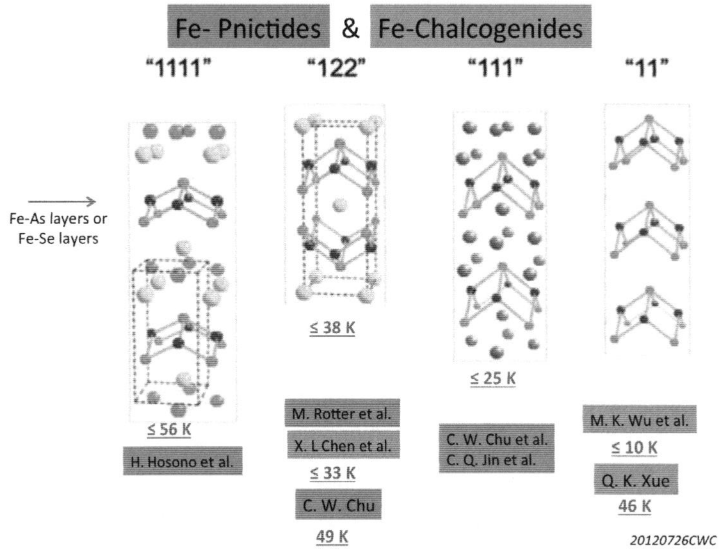

Figure 3. The schematic crystal structures of the four families of Fe-pnictide (green) and Fe-chalcogenides (blue) superconductors with their maximum T_cs.

3. Possible evidence for interface-enhanced T_c in rare-earth doped Ca122

Recently, non-bulk superconductivity was detected in Pr-doped single-crystalline $CaFe_2As_2$ [$(Ca_{1-x}Pr_x)Fe_2As_2$ or (Ca,Pr)122] with an onset-T_c (T_c^{on}) up to ~ 49 K [21]. Such a T_c^{on} is much higher than the T_c previously known for compounds at ambient or high pressure that contain one or more of the constituent elements Pr, Ca, Fe, and As. In addition, it exceeds that expected from the phonon-mediated superconductor [7]. These authors, therefore, conjectured that the enhanced T_c^{on} of 49 K may have occurred at the interfaces associated with strains and/or defects without further elaboration.

We have therefore examined systematically the rare-earth (R) doped single crystals of $(Ca_{1-x}R_x)Fe_2As_2$ [(Ca,R)122] with R = La, Ce, Pr, and Nd to determine whether the enhanced-T_c is a mere effect of doping in small segregated regions of the sample and, if not, what the morphology of the superconducting body is. Quality crystals with x up to their solubility limits were subsequently grown by the self-flux technique in an Ar inert atmosphere. Wavelength-dispersive-spectrometry (WDS) analysis shows they are chemically homogeneous. All (Ca,R)122 crystals display a T_c^{on} between 42 K for R = Nd and 49 K for R = Pr and show similar transport and magnetic behavior. They also display two sequential superconducting transitions resistively in the otherwise homogeneous single-crystalline

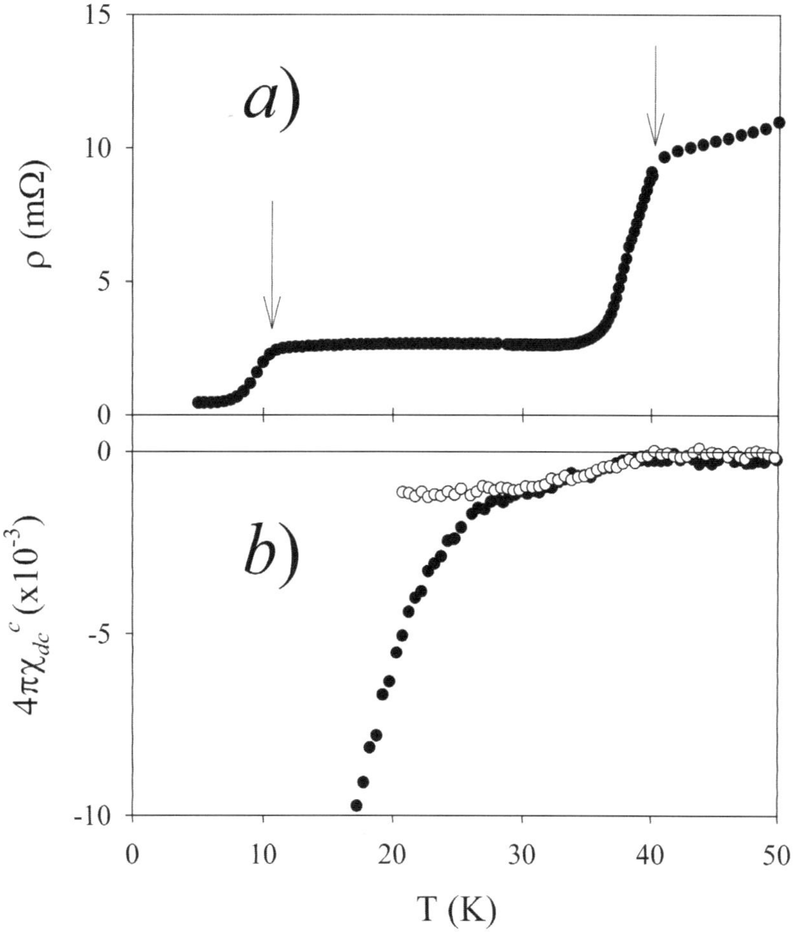

Figure 4. (a) ρ(T) at ambient and 3 mA and (b) χ_{dc}(T) at 1 Oe of single-crystalline (Ca,Nd)122, where arrows in (a) show the two superconducting transitions.

samples: one with a T_c^{on} at ~ 42 K and the other with a T_c^{on} ~ 12 K, as exemplified by the resistance ρ(T) and low field dc magnetic susceptibility χ_{dc}(T) for $Ca_{0.93}Nd_{0.07}Fe_2As_2$ shown in figures 4a and 4b, respectively. A dc magnetic field was found to broaden the ρ(T)-transition and shifts it toward lower temperature but cannot suppress it completely down to 4 K at 5 T, indicating a Type-II characteristic for the superconducting state, while the diamagnetic transition in χ_{dc}(T) was observed to disappear in less than 1 kOe, suggesting a large magnetic background of the sample, similar to [21]. Because of the slightly higher T_c^{on} at 49 K and 24 K for the two transitions, respectively, for R = Pr, we shall discuss only (Ca,Pr)122 hereafter. The two transitions were recovered in the ac-susceptibilities in an ac field H_{ac} of 0.5 Oe along the *c*-axis (χ_{ac}^{c}) and the *ab*-direction (χ_{ac}^{ab}) shown in figure 5. The two transitions exhibit drastically different responses to the magnetic fields as previously reported for R = Pr. This is clear from the rapid suppression of the diamagnetic transition at 47 K to below 20 K by a dc-bias field H_{dc} of only ~ 300 Oe (figure 5), suggesting that the superconducting body consists of Josephson Junction Arrays (JJAs). The magnetic anisotropy $\eta = \chi_{ac}^{c}/\chi_{ac}^{ab}$ is large and exhibits a drastic drop at ~ 20 K, coinciding with the entering into the low temperature phase, as displayed in the inset of figure 5.

We have determined the x-dependence of T_c^{on} on $(Ca_{1-x}R_x)Fe_2As_2$ single crystals for x up to its maximum solubility of 0.13. T_c^{on} is found to be x-insensitive to within ± 2 K throughout the superconducting region of x = 0.06–0.13, as shown in figure 6. In contrast to T_c^{on}(x), the upper-limit of the superconducting volume-fraction *f*, estimated as the zero-field-cooled dc diamagnetic $4\pi\chi_{dc}^{ab}$ at

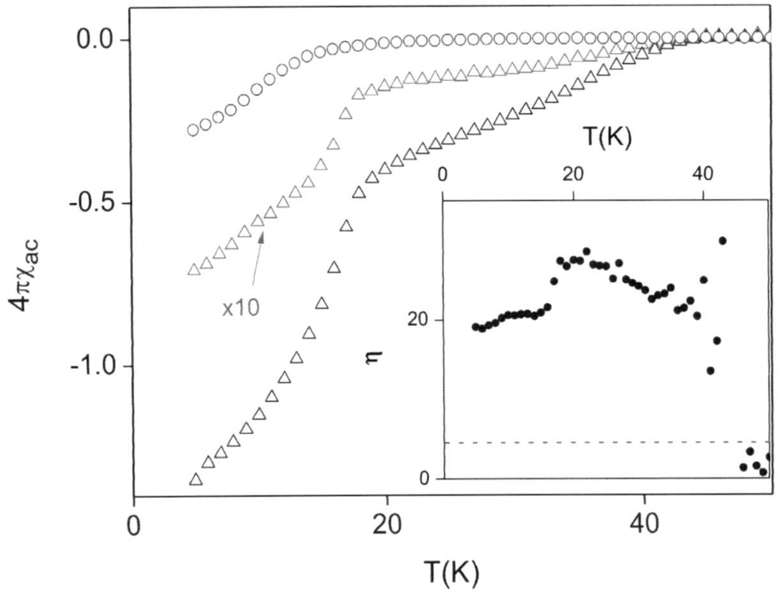

Figure 5. The temperature dependences of the magnetic susceptibilities for single-crystalline $Ca_{0.87}Pr_{0.13}Fe_2As_2$: red triangles— $10\times4\pi\chi_{ac}^{ab}$ at H_{ac} = 0.5 Oe and H_{dc} = 0; blue circles— $4\pi\chi_{ac}^{ab}$ at H_{ac} = 0.5 Oe and H_{dc} = 300 Oe; and black triangles— $4\pi\chi_{ac}^{c}$ at H_{ac} = 0.5 Oe and H_{dc} = 0. Inset: $\eta = \chi_{ac}^{c}/\chi_{ac}^{ab}$.

5 K and H_{dc} = 2 Oe, was also obtained and shown in the same figure. f increases rapidly with x but, even at the solubility limit of x = 0.13, is still far below 10% of that of a perfect superconductor, demonstrating that the superconducting state is indeed non-bulk, consistent with the previous report [21] and the negligible specific heat anomaly at the transition detected by us. The combined observations of the x-independent T_c^{on} and the rapidly increasing f with x in contrast to the large x-range of $0.06 < x \leq 0.13$ for superconductivity and the excellent chemical homogeneity of the samples suggest that the non-bulk superconductivity in (Ca,RE)122 cannot be caused by local chemical-doping, *i.e.* small inclusions of a bulk superconducting phase in the sample. These unusual observations, however, can be attributed to an x-dependent defect morphology, consistent with the picture of interface superconductivity.

To elucidate the morphology of such an unusual superconducting body in the single-crystalline (Ca,R)122, we have examined the field effects on χ_{ac}^{ab} in a total field $H_t = H_{ac} + H_{dc}$ between 10^{-3} and 10^3 Oe on another crystal. $4\pi\chi_{ac}^{ab}(H_t)$ and $4\pi\chi_{ac}^{c}(H_t)$ at 25 K without the interference of the low T_c^{on}-phase is shown in figure 7. It is evident that as H_t increases, the magnitude of the diamagnetic $4\pi\chi_{ac}^{ab}$ or f^{ab} remains constant at $f_1^{ab} \sim 1.6 \times 10^{-3}$ for $H_t \leq H_{t1} \sim 10$ Oe; decreases for $H_t > H_{t1}$, representing the beginning of flux penetration into the sample along the *ab*-direction; and finally levels off at $f_2^{ab} \sim 2 \times 10^{-4}$ for $H_{t2} \geq 500$ Oe , presumably $< H_{c1}$. The much larger $4\pi\chi_{ac}^{c}(H_t)$ or $f^c \sim 63 \times 4\pi\chi_{ac}^{ab}(H_t)$ exhibits a similar behavior: a constant $f_1^c \sim 100 \times 10^{-3}$ for $H_t \leq H_{t1} \sim 0.4$ Oe that deceases linearly until $H_t \sim 5$ Oe and nonlinearly beyond. An unusually large magnetic anisotropy $\eta \sim 63$ at low field and 25 K is thus evident, in contrast to the geometric anisotropy ~ 4 of the sample, suggesting that the superconducting body must be composed of sparsely populated domains made of Josephson-coupled thin superconducting platelets, which have a large aspect ratio ~ η. This is consistent with the low superconducting screening and flux penetration threshold H_{t1} observed along both directions. Similar

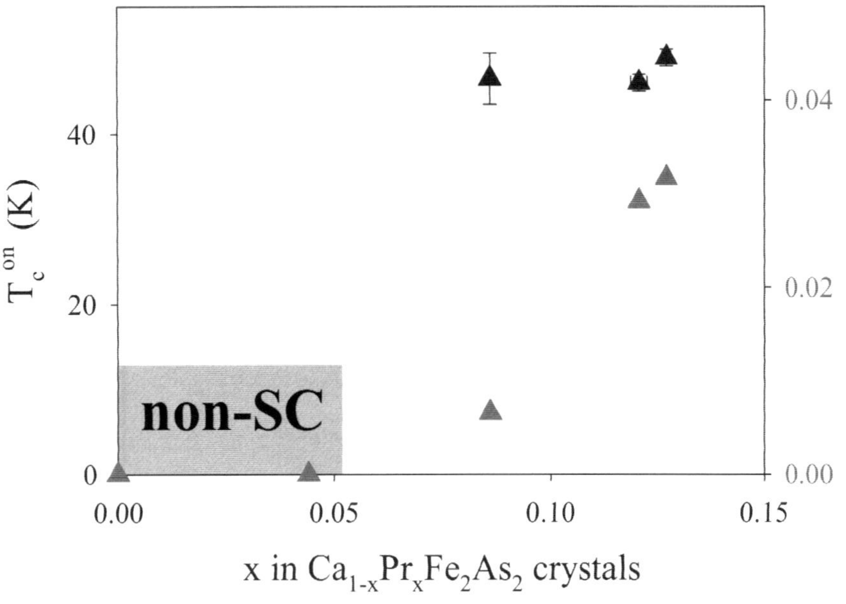

Figure 6. The x-dependences of T_c^{on} : black triangles—for the high T_c^{on} transition; and red triangles—for the superconducting volume-fraction, for x = 0–0.13 of $(Ca_{1-x}Pr_x)Fe_2As_2$ single crystals.

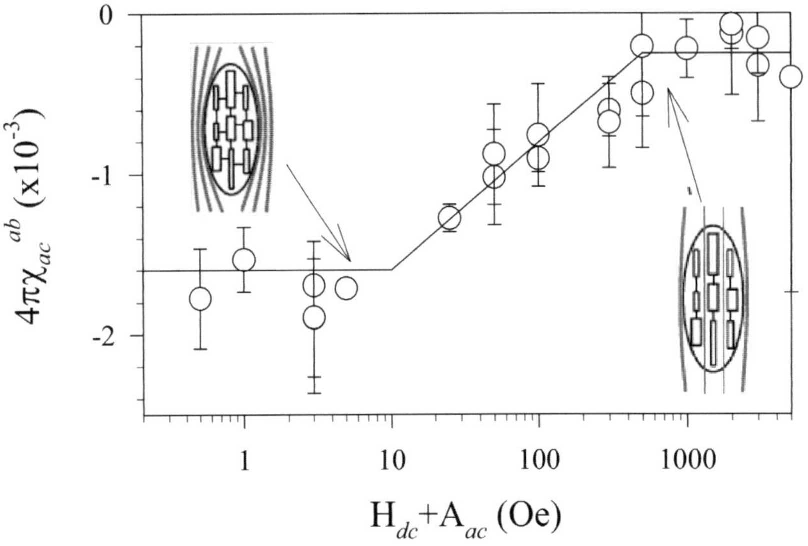

Figure 7. The field effects on $4\pi\chi_{ac}^{ab}$ at 25 K between 10^{-3} and 10^3 Oe.

field effects on $4\pi\chi_{ac}^{ab}$ (H_t) and $4\pi\chi_{ac}^{c}$ (H_t) at lower temperatures have also been observed.

The above field-dependences of $4\pi\chi_{ac}^{ab}$ (H_t) and $4\pi\chi_{ac}^{c}$ (H_t) are reminiscent of field-penetration into granular ceramic superconductors [30] and can thus be so understood in terms of a superconducting morphology in the sample proposed in figure 8. In such a sample, penetration occurs in two stages of vortex creation: first the Josephson vortices within the weak-links between the superconducting platelets in low field and then Abrikosov vortices at a higher field within the individual platelets. In our samples, as H_t increases, *f* is initially a constant for $H_t < H_{t1} \sim 10$ Oe when the sample is shielded by the JJAs; decreases as $H_t > H_{t1}$ when the Josephson vortices form and enter the sample; and finally stops decreasing at $H_{t2} \sim 500$ Oe when all weak-links between the superconducting platelets are decoupled and before the Abrikosov vortices can enter the platelets when H_{t2} is smaller than the effective lower critical field H_{c1} of the platelet, as presumably is the case.

In an attempt to estimate the thickness of the superconducting platelets, we examined the suppression of the superconducting screening in a thin long sample with the field parallel to the *ab*-

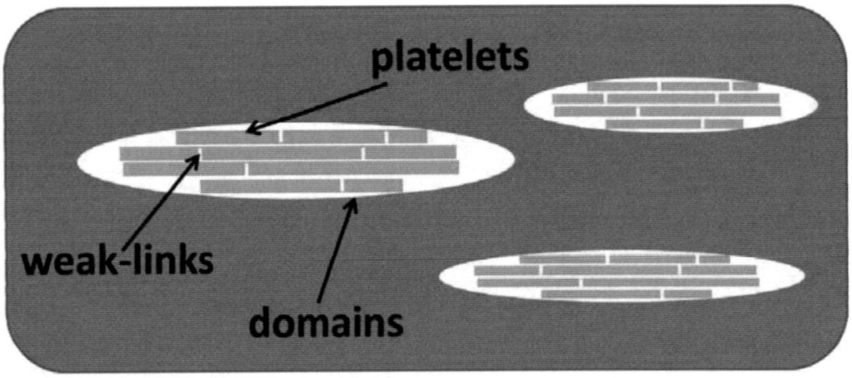

Figure 8. The proposed morphology for the non-bulk highly anisotropic superconducting body in the $(Ca_{0.87}Pr_{0.13})Fe_2As_2$ single crystal, which is composed of interfaces between alternately stacked perfect and defected Fe_2As_2 layers.

axis from that of an ideal sample. The expected suppression factor is $S = 1/\left[1-\frac{2\lambda}{d}\cdot\tanh\left(\frac{d}{2\lambda}\right)\right]$ for a thin sheet with a thickness d, where λ is the penetration depth [31]. The large $S \sim f(<H_{t1})/f(>H_{t2}) = 8$ for the superconducting platelets offers a model-insensitive way to estimate the thickness of the platelets d, leading to an upper limit of $d < 1.3\lambda$. For known bulk Fe-pnictide superconductors, the reported λ varies between 0.2 and 0.6 μm. The estimated d for the superconducting platelets is thus on the order of sub-microns, consistent with the dimension-induced, unusually large $H_{c1} > 500$ Oe along the *ab*-direction suggested by figure 7. Dense interfaces may also exist within the platelets that consist of alternately stacked perfect and defect-infested Fe_2As_2 layers, similar to the case of the $K_xFe_{2-y}Se_2$ system. This is in agreement with our preliminary finding that f is proportional to the superparamagnetic cluster density in the samples investigated [32]. It is therefore rather natural for us to attribute the enhanced T_c^{on} observed to the interfaces between the closely packed perfect and defect-infested Fe_2As_2 layers. The defects may arise from As-deficiencies that usually occur in pure or slightly doped $AeFe_2As_2$ samples for Ae = alkaline earth elements [32].

4. Conclusion

In the present study we have provided the most convincing evidence for interface-enhanced T_c to date. This may also be able to account for the non-bulk superconductivity previously observed in the undoped Ba122, Sr122, and Ca122 single crystals below ~ 20 K and other systems. The full deployment of the interface mechanism in other artificially made or naturally assembled material systems may lead to higher T_c. Detailed investigations on the nature of these defects will help further reveal the working of interfaces and are currently underway in our lab.

Acknowledgments

The work in Houston is supported in part by U.S. Air Force Office of Scientific Research contract FA9550-09-1-0656, U.S. Air Force Research Laboratory subcontract R15901 (CONTACT) through Rice University, the T. L. L. Temple Foundation, the John J. and Rebecca Moores Endowment, and the State of Texas through the Texas Center for Superconductivity at the University of Houston.

References

[1] Onnes H K 1911 *Leiden Comm.* **120b**; Onnes H K 1911 *Leiden Comm.* **122b**; Onnes H K 1912 *Leiden Comm.* **124c**
[2] Gavaler J R, Janocko M A and Jones C K 1974 *J. Appl. Phys.* **45** 3009; Testardi L R, Wernick J H and Royer W A 1974 *Solid State Commun.* **15** 1
[3] Bednorz J G and Müller K A 1986 *Z. Phys. B.* **64** 189
[4] Takagi H, Uchida S, Kitazawa K and Tanaka S 1987 *Jpn. J. Appl. Phys.* **26** L123
[5] Chu C W, Hor P H, Meng R L, Gao L, Huang Z J and Wang Y Q 1987 *Phys. Rev. Lett.* **58** 405
[6] Chu C W, Hor P H, Meng R L, Gao L and Huang Z J 1987 *Science* **235** 567–569
[7] McMillan W L 1968 *Phys. Rev.* **167** 331–344
[8] Chu C W 1988 *Physics in a Technological World*, ed A P French (New York: American Institute of Physics) p 211
[9] See for example, Chu C W 1987 (January 12) Patent application; Chu C W 1996 *History of Original Ideas and Basic Discoveries in Particle Physics* (*NATO Advanced Study Institute, Series B: Physics* vol 352) ed. H. B. Newman and T. Ypsilantis (New York: Plenum) p 793
[10] Wu M K, Ashburn J R, Torng C J, Hor P H, Meng R L, Gao L, Huang Z J, Wang Y Q and Chu C W 1987 *Phys. Rev. Lett.* **58** 908
[11] Hor P H et al. 1987 *Phys. Rev. Lett.* **58** 911
[12] Hazen R M, Finger L W, Angel R J, Prewitt C T, Ross N L, Mao H K, Hadidiacos C G, Hor P H, Meng R L and Chu C W 1987 *Phys. Rev.* B **35** 7238(R)
[13] Hor P H, Meng R L, Wang Y Q, Gao L, Huang Z J, Bechtold J, Forster K and Chu C W, *Phys. Rev. Lett.* **58** 1891

[14] Maeda H, Tanaka Y, Fukutomi M and Asano T 1988 *Jpn. J. Appl. Phys.* **27** L209
[15] Sheng Z Z and Hermann A M 1988 *Nature* **332** 138
[16] Schilling A, Cantoni M, Guo J D and Ott H R 1993 *Nature* **363** 56
[17] Gao L, Xue Y Y, Chen F, Xiong Q, Meng R L, Ramirez D, Chu C W, Eggert J H and Mao H K 1994 *Phys. Rev.* B **50** 4260(R)
[18] Kamihara Y, Watanabe T, Hirano M and Hosono H 2008 *J. Am. Chem. Soc.* **130** 3296
[19] Chen X H, Wu T, Wu G, Liu R H, Chen H & Fang D F 2008 *Nature* **453** 761; Ren Z A et al. 2008 *Phys. Lett.* **25** 2215
[20] Rotter M, Tegel M and Johrendt D 2008 *Phys. Rev. Lett.* **101** 107006; Sasmal K, Lv B, Lorenz B, Guloy A M, Chen F, Xue Y Y and Chu C W 2008 *Phys. Rev. Lett.* **101** 107007; Guo J G, Jin S F, Wang G, Wang S C, Zhu K X, Zhou T T, He M and Chen X L 2010 *Phys. Rev.* B **82** 180520(R)
[21] Lv B, Deng L Z, Gooch M, Wei F Y, Sun Y Y, Meen J, Xue Y Y, Lorenz B and Chu C W 2011 *Proc. Nat. Acad. Sci. USA* **108** 15705–15709
[22] Tapp J H, Tang Z J, Lv B, Sasmal K, Lorenz B, Chu C W and Guloy A M 2008 *Phys. Rev.* B **78** 060505(R); Wang X C, Liu Q Q, Lv Y X, Gao W B, Yang L X, Yu R C, Li F Y and Jin C Q 2008 *Solid State Comm.* **148** 538; Chu C W, Chen F, Gooch M, Guloy A M, Lorenz B, Lv B, Sasmal K, Tang Z J, Tapp J H and Xue Y Y 2009 *Physica C* **469** 326; Parker D R, Pitcher M J, Baker P J, Franke I, Lancaster T, Blundell S J and Clarke S J 2009 *Chem. Commun.* 2189–2191; Sasmal K, Lv B, Tang Z J, Chen F, Xue Y Y, Lorenz B, Guloy A M and Chu C W 2009 *Phys. Rev.* B **79** 184516.
[23] Hsu F C et al. 2008 *Proc. Natl. Acad. Sci. USA* **105** 14262
[24] Wang Q Y et al. 2012 *Chin. Phys. Lett.* **29** 037402
[25] Chu C W 2008 *AAPPS Bulletin* **18** 9
[26] Little W A 1964 *Phys. Rev.* **134** A1416; Ginzburg V L 1964 *JETP* **47** 2318
[27] Allender D, Bray J and Bardeen J 1973 *Phys. Rev.* B **7** 1020
[28] Inkson J C and Anderson P W 1973 *Phys. Rev.* B **8** 4429–4432; Allender D, Bray J and Bardeen J 1973 *Phys. Rev.* B **8** 4433–4434; Inkson J C 1975 *J. Phys. C: Solid State Phys.* **8** L164–L168
[29] For a review, see Pereiro J, Petrovic A, Panagopoulos C and Božović I 2011 Interface superconductivity: History, development and prospects *Physics Express* **1** 208–241
[30] Müller K H, Nikolo M and Driver R 1988 *Phys. Rev.* B **43** 7976; Clem J R 1988 *Physica C* **153–155** 50
[31] See for example, Porch A, Cooper J R, Zheng D N, Waldram J R, Campbell A M and Freeman P A 1993 *Physica C* **214** 350
[32] Y. Y. Xue et al., unpublished

Superconductivity and Magnetism in (Tl, K, Rb)Fe_xSe_2

Minghu Fang[1*], Hangdong Wang[1,2], Chiheng Dong[1], Q. Huang[3]
[1]Department of Physics, Zhejiang University, Hangzhou 310027, China
[2]Department of Physics, Hangzhou Normal University, Hangzhou 310036, China
[3]NIST Center for Neutron Research, National Institute of Standards and Technology, Gaithersburg, MD 20899, USA

E-mail: *mhfang@zju.edu.cn

Abstract. The iron selenide superconductors AFe_xSe_2 (A=Tl, K, Rb and Cs) discovered at the end of 2010 has attracted great interest because they display a variety of properties unprecedented for the cuprates and iron pnictides. Here, we present the motivation for our discovery on the superconductivity in (Tl,K,Rb)Fe_xSe_2 system, in which the onset superconducting transition temperature is as high as 40 K. We found that the compound with more Fe vacancies is an AFM insulator, which has been confirmed to be associated with the Fe-vacancy ordering in the crystals. Our discovery represents the first Fe-based superconductivity at the verge of an AFM insulator. A review on the recent results of Fe-vacancy super-lattice, AFM ordering, superconductivity and phase separation in (Tl,K,Rb)Fe_xSe_2 system is presented.

1. Introduction

Cuprates [1] and Fe-based compounds [2, 3, 4, 5, 6, 7, 8, 9, 10, 11, 12, 13] are two families with highest superconducting (SC) transition temperature. A common feature in both families is that superconductivity emerges as antiferromagnetic (AFM) long range order is suppressed. While the parent compound of cuprates is a Mott insulator where the electron repulsion is strong, the parent compound of Fe-based materials is metallic implying weak or moderate electron correlation. A key strategy to develop a unified picture for the Fe- and Cu-based high temperature superconductivity (HTSC) is to explore the possibility to tune the Fe-based compound into an insulator. But this has not been successful albeit the worldwide efforts before the reports on the superconductivity and magnetism in (Tl,K,Rb)Fe_xSe_2 system at the end of 2010 [14, 15, 16, 17].

As we know, all the iron-based superconductors share a common layered structure based on a square planar Fe^{2+} layer, as shown in Fig. 1(b), tetrahedrally coordinated pnictogen [18, 19, 20] (P, As) or chalcogen [12, 13, 21, 22] (S, Se, Te) anions. It is now widely believed that the common Fe^{2+} layers are responsible to the observed superconductivity, similar to the common CuO_2 layer in the cuprates. The crystal structure [23] of stoichiometric $TlFe_2Se_2$, which is shown in Fig. 1(a), is the same as that of the $BaFe_2As_2$ with the $ThCr_2Si_2$-type. However, Tl has a monovalence in the $TlFe_2Se_2$ compound and there are always vacancies on the Fe-square lattice sites [24], similar to the situation in the $TlFe_2S_2$ compound [23, 25]. Therefore, the actual chemical formula should be $TlFe_xSe_2$ with $1.3 \leq x < 2.0$. There are evidences that Fe-ions in $TlFe_xSe_2$ or $TlFe_xS_2$ with x = 1.5 exhibit long-range AFM orderings. Based on the Mössbauer and neutron diffraction studies, Haggstrom et al. [24] and Sabrowsky et al.[26] argued that the AFM order in $TlFe_xSe_2$ and $TlFe_xS_2$ compounds is associated with the crystallization of the Fe-vacancies, as shown in Fig. 1(c) and 1(d) for x = 1.5 and 1.6, respectively. From this viewpoint, we expect the AFM order is suppressed as Fe-content increases. At x = 2.0, Fe-ions occupy all the square lattice sites as shown in Fig. 1(b). The possible emergence of the superconductivity in this new 122-type iron-chalcogenide motivated us to grow $TlFe_xSe_2$ single crystals with a relatively high Fe-content.

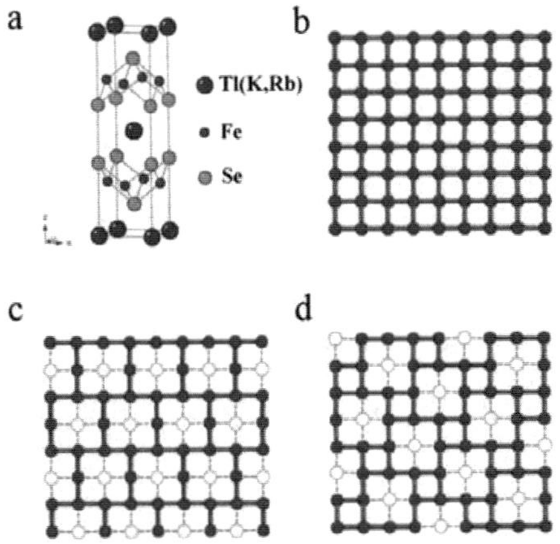

Figure 1. Crystal structure and super-lattice of Fe-vacancies in (Tl,K,Rb)Fe_xSe_2. (a) Crystal structure of stoichiometric (Tl,K,Rb)Fe_xSe_2. Tl/K or Tl/Rb- and Se-sites are fully occupied in the tetragonal $ThCr_2Si_2$ structure. Fe-sites are partially occupied. (b) Fe^{2+} square planar for stoichiometric (Tl,K,Rb)Fe_2Se_2, which is the common layer in all other Fe-based superconductors. (c) Fe-vacancy ordering patterns for (Tl,K,Rb)$Fe_{1.5}Se_2$, where a Fe-atom has either two or three Fe-neighbors. (d) Fe-vacancy ordering patterns for (Tl,K,Rb)$Fe_{1.6}Se_2$, where each Fe-atom has three equivalent neighbors, according to the proposition by M. Zabel et al.[25] Blue solid circle: Fe atom, Green hollow circle: Vacancy.

At the end 2010, we first found that the increase of Fe content in $TlFe_xSe_2$ indeed lead to an emergence of superconductivity, and the addition of the (K, Rb) elements can result in the increase of the Fe content in the grown crystals, the maximum Fe-content can reach to 1.88. Superconductivity above 30K, the onset superconducting transition temperature as high as 40K, emerges in the (Tl,K,Rb)Fe_xSe_2 system. The compounds with less Fe content is an AFM insulator, which represents the first Fe-based HTSC at the verge of an AFM insulator. A review on the recent results of Fe-vacancy super-lattice, block AFM ordering, superconductivity and phase separation in (Tl,K,Rb)Fe_xSe_2 system is presented.

2. From an insulator to a superconductor in (Tl,K,Rb)Fe_xSe_2 system

First, we discuss the resistivity and superconductivity of the $TlFe_xSe_2$ compounds. At the beginning, in order to get the crystals with more Fe content, we tried to grow $TlFe_xSe_2$ single crystals using various nominal compositions as the staring materials, *i.e.* $Tl_zFe_2Se_2$ ($0.4 \leq z \leq 1.0$), $TlFe_zSe_2$ ($1.5 \leq z \leq 3.0$). We have obtained a series of $TlFe_xSe_2$ ($1.30 \leq y \leq 1.70$) single crystals. Their compositions were determined by an Energy Dispersive X-ray Spectrometer (EDXS). The Tl and Se contents are almost of 1.0 and 2.0, respectively, but the Fe content is always less than 2.0. The temperature dependence of resistivity for the crystals with less Fe content exhibits a thermally activated semiconducting behavior with an activation energy of 80.2 meV and 57.7 meV, as shown by the fitting line in the inset in Fig 2(a) and (b), for x = 1.30 and 1.47, respectively. Superconductivity with a transition temperature T_c^{mid}=22.4 K and zero resistivity at T_c^{zero}=20 K is indeed first observed in the $TlFe_{1.70}Se_2$ crystal, as shown in Fig. 2(c), but the magnetic susceptibility measurement show its superconducting volume fraction being very small (< 1 %), see Fig 2(d). It was first found that the evolution from an insulator to a SC in $TlFe_xSe_2$ system with increasing Fe content. At the same time, we found that it is difficult to grow a $TlFe_xSe_2$ single crystal with x > 1.70.

Journal of Physics: Conference Series **449** (2013) 012015 doi:10.1088/1742-6596/449/1/012015

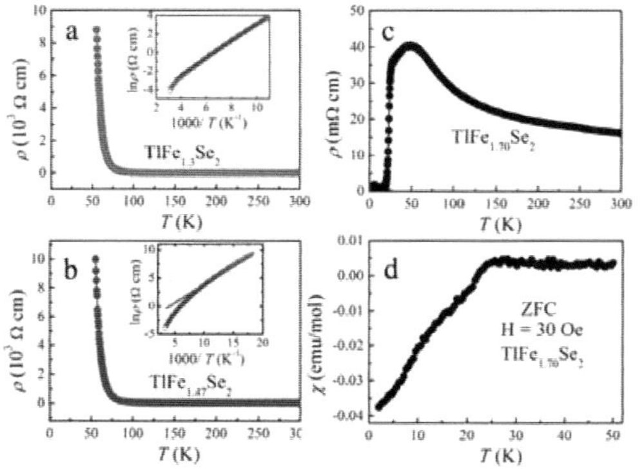

Figure 2. Temperature dependence of the in-plane resistivity, ρ and lnρ as a function of $1/T$ (in the inset) for both insulators. (a) $TlFe_{1.30}Se_2$ crystal. (b) $TlFe_{1.47}Se_2$ crystal. (c) Temperature dependence of the in-plane resistivity, ρ(T). (d) Magnetic susceptibility, χ(T), measured at 30Oe with zero-field-cooling (ZFC) process, for the $TlFe_{1.70}Se_2$ crystal.

The above results indicate that, in order to get bulk superconductivity in this system, it is necessary to further increase the Fe content. We tried to grow $Tl_{1-y}K_yFe_xSe_2$ single crystals by adding K in the starting materials and choosing three groups nominal composition as the starting material, *i.e.* $Tl_{0.5}K_zFe_2Se_2$ ($0.15 \leq z \leq 0.45$), $Tl_{0.4}K_zFe_2Se_2$ ($0.2 \leq z \leq 0.5$) and $Tl_{0.8-z}K_zFe_2Se_2$ ($0.1 \leq z \leq 0.4$). We have obtained a series of $Tl_{1-z}K_zFe_xSe_2$ ($1.50 \leq x \leq 1.88$, $0.14 \leq z \leq 0.57$) crystals. Figure 3(d) shows the photos of a crystal $Tl_{0.48}K_{0.52}Fe_{1.50}Se_2$ (before being cleaved) and a crystal $Tl_{0.64}K_{0.36}Fe_{1.83}Se_2$. Figure 3(a), 3(b) and 3(c) show the single crystal and powder x-ray diffraction (XRD) patterns for the $Tl_{0.64}K_{0.36}Fe_{1.83}Se_2$ and $Tl_{0.48}K_{0.52}Fe_{1.50}Se_2$. All the peaks in the powder XRD pattern can be well indexed with a $ThCr_2Si_2$-type structure (space group: *I4/mmm*). By fitting the XRD data, we obtain the lattice parameters a = 3.88 Å and c = 14.05 Å for $Tl_{0.64}K_{0.36}Fe_{1.83}Se_2$, and a = 3.90 Å, and c = 13.99 Å for $Tl_{0.48}K_{0.52}Fe_{1.50}Se_2$. Broader peaks in the powder XRD patterns might indicate the existence of some disorder in the crystals, and it is difficult to observe super-lattice peaks of Fe vacancy order. Only (00l) peaks were observed in the single crystal XRD patterns, indicating that the crystallographic *c* axis is perpendicular to the plane of the single crystal.

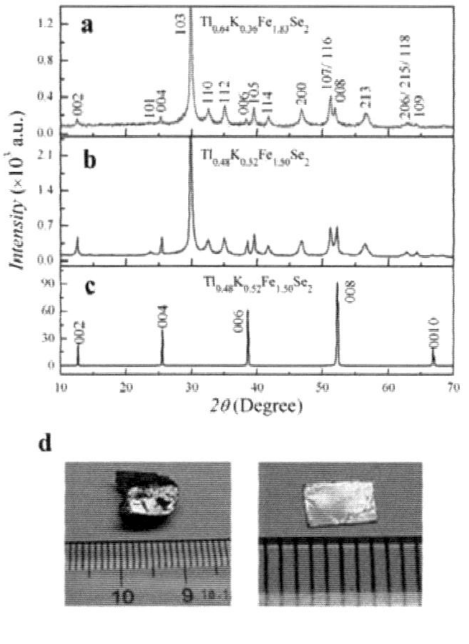

Figure 3. X-ray diffraction (XRD) pattern and photos for (Tl,K)Fe_xSe_2 crystals. (a) XRD pattern of $Tl_{0.64}K_{0.36}Fe_{1.83}Se_2$ powder obtained by grinding many pieces of crystals. (b) Powder XRD pattern of $Tl_{0.48}K_{0.52}Fe_{1.50}Se_2$. (c) Single crystals XRD pattern of $Tl_{0.48}K_{0.52}Fe_{1.50}Se_2$. (d) Photos of $Tl_{0.48}K_{0.52}Fe_{1.50}Se_2$ crystals (before being cleaved) and $Tl_{0.64}K_{0.36}Fe_{1.83}Se_2$ crystal.

Figure 4(a) and 4(b) show the in-plane resistivity as a function of temperature, ρ(T), for the (Tl,K)Fe_xSe_2 ($1.50 \leq x \leq 1.85$) crystals. We first discuss x = 1.50 and 1.64 samples, which are insulators. As shown in Fig. 4(a), the ρ at 300K is about 27.8 mΩ cm for x=1.50 and 27.2 mΩ cm for x=1.64. As T decreases, ρ(T) increases rapidly and shows thermally activated behavior: $\rho = \rho_0\exp(E_a/k_BT)$, where k_B is the Boltzmann constant. From the fitting, the activation energy E_a was estimated to be 36 meV for x = 1.50, as shown by the red line in Fig. 4(c), and 24 meV for x = 1.64. The ρ(T) for both crystals exhibits a typical semiconducting behavior.

We now discuss the superconductivity in the (Tl,K)Fe_xSe_2 crystals with higher Fe-contents. With the x value further increasing, ρ at lower T decreases rapidly. For example, at T=40K, ρ = 18.6 Ω cm for x=1.64 crystal, and drops to 0.46 Ω cm for x = 1.69 crystal. The preliminary indication of SC transitions were observed at 25.8 K and 27.6 K for x = 1.69 and 1.76 crystal, respectively. However, no zero resistivity is observed above 2 K in these Fe-content crystals, as shown in Fig. 4(a). In the x = 1.78 crystal, a sharp SC transition occurs at T_c^{mid} = 28.4 K, and reach to zero resistivity at T_c^{zero} = 27.4 K, confirmed by the existence of diamagnetism below T_c, as shown in Fig. 4(d). For the x = 1.83, *i.e.* $Tl_{0.64}K_{0.36}Fe_{1.83}Se_2$ crystal, T_c^{mid} increases to 31.1 K and T_c^{zero} = 30 K. And for the x = 1.84 ($Tl_{0.69}K_{0.31}Fe_{1.84}Se_2$) and 1.85 ($Tl_{0.75}K_{0.25}Fe_{1.85}Se_2$) crystals, although their T_c^{mid} values are almost the same as that in x = 1.83, they have a much larger SC volume fraction (SVF) (see Fig. 4(d)). We found that the SC transition temperatures, T_c^{mid} and T_c^{zero}, vary only with the Fe content in the crystals, and are independent of the K content in the crystals. Hall coefficient measurements have confirmed that charge carriers are n-type [16]. There is a tendency that T_c should increases to the highest in the crystal with an optimal Fe-content.

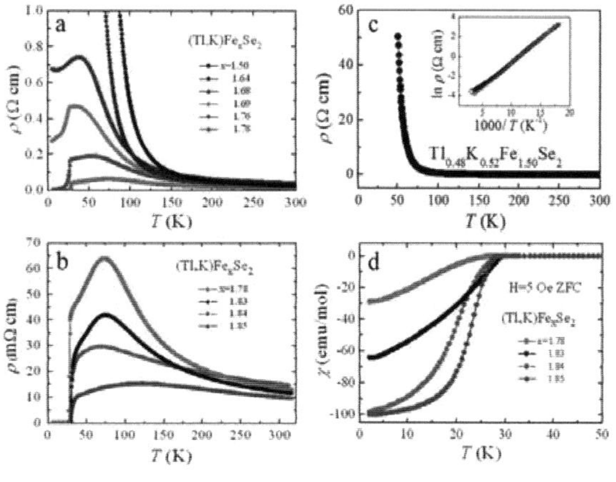

Figure 4. Temperature Dependence of Resistivity and Magnetic susceptibility of (Tl,K)Fe_xSe_2 ($1.50 \leq x \leq 1.85$) crystals. (a) In-plane resistivity, ρ(T), for (Tl,K)Fe_xSe_2 (x = 1.50, 1.64, 1.68, 1.69, 1.76 and 1.78). (b) ρ(T) for (Tl,K)Fe_xSe_2 (x = 1.78, 1.83, 1.84, and 1.85). (c) ρ(T) for $Tl_{0.48}K_{0.52}Fe_{1.50}Se_2$ crystal. Inset: $\ln\rho$ as a function of 1/T. (d) Zero field cooling χ measured at a field of 5 Oe for (Tl,K)Fe_xSe_2 (x = 1.78, 1.83, 1.84 and 1.85).

The SC transitions in ρ(T) for the (Tl,K)Fe_xSe_2 crystals with various x values are shown in Fig. 5. As we can see, there is only one sharp transition at T_c = 28.4 K for x = 1.78, but two transitions for x = 1.84, with one at T_c = 29.8 K and the other at 33.0 K. For $Tl_{0.75}K_{0.25}Fe_{1.88}Se_2$ crystal with the highest Fe-content in our study, three transitions are observed at T_c = 30.4, 34.1, and 40.4 K. The existence of SC phase with T_c = 40 K in the crystals with higher Fe contents is confirmed in the measurements of ρ(T) at various fields for the $Tl_{0.64}K_{0.36}Fe_{1.83}Se_2$ crystal, as shown in Fig. 3(d). all the SC transitions at magnetic field B = 0, 4 and 9 Tesla start at T = 40 K for x = 1.83. The multi-transition in ρ(T) was observed in all the crystals with $x \geq 1.83$, regardless of the K content. This indicates that the domains with various Fe-content, corresponding to different carrier concentration, likely exist in the crystals. In the majority domain, SC with T_c = 30 K emerges. The volume fraction of the domains with higher T_c value and higher Fe content may be small because no diamagnetic signal appears at higher temperature. To our knowledge, this phase separation behavior is first observed in Fe-based SCs. Bulk SC with T_c = 40 K may emerges in the crystal with an optimal Fe-content, but it is difficult to realize in this system due to the Fe^{2+} valence limitation. SC with T_c = 40 K emergence in the (Tl,K)Fe_xSe_2 crystals is reminiscent of the highest T_c = 38 K in the optimal doping $Ba_{1-x}K_xFe_2As_2$ compound [8].

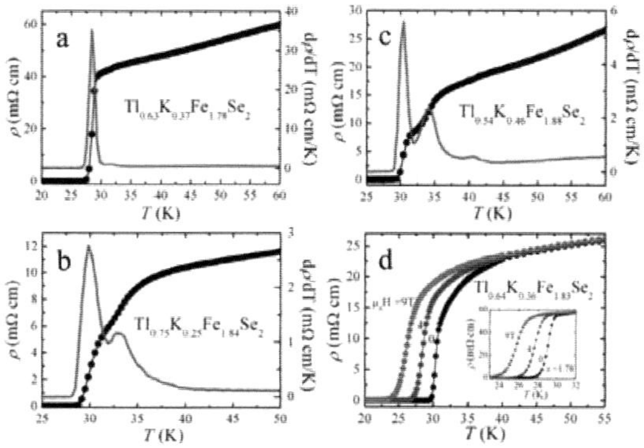

Figure 5. SC transitions of four typical (Tl,K)Fe_xSe_2 crystals. In-plane resistivity ρ(T) and its derivative dρ(T)/dT near the SC transitions. (a) for $Tl_{0.63}K_{0.37}Fe_{1.78}Se_2$ with T_c = 28.4 K. (b) for $Tl_{0.75}K_{0.25}Fe_{1.84}Se_2$ with two transitions at T_c = 29.8 K and 33.0 K, respectively. (c) for $Tl_{0.54}K_{0.46}Fe_{1.88}Se_2$, with three transitions at T_c = 30.4, 34.1, and 40.4 K, respectively. (d) ρ(T) near the SC transition in magnetic field of 0, 4, and 9 Tesla for $Tl_{0.64}K_{0.36}Fe_{1.83}Se_2$. The transition starts at 40 K. The inset shows the ρ(T) at 0, 4, and 9 Tesla for $Tl_{0.63}K_{0.37}Fe_{1.78}Se_2$ crystal, which has only one transition.

At the same time, we also tried to grow (Tl,Rb)Fe_xSe_2 single crystals by adding Rb in the starting materials. We have also obtained a series of (Tl,Rb)Fe_xSe_2 (1.30 $\leq$ x $\leq$ 1.72) crystals. Figure 6(a), (b) and (c) show their in-plane resistivity as a function of temperature, ρ(T). We first discuss x = 1.30 sample, which is an insulator. As shown in Fig. 6(a), the ρ at 300K is about 18.1 mΩ cm. As T decreases, ρ(T) increases rapidly and shows thermally activated behavior with an activation energy E_a of 51.7 meV (see the red fitting line in Fig 6(c)). With increasing Fe content, the bulk superconductivity with T_c = 32 K emerges in the crystals, which is evidenced by the susceptibility measurements as shown in Fig. 6(d). The evolution from an insulator to a superconductor with increase in Fe content has also been found in the (Tl, Rb)Fe_xSe_2 system.

Bulk superconductivity with T_c = 30 K emerges in TlFe_xSe_2 (x>1.7), (Tl,K)Fe_xSe_2 (x>1.72) and (Tl,Rb)Fe_xSe_2 (x>1.42) crystals. It raises a question: why the Fe content of the crystals bulk SC starting to appear is different for three systems. Nowadays, we can understand this by a phase separation discussed as follow. Two phases always coexist in the SC crystals below AFM transition temperature, T_N. In one phase, Fe sites are fully occupied, which should be responsible for bulk SC, while in another phase, Fe vacancies order into the $\sqrt{5} \times \sqrt{5}$ super-lattice, which should be an AFM insulator. Due to the difference in the ionic radius of K and Rb, the Fe content of the crystals a phase separation starting to appear is different. We indeed observed this phenomenon in the single crystal XRD experiments.

3. The Fe-vacancy super-lattice and AFM order in (Tl,K,Rb)Fe_xSe_2 system

As discussed above, we suggested [16] that there are two kinds of Fe-vacancy order in (Tl,K,Rb)Fe_xSe_2 system. One with the formula of (Tl,K,Rb)$Fe_{1.5}Se_2$, the iron atoms have two or three iron neighbors. Another with the formula of (Tl,K,Rb)$Fe_{1.6}Se_2$ (now called as the $A_2Fe_4Se_5$ phase), each iron atom has three iron neighbors. We pointed out [16] that the insulating phase mentioned above should be associated with these Fe-vacancy order. With these two kinds of Fe-vacancy order, the band structure calculations indicate a band gap with the value of about 0.3-0.5 eV for Tl$Fe_{1.5}Se_2$ [27] and 60 meV for Tl$Fe_{1.6}Se_2$ [28]. Of course, it is very important to determine which kind of the Fe vacancy and AFM order exists most likely in this system.

Based on the neutron diffraction experiments, Bao et al. [29, 30] concluded that the dominating phase had a composition of $A_{0.82}Fe_{1.62}Se_2$ (A=K, Tl/K and Tl/Rb) and Fe vacancies

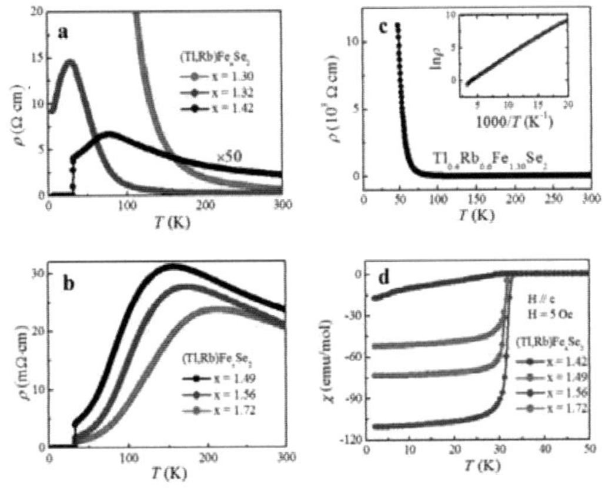

Figure 6. Temperature dependence of in-plane resistivity, ρ(T), for (Tl,Rb)Fe_xSe_2 crystals. (a) for x=1.30, 1.32 and 1.42. (b) for x=1.49, 1.56 and 1.72. (c) ρ(T) for x=1.30. Inset: $\ln\rho$ as a function of 1/T. (d) magnetic susceptibility, χ, measured at 5 Oe field with ZFC, for (Tl,Rb)Fe_xSe_2 (x = 1.72, 1.56, 1.49 and 1.42) crystals.

order into the $\sqrt{5}\times\sqrt{5}$ super-lattice between 460-580 K, corresponding to the second Fe-vacancy ordered structure suggested by us. At or slightly below this temperature, the spins align along the tetragonal c axis in a block-checkerboard antiferromagnetic structure (BCAF-c, where c indicates the alignment of the spin). The four nearby Fe ions, with a spin 3.31 μ_B/Fe, form a square block with the spins aligned ferromagnetically. The total magnetic moment of the 4-iron-block is 13.4 μ_B. This magnetic structure was later proved by another group at ORNL [31]. They concluded that this is a common crystalline and magnetic structure of superconducting $A_2Fe_4Se_5$ (A=Tl, K, Rb and Cs)[30]. The transmission electron microscopy experiments [32] have also confirmed this kind of Fe-vacancy order.

Very recently, A. F. May *et al.* [33] reexamined the Fe vacancy order and AFM structure by means of single crystal neutron diffraction, nuclear forward scattering, and transmission electron microscopy for $TlFe_{1.6}Se_2$ crystal with complete chemical/vacancy order. They confirmed the existence of $\sqrt{5}\times\sqrt{5}$ Fe-vacancy super-lattice, and found that the Fe moment orientation changes from parallel to c axis above 100K to perpendicular to c below 100K. And there may be two types of magnetic structure below 100K, one is a non-collinear, another is an in-plane, block-checker antiferromagnetic structure. They argued that, although the BCAF-c structure is the main magnetic structure observed in the alkali-metal compounds, the $TlFe_xSe_2$ has a limited compositional window and the Tl sites is always fully occupied, there may be additional magnetic phase transitions in the lower temperatures, which is consistent with our susceptibility data [16] and the results reported in [34].

Another, recently, J. Zhao *et al.* [35] re-checked also the Fe vacancy super-lattice and magnetic structure in both the insulating (E_a=500 meV) and semiconducting (E_a=40 meV) $K_yFe_xSe_2$ crystals. They found that there is only a $\sqrt{5}\times\sqrt{5}$ Fe-vacancy super-lattice, corresponding to the BCAF-c magnetic structure, in the insulating $K_{0.8}Fe_{1.6}Se_2$ crystal. But in the semiconducting $K_yFe_xSe_2$ crystals, in addition to the $\sqrt{5}\times\sqrt{5}$ Fe-vacancy structure, another a $\sqrt{2}\times\sqrt{2}$ Fe vacancy super-lattice was also found, which corresponds to the first one discussed above, in which the iron atoms have two or three iron neighbors, with the formula of $TlFe_{1.5}Se_2$ [16]. According to the Zhao's neutron diffraction experiments, the magnetic structure in this Fe vacancy super-lattice has a stripe type magnetic order with Néel temperature 280 K, which is the same as that in other iron pnictides. They concluded that this AFM semiconductor is a parent compound of the supercondicting AFe_xSe_2 compounds, interpolating between the antiferromagntic Mott insulator of the cuprates and the antiferromagnetic semimetal of the iron pnictides. It is important that this indicates the correlation physics playing an important role in the magnetism for this system, which is consistent with that suspected in our work [16]. Of course, now there are debates about which antiferromagnetic order being a parent of superconductivity emerging in this system.

Table 1. The fraction of $I4/m$ and $I4/mmm$ phases obtained by fitting to ND data measured at different temperatures for superconducting $Tl_{0.374}Rb_{0.374}Fe_{1.764}Se_2$ crystals.

	30K	300K	400K	550K
$I4/m$	70.8%	70.9%	70.0%	0%
$I4/mmm$	20.8%	21.5%	21.8%	91.56%

4. The phase separation in (Tl,K,Rb)Fe$_x$Se$_2$ system

Another important issue is the coexistence of magnetism and superconductivity at temperature up to $\sim$ 30K in (Tl,K,Rb)$_y$Fe$_x$Se$_2$ system. A lot of techniques have been employed to understand this interesting property of these materials. Now, the coexistence of superconductivity and magnetism is believed to be due to fine-scale phase separation, with each property being associated with a different composition and/or degree of order.[36, 37, 38, 39, 40, 41] All observations by transmission electron microscopy (TEM) [32, 42, 38], nano-focused x-ray diffraction [36] and Mössbauer spectroscopy [37, 43, 44] have confirmed the existence of the phase separation over length scales of 10-100 nm in the superconducting samples. Although superconducting samples with single phase (structural and chemically homogenous) have not produced up to now, it is not so difficult to grow the insulating single phase crystals with BCAF-c magnetic structure.

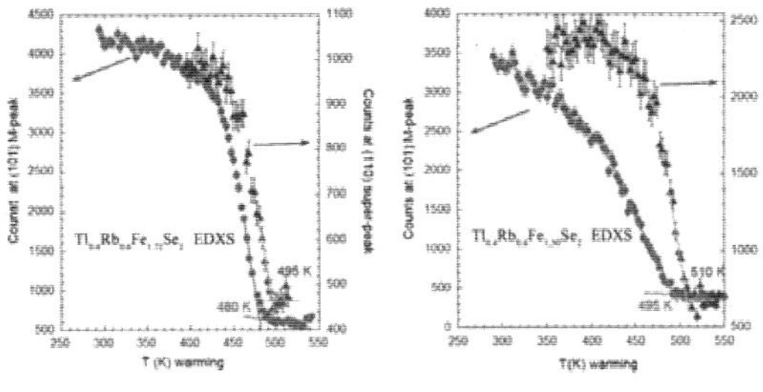

Figure 7. Temperature dependence of the intensity of (110) peak, corresponding to $\sqrt{5} \times \sqrt{5}$ super-lattice, and the intensity of (101) peak, corresponding to BCAF-c lattice, in the powder neutron diffraction pattern. (Left) for the superconducting $Tl_{0.4}Rb_{0.6}Fe_{1.72}Se_2$ crystals. (Right) for the insulator $Tl_{0.4}Rb_{0.6}Fe_{1.30}Se_2$ crystals.

We chosen two kinds of (Tl,Rb)Fe$_x$Se$_2$ crystals to determine their crystal and magnetic structure by neutron diffraction experiments. The first one is an insulator as shown in Fig.6(c), *i.e.* $Tl_{0.4}Rb_{0.6}Fe_{1.30}Se_2$. Above 510 K, the powder neutron diffraction (ND) data can be well fitted by a pure 122 structure phase with $I4/mmm$, and the composition with $Tl_{0.452}Rb_{0.452}Fe_{1.578}Se_2$. Below 510 K, the ND data can be well fitted by a pure 245 block phase with $I4/m$. At 495K, a BCAF-c magnetic structural transition occurs. The temperature dependence of the intensity of (110) peak, corresponding to $\sqrt{5} \times \sqrt{5}$ super-lattice, and the intensity of (101) peak, corresponding to BCAF-c lattice, in the powder neutron diffraction experiments, are shown in Fig.7(Right).

The second is a superconductor as shown in Fig. 6(b), *i.e.* $Tl_{0.4}Rb_{0.6}Fe_{1.72}Se_2$ crystals. Above 495 K, the powder neutron diffraction (ND) data can also be well fitted by a pure 122 phase with $I4/mmm$, with a composition $Tl_{0.374}Rb_{0.374}Fe_{1.764}Se_2$. But below 495 K, two phases emerge in these crystals. One is (Tl,Rb)$_{0.818}$Fe$_{1.563}$Se$_2$ with $I4/m$, which becomes BCAF-c below 480K shown in Fig.7 (Left), its volume fraction is about 70%, its refined structure parameters from ND data below T_N is shown in Table 3; Another is (Tl,Rb)$_{0.532}$Fe$_2$Se$_2$ with $I4/mmm$, its refined structure parameters from ND data below T_N is shown in Table 2, no magnetic order was discovered in this phase, its volume fraction is about 22%. Little change was found in the relative ratio of two phases between 490K and 30K, as listed in Table 1. We believed that (Tl,Rb)$_{0.534}$Fe$_2$Se$_2$ phase is responsible for the superconductivity in this compound.

Table 2. Refined structure parameters for the *I4/mmm* phase, obtained by fitting to ND data measured at 30, 300, 400 and 550 K for superconducting $Tl_{0.374}Rb_{0.374}Fe_{1.764}Se_2$. At 550K, Tl/Rb at $2a$ (0 0 0), Fe $4d$ (0 1/2 1/4) and Se at $4e$ (0 0 z) sites. 3.66% Fe and 4.78% FeSe as impurity were taken into account in the refinement.

		30K	300K	400K	550K
	a(Å)	3.8264(3)	3.8423(3)	3.8550(4)	3.9226(1)
	c(Å)	14.589(3)	14.820(2)	14.894(4)	14.4414(9)
Rb/Tl	n	**0.268(9)**	**0.267(8)**	**0.26(1)**	**0.374(5)**
Fe	n	**1**	**1**	**1**	**0.88(1)**
Se	n	**1**	**1**	**1**	**1**
	z	0.3503(5)	0.3515(4)	0.3543(7)	0.3532(2)

Table 3. Refined structure parameters for the *I4/m* phase obtained by fitting to ND data measured at below T_N for superconducting $Tl_{0.374}Rb_{0.374}Fe_{1.764}Se_2$.

		30K	300K	400K	500K
	a(Å)	8.6685(4)	8.7067(4)	8.7303(5)	8.7487(5)
	c(Å)	14.235(1)	14.335(1)	14.355(2)	14.405(2)
Rb1/Tl1	n	**0.408(6)**	**0.409(5)**	**0.37**	**0.37**
Rb2/Tl2	n	**0.408(6)**	**0.409(5)**	**0.37**	**0.37**
	x	0.399(2)	0.396(1)	0.394(2)	0.396(3)
	y	0.190(2)	0.194(2)	0.181(2)	0.184(3)
Fe1	n	**0.15(2)**	**0.10(1)**	**0.06(2)**	0.22(2)
Fe2	n	**0.92(1)**	**0.943(8)**	**1**	**1**
	x	0.1992(8)	0.1996(6)	0.1955(9)	0.198(1)
	y	0.0909(6)	0.0895(4)	0.0910(7)	0.0930(8)
	z	0.2491(9)	0.2477(9)	0.252(1)	0.252(1)
Se1	n	**1**	**1**	**1**	**1**
	z	0.1291(9)	0.1333(8)	0.132(1)	0.133(2)
Se2	n	**1**	**1**	**1**	**1**
	x	0.1077(9)	0.1080(7)	0.107(1)	0.111(1)
	y	0.288(1)	0.2946(10)	0.294(1)	0.298(2)
	z	0.1480(3)	0.1483(2)	0.1490(4)	0.1498(4)
	Rp (%)	5.50	5.45	6.08	7.23
	Rwp (%)	7.01	6.75	7.67	9.32
	χ^2	2.399	1.785	1.384	1.985

5. Summary and conclusions

In this article, we summarized our discovery about the evolution from an insulator to a superconductor with increasing Fe content in the (Tl,K,Rb)Fe_xSe_2 system. We first suggested that the AFM insulator in these systems is associated with the Fe-vacancy ordering in the crystals, which was confirmed by many experiments and the band calculations later. Superconductivity above 30K in (Tl,K,Rb)Fe_xSe_2 system was discovered, in which the onset superconducting transition temperature is as high as 40K. These results represent the first Fe-based superconductivity at the verge of an AFM insulator. But due to either chemical or electronic reason, the superconducting phase always coexists with the BCAF-c insulating phase in this system. The superconducting samples with single phase (structural and chemically homogenous) have not produced up to now. Whether BCAF-c insulator or other AFM insulating or semiconduting phase can be regarded as the parent compound of superconductivity in this system, remains unclear.

Acknowledgements

This work is supported by the National Basic Research Program of China (973 Program) under grant No. 2011CBA00103, 2012CB821404 and 2009CB929104, and the Nature Science Foundation of China (Grant No. 10974175, 10934005 and 11204059). We acknowledge Qimiao Si, Fuchun Zhang and Jianhui Dai for discussion and encouragement.

References

[1] Bednorz J G and Muller K A 1986 *Zeit. Fur Physik B-Condensed Matter* **64** 189
[2] Kamihara Y, Watanabe T, Hirano M and Hosono H 2008 *J. Am. Chem. Soc.* **130** 3296
[3] Chen X H, Wu T, Wu G, Liu R H, Chen H and Fang D F 2008 *Nature (London)* **453** 761
[4] Chen G F, Li Z, Wu D, Li G, Hu W Z, Dong J, Zheng P, Luo J L and Wang N L 2008 *Phys. Rev. Lett.* **100** 247002
[5] Ren Z A *et al* 2008 *Europhys. Lett.* **82** 57002
[6] Wen H H, Mu G, Fang L, Yang H and Zhu X Y 2008 *Europhys. Lett.* **82** 17009
[7] Wang C *et al* 2008 *Europhys. Lett.* **83** 67006
[8] Rotter M, Tegel M and Johrendt D 2008 *Phys. Rev. Lett.* **101** 107006
[9] Sefat A S, Jin R Y, McGuire M A, Sales B C, Singh D J and Mandrus D 2008 *Phys. Rev. Lett.* **101** 117004
[10] Li L J *et al* 2009 *New J. Phys.* **11** 025008
[11] Wang X C, Liu Q Q, Lv Y X, Gao W B, Yang L X, Yu R C, Li F Y and Jin C Q 2008 *Solid State Commun.* **148** 538
[12] Hsu K C *et al* 2008 *Proc. Natl. Acad. Sci. U.S.A.* **105** 14262
[13] Fang M H, Pham H M, Qian B, Liu T J, Vehstedt, Liu Y, Spinu L and Mao Z Q 2008 *Phys. Rev. B* **78** 224503
[14] Guo J, Jin S, Wang G, Wang S, Zhu K,Zhou T, He M and Chen X L 2010 *Phys. Rev.* B **82** 180520(R)
[15] Krzton-Maziopa A, Shermadini Z, Pomjakushina E, Pomjakushin V, Bendele M, Amato A, Khasanov R, Luetkens H and Conder K 2010 *J. Phys. Condens. Matter* **23** 052203
[16] Fang M H, Wang H D, Dong C H, Li Z J, Feng C M, Chen J and Yuan H Q 2011 *Europhys. Lett.* **94** 27009
[17] Wang H D, Dong C H, Li Z J, Mao Q H, Zhu S S, Feng C M, Yuan H Q and Fang M H 2011 *Europhys. Lett.* **93** 47004
[18] de la Cruz C *et al* 2008 *Nature* **453** 899
[19] Huang Q, Qiu Y, Bao W, Green M A, Lynn J W, Gasparovic Y C, Wu T, Wu G, and Chen X H 2008 *Phys. Rev. Lett.* **101** 257003
[20] Takahashi H, Igawa K, Arii K, Kamihara Y, Hirano M and Hosono H 2008 *Nature* **453** 376
[21] Dong C H, Wang H D, Yang J H, Qian B, Chen J, Li Z J, Yuan H Q, Fang M H, 2010 *SCIENCE CHINA* **53** 1216
[22] Bao W *et al* 2009 *Phys. Rev. Lett.* **102** 247001
[23] Klepp K and Boller H 1978 *Monatsh. Chem.* **109** 18
[24] Haggstrom L, Seidel A and Berger R A 1991 *J. Magn. Magn. Mater.* **98** 37
[25] Zabel M and Range K J 1984 *Rev. Chim. Miner.* **21** 139
[26] Sabrowsky H, Rosenberg M, Welz D, Deppe P and Schafer W 1986 *J. Magn. Magn. Mater.* **54-57** 1497
[27] Yan X W, Gao M, Lu Z Y, and Xiang T 2011 *Phys. Rev. Lett.* **106** 87005
[28] Cao C and Dai J 2011 *Phys. Rev. Lett.* **107** 056401
[29] Bao W, Huang Q, Chen G F, Green M A, Wang D M, He J B, Wang X Q and Qiu Y 2011 *Chin. Phys. Lett.* **28** 086104
[30] Ye F, Chi S, Bao W, Wang X F, Ying J J, Chen X H, Wang H D, Dong C H and Fang M H 2011 *Phys. Rev. Lett.* **107** 137003
[31] Wang M *et al* 2011 *Phys. Rev. B* **84** 094504
[32] Wang Z, Song W J, Shi H L, Wang Z W, Chen Z, Tian H F, Chen G F, Guo J G, Yang H X, and Li J Q 2011 *Phys. Rev. B* **83** 140505
[33] May A F, McGuire M A, Cao H, Sergueev I, Cantoni C, Chakoumakos B C, Parker D S and Sales B C 2012 *Phys. Rev. Lett.* **109** 077003
[34] Sales B C, McGuire M A, May A F, Cao H B, Chakoumakos and Sefat S 2011 *Phys. Rev. B* **83** 224510
[35] Zhao J, Cao H B, Bourret-Courchesne E, Lee D H, and Birgeneau 2012 *Phys. Rev. Lett.* **109** 267007

[36] Ricci A *et al* 2011 *Phys. Rev.* B **84** 060511
[37] Ksenofontov V, Eortmann G, Medvedev S A, Tsurkan V, Deisenhofer J, Loidl A and Felser C 2011 *Phys. Rev.* B **84** 180508
[38] Yuan R H, Dong T, Song Y J, Zheng P, Chen G F, Hu J P, Li J Q and Wang N L 2012 *Sci. Rep.* **2** 221
[39] Ma L, Ji G F, Dai J, He J B, Wang D M, Chen G F, Normand B and Yu W 2012 *Phys. Rev.* B **84** 220505
[40] Charnukha A, Cvitkovic A, Prokscha T, Pröpper D, Ocelic N, Suter A, Salman Z, Morenzoni Z, Deisenhofer J, Tsurkan V, Loid A, Keimer B, and Boris A V 2012 *Phys. Rev. Lett.* B **109** 017003
[41] Han F, Shen B, Wang Z Y and Wen H H 2012 *Phil. Mag.* B **92** 2553
[42] Song W J, Wang Z, Wang Z W, Shi H L, Chen Z, Tian H F, Chen G F, Guo J G, Yang H X, and Li J Q 2011 *Europhys. Lett.* **95** 37007
[43] Stadnik Z M, Wang P, Zukrowski J, Wang H D, Dong C H and Fang M H 2012 *J. Phys.: Condens. Matter* **24** 245701
[44] Stadnik Z M, Wang P, Zukrowski J, Wang H D, Dong C H and Fang M H 2013 *J. of Alloys and Compounds* **549** 288

Journal of Physics: Conference Series **449** (2013) 012016 doi:10.1088/1742-6596/449/1/012016

Coaligning arrays of air-sensitive single crystals for inelastic neutron scattering experiments

G. Friemel, M. Ohl, J. T. Park, B. Keimer and D. S. Inosov

Max Planck Institute for Solid State Research, Heisenbergstraße 1, 70569 Stuttgart, Germany.

E-mail: d.inosov@fkf.mpg.de

Abstract. The immense achievements of inelastic neutron scattering (INS) in recent years on the way to understanding the mechanism of high-temperature superconductivity were to a great extent enabled by the progress in fabricating large single-crystalline samples or coaligned mosaics consisting of up to several hundreds of individual single crystals. With the recent discovery of unconventional superconductivity in iron-based compounds, many of which can only be synthesized as relatively small crystals that rapidly decompose or deteriorate in air, we were faced with the technical challenge of preparing coaligned single-crystal arrays without exposing the samples to the ambient atmosphere. Here we describe a possible solution to this problem, which we have successfully employed to coalign both iron-pnictide and iron-chalcogenide single crystals in an argon atmosphere using a real-time digital x-ray Laue backscattering camera.

1. Introduction

For most of the high-temperature superconductors known today, including both copper oxides and iron-based compounds, high-quality single crystals can only be prepared in relatively small sizes, which are insufficient for inelastic neutron scattering (INS) measurements. Larger single crystals of both copper- and iron-based superconductors typically suffer from inclusions of foreign phases or doping inhomogeneities [1–3]. At the same time, the typical magnetic moments in these systems are small, of the order of one Bohr magneton or less. Therefore, to obtain reasonable intensity of the magnetic INS signal, sizable samples with a mass of several grams or more are desirable. To overcome this problem of sample size, INS measurements are often performed on large arrays of individual single crystals, coaligned on a sample plate (usually a thin aluminium grid or a silicon wafer) using x-ray Laue diffractometry [4–8]. Such crystal mosaics can include several tens or sometimes even more than a hundred [6, 8] of individual single crystals, mimicking a large single-domain crystal with a mosaicity determined by the accuracy of the coalignment. A typical mosaic spread of such arrays is of the order of 1–5°, defined as the full width at half maximum of the rocking curve measured across one of the structural Bragg peaks. Some of the largest cuprate samples prepared in this way for time-of-flight (TOF) spectroscopy measurements reached ~ 100 g in mass [9].

After the discovery of high-temperature superconductivity in iron-based compounds [10], this technique, which proved so successful for the copper oxides, also started to be applied to the new materials. Only for a few of the iron pnictides, such as electron-doped $Ba(Fe_{1-x}Co_x)_2As_2$, whole large-sized single crystals with a mass exceeding 1 g could be synthesized [11, 12]. Most other compounds, in particular hole-doped $Ba_{1-x}K_xFe_2As_2$ [13] or pure KFe_2As_2 [14] that naturally grow in the form of small platelets, had to be coaligned into large arrays to reach the required sample size. One of the first such arrays prepared by our group is shown in Fig. 1. For some other families, most notably those with the '1111'-type crystal structure, even such a coalignment turned out to be impossible due to the tiny sizes of the so far available single crystals. As a result, INS experiments on the '1111' family of iron pnictides until now remain limited to polycrystalline samples [15]. For a more complete overview of the INS results on iron pnictides, see Refs. 16–18.

2. Air sensitivity of iron-based superconductors

Different families of iron-based superconductors received disproportionate experimental attention, especially from INS, not only due to reasons related to the availability of sizeable single crystals, but also because of miscellaneous issues related to their chemical stability. As already mentioned, the '122'-type compounds have been investigated to a much greater extent than any other family, because these materials are typically stable in air and are easily available as large single crystals. However, even in these relatively stable compounds, changes in superconducting properties may occur after exposure of the samples to water vapor [19]. Other families of iron pnictides are more difficult to handle. For example, both $Li_{1-\delta}FeAs$ and $Na_{1-\delta}FeAs$ single crystals with the '111'-type structure, as well as their doped derivatives, are extremely sensitive to oxygen and air moisture [20, 21, 23] and decompose rapidly under ambient atmosphere. Therefore, meticulous care has to be taken to exclude any contact with air while handling the crystals over the entire process of sample preparation and measurements [24–26]. Some oxypnictides of the '1111' family can also react with water vapor in air with the formation of a rare-earth hydroxide phase [27], though much less violently.

Undoped iron chalcogenides are known to be hygroscopic and can reportedly acquire superconducting properties after exposure to air within several days [28]. This could possibly result from an intercalation of water molecules or oxygen in between the crystallographic layers [29]. The recently discovered alkali-metal iron chalcogenide superconductors of the $A_xFe_{2-y}Se_2$ type ($A = K, Rb, Cs$) [30–32] are even more air-sensitive [32]. Their freshly cleaved surface typically degrades in air within several minutes [33] with a characteristic change of color, which seriously complicates the surface-sensitive Laue diffraction measurements. Bulk samples usually decompose on the scale of hours and therefore can only be exposed to air for very short times during sample mounting, which is by far insufficient for assembling multiple single crystals into a well-aligned mosaic.

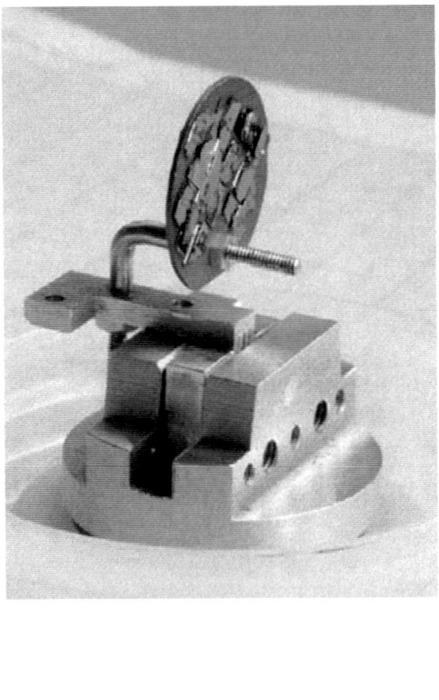

Figure 1: One of our first coaligned arrays of platelike optimally doped $Ba_{1-x}K_xFe_2As_2$ single crystals grown from Sn flux [34, 35], assembled on a silicon wafer and mounted on an aluminium sample holder.

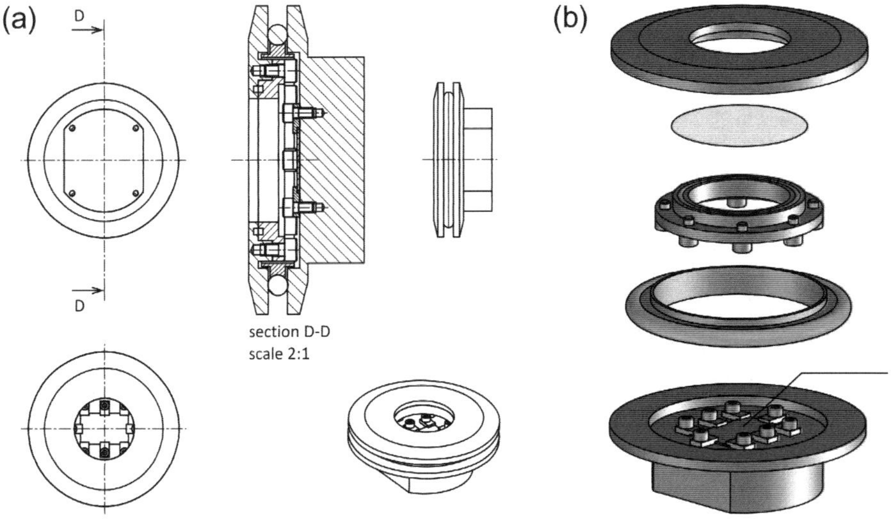

Figure 2: (a) technical drawing and (b) exploded view of the aluminium sample chamber for x-ray Laue backscattering measurements.

3. Airtight sample chamber for x-ray Laue measurements

To overcome this problem of air sensitivity, we have developed a simple and efficient way of manipulating air-sensitive samples during x-ray Laue backscattering measurements without exposing them to the ambient atmosphere. For this purpose, we have designed an airtight UFO-shaped chamber (see Fig. 2) that can be mounted vertically into the *MWL 110* real-time digital Laue-backscattering system (*Multiwire Laboratories, Ltd.*), as shown in Fig. 3. To minimize the weight of the chamber, most of its components are made of aluminium. The top and bottom parts of the chamber form a standard ISO KF50-type flange and are connected together using a standard KF-type o-ring gasket and a lightweight chain clamp. On the side of the chamber facing the detector, there is a thin transparent window made of an elastic polymer film, such as the one used in membrane boxes for sample storage. The window is sufficiently transparent for the

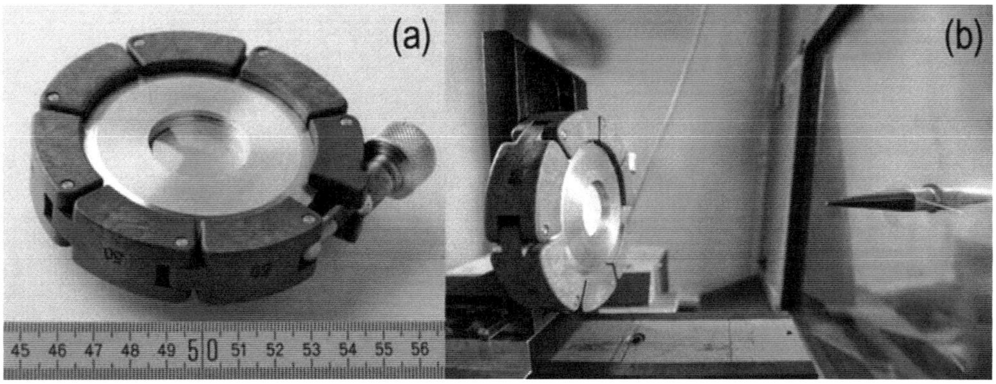

Figure 3: (a) the assembled sample chamber; (b) the same chamber installed on the *MWL 110* real-time digital Laue-backscattering system (*Multiwire Laboratories, Ltd.*).

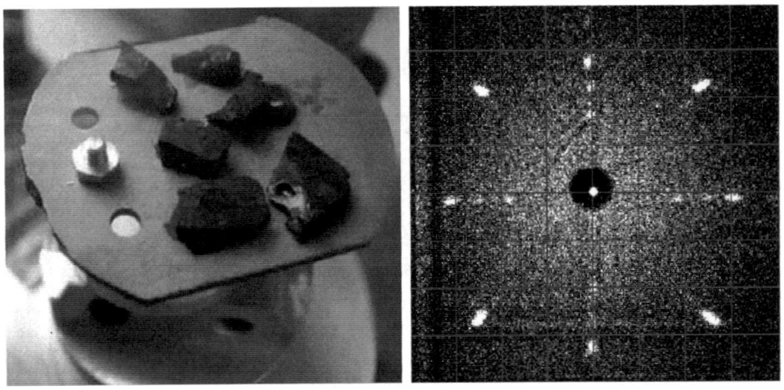

Figure 4: The left panel shows an array of six $Rb_xFe_{2-y}Se_2$ single crystals [36], coaligned on a silicon plate by means of the airtight sample chamber described in the text. The right panel presents a typical Laue diffraction pattern measured on one of these crystals.

x-ray beam, so that the Laue diffraction signal is visible on the detector. At the same time the elasticity of the film allows us to adjust the orientation of the samples inside the chamber directly through the membrane window without the need of opening the chamber at every iteration.

First, the sample plate is mounted inside the chamber as close as possible to the membrane window to allow enough space for sample mounting. If necessary, a spacer is placed below the sample plate to minimize the distance between the sample surface and the membrane window. Next, a single crystal is glued to the sample plate using a small amount of GE varnish inside a glove bag under a constant flow of argon. Immediately afterwards (before the varnish has time to harden), the chamber is closed and transferred to the Laue system, where the diffraction pattern from the sample is measured to estimate the misalignment of the crystal. Its orientation can be then adjusted immediately by pushing the crystal carefully through the membrane window with a soft ear-stick. Several iterations are usually required to reach an acceptably accurate orientation. After allowing some time for the varnish to solidify, the same procedure can be repeated for other single crystals until the complete mosaic is assembled. Finally, before the actual INS experiment, the assembled sample is removed from the chamber and mounted inside the cryostat in the usual way, without any extra foreign parts in the neutron beam.

Manipulating single crystals directly through the membrane without opening the chamber is a trick that allows us to save time and finish sample alignment before the varnish hardens, which would be much more difficult to achieve if the chamber had to be transferred to the glove bag and opened for every single adjustment of the crystal position. The elasticity of the membrane window also makes it possible to control the tightness of the chamber during measurements. A slight overpressure of argon is usually formed inside the chamber while the flange is being clamped, so that the membrane acquires a convex shape as can be seen in Fig. 3 (a). The loss of this overpressure would indicate a leak, possibly due to a puncture in the membrane, which can be thus immediately noticed.

4. Results

We have successfully used the described method for characterizing and orienting individual single crystals of air-sensitive '111'-type iron pnictide materials [26] and for assembling several arrays of both K- and Rb-based '245'-type iron selenides (see Fig. 4) [36, 37]. The resulting samples consisted of up to 20 single crystals and were used both for TOF neutron spectroscopy [38] and for conventional triple-axis INS experiments [39, 40]. A typical Laue diffraction pattern from one of the $Rb_xFe_{2-y}Se_2$ crystals is also depicted in Fig. 4. After the coalignment was completed, both $K_xFe_{2-y}Se_2$ and $Rb_xFe_{2-y}Se_2$ single crystals did not show any notable signs of surface degradation or any changes in their physical properties as compared to those constantly stored in argon atmosphere within a glove box.

Acknowledgments

This work has been supported, in part, by the DFG within the Schwerpunktprogramm 1458, under Grant No. BO3537/1-1. The authors are grateful to A. Ivanov, A. Piovano, O. Sobolev and E. Goremychkin for instrument support during INS experiments.

References

[1] H. A. Mook, P. C. Dai, and F. Doğan, Phys. Rev. Lett. **88**, 097004 (2002).
[2] R. Morinaga, K. Matan, H. S. Suzuki1, and T. J. Sato, Jpn. J. Appl. Phys. **48**, 013004 (2009).
[3] J. T. Park, D. S. Inosov, A. Yaresko, S. Graser, D. L. Sun, Ph. Bourges, Y. Sidis, Yuan Li, J.-H. Kim, D. Haug, A. Ivanov, K. Hradil, A. Schneidewind, P. Link, E. Faulhaber, I. Glavatskyy, C. T. Lin, B. Keimer, and V. Hinkov, Phys. Rev. B **82**, 134503 (2010).
[4] H. He, Y. Sidis, P. Bourges, G. D. Gu, A. Ivanov, N. Koshizuka, B. Liang, C. T. Lin, L. P. Regnault, E. Schoenherr, and B. Keimer, Phys. Rev. Lett. **86**, 1610–1613 (2001).
[5] V. Hinkov, S. Pailhès, P. Bourges, Y. Sidis, A. Ivanov, A. Kulakov, C. T. Lin, D. P. Chen, C. Bernhard, and B. Keimer, Nature **430**, 650–654 (2004).
[6] V. Hinkov, P. Bourges, S. Pailhes, Y. Sidis, A. Ivanov, C. D. Frost, T. G. Perring, C. T. Lin, D. P. Chen, and B. Keimer, Nat. Phys. **3**, 780–785 (2007).
[7] V. Hinkov, D. Haug, B. Fauque, P. Bourges, Y. Sidis, A. Ivanov, C. Bernhard, C. T. Lin, and B. Keimer, Science **319**, 597–600 (2008).
[8] H. He, P. Bourges, Y. Sidis, C. Ulrich, L. P. Regnault, S. Pailhes, N. S. Berzigiarova, N. N. Kolesnikov, and B. Keimer, Science **295**, 1045–1047 (2002).
[9] S. M. Hayden, G. Aeppli, R. Osborn, A. D. Taylor, T. G. Perring, S-W. Cheong, and Z. Fisk, Phys. Rev. Lett. **67**, 3622–3625 (1991).
[10] G. R. Stewart, Rev. Mod. Phys. **83**, 1589–1652 (2011).
[11] D. S. Inosov, J. T. Park, P. Bourges, D. L. Sun, Y. Sidis, A. Schneidewind, K. Hradil, D. Haug, C. T. Lin, B. Keimer, and V. Hinkov, Nat. Phys. **6**, 178–181 (2010).
[12] Y. Liu, D. L. Sun, J. T. Park, and C. T. Lin, Physica C **470**, S513–S515 (2010).
[13] C. Zhang, M. Wang, H. Luo, M. Wang, M. Liu, J. Zhao, D. L. Abernathy, K. Marty, M. D. Lumsden, S. Chi, S. Chang, J. A. Rodriguez-Rivera, J. W. Lynn, T. Xiang, J. Hu, and P. Dai, Scientific Reports **1**, 115 (2011).
[14] H. Kawano-Furukawa, C. J. Bowell, J. S. White, R. W. Heslop, A. S. Cameron, E. M. Forgan, K. Kihou, C. H. Lee, A. Iyo, H. Eisaki, T. Saito, H. Fukazawa, Y. Kohori, R. Cubitt, C. D. Dewhurst, J. L. Gavilano, and M. Zolliker, Phys. Rev. B **84**, 024507 (2011).
[15] S. I. Shamoto, M. Ishikado, A. D. Christianson, M. D. Lumsden, S. Wakimoto, K. Kodama, A. Iyo, and M. Arai, Phys. Rev. B **82**, 172508 (2010).
[16] J. W. Lynn and P. Dai, Physica C: Superconductivity **469**, 469–476 (2009).
[17] M. D. Lumsden and A. D. Christianson, J. Phys.: Condens. Matter **22**, 203203 (2010).
[18] D. S. Inosov, J. T. Park, A. Charnukha, Yuan Li, A. V. Boris, B. Keimer, and V. Hinkov, Phys. Rev. B **83**, 214520 (2011).
[19] H. Hiramatsu, T. Katase, T. Kamiya, M. Hirano, and H. Hosono, Phys. Rev. B **80**, 052501 (2009).
[20] M. J. Pitcher, D. R. Parker, P. Adamson, S. J. C. Herkelrath, A. T. Boothroyd, R. M. Ibberson, M. Brunellid, and S. J. Clarke, Chem. Commun. **2008**, 5918–5920 (2008).
[21] I. Todorov, D. Y. Chung, H. Claus, C. D. Malliakas, A. P. Douvalis, T. Bakas, J. He, V. P. Dravid, and M. G. Kanatzidis, Chem. Mater. **22**, 3916–3925 (2010).
[22] D. S. Inosov, J. S. White, D. V. Evtushinsky, I. V. Morozov, A. Cameron, U. Stockert, V. B. Zabolotnyy, T. K. Kim, A. A. Kordyuk, S. V. Borisenko, E. M. Forgan, R. Klingeler, J. T. Park, S. Wurmehl, A. N. Vasiliev, G. Behr, C. D. Dewhurst, V. Hinkov, Phys. Rev. Lett. **104**, 187001 (2010).
[23] M. A. Tanatar, N. Spyrison, K. Cho, E. C. Blomberg, G. Tan, P. Dai, C. Zhang, and R. Prozorov, Phys. Rev. B **85**, 014510 (2012).
[24] I. Morozov, A. Boltalin, O. Volkova, A. Vasiliev, O. Kataeva, U. Stockert, M. Abdel-Hafiez, D. Bombor, A. Bachmann, L. Harnagea, M. Fuchs, H.-J. Grafe, G. Behr, R. Klingeler, S. Borisenko, C. Hess, S. Wurmehl, and B. Büchner, Cryst. Growth Des. **10**, 4429–4432 (2010).
[25] A. E. Taylor, M. J. Pitcher, R. A. Ewings, T. G. Perring, S. J. Clarke, and A. T. Boothroyd, Phys. Rev. B **83**, 220514(R) (2011).
[26] J. T. Park, G. Friemel, T. Loew, V. Hinkov, Y. Li, B. H. Min, D. L. Sun, A. Ivanov, A. Piovano, C. T. Lin, B. Keimer, Y. S. Kwon, and D. S. Inosov, Phys. Rev. B **86**, 024437 (2012).
[27] A. D. Christianson, M. D. Lumsden, O. Delaire, M. B. Stone, D. L. Abernathy, M. A. McGuire, A. S. Sefat,

R. Jin, B. C. Sales, D. Mandrus, E. D. Mun, P. C. Canfield, J. Y. Y. Lin, M. Lucas, M. Kresch, J. B. Keith, B. Fultz, E. A. Goremychkin, and R. J. McQueeney, Phys. Rev. Lett. **101**, 157004 (2008).

[28] Y. Mizuguchi, K. Deguchi, S. Tsuda, T. Yamaguchi, and Y. Takano, Phys. Rev. B **81**, 214510 (2010).

[29] K. Deguchi, Y. Mizuguchi, Y. Kawasaki, T. Ozaki, S. Tsuda, T. Yamaguchi, and Y. Takano, Supercond. Sci. Technol. **24**, 055008 (2011).

[30] J. Guo, S. Jin, G. Wang, S. Wang, K. Zhu, T. Zhou, M. He, and X. Chen, Phys. Rev. B **82**, 180520(R) (2010).

[31] A. F. Wang, J. J. Ying, Y. J. Yan, R. H. Liu, X. G. Luo, Z. Y. Li, X. F. Wang, M. Zhang, G. J. Ye, P. Cheng, Z. J. Xiang, and X. H. Chen, Phys. Rev. B **83**, 060512(R) (2011).

[32] A. Krzton-Maziopa, Z. Shermadini, E. Pomjakushina, V. Pomjakushin, M. Bendele, A. Amato, R. Khasanov, H. Luetkens, and K. Conder, J. Phys.: Condens. Matter **23**, 052203 (2011).

[33] Y. Mizuguchi, H. Takeya, Y. Kawasaki, T. Ozaki, S. Tsuda, T. Yamaguchi, and Y. Takano, Appl. Phys. Lett. **98**, 042511 (2011).

[34] J. T. Park, D. S. Inosov, Ch. Niedermayer, G. L. Sun, D. Haug, N. B. Christensen, R. Dinnebier, A. V. Boris, A. J. Drew, L. Schulz, T. Shapoval, U. Wolff, V. Neu, X. Yang, C. T. Lin, B. Keimer, and V. Hinkov, Phys. Rev. Lett. **102**, 117006 (2009).

[35] G. L. Sun, D. L. Sun, M. Konuma, P. Popovich, A. Boris, J. B. Peng, K.-Y. Choi, P. Lemmens, and C. T. Lin, J. Supercond. Nov. Magn. **24**, 1773–1785 (2011).

[36] V. Tsurkan, J. Deisenhofer, A. Gunther, H.-A. Krug von Nidda, S. Widmann, and A. Loidl, Phys. Rev. B **84**, 144520 (2011).

[37] Y. Liu, Z. C. Li, W. P. Liu, G. Friemel, D. S. Inosov, R. E. Dinnebier, Z. J. Li, and C. T. Lin, Supercond. Sci. Technol. **25**, 075001 (2012).

[38] G. Friemel, W. P. Liu, E. A. Goremychkin, Y. Liu, J. T. Park, O. Sobolev, C. T. Lin, B. Keimer, and D. S. Inosov, EPL **99**, 67004 (2012).

[39] J. T. Park, G. Friemel, Y. Li, J.-H. Kim, V. Tsurkan, J. Deisenhofer, H.-A. Krug von Nidda, A. Loidl, A. Ivanov, B. Keimer, and D. S. Inosov, Phys. Rev. Lett. **107**, 177005 (2011).

[40] G. Friemel, J. T. Park, T. A. Maier, V. Tsurkan, Y. Li, J. Deisenhofer, H.-A. Krug von Nidda, A. Loidl, A. Ivanov, B. Keimer, and D. S. Inosov, Phys. Rev. B **85**, 140511(R) (2012).

A Short Review of the S_4 Symmetric Microscopic Model for Iron-Based High Temperature Superconductors

Jiangping Hu

Beijing National Laboratory for Condensed Matter Physics, Institute of Physics, Chinese Academy of Sciences, Beijing 100080, China & Department of Physics, Purdue University, West Lafayette, Indiana 47907, USA

E-mail: jphu@iphy.ac.cn & hu4@purdue.edu

Abstract. We briefly review the recently constructed two orbital microscopic model for iron-based superconductors based on S_4 symmetry[1]. With this faithful representation of the kinematics of the tri-layer FeAs or FeSe structure, the model provides answers and physical pictures to fundamental questions related to the robustness of superconductivity and pairing symmetry, unifies different families of iron-based superconductors, casts new insight into the connections to the other high T_c superconductors, cuprates, and reveals why an s-wave pairing can be stabilized by repulsive interactions. Further progresses include that the model provides a clean understanding of band reconstruction observed in magnetically ordered states, which is a strong support to the kinematics of the S_4 model, and captures the essential low energy physics of iron-based superconductors based on numerical results from unbiased quantum Monte Carlo simulation.

1. Introduction

In the past four years, the main theoretical difficulty in understanding iron-based high temperature superconductors[2, 3, 4, 5] stems from the fact that the materials are intrinsic multi-orbital systems due to the presence of all five 3d-orbitals of iron atoms. Several initial attempts to construct simplified models with fewer orbitals was made in an early stage of the research[6, 7, 8]. While these models could successfully explain some properties, they all had serious drawbacks, which resulted in a belief that any microscopic model composed of less than all five d-orbitals is insufficient[9, 10].

Recently, very surprisingly, with a proper consideration of the S_4 symmetry, the symmetry of the building block–the FeAs or FeSe trilayer, we demonstrate that an effective two-orbital model is sufficient to capture the kinematics responsible for low energy physics [1]. The two orbital model includes two nearly degenerate and weakly coupled single-orbital parts that can be mapped to each other under the S_4 transformation. Not only has the model greatly simplified kinematics which essentially can be described by only three important tight-binding hopping parameters, but the model also has manifested a clear physical picture to explore many intriguing fundamental properties of the materials and has drawn a deep connection to the other high temperature superconductors, cuprates. In this paper, we give a quick review of the model by emphasizing its unique aspects and summarizing recent numerical and analytic results[11, 12].

2. Motivation

Before proceeding to the S_4 model, we discuss several main initial motivations which lead to the construction of the model. The motivations stem from a few intriguing questions related to iron-based superconductors.

- *What is the relation between iron-based superconductors and cuprates?* This question has undoubtedly been asked by many researchers since iron-pnictides were discovered. In the past several years, it has been shown that the two classes of materials share many similar properties, including similar layered lattice structures, phase diagrams, strong antiferromagnetism in parental compounds, short superconducting coherent lengths and so on. However, it is still highly controversial that whether iron-based superconductors allow us to finally apply the induction method for high temperature superconductors.
- *How can the two different families of iron-based superconductors, iron-pnictides and iron chalcogenides, be unified?* Iron-based superconductors include two families, iron-pnictides[2, 3, 4] and iron-chalcogenides[5]. They share many intriguing common properties. They both have the highest T_cs around 50K[13, 3, 14, 15, 16]. The superconducting gaps are close to isotropic around Fermi surfaces[17, 18, 19, 20, 21, 22] and the ratio between the gap and T_c, $2\Delta/T_c$, are much larger than the BCS ratio, 3.52, in both families. However, the electronic structures in the two families, in particular, the Fermi surface topologies, are quite different in the materials reaching the highest T_c. The hole pockets are absent in iron-chalcogenides but present in iron-pnictides[17, 20, 21, 22]. The presence of the hole pockets has been a necessity for superconductivity in the majority of studies and models which deeply depend on the properties of Fermi surfaces[9, 13, 23, 24, 25, 26, 27, 28, 29, 30]. Therefore, the absence of the hole pockets in iron-chalcogenides causes a strong debate over whether both families belong to the same category that shares a common superconducting mechanism.
- *Why and how can a s-wave pairing state be stabilized if the superconductivity is originated from a repulsive interaction?* With a d-wave pairing symmetry, such as in cuprates, the positive and negative signs of the superconducting order parameter are equally distributed in both real and reciprocal space . Therefore, one can argue that the onsite repulsive interaction is naturally avoided in a d-wave pairing state. For a s-wave pairing symmetry, even in a sign changed $s^{\pm}$ state[24], the argument can not be held in general. While many of us use repulsive interactions to obtain a s-wave pairing or $s^{\pm}$, in iron-pnictides, this fundamental problem is largely unaddressed theoretically.
- *Why is the superconductivity so robust and how can a single energy scale dominate in pairing gap functions in iron-based superconductors?* Observed by angle-resolved photoemission microscopy (ARPES), a very intriguing property in the SC states of iron-pnictides is that the SC gaps around optimal doping on different Fermi surfaces are nearly proportional to a simple form factor $cosk_x cosk_y$ in reciprocal space. This form factor has been observed in both 122[17, 31, 18, 32] and 111[33, 34] families of iron-pnictides. Just like the *d*-wave form factor $cosk_x - cosk_y$ in cuprates, such a form factor indicates that the pairing between two next nearest neighbour (NNN) iron sites in real space dominates. In a multi-orbital model, many theoretical calculations based on weak coupling approaches have shown that the gap functions are very sensitive to detailed band structures and vary significantly when the doping changes[9, 26, 27, 28, 29, 30]. The robustness of the form factor has been argued to favor strong coupling approaches which emphasize electron-electron correlation or the effective next nearest neighbour (NNN) antiferromagnetic (AF) exchange coupling J_2[35, 36, 37, 38, 39, 40, 41, 42] as a primary source of the pairing force. However, realistically, it is very difficult to imagine there is a single energy scale in local exchange interactions if a multi *d*-orbital model is considered.

In order to answer the above questions, it is clear that we have to identify microscopic parameters that control the essential physics of iron-based superconductors at low energy.

3. Gauge mapping between the symmetries of superconductivity pairing and kinematics

The idea to address the robustness of possible s-wave pairing in iron-based superconductors stems from a paper by Berg, Scalapino and Kivelson[42], where they showed that the s-wave pairing in a two leg-ladder model for iron-pnictides is equivalent to a d-wave pairing in a similar model of cuprates. The s-wave and the d-wave can be mapped by a simple gauge transformation. Although their idea was carried out in an one-dimensional model, they raised an indispensable general point for any dimensional model, that is, *the pairing symmetry and kinematics must be considered together.* In particular, in a system where the short range pairing dominates, the phases or the signs of superconducting order parameters can be exchanged with those of the local hopping parameters.

3.1. Gauge Mapping

A gauge mapping can be generally described as follows. Given a general Hamiltonian including a kinetic energy $\hat{H}_0$ and an interaction energy $\hat{H}_I$, we can perform a gauge mapping that changes the sign of local Fermionic operators, namely,

$$c_i \rightarrow (-1)^{\theta_i} c_i \tag{1}$$

where $\theta_i = 0$, or 1. It is clear that such a general gauge mapping can be described by an unitary operator $\hat{U}$. Under the mapping, the hopping terms in $\hat{H}_0$ can change the sign so that the kinetic energy Hamiltonian is changed. However, the eigenvalues of $\hat{H}_0$ are not changed since it is an unitary transformation. It is also easy to see that the interaction Hamiltonian $\hat{H}_I$, if it includes standard interaction terms, such as local density interaction, spin exchange interaction and so on, in general, is not changed under the gauge transformation. Therefore, under the mapping,

$$\hat{H} = \hat{H}_0 + \hat{H}_I \rightarrow \hat{H}' = \hat{U}^+ \hat{H} \hat{U} = \hat{H}_0' + \hat{H}_I. \tag{2}$$

If the original Hamiltonian $\hat{H}$ has a superconducting ground state with an superconducting order parameter $< c_i c_j >= \Delta_{ij}$, the ground state of the new Hamiltonian $\hat{H}'$ must have $< c_i c_j >= (-1)^{\theta_i+\theta_j}\Delta_{ij}$. Therefore, if the pairing is rather local short range, it is possible to change order parameter symmetry by choosing a proper gauge mapping.

3.2. Gauge mapping between d-wave and extended s-wave pairing symmetries in cuprates

Let's use cuprates as an example to illustrate the point. For cuprates, the kinetic energy $\hat{H}_0$ is dominated by the nearest neighbour(NN) hopping in a square lattice, namely,

$$H_0 = -t \sum_{<ij>} c^+_{i\alpha} c_{j\alpha} + ... \tag{3}$$

where $< ij >$ represents the links between two NN sites, and the order parameter of the d-wave pairing symmetry in real space is also dominated by coupling two NN sites which is corresponding to the d-wave gap form $cosk_x - cosk_y$ in reciprocal space, namely,

$$< c_i c_{i+\vec{y}} >= - < c_i c_{i+\vec{x}} >= \Delta_0 \tag{4}$$

where $\vec{x}, \vec{y}$ are lattice unit vectors along x and y directions respectively. As shown in fig.1, if we set a gauge mapping as

$$c_{i\alpha} \rightarrow (-1)^{i_x} c_{i\alpha}, \tag{5}$$

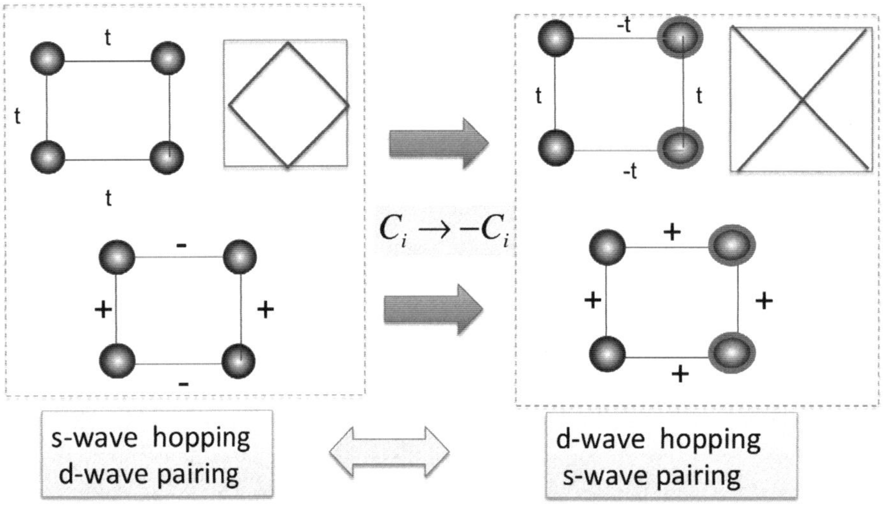

Figure 1. An illustration of the gauge mapping for cuprates. The left part of the figure shows the symmetry of the hopping parameters, d-wave pairing symemtry and the simplified Fermi surfaces at half filling. The right part shows their corresponding parts after the gauge mapping. The gague mapping changes the sign of Fermionic operators at the red sites.

where i_x labels the coordinate along the x direction of the ith site. After the mapping, we have

$$H_0' = -t \sum_{<ij>_y} c_{i\alpha}^+ c_{j\alpha} + t \sum_{<ij>_x} c_{i\alpha}^+ c_{j\alpha} \cdots \tag{6}$$

and

$$< c_i c_{i+\vec{y}} >=< c_i c_{i+\vec{x}} >= \Delta_0 \tag{7}$$

Therefore, the gauge mapping allows us to exchange the symmetries between the hopping parameters and the superconducting pairing order parameters. Before the gauge mapping, the hopping has a s-wave symmetry and the pairing has a d-wave symmetry. After the gauge mapping, the hopping has a d-wave symmetry and the pairing has a s-wave symmetry. It is also important to notice that the Fermi surface topology is changed after the gauge mapping as shown in fig.1. In a weak coupling approach, the Fermi surface topology has been the key ingredient to determine many properties, such as pairing symmetry. If the pairing is driven by repulsive interaction or antiferromagnetic fluctuations, the Fermi surface topology before the gauge mapping favors a d-wave symmetry while one can argue that the Fermi surface topology after the gauge mapping favors a s-wave symmetry. Therefore, the result obtained from a simple Fermi surface topology analysis is actually consistent with the gauge mapping result. We will revisit this point for iron-based superconductors in the following subsection.

3.3. Gauge mapping between extended s-wave and d-wave pairing symmetries in iron-based superconductors

The above analysis has been generalized to iron-based superconductors[1]. A similar gauge mapping which maps a s-wave to a d-wave pairing symmetry can be designed as shown in fig.2$(1, 2, 1', 2')$. We group four neighbour iron sites to form a super site and change the sign of Fermionic operators at the red iron sites in fig.2$(1')$. This transformation maps the A_{1g} *s*-wave $cos(k_x)cos(k_y)$ pairing symmetry in the original Fe lattice to a familiar *d*-wave $cosk_x' - cosk_y'$

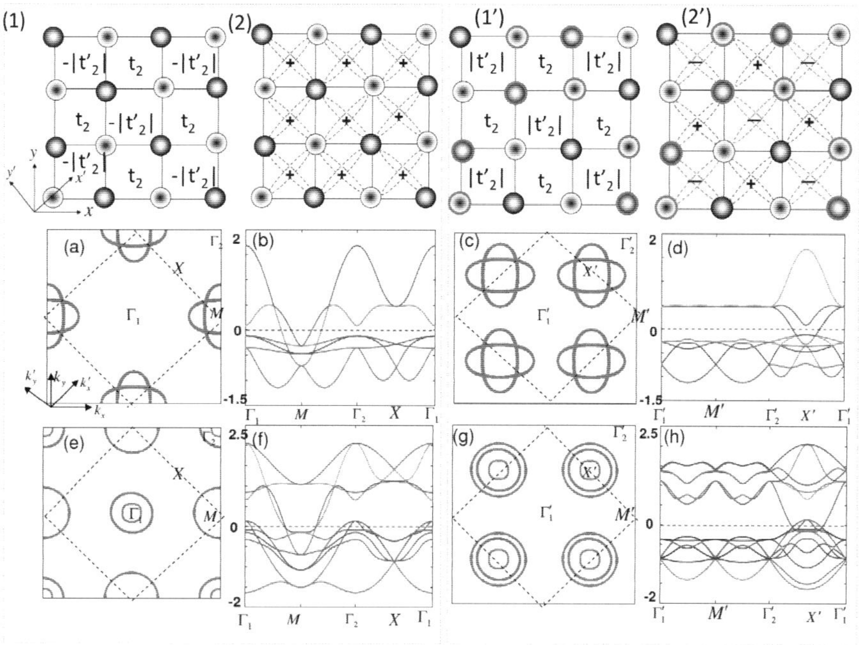

Figure 2. The illustration of the gauge mapping for iron-based superconductors: (1,2) and $(1', 2')$ sketch hopping parameters and pairing symmetries before and after the gauge mapping respectively; (a-h) band structures and Fermi surfaces for two models[36, 10] before and after the gauge mapping: (a,e) the Fermi surfaces, (b,f) the band dispersion along the high symmetry lines, (c,g) the Fermi surfaces after the gauge transformation, (d,h) the band dispersions along the high symmetry lines after the gauge transformation. The hopping parameters can be found in the above two references.

pairing symmetry defined in the two sublattices of the orginal lattice, where (k_x, k_y) and (k'_x, k'_y) label momentum in Brillouin zones of the origin lattice and sublattice respectively. This is actually a straightforward generalization of the transformation used in the study of the one dimensional iron ladder model[42].

Following the same procedure, we can ask how the kinetic energy changes. In ref.[1], we have shown that all the different tight binding models constructed for iron-based superconductors, after the gauge mapping, display very striking simplicity: (1) all Fermi surfaces after the gauge mapping are relocated around X', the anti-nodal points in a standard d-wave superconducting state in the sublattice Brillouin zone; (2) the bands previously located at the different places are magically linked in the new gauge setting. In particular, the two bands that contribute to electron pockets are nearly degenerate and the bands that contribute to hole pockets are remarkably connected to the former ones.

The first feature seriously challenges the consistency of previous weak coupling analysis depending on the difference of Fermi surface topologies for iron-pnictides and iron-chalcogenides. After the gauge mapping, based on the new Fermi surface topology, a standard weak coupling analysis would naturally favor a robust d-wave superconducting state with respect to the sublattice in the presence of repulsive interaction or NN AF coupling[43, 38]. Therefore, as shown in fig.3, if we reversely map to the original gauge setting, the original Hamiltonian must have a robust s-wave pairing symmetry. Namely, based on the new Fermi surface topology, both

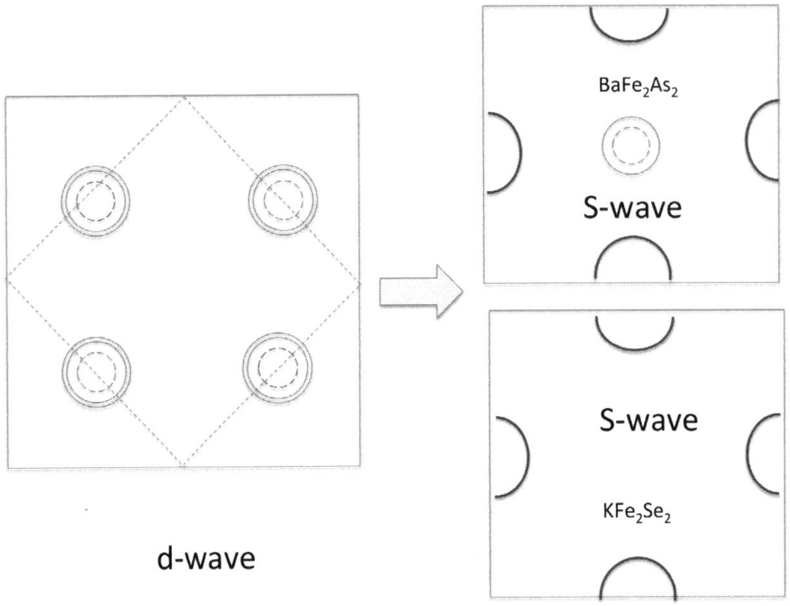

Figure 3. A sketch of a reverse gauge mapping from d-wave to s-wave for both iron-pnictides and iron-chalcogenides.

iron-pnictides and iron-chalcogenides should have s-wave pairing symmetry before the gauge mapping as long as the Fermi pockets are small.

The second feature is the key feature that leads us to believe that the low energy physics is controlled by a two orbital model which is characterized by two nearly degenerate bands. The emergence of the universal kinematics in different models suggest that the underlining electronic dynamics should be controlled by a much simplified set of parameters.

4. The construction of a two-orbital model with the S_4 symmetry

With above observations and the following important facts in the tri-layer FeAs or FeSe structure,

- The d-orbitals that form the bands near the Fermi surfaces are strongly hybridized with the p-orbitals of As(Se). Since the $d_{x'z}$ and $d_{y'z}$ have the largest overlap with the $p_{x'}$ and $p_{y'}$ orbitals, it is natural for us to use $d_{x'z}$ and $d_{y'z}$ to construct the model;
- The local lattice symmetry at an iron site is D_{2d} in which S_4 symmetry is the key symmetry operation;
- The two As(Se) planes are separated in space along c-axis in the tri-layer structure. Since there is little coupling between the p orbitals of the two planes and the hopping integrals through the p-orbitals are expected to dominate over the direct exchange couplings between the d-orbitals themselves, the two orbital model essentially could be decoupled into two nearly degenerate one orbital models;
- The decoupling must be controlled by the S_4 symmetry since S_4 symmetry links two As(Se) planes;

we draw a natural conclusion that the two-orbital model should be classified according to the S_4 symmetry and each of the two nearly degenerate one-orbital models should have a unit cell with

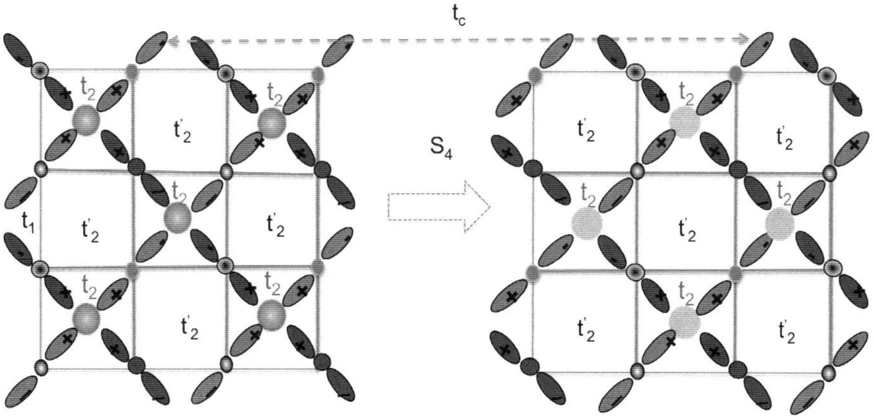

Figure 4. A sketch of the S_4 kinematics with $d_{x'z}$ and $d_{y'z}$ orbitals. The green balls represent As(Se) in the top layer and the grey balls represent those in the bottom layer. The blue and red orbitals represent the $d_{x'z}$ and $d_{y'z}$ orbitals.

two iron sites. A picture of the effective two-orbital model with the S_4 symmetry is illustrated in fig.4. The two orbital model is divided into two weakly and coupled one-orbital models. The orbital degree of freedom in the two one-orbital models can be called as two components of an S_4 iso-spin. The first one includes the $d_{x'z}$ in the A sublattice and the $d_{y'z}$ in the B sublattice if we use A and B to label two sublattices, and the second one includes the $d_{x'z}$ in the B sublattice and the $d_{y'z}$ in the A sublattice. The first strongly couples to the top As(Se) layer and the second couples to the bottom As(Se) layer. We denote $\hat{c}_{i\sigma}$ to $\hat{d}_{i\sigma}$ to label Fermionic operators of the two models that are connected by the S_4 transformation. Then, a general Hamiltonian that describes the kinematics is given by,

$$\hat{H}_{0,S_4} = \hat{H}_0 + \hat{H}_{0,c}, \tag{8}$$

where H_0 includes the two decoupled one-orbital models and $H_{0.c}$ which also obey the S_4 symmetry describes the weak coupling between the two models. As shown in[1], if we keep the leading hopping parameters,

$$\begin{aligned}
\hat{H}_0 &= \sum_{k\sigma}[4t_{2s}cosk_x cosk_y - \mu](\hat{c}^+_{k\sigma}\hat{c}_{k\sigma} + \hat{d}^+_{k\sigma}\hat{d}_{k\sigma}) \\
&+2t_1(cosk_x + cosk_y)(\hat{c}^+_{k\sigma}\hat{c}_{k\sigma} - \hat{d}^+_{k\sigma}\hat{d}_{k\sigma}) \\
&+2t_{1d}(cosk_x - cosk_y)(\hat{c}^+_{k\sigma}\hat{c}_{k\sigma} + \hat{d}^+_{k\sigma}\hat{d}_{k\sigma}) \\
&+4t_{2d}sink_x sink_y(\hat{c}^+_{k\sigma}\hat{c}_{k+Q\sigma} - \hat{d}^+_{k\sigma}\hat{d}_{k+Q\sigma}) \\
&+2t_3(cos2k_x + cos2k_y)(\hat{c}^+_{k\sigma}\hat{c}_{k\sigma} + \hat{d}^+_{k\sigma}\hat{d}_{k\sigma}) \\
&+...
\end{aligned} \tag{9}$$

and

$$\hat{H}_{0,c} = \sum_k 2t_c(cosk_x + cosk_y)(\hat{c}^+_{k\sigma}\hat{d}_{k\sigma} + h.c.). \tag{10}$$

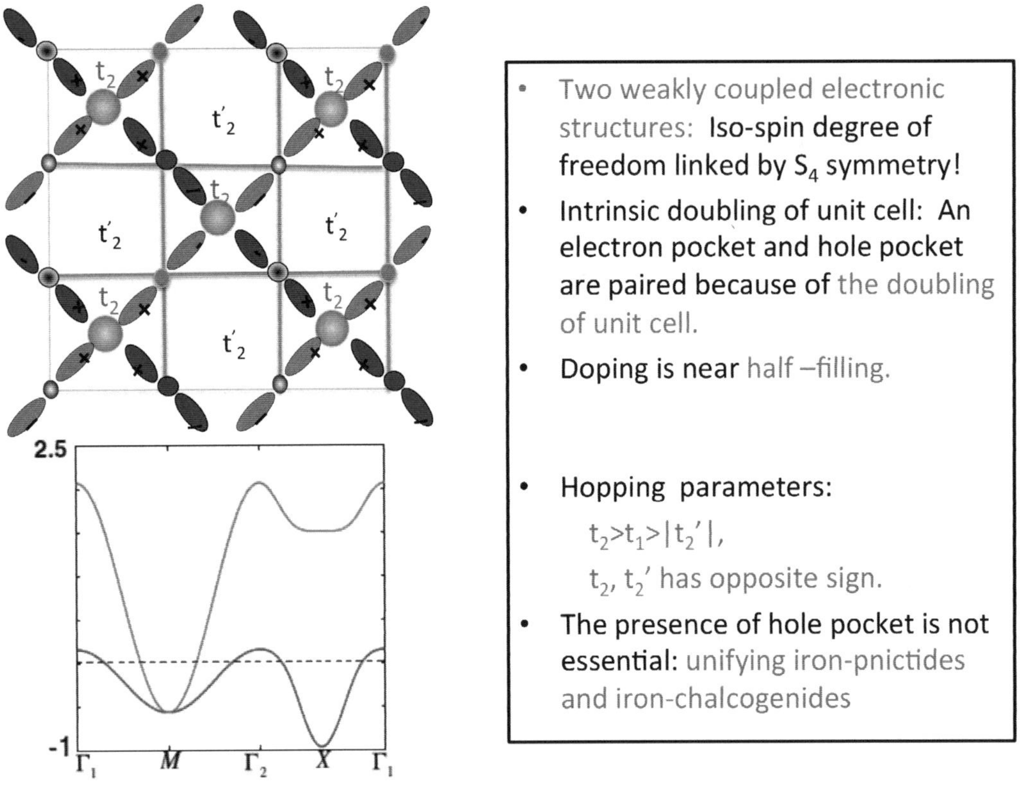

Figure 5. A sketch of the underlining kinematics and band structure of the S_4 model, and a list of the unique and important properties of the model.

where we limit the hopping parameters up to the third NN (TNN), including NN hoppings, t_{1x}, t_{1y}, NNN hoppings, t_2, t_2', and TNN hoppings, t_3. The longer range hoppings can be included if needed. For convenience, we have defined $t_{1s} = (t_{1x}+t_{1y})/2$, $t_{1d} = (t_{1x}-t_{1y})/2$, $t_{2s} = (t_2+t_2')/2$ and $t_{2d} = (t_2 - t_2')/2$.

Ignoring t_c, each S_4 iso-spin components provides the following two paired band energy dispersions due to unit cell doubling,

$$E_e = \epsilon_k + 4\sqrt{t_{2d}^2 sin^2 x sin^2 y + [\frac{t_{1s}(cosk_x + cosk_y) \pm t_{1d}(cosk_x - cosk_y)}{2}]^2}, \qquad (11)$$

$$E_h = \epsilon_k - 4\sqrt{t_{2d}^2 sin^2 x sin^2 y + [\frac{t_{1s}(cosk_x + cosk_y) \pm t_{1d}(cosk_x - cosk_y)}{2}]^2}, \qquad (12)$$

where $\epsilon_k = 4t_{2s}cosk_x cosk_y + 2t_3(cos2k_x + cos2k_y) - \mu$. E_e can capture the electron pocket at M points and E_h can capture the hole pocket at Γ point. t_{1s}, t_{2s} and t_{2d} are the key parameters. By just keeping these three parameters, the model is already good enough to capture the main characters of the bands contributing to the Fermi surface. Including interactions, the general

Hamiltonian is given by

$$H = H_{0,s_4} + U \sum_{i,\alpha=1,2} \hat{n}_{i,\alpha\uparrow}\hat{n}_{i,\alpha\downarrow} + U' \sum_i \hat{n}_{i,1}\hat{n}_{i,2} + J'_H \sum_i \hat{S}_{i,1} \cdot \hat{S}_{i,2} \tag{13}$$

where $\alpha = 1, 2$ labels the S_4 iso-spin, U describes the effective Hubbard repulsion interaction within each iso-spin, U' describes the one between them and J'_H describes the effective Hund's coupling. Since the two components couple weakly, we may expect U dominates over U' and J'_H. Then, in the first order approximation, the model could become a single orbital-Hubbard model near half-filling.

In the original paper[1], we have suggested that t_{1s} breaks D_{2d} symmetry. This statement is not absolutely correct because t_{1s} and t_{1d} can be exchanged by choosing different orbital configurations. With the orbital configuration shown in fig.4, t_{1s} does not break D_{2d} symmetry. It is also important to note that the Γ point and the (π, π) point in Brillouin zone here are exchangeable because of the doubling of the unit cell.

5. Numerical results and supporting experimental evidence for the S_4 model

In fig.5, we summarize the important properties in the S_4 model. These properties include

- For each S_4 iso-spin, there is a pair of bands generated by unit cell doubling: one contributes a hole pocket and the other contributes an electron pocket.
- For each S_4 iso-spin, the filling factor is close to half-filling, an important property to allow strong spin fluctuation.
- The unit cell is intrinsically doubled, which is a very different starting point of kinematics from previous models.
- The hole pockets can be suppressed by adjusting t_{1s} or t_{2s}. Fixing t_{2d} and t_{2s}, reducing t_{1s} leads to flatter dispersion of the band responsible for the hole pockets. Thus, it suppresses the hole pockets. Therefore, the model unifies iron-pnictides and iron-chalcogenides.
- From orbital configuration, we have $t_2 > |t_{1s}| > |t'_2|$ and the signs of t_2 and t'_2 are opposite.

These characters have important physical consequences. In the following, we discuss some recent progresses associated with them.

5.1. Quantum Monte Carlo (QMC) Simulation on the S_4 model

The microscopic understanding provides a new opportunity to make use of highly controllable and unbiased numerical methods to study iron-based superconductors. In a recent paper[11], we have demonstrated that the model in the presence of electron-electron interaction is characterized by strong $(0, \pi)$ collinear antiferromagentic (C-AFM) fluctuations as observed in iron-based superconductors[44, 45, 46]. Moreover, the pairing with an extended s-wave symmetry robustly dominates over other pairings at low temperature in reasonable parameter region regardless of the change of Fermi surface topologies, namely, the presence or absence of hole pockets at Γ point. For both iron-pnictides and iron-chalcogenides, the pairing correlation, the effective pairing interaction and the antiferromagnetic correlation strongly increase as the on-site Coulomb interaction increases.

QMC results are summarized in fig.6. The competition between the s-wave and the d-wave in the S_4model is manifested by t_1 and t_2 where t_1 favors the d-wave while t_2 favors the s-wave. A simple phase diagram from QMC simulation is given as the inset in fig. 6(b). For iron-based superconductors, t_1 should be roughly equal to $t_2/2$ and $|t'_2|$ is much smaller than t_1 and t_2. Therefore, the s-wave should be robustly favored.

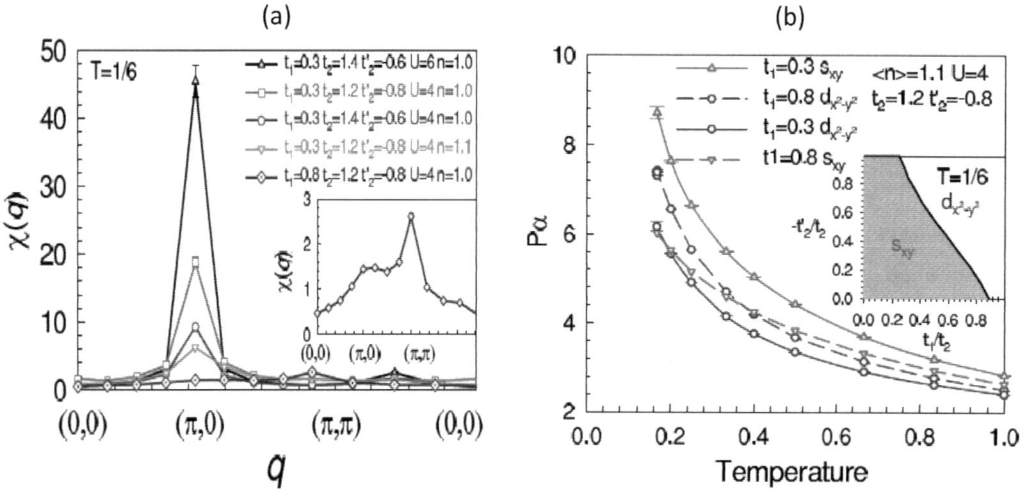

Figure 6. (a) Quantum Monte Carlo calculations about the static spin susceptibility for the S_4 model. The inset shows a significant large t_1 can change the peak from $(0, \pi)$ to (π, π), which signals the competition between NN hopping and NNN hopping. (b) The effective pairing interaction as a function of temperature at $< n >$=1.1, $U = 4.0$, $t_2 = 1.2$, $t_2' = -0.8$ for different t_1. Inset: the competition between s_{xy} and $d_{x^2-y^2}$ depends on t_1/t_2 and $-t_2'/t_2$.

5.2. Oriented gap opening in the C-AFM state: an impact of intrinsic unit cell doubling

As we have discussed above, one unique feature in the S_4 symmetry construction is that the unit cell includes 2 irons. Most of previous studies are based on models with one iron per unit cell. In principle, one can argue that models based on both representations should be equivalent. However, since the search of a meanfield result depends on the starting point of kinematics, a proper starting point is crucial to obtaining a correct meanfield result.

In a recent paper, we have demonstrated an important impact of the intrinsic unit cell doubling[12]. Unlike in a conventional spin density wave (SDW) state where gaps at any two points on the Fermi surface that are connected by the ordered SDW wave vector should be opened, the band structure in the C-AFM state appears to be reconstructed by the development of the magnetic order. ARPES experiments [47, 48, 49, 50, 51, 52] observe that small pockets or hot spots develop around Fermi surfaces. It is fair to say that the scenario of the gap opening due to a magnetic order in a multi-orbital system, such as iron-pnictides, is complex. For example, it has been argued that the entire Fermi surfaces can be gaped out[53]. Nevertheless, the surprising band reconstruction is never fully understood in previous models.

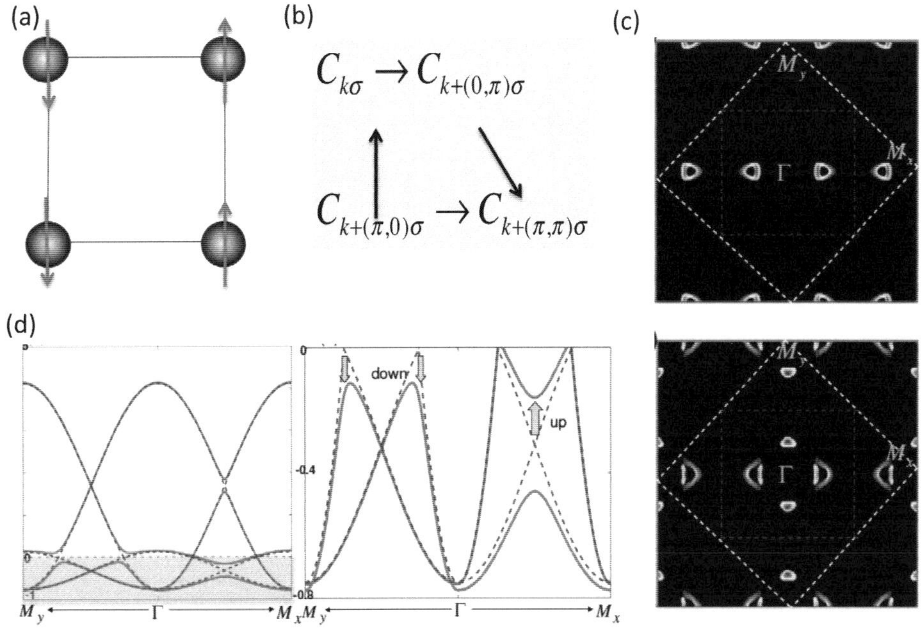

Figure 7. The illustration of oriented gap opening in magentically ordered state of iron-based superconductors: (a) The C-AFM magnetic state; (b) The illustration of the connections of four Fermionic operators in reciprocal space by the C-AFM order; (c) The typical Fermi surfaces in the C-AFM state; (d) The band reconstruction and gap opening in the C-AFM state.

As summarized in fig.7, the complicated band reconstruction near Fermi surfaces in the magnetically ordered state of iron-pnictides observed by ARPES can be understood in a meanfield level by the S_4 model. The $(0,\pi)$ or $(\pi,0)$ C-AFM order does not open gaps between two points on theFermi surfaces linked by the ordered wave vector but forces a band reconstruction involving four points in unfolded Brillouin zone and gives rise to small pockets or hot spots.

The doubling of the unit cell leads to an orientation-dependent gap opening in reciprocal space. In one $\Gamma - M$ direction, the magnetic order splits two folded bands while in the other direction the gap can be opened between the pair of electron and hole pocket bands. Such an oriented gap opening generates the hot spots or small pockets as observed by ARPES as shown in fig.7(c)..

5.3. Sign change between the top and bottom As(Se) planes in the s-wave state

The NNN hopping parameters t_2 and t_2' have opposite signs in the S_4 model. This sign difference stems from the orbital symmetries. If we take the gauge mapping mentioned earlier, after the mapping, the two hopping parameters in the new model have the same signs. Therefore, the gauge mapping exactly switches the symmetries between the pairing and the underlining

properties	iron SCs	cuprates
pairing symmetry	s-wave	d-wave
underlining hopping symmetry	d-wave	s-wave
dominant pairing form	$cosk_xcosk_y$	$cosk_x - cosk_y$
pairing classification symmetry	S_4	C_{4v}
unit cell	2 Fe (doubled)	1 Cu
AF coupling	NNN J_2	NN J_1
sign change in real space	c-axis	a-b plane
filling density	half-filling	half-filling

Table 1. A list of the close connections between iron-based superconductors (iron SCs) and cuprates.

kinematics as shown in fig.2$(1, 2, 1', 2')$.

In the view of the d-wave picture after the gauge mapping, the superconducting phase for each S_4 iso-spin alternates between neighboring squares, which is corresponding to the sign change between the top and bottom As/Se planes in the view of the S_4 symmetry. Such an intrinsic sign change between the top and bottom As planes along c-axis explains why a s-wave before gauge mapping can be stabilized even if the repulsive interaction is responsible for superconductivity.

The relative phase between the two S_4 iso-spins can take two different values, 0 or π, in principle. The difference can be classified as the irreducible representation of the S_4 symmetry. The S_4 point group has four one-dimensional representations, including A, B and $2E$. In the A state, the S_4 symmetry is maintained. In the B state, the state changes sign under the S_4 transformation. Therefore, in the A phase, since the S_4 symmetry is not violated, the relative phase between the two iso-spin are equal to π in space, while in the B phase, the relative phase is zero. The A phase in the presence of the hole pockets is an sign-changed s-wave or $s^{\pm}$ state. Therefore, the model predicts a new superconducting state, B phase state. However, it is still an open question whether iron-base superconductors can carry both states or not.

6. Discussion and perspectives

6.1. Comparison between cuprates and iron-based superconductors

By revealing the underlining electronic structure, we can make a close comparison between cuprates and iron-based superconductors. In Table.1, we list the close relations between two high T_c superconductors. From the table, it is clear that by determining these physical properties of iron-based superconductors listed in the table can help to determine the high T_c superconducting mechanism.

6.2. More supporting evidence for the S_4 symmetry

The S_4 model provides a microscopic understanding to two measured universal behaviors: the superconducting gap function that is dominated by the single $cosk_xcosk_y$ form and the almost universal NNN antiferromagnetic exchange coupling strength J_2 measured by neutron scattering[45, 54, 55] . The existence of such two single energy scale quantities is also a strong support for the S_4 model. If we treat the S_4 model in the presence of strong electron-electron correlation, both J_2 and the gap function in the s-wave state is determined by the NNN hopping parameters.

The S_4 model also suggests that the properties of an iron-based superconductor can be understood by simply observing bands responsible for electron pockets. Since the physics is dominated by the pair of bands that are responsible for a pair of hole and electron pockets. The third hole pockets at the Γ point observed in iron-pnictides is not expected to play a significant role. However, the existence of the third hole pocket may change the doping concentration for the S_4 model in a stoichiometric material. Namely, in the view of the S_4 model, the material could be self electron doped. For example, if a large third hole pocket is present, we would expect that the S_4 model could be heavily electron-doped. This expectation naturally explains puzzled physical properties of $LiFeAs$ where a large third hole pocket was observed[56]. No static magnetic order is observed in the stoichiometric $LiFeAs$ with $T_c \sim 18K$[57, 58]. In the view of the S_4 model, the $LiFeAs$ is self electron doped so that many physical properties of the material should be understood from the heavy electron doped side of phase diagram[59].

6.3. Perspectives
The S_4 model provides a fresh starting point to explore novel physics in iron-based superconductors. Here, we discuss several perspectives for the future research.

First, since the S_4 symmetry stems from the fact that As stands out of Fe plane, obvious interesting physics may stem from c-axis coupling. Recently, ARPES has provided the existence of superconducting gap nodes along c-axis in $BaFe_2As_{2-x}P_x$[60, 61]. It is very likely that the S_4 symmetry may be linked to such a development of nodes since the superconducting phase along c-axis in the S_4 model has very intriguing sign change as we discussed earlier.

Second, electron nematism observed in iron-pnictides[62] can be fully investigated in the S_4 model. Namely, the S_4 model provides a microscopic base to understand electron nematism. The S_4 symmetry breaking can lead to an electron nematic state. In the previous theoretical studies, electron nematism is obtained through phenomenological models[63, 64, 65, 66].

Third, with only a set of three parameters describing kinematics, it becomes possible to obtain the dependence of these parameters on lattice parameters. Such an dependence may help us to explain empirical observations of the connection between the maximum T_cs and the As-Fe-As angles[13].

Fourth, it is interesting to investigate physical properties on surfaces where the S_4 symmetry, in principle, is naturally broken. A natural consequence of such a symmetry breaking is that symmetry broken states may be more stable on surfaces than in bulk.

Finally, the doping phase diagram and its parameter dependence, as well as many other physical quantities need to be investigated in the S_4 model.

6.4. Summary
The S_4 model provides great simplicity to study and unify the physics of iron-based superconductors. It reveals the deep connections between curpates and iron-based superconductors. Since the uniqueness of the model stems from the unique structure of FeAs or FeSe tri-layer, revealing such an underlining microscopic kinematics cast a new light on the search of new structures that can produce high superconducting transition temperatures.

7. Acknowledgments
The author acknowledges NN Hao, T. Ma, H. Lin, Y. Wang for collaborations and H. Ding, DL Feng, PC Dai, T. Xiang, X. Dai, S. Kivelson, D. H. Lee, F. Wang for useful discussions. The work is supported by the Ministry of Science and Technology of China 973 program(2012CB821400) and NSFC-1190024.

[1] Hu J and Hao N 2012 *Phys. Rev X* **2** 021009
[2] Kamihara Y, Watanabe T, Hirano M and Hosono H 2008 *J. Am. Chem. Soc.* **130** 3296
[3] Chen X H, Wu T, Wu G, Liu R H, Chen H and Fang D F 2008 *Nature* **453** 761
[4] Chen G F, Li Z, Wu D, Li G, Hu W Z, Dong J, Zheng P, Luo J L and Wang N L 2008 *Phys. Rev. Let.* **100**
[5] Guo J, Jin S, Wang G, Wang S, Zhu K, Zhou T, He M and Chen X 2010 *Phys. Rev. B* **82** 180520
[6] Daghofer M, Nicholson A, Moreo A and Dagotto E 2010 *Phys. Rev. B* **81** 14511
[7] Lee P A and Wen X G 2008 *Phys. Rev. B* **78** 144517

[8] Raghu S, Qi X L, Liu C X, Scalapino D J and Zhang S C 2008 *Phys. Rev. B* **77** 220503
[9] Hirschfeld P J, Korshunov M M and Mazin I I 2011 *Rep. Prog. Phys.* **74** 124508
[10] Kuroki K, Onari S, Arita R, Usui H, Tanaka Y, Kontani H and Aoki H 2008 *Phys. Rev. Lett.* **101** 087004
[11] Ma T, Lin H and Hu J 2012 *arXiv:1206.6277*
[12] Hao N N, Wang Y and Hu J 2012 *arXiv:1207.2798*
[13] Johnston D 2010 *Advances in Physics* **59** 803–1061
[14] Liu D and et al. 2012 *arXiv:1202.5849*
[15] Wang Q Y and et al. 2012 *arXiv:1201.5694*
[16] Sun L and et al. 2012 *Nature* **10.1038/nature10813**
[17] Ding H and et al 2008 *EPL (Europhysics Lett.)* **83** 47001
[18] Zhao L and et al. 2008 *Chin. Phys. Lett* **25** 4402
[19] Zhang Y, Yang L X, Chen F, Zhou B, Wang X F, Chen X H, Arita M, Shimada K, Namatame H, Taniguchi M, Hu J P, Xie B P and Feng D L 2010 *Phys. Rev. Lett.* **105** 1117003
[20] Wang X P, Qian T, Richard P, Zhang P, Dong J, Wang H D, Dong C H, Fang M H and Ding H 2011 *Europhysics Lett.* **93** 57001
[21] Zhang Y, Yang L X, Xu M, Ye Z R, Chen F, He C, Xu H C, Jiang J, Xie B P, Ying J J, Wang X F, Chen X H, Hu J P, Matsunami M, Kimura S and Feng D L 2011 *Nature Materials* **10** 273–277
[22] Mou D, Liu S, Jia X, He J, Peng Y, Zhao L, Yu L, Liu G, He S, Dong X, Zhang J, Wang H, Dong C, Fang M, Wang X, Peng Q, Wang Z, Zhang S, Yang F, Xu Z, Chen C and Zhou X J 2011 *Phys. Rev. Lett.* **106** 107001
[23] Dong J, Zhang H J, Xu G, Li Z, Li G, Hu W Z, Wu D, Chen G F, Dai X, Luo J L, Fang Z and Wang N L 2008 *Europhysics Lett.* **83** 27006
[24] Mazin I I, Singh D J, Johannes M D and Du M H 2008 *Phys. Rev. Lett.* **101** 57003
[25] Kuroki K, Onari S, Arita R, Usui H, Tanaka Y, Kontani H and Aoki H 2008 *Phys. Rev. Lett.* **101** 087004
[26] Wang F, Zhai H, Ran Y, Vishwanath A and Lee D H 2009 *Phys. Rev. Lett.* **102** 47005
[27] Thomale R, Platt C, Hu J P, Honerkamp C and Bernevig B A 2009 *Physical Review B* **80** 180505
[28] Thomale R, Platt C, Hanke W, Hu J and Bernevig B A 2011 *Phys. Rev. Lett.* **107** 117001
[29] Chubukov A V, Efremov D V and Eremin I 2008 *Phys. Rev. B* **78** 134512
[30] Cvetkovic V and Tesanovic Z 2009 *Phys. Rev. B* **80** 24512
[31] Richard P, Sato T, Nakayama K, Takahashi T and H D 2011 *Rep. Prog. Phys.* **74** 124512
[32] Nakayama K, Sato T, Richard P, Xu Y M, Sekiba Y, Souma S, Chen G F, Luo J L, Wang N L, Ding H and Takahashi T 2009 *Europhysics Lett.* **85** 67002
[33] Umezawa K, Li Y, Miao H, Nakayama K, Liu Z H, Richard P, Sato T, He J B, Wang D M, Chen G F, Ding H, Takahashi T and Wang S C 2012 *Phys. Rev. Lett.* **108**
[34] Liu Z H, Richard P, Nakayama K, Chen G F, Dong S, He J B, Wang D M, Xia T L, Umezawa K, Kawahara T, Souma S, Sato T, Takahashi T, Qian T, Huang Y B, Xu N, Shi Y B, Ding H and Wang S C 2011 *Phys. Rev. B* **84**
[35] Seo K J, Bernevig B A and Hu J P 2008 *Phys. Rev. Lett.* **101** 206404
[36] Fang C, Wu Y L, Thomale R, Bernevig B A and Hu J 2011 *Phys. Rev X* **1** 011009
[37] Si Q and Abrahams E 2008 *Phys. Rev. Lett.* **101** 76401
[38] Hu J and Ding H 2012 *Scientific Reports* **2** 381
[39] Hu J, Xu B, Liu W, Hao N and Wang Y 2012 *Phys. Rev. B* **85** 144403
[40] Lu X, Fang C, Tsai W F, Jiang Y and Hu J 2012 *Phys. Rev B* **85** 054505
[41] Berg E, Kivelson S A and Scalapino D J 2010 *Phys. Rev. B* **81** 172504
[42] Berg E, Kivelson S A and Scalapino D J 2009 *N. J. Phys.* **11** 085007
[43] Scalapino D J 1999 *Science* **284** 1282–1283
[44] de La Cruz C, Huang Q, Lynn J W, Li J, W Ratcliff I, Zarestky J L, Mook H A, Chen G F, Luo J L, Wang N L and Dai P 2008 *Nature* **453** 899–902
[45] Zhao J, Adroja D T, Yao D X, Bewley R, Li S L, Wang X F, Wu G, Chen X H, Hu J P and Dai P C 2009 *Nature Physics* **5** 555–560
[46] Dai P, Hu J and Dagotto E 2012 *Nature Physics* **In Press**
[47] Yi M, Lu D H, Analytis J G, Chu J H, Mo S K, He R H, Hashimoto M, Moore R G, Mazin I I, Singh D J, Hussain Z, Fisher I R and Shen Z X 2009 *Phys. Rev.B* **80** 174510
[48] Richard P, Nakayama K, Sato T, Neupane M, Xu Y M, Bowen J H, Chen G F, Luo J L, Wang N L, Ding H and Takahashi T 2010 *Phys. Rev. Lett* **104** 0909.0574
[49] Yi M, Lu D H, Chu J H, Analytis J G, Sorini A P, Kemper A F, Mo S K, Moore R G, Hashimoto M, Lee W S, Hussain Z, Devereaux T P, Fisher I R and Shen Z X 2011 *PNAS* **108** 6878
[50] Zhang Y, He C, Ye Z R, Jiang J, Chen F, Xu M, Ge Q Q, Xie B P, Wei J, Aeschlimann M, Cui X Y, Shi M, Hu J P and Feng D L 2012 *Phys. Rev. B* **85**

[51] Liu H and et a 2008 *Phys. Rev B* **78** 184514
[52] Liu G and et al. 2009 *Phys. Rev B* **80** 134519
[53] Ran Y, Wang F, Zhai H, Vishwanath A and Lee D H 2009 *Phys. Rev B* **79** 014505
[54] Lipscombe O J, Chen G F, Fang C, Perring T G, Abernathy D L, Christianson A D, Egami T, Wang N, Hu J and Dai P 2011 *Phys. Rev. Lett.* **106** 57004
[55] Wang M, Fang C, Yao D X, Tan G, Harriger L W, Song Y, Netherton T, Zhang C, Wang M, Stone M B, Tian W, Hu J and Dai P 2011 *Nature Comm.* **2** 580
[56] Putzke C, Coldea A I, Guillamn I, Vignolles D, McCollam A, Leboeuf D, Watson M D, Mazin I I, Kasahara S, Terashima T, Shibauchi T, Matsuda Y and Carrington A 2012 *Phys. Rev. Lett.* **108** 47002
[57] Tapp J H, Tang Z, Lv B, Sasmal K, Lorenz B, Chu P C W and Guloy A M A 2008 *Phys. Rev. B* **78** 60505
[58] Wang X C, Liu Q Q, Lv Y X, Gao W B, Yang L X, Yu R C, Li F Y and Jin C Q 2008 *Sol. Stat. Comm.* **148** 538
[59] Wang M, Wang M, Miao H, Carr S V, Abernathy D L, Stone M B, Wang X C, Xing L, Jin C Q, Zhang X, Hu J, Xiang T, Ding H and Dai P 2012 arXiv:1208.0909
[60] Zhang Y, Ye Z R, Ge Q Q, Chen F, Jiang J, Xu M, Xie B P and Feng D L 2012 *Nature Physics* **8** 371
[61] Qiu X, Zhou S Y, Zhang H, Pan B Y, Hong X C, Dai Y F, Eom M J, Kim J S, Ye Z R, Zhang Y, Feng D L and Li S Y 2012 *Phys. Rev. X* **2** 11010
[62] Fisher I R, Degiorgi L and Shen Z X 2011 *Rep. Prof. Phys.* **74** 124506
[63] Fang C, Yao H, Tsai W F, Hu J P and Kivelson S A 2008 *Phys. Rev. B* **77** 224509
[64] Xu C, Mller M and Sachdev S 2008 *Phys. Rev. B* **78** 20501
[65] Eremin I and Chubukov A V 2010 *Phys. Rev. B* **81** 24511
[66] Fernandes R M, Vanbebber L H, Bhattacharya S, Chandra P, Keppens V, Mandrus D, McGuire M A, Sales B C, Sefat A S and Schmalian J 2010 *Phys. Rev. Lett.* **105** 157003

Journal of Physics: Conference Series **449** (2013) 012018 doi:10.1088/1742-6596/449/1/012018

Impurity bound-state as a probe of order-parameter symmetry in iron-pnictide superconductors: T-matrix approach

Masao Ogata and Toshikaze Kariyado

Department of Physics, University of Tokyo, Hongo, Bunkyo-ku, Tokyo 113-0033, Japan

E-mail: ogata@phys.s.u-tokyo.ac.jp

Abstract. The impurity induced in-gap bound state in multi-orbital superconductors is discussed based on Bogoliubov-de Gennes equation and T-matrix approximation. It is shown that multiple bound states appear depending on the strength of the impurity potential in the case of unconventional s-wave (or s_{+-}-wave) superconductivity. This result can be used to distinguish between the conventional s-wave and unconventional s-wave states. T-matrix approximation using a simplified model supports the numerical results obtained in the Bogoliubov-de Gennes equation. It is shown that the T-matrix can be block diagonalized when we use orbital representation. As a result, each orbital can make the impurity bound state independently, leading to the multiple structure of the bound states. From the analysis of the T-matrix approximation, we understand the reasons why the bound states always appear pair-wise at the positions of $\pm\omega$ and why there is a large asymmetry of the two peaks. We also show that the particle-hole asymmetry of the density of states gives rise to the asymmetry with respect to the strength of the impurity potential. The connection between the obtained results and the iron-pnictide superconductors is also provided.

1. Introduction

The discovery of iron-based superconductors [1] shed light on the high-temperature superconductivity in multi-orbital systems and the possibility of unconventional s-wave superconducting states [2]. Here, unconventional s-wave superconductor means that its gap function changes its sign but has s-wave symmetry. In multi-orbital superconductors, the gap functions are represented by matrices which represent orbital degrees of freedom. It is very interesting to develop a theory to characterize the unconventional s-wave superconductivity in multi-orbital superconductors.

In iron-pnictide superconductors, the order parameter is now controversial, i.e., simple s-wave (s_{++}-wave) or unconventional s_{+-}-wave superconductivity. Usually the latter is believed to be realized next to the magnetic phase [2]. Actually it is supported by the quasi-particle interference experiment [3]. However, the former s_{++}-wave state is claimed based on the fact that the superconductivity is robust against nonmagnetic impurities [4]. For this s_{++} state to be stabilized, the importance of lattice and orbital degrees of freedom has been discussed [5].

Having these in mind, it is necessary to have a theory to distinguish between the two candidates of superconductivity, *i.e.*, s_{+-}-wave or s_{++}-wave. In contrast to the high-T_c cuprate superconductivity with $d_{x^2-y^2}$-wave, the phase sensitive experiment will be difficult in iron-pnictide since the s_{+-}- and s_{++}-wave states have the same crystallographic symmetry. However, we noticed that the single impurity problem is suitable for distinguishing between the two states [6]. It is well known that there is no bound state around a nonmagnetic impurity in a simple s-wave state, while a bound state appears in unconventional superconductors, such as $d_{x^2-y^2}$-wave superconductivity due to the phase change of the order parameter along the quasi-particle trajectory. However, it was not known what happens in the case of s_{+-}-wave state.

In the previous paper [6], we studied a single-impurity problem by solving Bogoliubov-de Gennes equation in real-space with a nonmagnetic impurity using a realistic five-orbital model for iron-pnictides and discussed the impurity bound-states. We found that multiple bound states appear depending on the strength of the impurity potential. Our results can be used to distinguish between the s_{+-}-wave and s_{++}-wave state. Namely, if any bound state around an nonmagnetic impurity is observed, for example, in STS experiments, we can exclude the s_{++}-wave state which should not give any impurity bound state.

In the next section, we summarize the essence of the previous work by pointing out several important numerical results. In section 3, we develop T-matrix approach for the multi-orbital superconductors and study the impurity-induced in-gap bound state in section 4.

2. Impurity bound state obtained in Bogoliubov-de Gennes equations

In the previous paper [6], we studied numerically a single-impurity by solving Bogoliubov-de Gennes equation for five-orbital iron-pnictide model. The model and formalism are as follows.

(i) For the kinetic energy part, we use five band Hubbard model developed by Kuroki *et al* [7].
(ii) We construct a simple effective attractive interaction which reproduces the RPA results obtained in the multi-orbital Hubbard model with the above kinetic energy. For this purpose, we notice that using the orbital representation (not the band representation) and using the real-space picture are essential for making a simple effective interaction [8]. Actually, the multi-orbital gap functions obtained in RPA can be reproduced quite well by real-space pairings up to the next-nearest-neighbor sites when we use the orbital representation. Accordingly, our effective attractive interaction is up to the next-nearest-neighbor sites.
(iii) By using this simple effective interaction, we can study Bogoliubov-de Gennes equation that is basically a real-space mean-field approximation of the BCS theory. Note that the multi-orbital Hubbard model can not be used in the Bogoliubov-de Gennes equation, and it is essential to use the simplified model. We solve Bogoliubov-de Gennes equation in a unit cell with up-to 28×28 lattice sites with a single impurity at the center. We use a super-cell method which was used before for the vortex problem in the t-J model [9].
(iv) As an impurity potential we assume

$$\mathcal{H}_{\mathrm{imp}} = \sum_{a\sigma} I_a c^{\dagger}_{\boldsymbol{r}_0 a\sigma} c_{\boldsymbol{r}_0 a\sigma}, \tag{1}$$

where a represents the five orbitals and $\boldsymbol{r}_0$ is the impurity site. Generally, we can choose orbital-dependent strengths of the impurity potential, I_a. However, in the numerical simulations we assumed $I_a = I$ for simplicity. Even then, there appear many interesting phenomena as shown below.

The obtained results are summarized as follows [6].

(i) For the s_{+-}-wave state obtained in the RPA in multi-orbital Hubbard model, we found that the impurity-bound states near the Fermi energy become prominent when I = 1-2eV. They give relatively sharp peaks in local density of states around the impurity, which can be directly checked by STS experiments. We also found that bound-states appear near the gap edge when I = 0.5eV, I = 8eV or I is negative. These results indicate that the impurity bound-state is sensitive to the impurity strength, I. This situation is different from the d-wave case in which

Journal of Physics: Conference Series **449** (2013) 012018 doi:10.1088/1742-6596/449/1/012018

the bound state is a resonance state due to the finite density of states inside the gap and it is sometimes difficult to be observed.
(ii) We found multiple bound states for the cases with I = 1-2eV. When we checked the components of these bound states, we can see that the multiplicity comes from the orbital degrees of freedom. In other words, the positions of the bound states are orbital-dependent.
(iii) The bound states always appear pair-wise at the positions of $\pm\omega$, with ω being the energy measured from the Fermi energy. However there is a large asymmetry of these two peaks. Depending on the parameter I and the related orbital, the wights of the two peaks change.
(iv) The impurity bound state is localized around the impurity site. The spatial extension is about up-to the next-nearest-neighbor sites. Also the peak height oscillates as a function of the distance from the impurity.
(v) We checked that there appears no bound state for the s_{++}-wave case.

This kind of variety of bound-state spectra is a new aspect for the unconventional multi-orbital superconductivity. A recent experiment of STS by Hanaguri *et al* [10] supports this phenomenon. To repeat our claim, our results can be used to distinguish between the s_{+-}-wave and s_{++}-wave state. Namely, if any bound state around a nonmagnetic impurity is observed, we can exclude the s_{++}-wave state which does not give any impurity bound states.

3. T-matrix approximation in multi-orbital superconductors

In order to understand the impurity bound states discussed in the previous section, we develop a T-matrix approach for the multi-orbital superconductivity. In particular, we show the reason of the appearance of the orbital-dependent multiple-peaks in the local density of states as well as the characteristic I-dependence observed numerically in the Bogoliubov-de Gennes equation.

For the impurity potential, we use eq. (1) in which the impurity is treated as a simple potential shift, and assume that the potential strength I_a depends on the orbital index a. This on-site potential means that the impurity potential is short-ranged and affects only a single site. Although this assumption is not generally correct, it will capture most of the interesting phenomena. It is also assumed that there are no inter-orbital scatterings. This assumption is not general. However, inter-orbital components will vanish when the impurity scattering is isotropic and the related orbital wave functions are orthogonal to each other. Note that the orbital representation is essential for this argument. As we will see shortly, these assumptions give more benefits than their drawbacks.

Under these assumptions, we obtain Green's function within T-matrix approximation [11] as

$$\left[\mathcal{G}(\boldsymbol{k},\boldsymbol{k}',i\omega_n)_{\rm imp}\right] = \delta_{\boldsymbol{k},\boldsymbol{k}'}\left[\mathcal{G}(\boldsymbol{k},i\omega_n)\right] + \left[\mathcal{G}(\boldsymbol{k},i\omega_n)\right]\left[T(i\omega_n)\right]\left[\mathcal{G}(\boldsymbol{k}',i\omega_n)\right], \tag{2}$$

where $\left[\cdots\right]$ represents a $2L\times 2L$ matrix with L being the number of orbitals ($L=5$ for the model of iron-pnictide), and the Green's function $\left[\mathcal{G}(\boldsymbol{k},i\omega_n)\right]$ without impurity is expressed as

$$\left[\mathcal{G}(\boldsymbol{k},i\omega_n)\right] = \begin{pmatrix} G_{ab}(\boldsymbol{k},i\omega_n) & F_{ab}(\boldsymbol{k},i\omega_n) \\ \bar{F}_{ab}(\boldsymbol{k},i\omega_n) & -G_{ab}(-\boldsymbol{k},-i\omega_n)\end{pmatrix}. \tag{3}$$

The T-matrix is defined as

$$\left[T(i\omega_n)\right] = \left(\left[1\right] - \left[I\right]\left[\mathcal{G}^{\rm loc}(i\omega_n)\right]\right)^{-1}\left[I\right]. \tag{4}$$

Here, $\left[\mathcal{G}^{\rm loc}(i\omega_n)\right]$ represents a local Green's function

$$\left[\mathcal{G}^{\rm loc}(i\omega_n)\right] \equiv \frac{1}{N}\sum_{\boldsymbol{k}}\left[\mathcal{G}(\boldsymbol{k},i\omega_n)\right], \tag{5}$$

and

$$[I] = \begin{pmatrix} \tilde{I} & 0 \\ 0 & -\tilde{I} \end{pmatrix}, \quad \tilde{I} = \begin{pmatrix} I_1 & & & \\ & I_2 & & \\ & & \ddots & \\ & & & I_L \end{pmatrix}. \tag{6}$$

Although the form of T-matrix is complicated in general, there appears a large simplification when we use the orbital representation. First, we consider the symmetry of Green's function $G_{ab}(\boldsymbol{k}, i\omega_n)$ in multi-orbital superconductors. It is determined from the symmetries of the wave functions for a and b orbitals together with the symmetry of the lattice. Generally we find

$$\frac{1}{N}\sum_{\boldsymbol{k}} G_{ab}(\boldsymbol{k}, i\omega_n) = 0 \qquad \text{(for } a \neq b\text{)}, \tag{7}$$

when the basis wave functions are orthogonal to each other. This is because $\frac{1}{N}\sum_{\boldsymbol{k}} G_{ab}(\boldsymbol{k}, i\omega_n)$ represents a kind of on-site hopping between orbital a and b.

For the anomalous Green's function, $F_{ab}(\boldsymbol{k}, i\omega_n)$, its symmetry is determined as

$$(\text{symm. of } F_{ab}(\boldsymbol{k}, i\omega_n)) = (\text{symm. of } G_{ab}(\boldsymbol{k}, i\omega_n)) \times (\text{symm. of } \Delta_{bb}(\boldsymbol{k})), \tag{8}$$

when the symmetries of the diagonal elements of the gap functions, $\Delta_{bb}(\boldsymbol{k})$, are specified. This can be understood from the relation

$$F_{ab}(\boldsymbol{k}, i\omega_n) = G_{ab}(\boldsymbol{k}, i\omega_n)\Delta_{bb}(\boldsymbol{k})G_{bb}(\boldsymbol{k}, i\omega_n) + (\text{other terms}). \tag{9}$$

When $\Delta_{bb}(\boldsymbol{k})$ has s-wave-like symmetry, eq. (8) indicates that the symmetry of $F_{ab}(\boldsymbol{k}, i\omega_n)$ with respect to the direction of $\boldsymbol{k}$ is same as that of $G_{ab}(\boldsymbol{k}, i\omega_n)$. As a result, together with eq. (7), we have

$$\frac{1}{N}\sum_{\boldsymbol{k}} F_{ab}(\boldsymbol{k}, i\omega_n) = 0 \qquad \text{(for } a \neq b\text{)}, \tag{10}$$

again when the basis wave functions are orthogonal, e.g., in the 3d five orbital model. Actually we confirm this relation in the case of iron-pnictide superconductors.

From eqs. (7) and (10), we find that many components of $[\mathcal{G}^{\text{loc}}(i\omega_n)]$ vanish in multi-orbital superconductors, making the form of the T-matrix very simple. For the s_{+-} and s_{++}-wave cases, we have

$$[\mathcal{G}^{\text{loc}}(i\omega_n)] = \begin{pmatrix} G_{11}(i\omega_n) & 0 & \cdots & F_{11}(i\omega_n) & 0 & \cdots \\ 0 & G_{22}(i\omega_n) & \cdots & 0 & F_{22}(i\omega_n) & \cdots \\ 0 & 0 & \ddots & 0 & 0 & \ddots \\ \bar{F}_{11}(i\omega_n) & 0 & \cdots & -G_{11}(-i\omega_n) & 0 & \cdots \\ 0 & \bar{F}_{22}(i\omega_n) & \cdots & 0 & -G_{22}(-i\omega_n) & \cdots \\ 0 & 0 & \ddots & 0 & 0 & \ddots \end{pmatrix}. \tag{11}$$

After simple exchanges of rows and columns, we can see that the T-matrix is block diagonalized for each orbital, *i.e.*,

$$[T](i\omega_n) = \begin{pmatrix} \hat{T}_1(i\omega_n) & 0 & 0 \\ 0 & \hat{T}_2(i\omega_n) & 0 \\ 0 & 0 & \ddots \end{pmatrix}, \quad \hat{T}_a(i\omega_n) = (\hat{1} - \hat{I}_a\hat{\mathcal{G}}_a^{\text{loc}}(i\omega_n))^{-1}\hat{I}_a \tag{12}$$

where

$$\hat{\mathcal{G}}_a^{\mathrm{loc}}(i\omega_n) = \begin{pmatrix} G_{aa}(i\omega_n) & F_{aa}(i\omega_n) \\ \bar{F}_{aa}(i\omega_n) & -G_{aa}(-i\omega_n) \end{pmatrix}, \quad \hat{I}_a = \begin{pmatrix} I_a & 0 \\ 0 & -I_a \end{pmatrix}. \tag{13}$$

It is remarkable that a multi-orbital problem now becomes as simple as a single-orbital problem. This is due to the assumptions about the impurity potential and due to the symmetry of the Green's functions. Equation (12) also means that the poles of T-matrix, or the impurity bound states are determined independently for each orbital. This is the reason why the multiple bound states (orbital dependent) are obtained numerically [6] as explained in the previous section.

4. Bound State Formation in Some Simple Models

Next question is why we have bound states for the s_{+-}-wave in particular for I = 1-2eV.

Applying the analytic continuation $i\omega_n \to \omega + i\delta$, T-matrix for each orbital becomes

$$\begin{aligned} \hat{T}_a(\omega) &= \frac{1}{(1-I_a g_a(\omega))(1-I_a\bar{g}_a(-\omega)) + I_a^2\bar{f}_a(\omega)f_a(\omega)} \begin{pmatrix} 1-I_a\bar{g}_a(-\omega) & I_a f_a(\omega) \\ -I_a\bar{f}_a(\omega) & 1-I_a g_a(\omega) \end{pmatrix} \\ &= \frac{1}{D_a(\omega)} \begin{pmatrix} c_a-\bar{g}_a(-\omega) & f_a(\omega) \\ -\bar{f}_a(\omega) & c_a-g_a(\omega) \end{pmatrix} \end{aligned} \tag{14}$$

where

$$g_a(\omega) \equiv G_{aa}(\omega+i\delta), \quad \bar{g}_a(\omega) \equiv G_{aa}(-\omega-i\delta) \neq G_{aa}(-\omega+i\delta), \tag{15}$$
$$f_a(\omega) \equiv F_{aa}(\omega), \quad c_a \equiv 1/I_a, \tag{16}$$
$$D_a(\omega) \equiv (c_a-g_a(\omega))(c_a-\bar{g}_a(-\omega)) + \bar{f}_a(\omega)f_a(\omega). \tag{17}$$

Then, the positions of the poles are determined from zeros of $D_a(\omega)$, i.e., $D_a(\omega)=0$.

In the following, we suppress the orbital index a. $g(\omega)$ and $f(\omega)$ are represented as

$$g(\omega) \equiv \frac{1}{N}\sum_{\boldsymbol{k}} G(\boldsymbol{k},\omega) = \frac{1}{N}\sum_{\boldsymbol{k}} \frac{\omega+\xi_{\boldsymbol{k}}}{(\omega+i\delta)^2-\xi_{\boldsymbol{k}}^2-|\Delta_{\boldsymbol{k}}|^2}, \tag{18}$$

and

$$f(\omega) \equiv \frac{1}{N}\sum_{\boldsymbol{k}} F(\boldsymbol{k},\omega) = \frac{1}{N}\sum_{\boldsymbol{k}} \frac{\Delta_{\boldsymbol{k}}}{(\omega+i\delta)^2-\xi_{\boldsymbol{k}}^2-|\Delta_{\boldsymbol{k}}|^2}. \tag{19}$$

For the simple s-wave case with $\Delta_{\boldsymbol{k}} = \Delta$ constant, we have

$$g(\omega) = \omega S_1(\omega) + S_2(\omega), \tag{20a}$$
$$f(\omega) = \Delta S_1(\omega), \tag{20b}$$

with

$$S_1(\omega) = \int \mathrm{d}\xi\, N(\xi)\frac{1}{\omega^2-\xi^2-\Delta^2}, \tag{21a}$$
$$S_2(\omega) = \int \mathrm{d}\xi\, N(\xi)\frac{\xi}{\omega^2-\xi^2-\Delta^2}, \tag{21b}$$

where ω is assumed to be inside the gap ($|\omega| < \Delta$), and $N(\xi)$ is the density of states. Note that the relations $S_1(\omega) = S_1(-\omega), S_2(\omega) = S_2(-\omega)$ hold. [In contrast, for the $\mathrm{d}_{x^2-y^2}$-wave case, we have $f(\omega) = 0$.]

In the s_{+-}-wave case, on the other hand, the gap function $\Delta_{\boldsymbol{k}}$ has sign reversal between Fermi surfaces, *i.e.*, $\Delta_{\boldsymbol{k}}$ is positive at some parts on the Fermi surface and negative on other parts. This kind of gap function should lead to the reduction of $f(\omega)$ compared with the simple s-wave case owing to the cancellation of the positive part and negative part. Thus, we assume

$$f(\omega) = \alpha\Delta S_1(\omega), \qquad (0 \leq \alpha < 1), \tag{22}$$

for the s_{+-}-wave case [12]. Although this is a little rough approximation, we think that it captures the essential part of the s_{+-}-wave case.

With this approximation, the denominator of the T-matrix becomes

$$D(\omega) = (c - S_2(\omega))^2 - (\omega^2 - \alpha^2\Delta^2)S_1(\omega)^2, \tag{23}$$

with c being $c = 1/I$. In a simple estimation, we have

$$S_1(\omega) = -\frac{\pi N(0)}{\sqrt{\Delta^2 - \omega^2}}, \tag{24}$$

for $|\omega| < \Delta$ and $S_2(\omega)$ is nearly constant. In this case, we can see the following features of the positions of poles in the T-matrix. (i) We have a solution at ω_0 satisfying $\alpha\Delta < \omega_0 < \Delta$. (ii) Since eq. (23) is even with respect to ω, we have always solutions at $\pm\omega_0$. This means that the bound states always appear as a pair. (iii) The weights of the poles at $\pm\omega_0$ are different, leading to the asymmetry of the local density of states. These properties are observed numerically as shown in the previous section. Note that, in the d-wave case, only one bound state appears since the weight of one of the poles always vanishes.

Next we discuss the asymmetry with respect to the sign of I which was also observed numerically. For this purpose we consider a special case with $\alpha = 0$. In this case the two solutions are

$$\omega = \pm\frac{\Delta(c - S_2)}{\sqrt{\pi^2 N(0)^2 + (c - S_2)^2}}. \tag{25}$$

However, by checking the numerator of the T-matrix, we find that one of the bound states looses its weight because the numerator vanishes. As a result, we only have a solution at

$$\omega = -\frac{\Delta(c - S_2)}{\sqrt{\pi^2 N(0)^2 + (c - S_2)^2}}, \tag{26}$$

which satisfies a relation $\omega S_1(\omega) = c - S_2$. In this sense, the case with $\alpha = 0$ is rather special, although the following argument becomes simple.

The solution of the pole of T-matrix can be obtained diagrammatically as shown in Fig. 1. Here $\omega S_1(\omega) + S_2$ is plotted as a function of ω for the case with $S_2 > 0$. The pole of the T-matrix is given at the point where this function has a value of c as discussed above. Note that $S_2(\omega)$ vanishes when $N(\xi)$ is an even function of ξ. In other words, $S_2(\omega)$ comes from the asymmetry of the density of states near the Fermi energy. In the iron-pnictide cases, LDA calculations show that the density of states is larger when $\varepsilon < \varepsilon_{\mathrm{F}}$. In such cases, we have $S_2 > 0$ for $|\omega| < \Delta$, which corresponds to the situation in Fig. 1. We can see that, when I is positive (or when c is positive), we have a solution near $\omega = 0$ with a large amplitude. On the other hand, when I is negative (or when c is negative), the solution approaches to $\omega = \Delta$. In this case we can see that the amplitude of the pole is quite small. This is the reason why there is an asymmetry with respect to the impurity strength I.

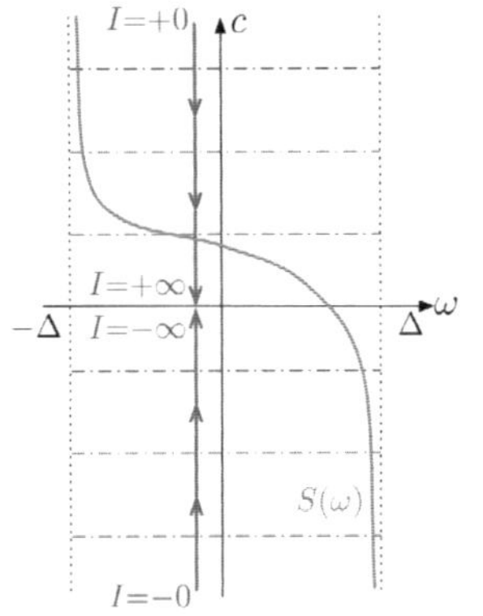

Figure 1. Schematic picture of the search for the solution of $D(\omega) = 0$.

5. Summary and Discussions

In this paper we show that the T-matrix can be block diagonalized in the multi-orbital superconductors when we use orbital representation. This is due to the symmetry of the normal and anomalous Green's functions. This result greatly simplifies the understanding of the bound state formation in multi-orbital superconductors. Since the T-matrix for each orbital can give poles independently, there is a chance to observe the multiple peak structure in the local density of states. In other words, the bound-state energy can take different values for each orbital.

The story becomes more interesting when there is remaining density of state inside the gap, i.e., the case that the gap function has nodes. Actually, the gap function with s-wave symmetry is not necessarily being nodeless, but it is possible to be nodal. In fact, nodal s-wave gap functions are predicted in some kind of iron-pnictide superconductors. In such cases, the bound states usually acquire finite life-time and becomes a virtual bound state. However, when the orbital providing the remaining density of states and that giving the bound state are different, the bound state remains to be sharp even if there is the intrinsic in-gap density of states.

The present analysis of the T-matrix approximation is consistent with the previous numerical study on the single impurity problem [6]. Numerical study shows that the multiple peak structure appears in a certain parameter range with $I > 0$ even if I_a have a common value of I. This is probably because the partial density of states for each orbital are different.

In the s_{+-}-wave case, the bound states can appear because $f(\omega)$ is reduced by a factor $\alpha < 1$, compared with the simple s-wave case. The case with $\alpha = 0$ is special in which only one bound state appears due to the cancellation of the denominator and the numerator of the T-matrix. On the other hand, for the general case with $0 < \alpha < 1$ there appear two bound states at $\omega = \pm\omega_0$ for each orbital. We have also shown that the strong asymmetry with respect to I can be well understood in the present formalism. This is due to the asymmetry of the density of states around the Fermi energy, *i.e.*, S_2-term. We expect that the effect of S_2, which is usually neglected for simplicity, plays an important role in iron-pnictide.

Acknowledgments

We would like to thank H. Hosono, T. Hanaguri, H. H. Wen, Y. K. Bang and A. V. Balatsky for stimulating discussions and comments. T.K. was supported by JSPS Research Fellowship.

References

[1] Kamihara Y, Watanabe T, Hirano M and Hosono H 2008 *J. Am. Chem. Soc.* **130** 3296
[2] Ishida K, Nakai Y and Hosono H 2009 *J. Phys. Soc. Japan* **78** 062001

[3] Hanaguri T, Niitaka S, Kuroki K and Takagi H 2010 *Science* **328** 474–476
[4] Sato M, Kobayashi Y, Lee S C, Takahashi H, Satomi E and Miura Y 2010 *J. Phys. Soc. Japan* **79** 014710
[5] Onari S and Kontani H 2009 *Phys. Rev. Lett.* **103**(17) 177001
[6] Kariyado T and Ogata M 2010 *J. Phys. Soc. Japan* **79** 083704
[7] Kuroki K, Usui H, Onari S, Arita R and Aoki H 2009 *Phys. Rev. B* **79**(22) 224511
[8] Kariyado T and Ogata M 2010 *J. Phys. Soc. Japan* **79** 033703
[9] Himeda A, Ogata M, Tanaka Y and Kashiwaya S 1997 *J. Phys. Soc. Japan* **66** 3367–3370
[10] Hanaguri T 2012 *private communication*
[11] Balatsky A V, Vekhter I and Zhu J X 2006 *Rev. Mod. Phys.* **78**(2) 373–433
[12] Preosti G and Muzikar P 1996 *Phys. Rev. B* **54**(5) 3489

Journal of Physics: Conference Series **449** (2013) 012019 doi:10.1088/1742-6596/449/1/012019

Direct observation of superconducting gaps in iron-based superconductors by laser ARPES

Kozo Okazaki

Institute for Solid State Physics (ISSP), University of Tokyo, Kashiwa, Chiba 277-8581, Japan

E-mail: okazaki@issp.u-tokyo.ac.jp

Abstract. We have developed a new apparatus for carrying out ultra-high energy resolution laser-excited angle-resolved photoemission spectroscopy (laser ARPES). The achieved energy resolution is $\sim$ 70 μeV at a sample temperature $\sim$ 1.5 K. As a test case, we have measured the superconducting gap of elemental metal Sn (T_c = 3.7 K). We have used this laser-ARPES apparatus, to directly observe the superconducting gap of various iron-based superconductors. The spectra of $FeTe_{0.6}Se_{0.4}$ below T_c show a very clear temperature dependent superconducting coherence peak. Preliminary momentum-dependent results indicate an anisotropic superconducting gap in $FeTe_{0.6}Se_{0.4}$.

1. Introduction

Since the discovery of the iron-based superconductors (FeSCs) [1], the pairing mechanism of their superconductivity has been extensively studied. One of the most intriguing aspects of FeSCs is their multi-band and multi-orbital nature and how they relate with various properties. Most of the FeSCs have characteristic electronic structures composed of hole-like Fermi surfaces (FSs) around Brillouin-zone (BZ) center and electron-like FSs at the BZ corner. The interband scattering between these hole-like and electron-like FSs is generally believed to be important for the pairing interaction based on the experimental observations that many FeSCs show superconductivity only upon carrier doping in the antiferromagnetic parent compounds [2]. Theoretically, spin-fluctuation mediated $s_{\pm}$ superconductivity, where the hole-like and electron-like FSs have opposite signs of the superconducting order parameter, have been proposed[3, 4]. However, there exist exceptional examples, such as: (i) Fe(Te,Se), whose parent compound $Fe_{1+y}Te$ is antiferromagnetic [5], but Fe(Te,Se) shows superconductivity upon isovalent substitution i.e. without carrier doping [6], (ii) KFe_2As_2, which is an extremely hole-doped compound of the series $(Ba,K)Fe_2As_2$ and has only hole FSs [7] but still shows superconductivity [8], and (iii) LiFeAs, which can be regarded as a parent compound, but also shows superconductivity without substitution [9]. To reveal the details of the superconducting pairing interactions, it should be crucial to investigate their superconducting (SC) order parameters or SC gaps including anisotropy in the momentum space and node structures.

Angle-resolved photoemission spectroscopy (ARPES) is the most powerful tool to investigate the magnitude of SC gaps. We can directly observe them utilizing this technique, although their phase cannot be detected. The direct observation of SC-gap anisotropy and nodes would provide us indispensable informations for understanding the superconducting pairing mechanism. Recently, we have developed a new laser-based ARPES (laser ARPES) apparatus, where a vacuum ultra violet (VUV) laser is adopted for an excitation light source. We have achieved an unprecedented energy resolution of 70 μeV at a low temperature of 1.5 K. Utilizing this laser-ARPES apparatus, we can directly observe the SC gaps of various FeSCs including above mentioned interesting materials, $FeTe_{0.6}Se_{0.4}$ (T_c = 14.5 K) [10], KFe_2As_2 (T_c = 3.4 K) [11], and LiFeAs (T_c = 18 K) [12].

2. Performance of new laser ARPES apparatus

Figure 1(a) shows a schematic description of our laser-ARPES apparatus. This apparatus mainly consists of a VG-Scienta HR8000 hemispherical electron analyzer, an upright helium-impounding cryostat, and a Nd:YVO_4 laser system with a Fabry-Perot etalon and a non-linear optical crystal $KBe_2BO_3F_2$ [13]. Because the hemisphere radius of the electron analyzer R = 200 mm and the smallest width of the installed slit is w = 50 μm, the highest energy resolution that can be achieved with this spectrometer, evaluated as $2R/w$, can be as high as 8000. Because the lowest pass energy of this spectrometer is 0.5 eV, we can expect the best energy resolution of this electron analyzer ΔE_a = 0.5 eV / 8000 = 62.5 μeV. The linewidth of the incident laser photons can be narrowed with the Fabry-Perot etalon down to ΔE_l = 25 μeV. Thus, we can expect the best total energy resolution of this laser-ARPES apparatus to

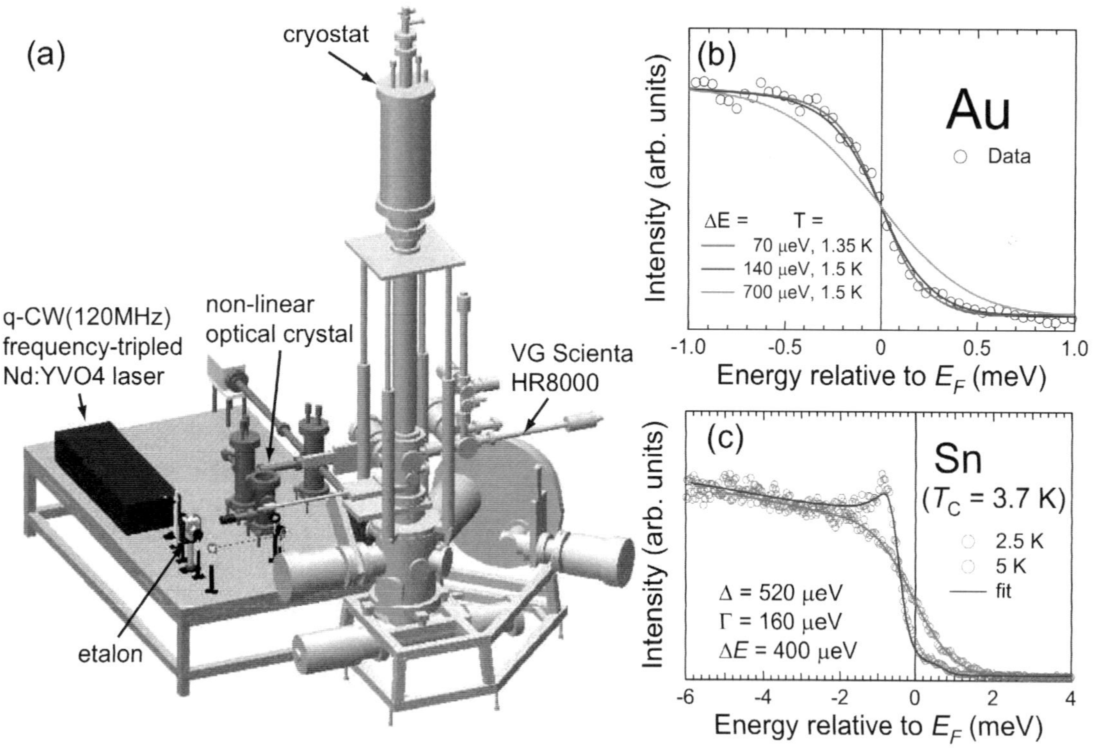

Figure 1. Performance of our recently developed laser-ARPES apparatus. (a) Schematic description of our laser-ARPES apparatus. (b) Photoemission spectrum of an evaporated gold film measured with the laser-ARPES apparatus, compared with the several fitting results. (c) Photoemission spectra of Sn (T_c = 3.7 K) below and above T_c. The superconducting coherence peak is clearly observed for the spectrum at 2.5 K.

be $\Delta E_{total} = \sqrt{\Delta E_a^2 + \Delta E_l^2} = \sqrt{62.5^2 + 25^2} \sim 67.5$ μeV. Figure 1(b) shows the spectrum of an evaporated gold film for the best performance of our new laser-ARPES apparatus. We have carried out a least-squares fitting of the gold spectrum to the Fermi-Dirac function convoluted with a Gaussian of FWHM ΔE to obtain the total energy resolution. If we fixed ΔE to be 70 μeV, the value of temperature T was obtained to be T = 1.35 K by the least-squares fitting. Accordingly, the quality of the fit in terms of the residual error was the best when we used ΔE = 70 μeV and T = 1.35 K, while it became worse for ΔE = 70 μeV and T = 1.5 K. For comparison, we show fits using ΔE = 70 μeV, T = 1.35 K, ΔE = 140 μeV, T = 1.5 K and ΔE = 700 μeV, T = 1.5 K. Figure 1(c) shows the spectrum of a fractured polycrystalline tin(Sn) sample (T_c = 3.7 K) measured at 2.5 K ($< T_c$) and 5 K ($> T_c$). The leading-edge shift below T_c clearly indicates the SC-gap opening and the superconducting coherence peak is also observed.

3. Direct observation of the superconducting gap in $FeTe_{0.6}Se_{0.4}$

Single crystals of $FeTe_{0.6}Se_{0.4}$ were prepared by a melt-growth technique. Chemical composition of the grown crystals was determined by electron probe microanalysis (EPMA) and inductively coupled plasma (ICP) atomic emission spectrometry. Details have been described in Ref. [10]. ARPES data were collected using the above-mentioned laser-ARPES apparatus. The overall energy resolution was set to $\sim$ 1.2 meV for the SC-gap measurements and the angular resolution was set to 0.1° to obtain a reasonable count rate for the angle-resolved measurements. The Fermi edge of an evaporated gold film was measured to calibrate the positions of Fermi level (E_F).

Figure 2(a) shows the FS map of $FeTe_{0.6}Se_{0.4}$ measured at 25 K with circularly polarized light. The image was made by integrating spectral intensity within $\pm$ 5 meV from E_F and symmetrized assuming the tetragonal crystal structure. The symbols indicate the positions of Fermi momenta (k_F) determined from the momentum distributions curves (MDCs) at E_F. The solid line corresponds to the shape of FS determined by fitting the positions of k_F to the model function having a fourfold symmetry. We have then measured a detailed temperature dependence of energy distribution curves (EDCs) at k_F with FS angle φ of 0°. The results are shown in Fig. 2(b). While the spectra above T_c have a cutoff of the Fermi-Dirac function at E_F, the spectrum at the lowest temperature (2.5 K) shows a very clear superconducting coherence

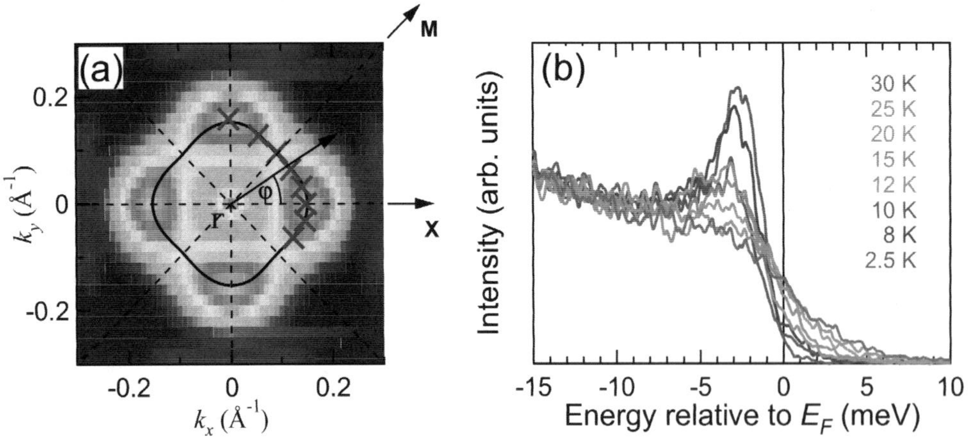

Figure 2. (a) FS intensity map of $FeTe_{0.6}Se_{0.4}$ measured at 25 K with circularly polarized light. k_F positions are indicated by the symbols. (b) Temperature-dependent EDCs at k_F with FS angle φ of 0°. The evolution of the superconducting gap with temperature is recognized very clearly.

Journal of Physics: Conference Series **449** (2013) 012019 doi:10.1088/1742-6596/449/1/012019

peak. The gap at the lowest temperature can be nicely fitted to a BCS spectral function and provides a gap value of $\Delta_0 \sim 2$ meV. Preliminary results of the momentum dependence of the gap indicate an anisotropic *s*-wave behavior for the Γ-centered hole FS. This result is in contrast to earlier ARPES studies [14, 15] but consistent with thermodynamic results [16]. The origin of the anisotropy has been theoretically discussed recently [17].

Acknowledgments

The development of the laser-ARPES apparatus has been done in collaboration with Y. Ota, Y. Kotani, T. Shimojima, T. Kiss, C.-T. Chen, S. Watanabe, A. Chainani and S. Shin. The work on $FeTe_{0.6}Se_{0.4}$ has been done in collaboration with Y. Ito, S. Niitaka, T. Hanaguri, and H. Takagi. This work is supported by JSPS through its FIRST Program.

References

[1] Kamihara Y, Watanabe T, Hirano M and Hosono H 2008 *J. Am. Chem. Soc* **130** 3296–3297
[2] de la Cruz C, Huang Q, Lynn J W, Li J, II W R, Zarestky J L, Mook H A, Chen G F, Luo J L, Wang N L and Dai P 2008 *Nature* **453** 899–902
[3] Mazin I I, Singh D J, Johannes M D and Du M H 2008 *Phys. Rev. Lett.* **101** 057003
[4] Kuroki K, Onari S, Arita R, Usui H, Tanaka Y, Kontani H and Aoki H 2008 *Phys. Rev. Lett.* **101** 087004
[5] Li S, de la Cruz C, Huang Q, Chen Y, Lynn J W, Hu J, Huang Y L, Hsu F C, Yeh K W, Wu M K and Dai P 2009 *Phys. Rev. B* **79** 054503
[6] Yeh K W, Huang T W, lin Huang Y, Chen T K, Hsu F C, Wu P M, Lee Y C, Chu Y Y, Chen C L, Luo J Y, Yan D C and Wu M K 2008 *Europhys. Lett.* **84** 37002
[7] Sato T, Nakayama K, Sekiba Y, Richard P, Xu Y M, Souma S, Takahashi T, Chen G F, Luo J L, Wang N L and Ding H 2009 *Phys. Rev. Lett.* **103** 047002
[8] Rotter M, Pangerl M, Tegel M and Johrendt D 2008 *Angewandte Chemie International Edition* **47** 7949–7952 ISSN 1521-3773
[9] Tapp J H, Tang Z, Lv B, Sasmal K, Lorenz B, Chu P C W and Guloy A M 2008 *Phys. Rev. B* **78** 060505
[10] Hanaguri T, Niitaka S, Kuroki K and Takagi H 2010 *Science* **328** 474–476
[11] Kihou K, Saito T, Ishida S, Nakajima M, Tomioka Y, Fukazawa H, Kohori Y, Ito T, Uchida S, Iyo A, Lee C H and Eisaki H 2010 *J. Phys. Soc. Jpn.* **79** 124713
[12] Imai Y, Takahashi H, Kitagawa K, Matsubayashi K, Nakai N, Nagai Y, Uwatoko Y, Machida M and Maeda A 2011 *Journal of the Physical Society of Japan* **80** 013704
[13] Chen C, Xu Z, Deng D, Zhang J, Wong G K L, Wu B, Ye N and Tang D 1996 *Appl. Phys. Lett.* **68** 2930–2932 ISSN 00036951
[14] Nakayama K, Sato T, Richard P, Kawahara T, Sekiba Y, Qian T, Chen G F, Luo J L, Wang N L, Ding H and Takahashi T 2010 *Phys. Rev. Lett.* **105** 197001
[15] Miao H, Richard P, Tanaka Y, Nakayama K, Qian T, Umezawa K, Sato T, Xu Y M, Shi Y B, Xu N, Wang X P, Zhang P, Yang H B, Xu Z J, Wen J S, Gu G D, Dai X, Hu J P, Takahashi T and Ding H 2012 *Phys. Rev. B* **85**(9) 094506
[16] Zeng B, Mu G, Luo H, Xiang T, Mazin I, Yang H, Shan L, Ren C, Dai P and Wen H H 2010 *Nat. Commun.* **1** 112
[17] Maiti S, Korshunov M M and Chubukov A V 2012 *Phys. Rev. B* **85**(1) 014511

Dome - like variation of the superconducting gap anisotropy in Fe-based superconductors

R Prozorov, K Cho, H Kim and M A Tanatar

The Ames Laboratory and Department of Physics & Astronomy, Iowa State University, USA

E-mail: prozorov@ameslab.gov

Abstract. Experiments performed on different iron - based superconductors suggest a variety of possible structures of the superconducting energy gap, both nodeless and nodal. To understand the pairing mechanisms, it is important to identify common features in the behavior of different materials. Measurements of the temperature - dependent London penetration depth provide important information on the structure of the superconducting gap. We show that despite significant differences between different iron - based superconductors, there is a universal trend: the gap is least anisotropic at the optimal doping and its anisotropy increases upon the departure towards underdoped and overdoped ends of the "superconducting dome". This trend is not related to the presence of the long - range magnetic order in the underdoped state.

The Fe-based superconductors (FeSCs) represent chemically and structurally diverse group of materials with significantly different transition temperatures, T_c, morphology and electronic properties [1–7]. Yet they are viewed as unique class of superconductors in which long - range magnetic order is in close proximity (and often coexists) with superconductivity. The superconducting phase forms a "dome" in the $T_c(x)$ transition temperature versus doping, x, phase diagram. The proximity of magnetism and superconductivity is considered to be instrumental for understanding the pairing mechanism in the FeSCs [1]. This might look similar to the high$-T_c$ cuprates, but with two significant distinctions. In the cuprates, robust nodal $d-$wave pairing is observed for all doping types and levels and there is a clear separation of the long - range magnetism from the domain of superconductivity. In Fe - based superconductors, the gap structure is strongly doping-dependent and varies between different sub-classes from isotropic to highly anisotropic and even nodal [4–11]. In view of this diversity, it is important to search for possible common trends in the structure of the superconducting gap in various compositions across the family of FeSCs.

Most of the FeSCs exhibit nearly isotropic full gaps at the optimal doping and the gap anisotropy increases towards the edges of the superconducting dome as the function of doping. This is observed in both electron-doped $Ba(Fe_{1-x}T_x)_2As_2$ ("BaT122", T = transition metal) [8–14] and hole-doped $(Ba_{1-x}K_x)Fe_2As_2$ ("BaK122") [15–18]. These materials also exhibit coexistence of superconductivity and magnetism in the underdoped regime, which might be responsible for increasing anisotropy at least on that side of the dope. On the other hand, the isovalently substituted LiFe(As,P) [19, 20] and layered $Ca_{10}(Pt_3As_8)[(Fe_{1-x}Pt_x)_2As_2]_5$ ("Ca10-3-8") compounds [21] show no coexistence, but obey this trend as well. The exception is the isovalently substituted $BaFe_2(As_{1-x}P_x)_2$ ("BaP122") [22, 23] that shows response consistent with the nodal gap even at the optimal doping. Such diversity is not well understood and

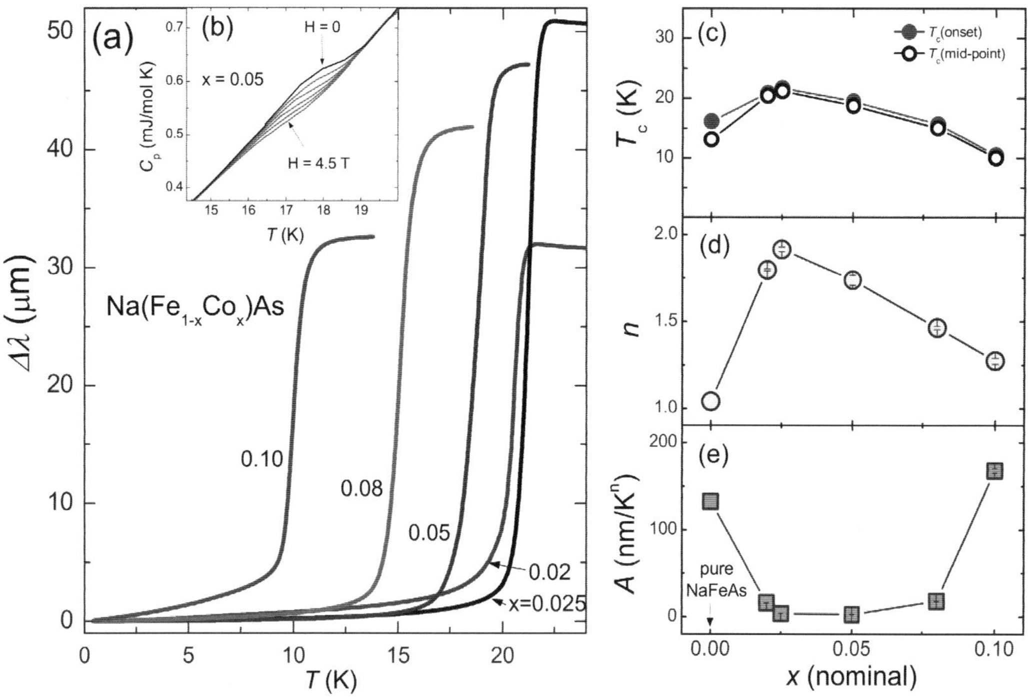

Figure 1. **(a)** Variation of London penetration depth, $\Delta\lambda(T)$ with temperature through the superconducting transition for single crystals of Na(Fe$_{1-x}$Co$_x$)As for x =0.02, 0.025, 0.05, 0.08 and 0.10. Inset **(b)** shows total specific heat measured at different values of applied magnetic field in a sample with x =0.05 demonstrating bulk nature of superconductivity. **(c)** Variation of the superconducting transition temperature, T_c, with Co - doping level, x, determined as the onset of a diamagnetic signal as well as the middle of the transition. **(d)** and **(e)** Parameters of the power - law fit, $\Delta\lambda(T) = A\,(T/T_c)^n$. The exponent n is shown on panel **(d)** and the pre-factor A is shown on panel **(e)**.

many theories were put forward, suggesting the importance of the doping-dependent electronic structure taking into account different orbital contents in each band, and the competition between intraband and interband interactions [4, 6, 24].

In this paper, we summarize the results of the penetration depth measurements on several FeSCs and show that the universal evolution from more to less isotropic gap structure upon the departure from the optimal doping persists in all compounds. To illustrate the analysis of the London penetration depth we use NaFe$_{1-x}$Co$_x$As as a model FeSC system spanning the whole superconducting dome. This material was studied by several probes that have produced conflicting results ranging from isotropic [25] to highly anisotropic, perhaps even nodal, gaps at the optimal doping [26] as well as the results compatible with both scenarios [27, 28].

The variation of the in-plane London penetration depth, $\Delta\lambda(T)$, was measured using a self-oscillating tunnel-diode resonator (TDR) [29–32]. The sample was placed with its c-axis parallel to an excitation field, $H_{ac} \sim 20$ mOe, much smaller than lower critical field, H_{c1}. The measured shift of the resonant frequency is converted to the change in London penetration depth, $\Delta f(T) = -G((\lambda/R)\tanh(R/\lambda) - 1)$, where G is the geometric factor and R is the effective sample dimension (for details of the calibration, see [7, 31, 33]).

Figure 1 shows summary of the measurements on single crystals of Na(Fe$_{1-x}$Co$_x$)As for

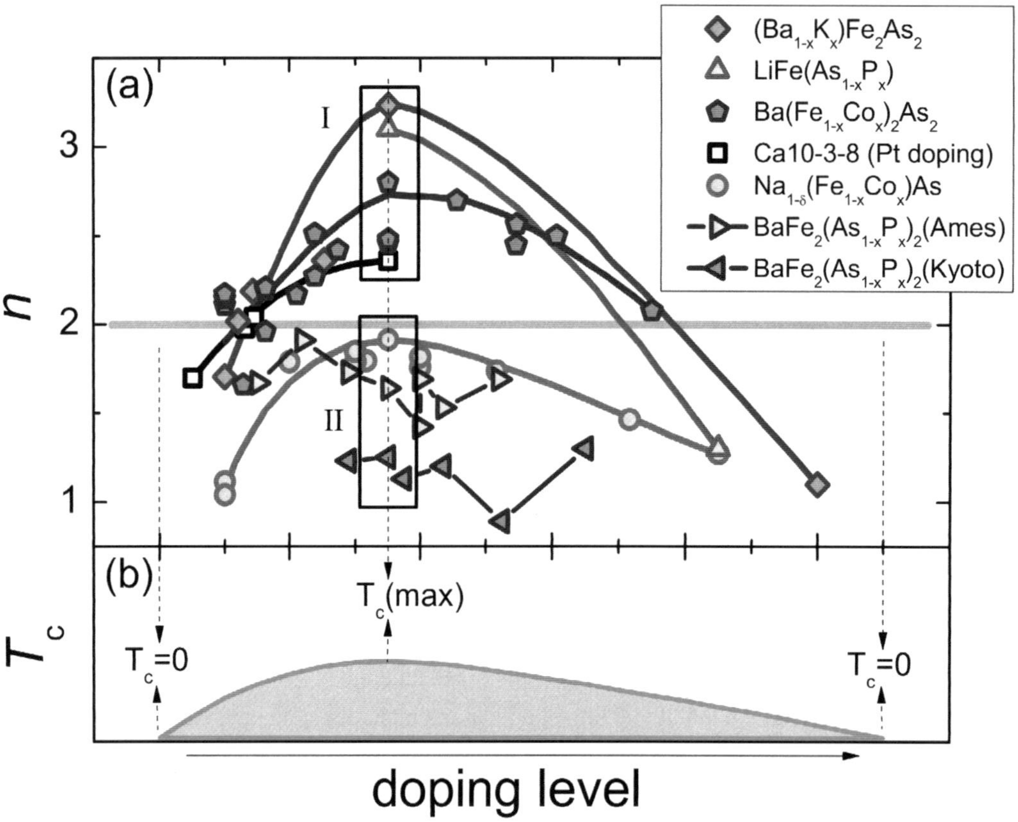

Figure 2. **(a)** Summary of the low - temperature exponent n for different families of iron - based superconductors as a function of doping. The maximum exponent is achieved at the optimal doping. For group **I** it is higher than 2 indicating nodeless gap. For group **II** the exponents lie below the n =2 line indicating significant anisotropy and, perhaps, line nodes in case of isovalently substituted $BaFe_2(As_{1-x}P_x)_2$. In all cases the anisotropy increases towards the edges of the superconducting dome, schematically shown in panel **(b)**.

x =0.02, 0.025, 0.05, 0.08 and 0.10. Full temperature sweeps are shown in panel (a) and the inset, panel (b), shows specific heat measured at several values of the applied magnetic field demonstrating the bulk nature of superconductivity in this material. Superconducting transition temperature, T_c, determined either as the onset of the diamagnetic signal (panel (c) - full symbols) or the mid - point of the transition (peak in the derivative, $d\Delta\lambda(T)/dT$, - panel (c) - open symbols) shows dome - like variation as the function of doping. Analysis of the low - temperature behavior of London penetration depth shows that, similar to most FeSCs, $\Delta\lambda(T)$ is best described by a power-law function, $\Delta\lambda(T) = A(T/T_c)^n$, with the exponent n (panel (d)) and pre-factor A (panel (e)) varying systematically with doping in a dome-like fashion. The values of n in the underdoped ($n \sim 1.1$) and heavily overdoped ($n \sim 1.3$) compositions are compatible with the superconducting gap with line nodes. Even at the optimal doping the exponent $n \sim 1.9$ is notably lower than in the other charge - doped FeSCs [7, 26].

Similar measurements conducted on a variety of Fe - based superconductors, electron - doped, hole - doped and isovalently - substituted, have shown that the trend that gap anisotropy increases towards the edges of the superconducting dome is universal. At the same time, the gap at the optimal doping is isotropic in most compounds, but can also be anisotropic as in described here $Na(Fe_{1-x}Co_x)As$ or even nodal as in $BaFe_2(As_{1-x}P_x)_2$ [22, 23]. These conclusions are represented graphically in Fig. 2. Panel (a) shows the variation of the power - law exponent, n, and panel (b) shows the pre - factor, A. The dome - like variation is correlated with the generic $T_c(x)$ shown in panel (c). While most charge - doped pnictides show full isotropic gap at the optimal doping (group I), charge - doped $Na(Fe_{1-x}Co_x)As$ is clearly more anisotropic and falls in group II together with isovalently - substituted $BaFe_2(As_{1-x}P_x)_2$).

Observation of the universal increase of the gap anisotropy towards the edges of the superconducting dome irrespective of whether the system shows coexistence of long - range magnetic order and superconductivity or not, makes it unlikely that long - range magnetism is responsible for the observed trend. Within the electronic structure of FeSCs, the long - range magnetic order can coexist with superconductivity only if the pairing is of the $s_\pm$ type [34, 35] and it is unlikely that different materials would have different pairing mechanisms. It is quite plausible that the pairing in iron pnictides is mediated by spin fluctuations that persist both in underdoped and overdoped regimes. In this case, the gap anisotropy has purely electronic origin and is determined by the competition between intra-band and inter-band interactions and depends sensitively on the orbital character of the bands [4, 35–45]. Unfortunately, experimental studies of the gap anisotropy in $s_\pm$ superconductors are hindered by the pair - breaking character of non - magnetic scattering that plays an important role in determining the low - temperature behavior of all thermodynamic quantities [7, 46–48]. This scattering is most likely doping - dependent, which necessitates further studies, for example, on samples with deliberately - introduced disorder.

Acknowledgments: We thank Hai-Hu Wen, Bing Shen, Pengcheng Dai, C. L. Zhang, R. J. Cava, Ni Ni, Y. Matsuda and S. Kasahara for samples used in this study. Work at the Ames Laboratory was supported by the U.S. Department of Energy, Office of Basic Energy Sciences, Division of Materials Sciences and Engineering under contract No. DE-AC02-07CH11358.

References

[1] Mazin I I 2010 *Nature* **464** 183–186
[2] Johnston D C 2010 *Adv. Phys.* **59** 803–1061
[3] Paglione J and Greene R L 2010 *Nat. Phys.* **6** 645–658
[4] Chubukov A 2012 *Ann. Rev. Cond. Matt. Phys.* **3** 57–92
[5] Stewart G R 2011 *Rev. Mod. Phys.* **83** 1589–1652
[6] Hirschfeld P J, Korshunov M M and Mazin I I 2011 *Rep. Prog. Phys.* **74** 124508
[7] Prozorov R and Kogan V G 2011 *Rep. Prog. Phys.* **74** 124505
[8] Gordon R T, Martin C, Kim H, Ni N, Tanatar M A, Schmalian J, Mazin I I, Bud'ko S L, Canfield P C and Prozorov R 2009 *Phys. Rev. B* **79** 100506
[9] Martin C, Kim H, Gordon R T, Ni N, Kogan V G, Bud'ko S L, Canfield P C, Tanatar M A and Prozorov R 2010 *Phys. Rev. B* **81** 060505
[10] Tanatar M A, Reid J P, Shakeripour H, Luo X G, Doiron-Leyraud N, Ni N, Bud'ko S L, Canfield P C, Prozorov R and Taillefer L 2010 *Phys. Rev. Lett.* **104** 067002
[11] Reid J P, Tanatar M A, Luo X G, Shakeripour H, Doiron-Leyraud N, Ni N, Bud'ko S L, Canfield P C, Prozorov R and Taillefer L 2010 *Phys. Rev. B* **82** 064501
[12] Ren C, Wang Z S, Wang Z Y, Luo H Q, Lu X Y, Sheng B, Li C H, Shan L, Yang H and Wen H H 2011 *arXiv:1106.2891*
[13] Bud'ko S L, Ni N and Canfield P C 2009 *Phys. Rev. B* **79** 220516
[14] Mu G, Tang J, Tanabe Y, Xu J, Heguri S and Tanigaki K 2011 *Phys. Rev. B* **84** 054505
[15] Kim H, Tanatar M A, Shen B, Wen H H and Prozorov R 2011 *arXiv:1105.2265*
[16] Reid J P, Tanatar M A, Luo X G, Shakeripour H, Rene de Cotret S, Doiron-Leyraud N, Chang J, Shen B, Wen H H, Kim H, Prozorov R and Taillefer L 2011 *arXiv:1105.2232*
[17] Dong J K, Zhou S Y, Guan T Y, Zhang H, Dai Y F, Qiu X, Wang X F, He Y, Chen X H and Li S Y 2010 *Phys. Rev. Lett.* **104** 087005
[18] Hashimoto K, Serafin A, Tonegawa S, Katsumata R, Okazaki R, Saito T, Fukazawa H, Kohori Y, Kihou K,

Lee C H, Iyo A, Eisaki H, Ikeda H, Matsuda Y, Carrington A and Shibauchi T 2010 *Phys. Rev. B* **82** 014526
[19] Kim H, Tanatar M A, Song Y J, Kwon Y S and Prozorov R 2011 *Phys. Rev. B* **83** 100502
[20] Hashimoto K, Kasahara S, Katsumata R, Mizukami Y, Yamashita M, Ikeda H, Terashima T, Carrington A, Matsuda Y and Shibauchi T 2012 *Phys. Rev. Lett.* **108** 047003
[21] Cho K, Tanatar M A, Kim H, Straszheim W E, Ni N, Cava R J and Prozorov R 2012 *Phys. Rev. B* **85** 020504
[22] Hashimoto K, Yamashita M, Kasahara S, Senshu Y, Nakata N, Tonegawa S, Ikada K, Serafin A, Carrington A, Terashima T, Ikeda H, Shibauchi T and Matsuda Y 2010 *Phys. Rev. B* **81** 220501
[23] Yamashita M, Nakata N, Senshu Y, Tonegawa S, Ikada K, Hashimoto K, Sugawara H, Shibauchi T and Matsuda Y 2009 *Phys. Rev. B* **80** 220509
[24] Thomale R, Platt C, Hanke W, Hu J and Bernevig B A 2011 *Phys. Rev. Lett.* **107** 117001
[25] Liu Z H, Richard P, Nakayama K, Chen G F, Dong S, He J B, Wang D M, Xia T L, Umezawa K, Kawahara T, Souma S, Sato T, Takahashi T, Qian T, Huang Y, Xu N, Shi Y, Ding H and Wang S C 2011 *Phys. Rev. B* **84** 064519
[26] Cho K, Tanatar M A, Spyrison N, Kim H, Song Y, Dai P, Zhang C L and Prozorov R 2012 *Phys. Rev. B* **86** 020508
[27] Yang H, Wang Z, Fang D, Kariyado T, Chen G, Ogata M, Das T, Balatsky A V and Wen H H 2012 *arXiv:1203.3123*
[28] Wang A F, Luo X G, Yan Y J, Ying J J, Xiang Z J, Ye G J, Cheng P, Li Z Y, Hu W J and Chen X H 2012 *Phys. Rev. B* **85** 224521
[29] Degrift C T V 1975 *Rev. Sci. Instrum.* **46** 599
[30] Carrington A, Bonalde I J, Prozorov R, Giannetta R W, Kini A M, Schlueter J, Wang H H, Geiser U and Williams J M 1999 *Phys. Rev. Lett.* **83** 4172–4175
[31] Prozorov R, Giannetta R W, Carrington A and Araujo-Moreira F M 2000 *Phys. Rev. B* **62** 115–118
[32] Prozorov R, Giannetta R W, Carrington A, Fournier P, Greene R L, Guptasarma P, Hinks D G and Banks A R 2000 *Appl. Phys. Lett.* **77** 4202–4204
[33] Prozorov R and Giannetta R W 2006 *Supercond. Sci. Technol.* **19** R41
[34] Fernandes R M, Pratt D K, Tian W, Zarestky J, Kreyssig A, Nandi S, Kim M G, Thaler A, Ni N, Canfield P C, McQueeney R J, Schmalian J and Goldman A I 2010 *Phys. Rev. B* **81**
[35] Fernandes R M and Schmalian J 2010 *Phys. Rev. B* **82**
[36] Chubukov A V, Efremov D V and Eremin I 2008 *Phys. Rev. B* **78** 134512
[37] Sknepnek R, Samolyuk G, bin Lee Y and rg Schmalian J 2009 *Phys. Rev. B* **79** 054511
[38] Mazin I and Schmalian J 2009 *Physica C: Superconductivity* **469** 614 – 627
[39] Mishra V, Boyd G, Graser S, Maier T, Hirschfeld P J and Scalapino D J 2009 *Phys. Rev. B* **79** 094512
[40] Chubukov A V, Vavilov M G and Vorontsov A B 2009 *Phys. Rev. B* **80** 140515
[41] Parker D, Vavilov M G, Chubukov A V and Mazin I I 2009 *Phys. Rev. B* **80** 100508
[42] Vorontsov A B, Vavilov M G and Chubukov A V 2009 *Phys. Rev. B* **79** 140507
[43] Daghofer M, Nicholson A, Moreo A and Dagotto E 2010 *Phys. Rev. B* **81** 014511
[44] Maiti S, Korshunov M M, Maier T A, Hirschfeld P J and Chubukov A V 2011 *Phys. Rev. B* **84** 224505
[45] Thomale R, Platt C, Hanke W and Bernevig B A 2011 *Phys. Rev. Lett.* **106** 187003
[46] Bang Y 2009 *Europhys. Lett.* **86** 47001
[47] Kogan V G 2009 *Phys. Rev. B* **80** 214532
[48] Onari S and Kontani H 2009 *Phys. Rev. Lett.* **103** 177001

Journal of Physics: Conference Series **449** (2013) 012021 doi:10.1088/1742-6596/449/1/012021

What have we learned from high-pressure experiments on Cu-oxide and Fe-based superconductors?

James Schilling, Narelle Hillier and Neda Foroozani
Department of Physics, Washington University, One Brookings Dr, St Louis, Missouri 63130, USA

E-mail: jss@wuphys.wustl.edu

Abstract. High-pressure experiments have made many important contributions to the field of superconductivity, including studies on the Cu-oxide and Fe-based high-temperature superconductors (HTS). However, for detailed quantitative understanding, the nature of the high-pressure environment acting on the sample must be well defined, a requirement often not met. Since HTS are quasi-2D, all properties are anisotropic. As a result, only a combination of fully hydrostatic and uniaxial pressure experiments are capable of providing the detailed information required. Unfortunately, very few such experiments have been carried out. For both Cu-oxide and Fe-based HTS at optimal doping, T_c is found to be enhanced by either *increasing* the separation of the superconducting planes or *decreasing* their area. Future experiments will extend such hydrostatic/uniaxial studies over the entire doping range.

1. Introduction

Since the discovery of the high-T_c cuprates in 1986 [1] and the Fe-based pnictides in 2008 [2], an enormous volume of experimental and theoretical work has been carried out. Unfortunately, the essential question – why these materials superconduct at such high temperatures in the first place – has yet to be resolved. Besides varying the crystal structure and stoichiometry of the materials, experimentalists have availed themselves of the thermodynamic variables temperature, magnetic field, and high pressure in an effort to further understanding.

In this paper we will attempt to illustrate a few examples of what high-pressure studies have, and have not, taught us about high-temperature superconductors (HTS). But first we would like to outline five different ways high-pressure experiments have advanced our understanding of the superconducting state in general.

(i) **Create new superconductors.** Many materials only superconduct under high pressure. Of the 92 naturally occurring elements in the periodic table, more than half (53) are known to superconduct as elemental solids – 30 at ambient pressure and 23 more under high pressure, as illustrated in Fig 1 below [3]. A number of distinct mechanisms may prompt a material to become superconducting under pressure, for example:

  (a) an insulator- or semiconductor-to-metal transition, as in O, Si or Ge;
  (b) a weakening or complete suppression of magnetism, as in Fe, Ce or Eu;

(c) a substantial change in the character of the conduction electrons, as in Ca, Sr, Ba, Sc, Y and the alkali metal Cs.

(ii) **Expose new phases, reveal new phenomena**. In 1999 Neaton and Ashcroft [4] pointed out that subjecting the highly compressible alkali metal Li to extreme pressures would bring neighboring atoms so close together that their core electrons begin to overlap, thus trapping the conduction electrons in interstitial sites. The result is a strong deviation in the conduction electrons from nearly-free-electron behavior and highly anomalous electronic properties, including possible superconductivity at relatively high temperatures, followed by a metal-insulator transition at higher pressure. Indeed, as seen in Fig 2, Li metal exhibits a superconducting transition temperature T_c which increases to a temperature as high as 14 K at 30 GPa [5]. This alone is a clear indication that Li's conduction electrons are no longer free-electron-like since, for all known simple-metal superconductors, T_c decreases under pressure [6]. At 67 GPa it is seen in Fig 2 that superconductivity abruptly disappears. Matsuoka *et al.* [7] demonstrated that this disappearance of superconductivity is the result of the fact that Li suffers a metal-to-semiconductor transition. A transition back to metallic behavior has been predicted at even higher pressures [8]. In our opinion these highly anomalous properties constitute some of the most interesting physics ever explored using the high pressure technique.

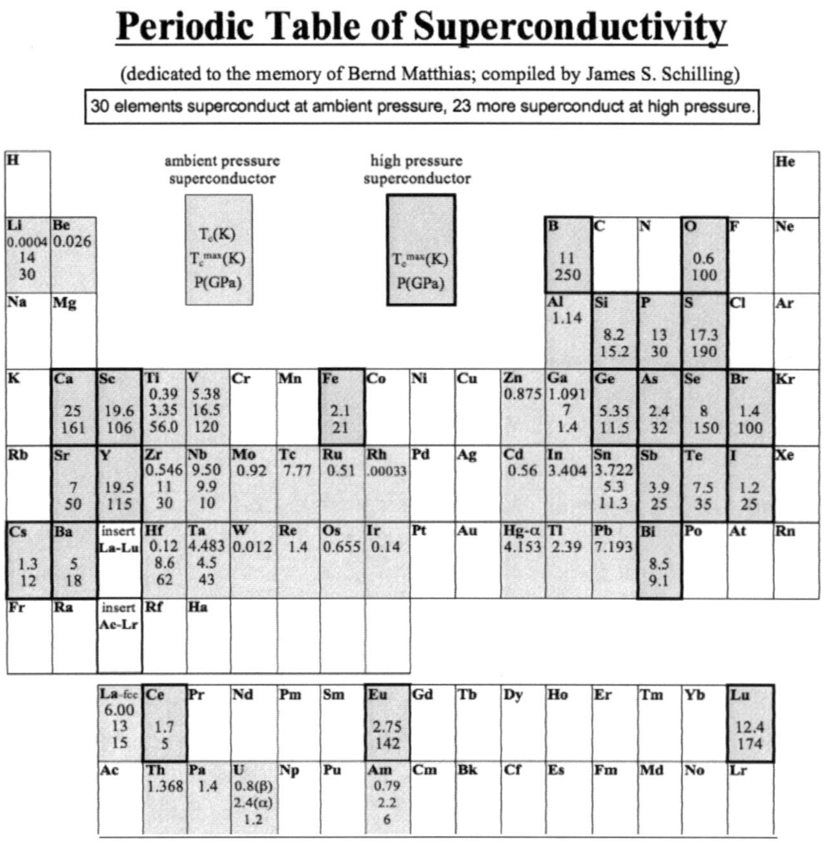

Figure 1. Periodic Table from Ref. [3] listing elements which superconduct at ambient pressure (yellow) or only under high pressure (light green). For each element the upper position gives the value of T_c(K) at ambient pressure; middle position gives maximum value T_c^{max}(K)in a high-pressure experiment at the pressure P(GPa) (lower position). If T_c decreases under pressure, only the ambient pressure value of T_c is given.

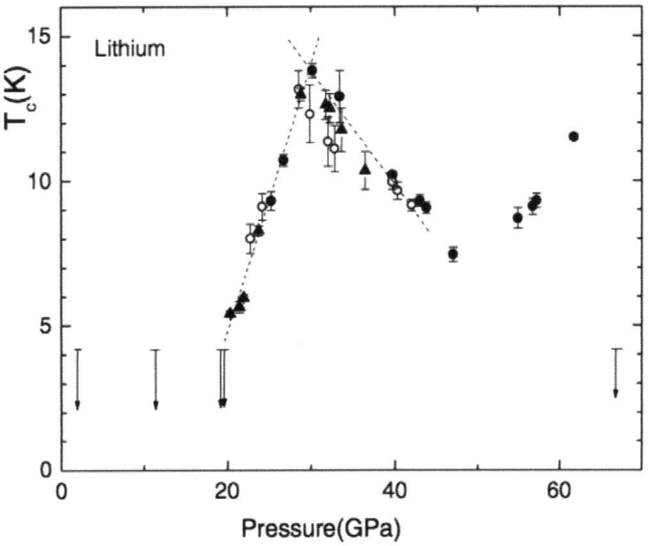

Figure 2. Dependence of T_c of Li metal on hydrostatic (He-gas) pressure from Ref [5].

With reference to HTS: (a) Sun *et al.* [9] have recently reported that in the iron chalcogenides a second superconducting phase suddendly reemerges near 12 GPa after T_c decreases in monotonic fashion from 32 K to 0 K at 10 GPa. In this reemergent phase T_c reaches values as high as 49 K. In addition, in resistivity studies on $Ba_{1-x}K_xFe_2As_2$ Hassinger *et al.* [10] find evidence for a new phase between the antiferromagnetic and the superconducting states, possibly a second spin-density-wave phase, which appears to compete with superconductivity. Further studies are required to identify the nature of this new phase.

(iii) **Synthesize at high-pressure/temperature.** As pointed out by Iimura *et al.* [11], the T_c versus x phase diagram for $Ln\mathrm{FeAsO}_{1-x}\mathrm{F}_x$, where Ln = *lanthanide,* is limited to $x \leq 0.2$ by the low solubility of F. Substituting instead H for O in this compound allows one to extend the $T_c(x)$ phase diagram to the much higher doping level $x \leq 0.5$. To synthesize these hydrogen-doped superconductors, simultaneous pressures/temperatures as high as 5 GPa/1200°C are necessary [11, 12]. The results of these studies are of considerable significance.

(iv) **Help push T_c to higher temperatures at ambient pressure.** Here we only outline three of the many examples.

(a) The T_c-value of the original HTS LaBaCuO, discovered by Bednorz and Müller [1] in 1986, was soon found to increase rapidly with pressure. Soon thereafter Wu *et al.* [13] availed themselves of chemical pressure by substituting the smaller cation Y for La, whereupon they discovered the 123-compound $YBa_2Cu_3O_7$, the first superconductor with a T_c value above liquid nitrogen temperatures.

(b) With the rapid increase in T_c with pressure for LaBaCuO and LaSrCuO in mind, Locquet *et al.* [14] grew a film of LaSrCuO epitaxially on a suitable substrate which compressed the CuO_2 basal plane, whereupon the separation between these planes increased. As a result the value of T_c nearly doubled to 49 K, a record high value for this cuprate.

(c) Early high-pressure studies on the Fe-based superconductor $LaFeAsO_{1-x}F_x$ by Takahashi *et al.* [15] showed that T_c initially increases rapidly with pressure. Ren *et al.* [16] increased the chemical pressure in this compound by substituting the smaller Sm cation for La and found that T_c nearly doubled to 55 K, the highest known value for any Fe-based superconductor.

(v) **Use high pressure as a quantitative tool to further understanding and test theory.** Compressing a superconducting solid through high pressure changes its lattice, electronic and magnetic properties, and thus also the interactions responsible for the attractive interaction binding the two electrons together in a Cooper pair. The universal decrease in T_c under pressure observed in simple-metal superconductors comes about because the electron-phonon interaction weakens as the crystal lattice stiffens under pressure, the change in electronic properties playing only a minor role [6]. In this way high-pressure studies serve to test the suitability of the BCS theory of superconductivity where the lattice vibrational modes form the basis for the pairing interaction. The isotope effect constitutes a more direct probe; however, in contrast to the isotope effect, the changes generated by high pressure are relatively large since they occur in the exponent of the BCS expression for T_c.
In the cuprate and Fe-based superconductors it appears likely that magnetic or electronic interactions lead to the pairing. In these quasi-2D materials a basic question is whether the superconducting state is strengthened or weakened as the planar area or the separation between planes decreases under pressure. In one theoretical model it is predicted that T_c should increase as the separation between the CuO_2 planes is reduced [17].

2. High Pressure as a Variable

In contrast to the 'clean' thermodynamic variables temperature T and magnetic field H, pressure is often relatively poorly defined. The reason for this is that some type of pressure medium is necessary when applying pressure to a solid. This pressure may be a gas, liquid, solid or perhaps even the sample itself. For cubic solids the ideal pressure medium is a gas or liquid which exerts hydrostatic pressure. A solid pressure medium, on the other hand, is unable to exert purely hydrostatic pressure but also subjects the sample to shear stresses which can cause nonuniform compression and even plastic deformation and extensive defect formation, i.e. the sample gets squashed. These shear stresses themselves may drastically change the way the sample's properties change compared to if hydrostatic pressure is applied. Helium is by far the most satisfactory hydrostatic pressure medium since it only becomes solid (freezes) above 27 bar at 0 K, 0.18 GPa at 20 K, 2.2 GPa at 100 K or 12 GPa at room temperature. All other fluids freeze at much lower pressures.

If the sample is noncubic, as for the cuprate and Fe-based superconductors, which are quasi-2D, then all properties, including the compressibility, are anisotropic. In this case hydrostatic pressure experiments alone are insufficient. Both hydrostatic and uniaxial pressures must be applied to single-crystalline samples to gain full, unequivocal, quantitative information on the response of the sample to changes in the respective lattice parameters. Unfortunately, uniaxial pressure measurements are normally restricted to pressures sufficiently weak as to not destroy the sample, i.e. approximately 0.1 GPa. Unequivocal hydrostatic/uniaxial information is, therefore, confined to a relatively narrow pressure range.

3. Cuprate Superconductors

The parent compounds for the high-T_c cuprates are antiferromagnetic insulators. As a function of either doping level or pressure, the cuprates becomes metallic as the antiferromagnetic phase is suppressed in favor of the superconducting phase, as represented in Fig 3; thus far the entire phase diagram has been traversed only through doping. In the superconducting phase $T_c(P)$ typically follows a bell-shaped curve. In the hole-doped cuprates, the carrier concentration in the CuO_2 planes normally increases with pressure [6]. Were this the only effect, one would then expect T_c to increase (decrease) with pressure for underdoped (overdoped) samples, but remain constant at optimal doping, as indicated by the respective black arrows in Fig 3.

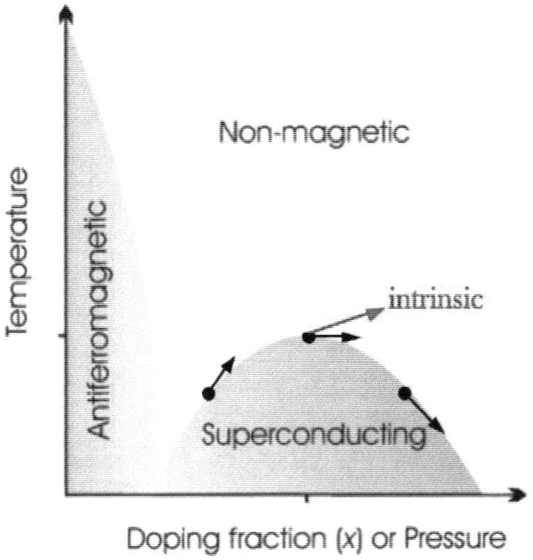

Figure 3. Generic phase diagram for cuprates and Fe-based pnictides, adapted from Ref [18].

This is not the behavior obtained in experiment, as seen in Fig 4(a), where the initial pressure dependence $T_c(P)$ is seen to be positive with an identical slope to 0.9 GPa for the one- ($HgBa_2CuO_4$ (Hg-1201)), two- ($HgBa_2CaCu_2O_6$ (Hg-1212)), and three-layer ($HgBa_2Ca_2Cu_3O_8$ (Hg-1223)) mercury cuprates at optimal doping [19]. The parallelism of $T_c(P)$ continues to pressures above 40 GPa, as illustrated in Fig 4(b) where T_c is seen to reach values near 160 K at 30 GPa for Hg-1223 [20]. From these data and those on other systems it is clear that for the cuprates the measured change in T_c under pressure is a function of both 'intrinsic' and doping dependent terms, i.e.

$$\left(\frac{dT_c}{dP}\right)_{measured} = \left(\frac{dT_c}{dP}\right)_{doping} + \left(\frac{dT_c}{dP}\right)_{intrinsic}, \quad (1)$$

as illustrated in Fig 3 by the black and red arrows. At optimal doping $(dT_c/dP)_{doping} = 0$. For a wide range of optimally doped cuprates with T_c values above 90 K it is found that $(dT_c/dP)_{measured} = (dT_c/dP)_{intrinsic} = +1$ to $+2$ K/GPa, consistent with the data in Fig 4.

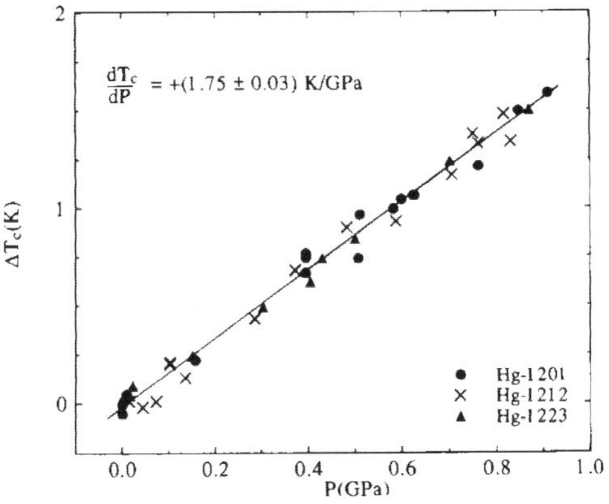

Figure 4(a). T_c versus pressure to 0.9 GPa [19].

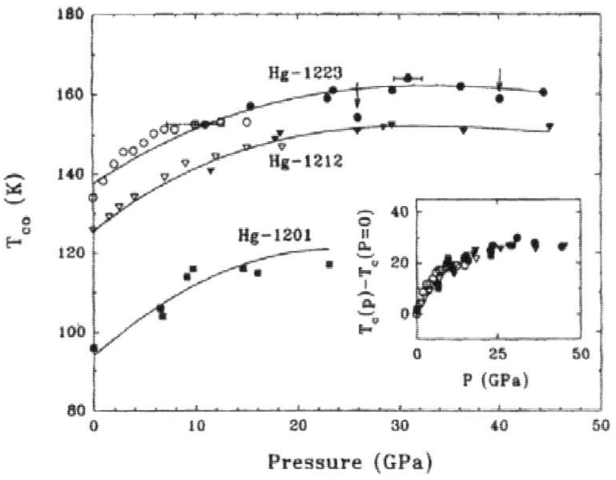

Figure 4(b). T_c versus pressure to 45 GPa [20].

The question remaining is: why does T_c here initially increase under pressure at the relatively modest rate of approximately 1-2 %/GPa? Since the cuprates are quasi-2D (planar) compounds, the first step in obtaining a quantitative answer to this question is to carry out parallel uniaxial pressure experiments on single crystalline samples. Hardy *et al.* [21] have determined the uniaxial pressure derivatives along the tetragonal a- and c-axes for the nearly optimally doped single-layer cuprate Hg-1201 and obtain $dT_c/dP_a = +2.3(2)$ K/GPa and $dT_c/dP_c = -3.6(3)$ K/GPa. Bringing the CuO_2 planes closer together through pressure along the c-axis is thus seen to rapidly decrease the value of T_c whereas it is increased by reducing the area of the CuO_2 plane. This contradicts the prediction of the theoretical model of Anderson referred to above [17].

A consistency check can be carried out by comparing the anticipated hydrostatic pressure derivative $dT_c/dP = 2dT_c/dP_a + dT_c/dP_c \simeq +1.0(5)$ K/GPa to the value obtained in a direct He-gas hydrostatic pressure experiment on the same single crystal, $dT_c/dP = +1.48(5)$ K/GPa, which is in reasonable agreement.

In Fig 5 these values of the partial pressure derivatives are compared to those for several other optimally doped cuprates. The main result remains, namely, that T_c increases both by increasing the separation between the CuO_2 planes and decreasing their area. In addition, the partial pressure derivatives are seen to be much larger than the hydrostatic derivative because they are opposite in sign and nearly cancel. It would be interesting to determine to what extent these results hold for the high-T_c cuprates in the over- and underdoped regions.

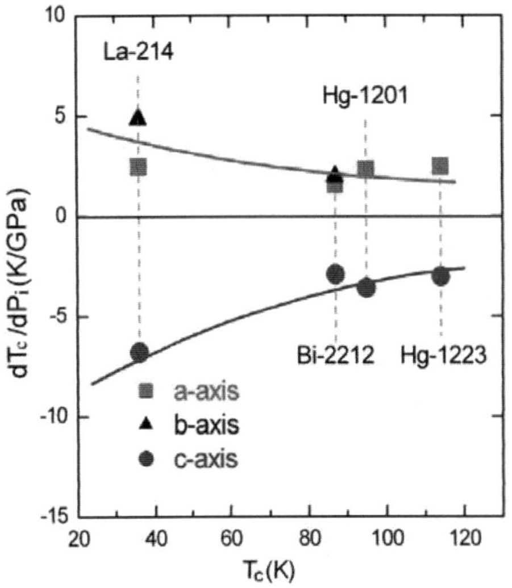

Figure 5. Partial pressure derivatives for several optimally doped cuprates from Ref [21].

Perhaps the first study which took advantage of the above results were by Locquet *et al.* [14] who grew an epitaxial film of $La_{1-x}Sr_xCuO_4$ (La-214) on a suitable substrate which compressed the CuO_2 planes and, by virtue of the Poisson effect, expanded the lattice in the c-direction, both effects serving to increase T_c. As a result, T_c was found to almost double in value from 25 K to 49 K in the epitaxially grown thin films!

4. Fe-Based Superconductors

Following the discovery in 2008 of high-temperature superconductivity in the Fe-based compound $LaFeAsO_{1-x}F_x$ [2], a veritable avalanche of research has been carried out, including many studies under high pressure conditions which have been extensively reviewed [22, 23, 24, 25, 26, 27], the most recent being the excellent review by Athena Sefat [18]. As indicated in Fig 3, the antiferromagnetic phase of the parent compound is suppressed through doping x or high pressure P, making way for a superconducting phase. T_c as a function of x or P exhibits a dome-like dependence, similar to that for the cuprates. Very recently Hosono's group has utilized temperatures to 1200°C and pressures to 5 GPa to synthesize the H-doped compounds LnFeAsO$_{1-x}$H$_x$ (Ln = La, Ce, Sm, Gd) [11]. The advantage is that a significantly higher doping level $x = 0.5$ can be reached with hydrogen, whereas only $x = 0.2$ is possible with fluorine. Remarkably, a double-dome $T_c(x)$ dependence is obtained to $x = 0.5$ for the La-compound, but only a single, broad dome appears for Ln = Ce, Sm, and Gd. High pressure has also successfully induced superconductivity in hydrogen substituted CaFeAsF$_{1-x}$H$_x$ [28].

The first high-pressure experiments on the Fe-doped superconductors were carried out by Takahashi *et al.* [15] on LaFeAsO$_{0.89}$F$_{0.11}$ who found that the ambient-pressure value $T_c \simeq 25$ K initially increases sharply with pressure, passing through a maximum of 43 K at 5 GPa before decreasing gradually to 8 K at 30 GPa.

As pointed out extensively in the reviews cited above for the Fe-based superconductors, however, the dependence of T_c on pressure differs widely from one experimenetal group to another, depending to a considerable extent on the pressure medium used and whether the sample is single- or polycrystalline. This situation is brought out clearly by high-pressure experiments on non-crystalline and single-crystalline LaFePO samples where $T_c \approx 6$ K. In piston-cylinder studies to 1.5 GPa by Igawa *et al.* [29] using Fluorinert pressure medium T_c is observed to rise sharply with pressure and pass through a maximum near 9 K at 0.8 GPa. Hamlin *et al.* [30] studied LaFePO in a diamond-anvil cell without pressure medium and found T_c to pass through a maximum of 10 K at 5 GPa. From these results one might conclude that a determination of $T_c(P)$ under purely hydrostatic pressure in a He-gas system would find that T_c rises very rapidly with pressure, perhaps passing through a maximum at a pressure below 0.8 GPa.

That this is not at all the case is seen for the He-gas measurements in Fig 6 where T_c is seen to *decrease rapidly* with pressure, irrespective of whether single- or polycrystalline LaFePO samples are studied, nor whether resistivity or magnetic susceptibility measurements are used to determine T_c; only when a polycrystalline sample was crushed in a diamond-anvil cell without pressure medium is a dome-shaped $T_c(P)$ dependence found for LaFePO with a T_c-maximum near 0.6 GPa [31].

The sensitivity of Fe-based HTS to strain also means that high-pressure studies should preferably be carried out only on samples in single-crystalline form rather than polycrystalline. The reason for this is that the compressibility of these materials is highly anisotropic; since the individual crystallites in a polycrystalline material are 'glued' together at the grain boundaries, applying even a completely hydrostatic pressure to a polycrystalline sample may result in a significant change in the strain field within the sample. This induced strain field may depend on the temperature at which the pressure is changed since the sample's elastic properties are temperature dependent. This may be the reason that Bi *et al.* [32] observed significant pressure/temperature history effects in He-gas $T_c(P)$ studies on polycrystalline LaO$_{1-x}$F$_x$FeAs.

The unusually strong dependence of the pressure dependence $T_c(P)$ on the pressure medium used is also exhibited by other Fe-doped systems. In an experiment using Daphne Oil 7373 or Fluorinert FC-75 pressure medium, Alireza *et al.* [33] find that superconductivity is induced under pressure in $CaFe_2As_2$ near 0.3 GPa pressure and in $SrFe_2As_2$ and $BaFe_2As_2$ near 3 GPa, all single crystals. However, in a later experiment on a $CaFe_2As_2$ single crystal using a He-gas pressure system, Yu *et al.* [34] can find no superconductivity to 0.7 GPa!

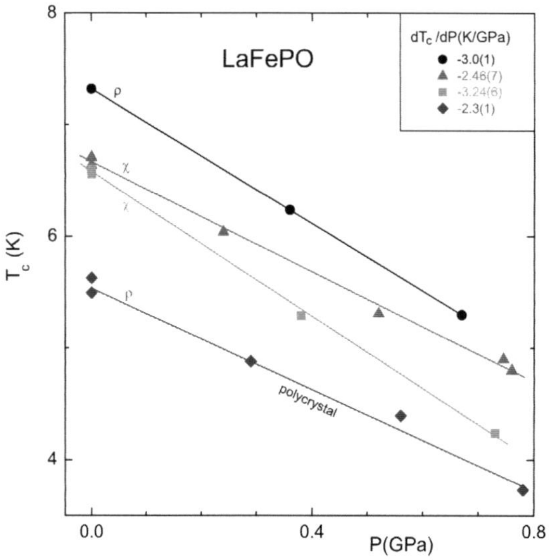

Figure 6. Initial dependence of T_c on pressure in He-gas system for three single-crystals and one polycrystal of LaFePO from both resistivity ρ and ac susceptibility χ measurements.

Comparing three different pressure media, two fluid at room temperature (Daphne 7474, Daphne 7373) and one solid (Stycast 1266), Kotegawa *et al.* [35] find that for $SrFe_2As_2$ crystals that the less hydrostatic the pressure medium, the lower the pressure needed to suppress the antiferromagnetism and the associated tetragonal-to-orthorhombic phase transition, thus allowing the superconducting state to appear.

Kirshenbaum *et al.* [36] find that some $SrFe_2As_2$ single crystals show evidence for superconductivity near 21 K at ambient pressure if they are measured before a strain-relieving anneal is carried out; they find the superconductivity disappears after annealing. Cleaving or quenching the crystal in liquid nitrogen is not able to resurrect the superconductivity, but severe deformation of the crystal does reinstate superconductivity near 21 K.

Saha *et al.* [37] suggest that there exists "an intimate relationship among superconductivity, magnetism, and crystallographic strain in this system of materials" and Duncan *et al.* [38] state that "an increasing uniaxial pressure component in $BaFe_2As_2$ quickly reduces spin-density-wave order and favors the appearance of superconductivity, similar to what is seen in $SrFe_2As_2$."

It thus seems clear for these Fe-based systems that the more highly strained the sample is, the more likely it is that the sample is superconducting! In fact, the strains and disorder associated with doping into the Fe_2As_2 planes, for example with Co, may make superconductivity more likely, in contrast to the situation with the high-T_c cuprates where in-plane doping quickly quenches the superconductivity.

To get a quantitative handle on the mechanisms behind the extreme sensitivity of the properties of the Fe-based superconductors to strain, uniaxial pressure experiments are essential. Hardy *et al.* [39] find the uniaxial pressure derivatives for a nearly optimally doped $Ba(Fe_{0.92}Co_{0.08})_2As_2$ crystal to be $dT_c/dP_a = +3.1$ K/GPa and $dT_c/dP_c = -7.0$ K/GPa using an ultrahigh resolution dilatometer. A direct uniaxial pressure measurement along the c-axis of a crystal with the same composition yields the surprisingly large negative value $dT_c/dP_c = -15$ K/GPa [40]. Note that the sign of the uniaxial pressure derivatives is the same as for the optimally doped cuprates.

For a $Ba(Fe_{0.926}Co_{0.074})_2As_2$ single crystal, Bud'ko *et al.* [41] find the derivatives $dT_c/dP_a = +3$ K/GPa and $dT_c/dP_c = -26$ K/GPa, confirming the sign of the uniaxial derivatives reported by Hardy *et al.* [39]. However, for $Ba(Fe_{0.962}Co_{0.038})_2As_2$, Bud'ko *et al.* find $dT_c/dP_a = -41$

K/GPa and $dT_c/dP_c = +17$ K/GPa, revealing that the uniaxial pressure derivatives of T_c both change sign at the lower doping level [41].

Recent extensive dilatometer experiments by the Meingast group [42] over the Co concentration range from 4% to 12% confirm this sign change, finding, in fact, that both derivatives dT_c/dP_a and dT_c/dP_c abruptly change sign at the 6% Co doping level; interestingly, this is the doping level above which the orthorhombic-to-tetragonal phase transition is fully suppressed.

5. Conclusions

With regard to the Cu-oxide and Fe-based superconductors, the high-pressure technique has proven to be a valuable resource to: (a) synthesize new materials, (b) induce normal-to-superconductor transitions, (c) expose new phases and reveal new phenomena, and (d) help develop strategies to push T_c to higher temperatures at ambient pressure. However, the high-pressure variable has not yet lived up to its potential to make substantial contributions as a quantitative tool to further our understanding and test theoretical models. The difficulty is that 'not all pressures are created equal' since the nature of high-pressure to which the sample is subjected (degree of hydrostaticity, size and direction of shear stresses, etc) depends on many factors, particularly the pressure medium used and the temperature at which the pressure is changed. This uncertainty is particularly worrisome for the Fe-based materials since they appear to be exquisitely sensitive to shear stresses which are, surprisingly, found to favor superconductivity.

To make substantial contributions to our quantitative understanding, high-pressure experiments on both the Cu-oxide and Fe-based superconductors should ideally be carried out under both fully hydrostatic and uniaxial pressure conditions, or perhaps both combined. Helium is the hydrostatic pressure medium of choice since it does not solidify at room temperature until pressures exceeding 12 GPa have been reached. Even as a solid helium is relatively soft, able to exert only minor shear stresses on the sample embedded in it. However, due to the quasi-2D nature of both the cuprate and Fe-based HTS, the changes in the normal and superconducting state properties under hydrostatic pressure do not provide sufficient information for quantitative interpretation. In fact, since the uniaxial pressure dependences are large and opposite in sign, the dependence of T_c on hydrostatic pressure alone has no direct physical significance! Uniaxial pressure experiments are thus absolutely essential. Only through combined hydrostatic/uniaxial pressure experiments can the most basic questions be answered, such as whether superconductivity is weakened or strengthened as the relevant quasi-2D planes are compressed or brought closer together. Such information would allow definitive tests of theoretical models. Ultimately, such quantitative experiments on the Cu-oxide and Fe-based superconductors have the potential to reveal the nature of the pairing mechanism, much as the universal suppression of T_c in simple-metal superconductors along with measurements of the lattice Grüneisen parameter clearly point to the importance of the electron-phonon interaction.

Clearly, carefully controlled hydrostatic AND uniaxial pressure experiments on high quality, high-T_c single crystals over a wide doping range are necessary to further our quantitative understanding. The uniaxial pressure experiments carried out thus far have concentrated on optimally doped samples and have, as we have discussed above, revealed that the superconducting state is strengthened (T_c increases) in one of two ways: (1) by reducing the area of the superconducting planes, or (2) by increasing the separation between these planes. Whether or not this result also holds for under- or overdoped samples is the subject of future experimentation.

Acknowledgments. The authors are grateful to Jinhyuk Lim for technical assistance. This work is supported by the National Science Foundation through grant DMR-1104742.

References

[1] Bednorz J G and Müller K A 1986 Z. Physik B **64** 189
[2] Kamihara Y, Watanabe T, Hirano M and Hosono H 2008 J. Am. Chem. Soc. **130** 3296
[3] Debessai M, Matsuoka T, Hamlin J J, Bi W, Meng Y, Shimizu K and Schilling J S 2010 J. Phys.: Conf. Ser. **215** 012034
[4] Neaton J B and Ashcroft N W 1999 Nature **400** 141
[5] Deemyad S and Schilling J S 2003 Phys. Rev. Lett. **91** 167001
[6] See, for example, Schilling J S 2007 *Handbook of High-Temperature Superconductivity: Theory and Experiment* ch 11, ed J R Schrieffer, assoc ed J S Brooks (Springer, New York) p 427
[7] Matsuoka T and Shimizu K 2009 Nature Physics **7** 211
[8] Marqués M, McMahon M I, Gregoryanz E, Hanfland M, Guillaume C L, Pickard C J, Ackland G J and Nelmes R J 2011 Phys. Rev. Lett. **106** 095502
[9] Sun L, Chen X-J, Guo J, Gao P, Huang Q-Z, Wang H, Fang M, Chen X, Chen G, Wu Q, Zhang C, Gu D, Dong X, Wang L, Yang L. Yang K, Li A. Dai X., Mao H-k and Zhao Z 2012 Nature **483** 67
[10] Hassinger E, Valade F, René de Cotret S, Juneau-Fecteau A, Reid J-Ph, Kim H, Tanatar M A, Prozorov R, Shen B, Wen H-H, Doiron-Leyraud N and Taillefer L 2012 New phase Induced by Pressure in the Iron-Arsenide Superconductor $Ba_{1-x}K_xFe_2As_2$ *Preprint* arXiv:1205.6730v1
[11] Iimura S, Matuishi S, Sato H, Hanna T, Muraba Y, Kim S W, Kim J E, Takata M and Hosono H 2012 Two-Dome Structure in Electron-Doped Iron Arsenide Superconductors *Preprint* arXiv:1207.0583
[12] Hanna T, Muraba Y, Matsuishi S, Igawa N, Kodama K, Shamoto S-i and Hosono H 2011 Phys. Rev. B **84** 024521
[13] Wu M K, Ashburgn J R, Torng C J, Hor P H, Meng R L, Gao L, Huang Z J, Wang Y Q and Chu C W 1987 Phys. Rev. Lett. **58** 908
[14] Locquet J-P, Perret J, Fompeyrine J, Mächler E, Seo J W and Van Tendeloo G 1998 Nature **394** 453
[15] Takahashi H, Igawa K, Arii K, Kamihara Y, Hirano M and Hosono H 2008 Nature **453** 376
[16] Ren Z A, Lu W, Yang J, Yi W, Shen X-L, Li Z-C, Che G-C, Dong X-L, Sun L-L, Zhou F and Zhao Z-X 2008 Chin. Phys. Lett. **25** 2215
[17] Anderson P W 1998 Science **279** 2296
[18] Sefat A S 2011 Rep. Prog. Phys. **74** 124502
[19] Klehe A-K, Gangopadhyay A K, Diederichs J and Schilling J S 1993 Physica C **213** 266; Klehe A-K, Schilling J S, Wagner J L and Hinks D G 1994 Physica C **223** 313
[20] Gao L, Xue Y Y, Chen F, Xiong Q, Meng R L, Ramirez D, Chu C W, Eggert J H and Mao H-K 1994 Phys. Rev. B **50** 4260
[21] Hardy F, Hillier N J, Meingast C, Colson D, Li Y, Barišić N, Yu G, Zhao X, Greven M and Schilling J S 2010 Phys. Rev. Lett. **105** 167002
[22] Aswathy P M, Anooja J B, Sarun P M and Syamaprasad U 2010 Supercond. Sci. Technol. **23** 073001
[23] Paglione J and Greene R L 2010 Nature Phys. **6** 645
[24] Chu C W and Lorenz B 2009 Physica C **469** 385
[25] Ren Z-A and Zhao Z-X 2009 Adv. Mater. **21** 4384
[26] Canfield P C, Bud'ko S L, Ni N, Kreyssig A, Goldman A I, McQueeney R J, Torikachvili M S, Argyriou D N, Luke G and Yu W 2009 Physica C **469** 404
[27] Takahashi H, Okada H, Igawa K, Kamihara Y, Hirano M and Hosono H 2009 Physica C **469** 413
[28] Okuma K, Ebata M, Tomita T, Takahashi H, Hanna T, Muraba Y, Matsuishi S and Hosono H 2011 High-Pressure Studies for Hydrogen Substituted $CaFeAsF_{1-x}H_x$ *Preprint Proceedings of 26th International Conference on Low Temperature Physics*, Aug 10-17, 2011, Beijing, China (to appear in J. Phys. Conf. Series)
[29] Igawa K, Okada H, Ani K, Takahashi H, Kamihara Y, Hirano M, Hosono H, Nakano S and Kikegawa T 2009 J. Phys. Soc. Japan **78** 023701
[30] Hamlin J J, Baumbach R E, Zocco D A, Sayles T A and Maple M B 2008 J. Phys.: Condens. Matter **20** 365220
[31] Hillier N J, Foroozani N, Zocco D A, Hamlin J J, Baumbach R E, Lum I K and Maple M B (unpublished)
[32] Bi W, Banks H B, Schilling J S, Takahashi H, Okada H, Kamihara Y, Hirano M and Hosono H 2010 New J. Phys. **12** 023005
[33] Alireza P L, Ko Y T C, Gillett J, Petrone C M, Cole J M, Lonzarich G G and Sebastian S E 2009 J. Phys.: Condens. Matter **21** 012208
[34] Yu W, Aczel A A, Williams T J, Bud'ko S L, Ni N, Canfield P C and Luke G M 2009 Phys. Rev. B **79** 020511(R)
[35] Kotegawa H, Kawazoe T, Sugawara H, Murata K and Tou H 2009 J. Phys. Soc. Japan **78** 083702
[36] Kirshenbaum K C, Saha S R, Butch N P, Magill J D and Paglione J 2009 Superconductivity in the Iron-

Pnictide Parent Compound $SrFe_2As_2$ *Preprint* arXiv: 0907.4141v1
[37] Saha S R, Butch N P, Kirshenbaum K, Paglione J and Zavalij P Y 2009 Phys. Rev. Lett. **103** 037005
[38] Duncan W J, Welzel O P, Harrison C, Wang X F, Chen X H, Grosche F M and Niklowitz P G 2010 J. Phys.: Condens. Matter **22** 052201
[39] Hardy F, Adelmann P, Wolf T, v. Löhneysen H and Meingast C 2009 Phys. Rev. Lett. **102** 187004
[40] Nakashima Y, Yui H and Sasagawa T 2010 Physica C **470** 1063
[41] Bud'ko S L, Ni N, Nandi S, Schmiedeshoff G M and Canfield P C 2009 Phys. Rev. B **79** 054525
[42] Meingast C (private communication)

Quantum oscillations in iron-based superconductors: $BaFe_2As_2$ vs. KFe_2As_2

Taichi Terashima[1,2], Nobuyuki Kurita[2,3], Motoi Kimata[2,4], Megumi Tomita[1], Satoshi Tsuchiya[1], Hidetaka Satsukawa[1], Atsushi Harada[1], Kaori Hazama[1], Motoharu Imai[1,2], Akira Sato[1], Shinya Uji[1,2], Kunihiro Kihou[2,5], Chul-Ho Lee[2,5], Hijiri Kito[2,5], Yasuhide Tomioka[2,5], Toshimitsu Ito[2,5], Akira Iyo[2,5], Hiroshi Eisaki[2,5], Tian Liang[2,6], Masamichi Nakajima[2,6], Shigeyuki Ishida[2,6], Shin-ichi Uchida[2,6], Taku Saito[7], Hideto Fukazawa[2,7], Yoh Kohori[2,7] and Hisatomo Harima[2,8]

[1]National Institute for Materials Science, Tsukuba, Ibaraki 305-0003, Japan
[2]JST, Transformative Research-Project on Iron Pnictides (TRIP), Chiyoda, Tokyo 102-0075, Japan
[3]Department of Physics, Tokyo Institute of Technology, Meguro, Tokyo 152-8551, Japan
[4]Institute for Solid State Physics, University of Tokyo, Kashiwa, Chiba 277-8581, Japan
[5]National Institute of Advanced Industrial Science and Technology (AIST), Tsukuba, Ibaraki 305-8568, Japan
[6]Department of Physics, University of Tokyo, Bunkyo-ku, Tokyo 113-0033, Japan
[7]Department of Physics, Chiba University, Chiba 263-8522, Japan
[8]Department of Physics, Graduate School of Science, Kobe University, Kobe, Hyogo 657-8501, Japan

E-mail: `TERASHIMA.Taichi@nims.go.jp`

Abstract. We present results of Shubnikov-de Haas oscillation measurements on detwinned $BaFe_2As_2$ and de Haas-van Alphen oscillation measurements on KFe_2As_2. The Fermi surface of $BaFe_2As_2$ in the antiferromagnetic phase is found to consist of one hole and two electron pockets, all of which are three-dimensional and closed, and can reasonably be accounted for by LSDA band calculations. We find only moderate mass enhancements m^*/m_{band} of 2–3. In the case of KFe_2As_2, four quasi-two-dimensional Fermi surface cylinders ϵ, α, ζ, and β are observed in qualitative agreement with previous ARPES data. In sharp contrast to $BaFe_2As_2$, agreement between the observed and LDA-calculated Fermi surface is poor: LDA calculations seem to predict wrong crystal-field splitting of Fe 3*d* states. Large effective masses up to 20 m_e, m_e being the free electron mass, are found. The Sommerfeld coefficient estimated from the observed Fermi surface and effective masses is consistent with the measured value of 93 mJ/K^{2}mol [H. Fukazawa *et al.*, J. Phys. Soc. Jpn. **80**, SA118 (2011)] and is 8–9 times larger than the band value, indicating strong electronic correlations in KFe_2As_2.

1. Introduction

Since the discovery of superconductivity at $T_c = 26$ K in LaFeAs(O, F) [1], extensive efforts have been devoted to the elucidation of the mechanism and paring symmetry of the iron-based high-T_c superconductivity (for reviews, see Refs. [2, 3, 4]). At the beginning, it looked like

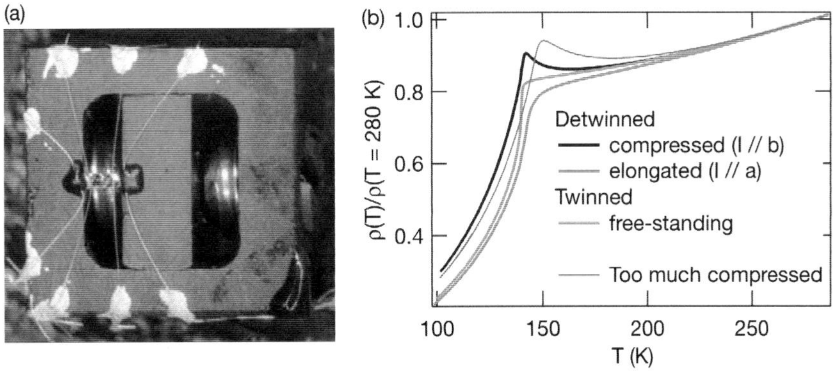

Figure 1. (a) Detwinning device. (b) Temperature dependences of resistivity near the structural/magnetic phase transition measured in compressed, elongated, and free-standing $BaFe_2As_2$ samples. If a sample is too much compressed or elongated, the transition shifts to higher temperatures.

spin-fluctuations mediated $s_\pm$ paring is the only answer. However, as more and more iron-based superconductors have been discovered, considerable diversity in their superconducting properties has become apparent. There is strong evidence for nodal gap structures in some compounds; nodes may be accidental ones compatible with the s-wave symmetry, but there is also possibility of d-wave paring. For the paring mechanism, orbital fluctuations are now seriously considered as possible alternative glue. In this situation, it is necessary to study electronic structures in various iron-based superconductors in detail and systematically to reveal relation between the electronic structures and superconducting properties. $(Ba_{1-x}K_x)Fe_2As_2$ [5] is a suitable system for such investigations. High-quality single crystals are available for the whole range of composition. Its phase diagram is typical of the iron-based high-T_c: The parent compound undergoes an antiferromagnetic transition and does not exhibit superconductivity. As x is increased, the antiferromagnetism is suppressed and superconductivity emerges. At the optimal doping ($x \sim 0.4$), T_c reaches 38 K, and the superconducting gap is a full gap. At $x = 1$, KFe_2As_2 still has a T_c of 3.4 K but the gap is nodal (see a review [4]). This contrast makes it fascinating to study how the electronic structure evolves with x in $(Ba_{1-x}K_x)Fe_2As_2$. We here show results of quantum oscillation measurements on the two end members, $BaFe_2As_2$ [6] and KFe_2As_2 [7].

2. $BaFe_2As_2$

Single crystals were grown by a self-flux method and were annealed [8, 9]. The residual resistivity ratios were 40–60. Because of a structural phase transition from tetragonal to orthorhombic, if samples are cooled without stress, they become a mixture of twin domains [10, 11]. To avoid that, we used a device shown in Fig. 1(a): A single crystal sample with the length along a tetragonal [110] axis is fixed to the outer frame and mobile plate by epoxy glue. By adjusting the screws, the sample can be compressed or elongated. If compressed (elongated), the length direction becomes the orthorhombic b (a) axis. Appropriate strength of stress needs to be found by trial and error so as to avoid appreciable shift in the transition temperature [Fig. 1(b)].

Figure 2 shows magnetoresistance and its oscillatory part for $B \parallel c$ in the compressed sample. SdH oscillations are already visible at $B = 5$ T, indicating high-quality of the sample. Figure 3(a) shows the field-angle dependence of the SdH frequencies. Three fundamental frequencies α, γ and δ are identified. An LSDA band structure calculation indicates that the Fermi surface is composed of relatively large hole and electron pockets and lines of tiny pockets. After

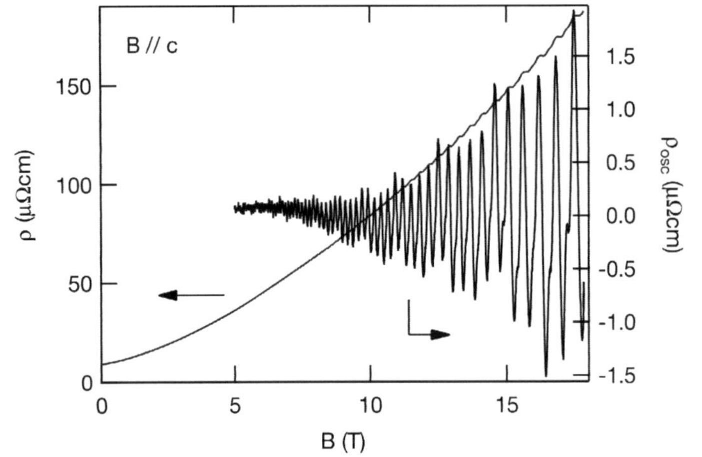

Figure 2. Magnetoresistance and its oscillatory part for $B \parallel c$ in a compressed $BaFe_2As_2$ sample.

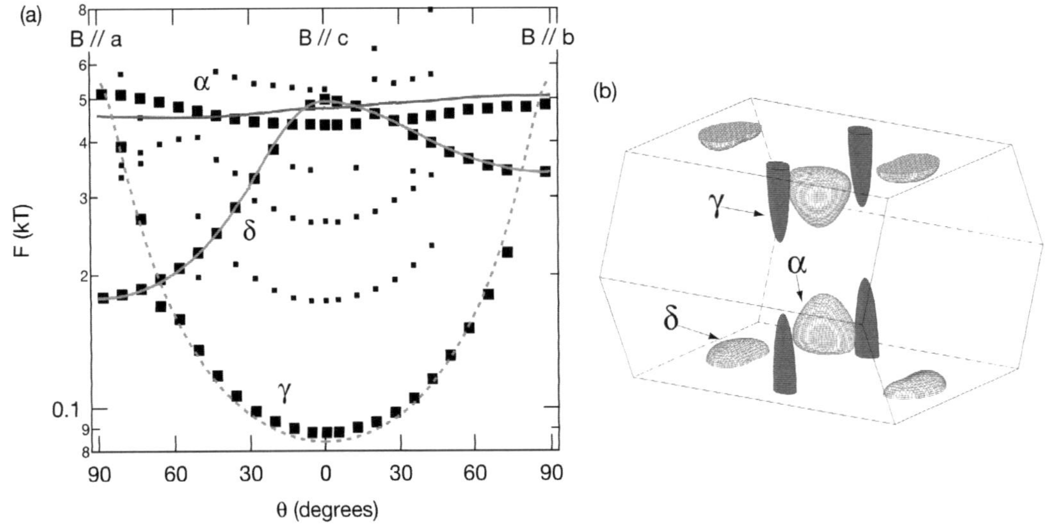

Figure 3. (a) Angle dependence of SdH frequencies in $BaFe_2As_2$. The filled squares indicate fundamental frequencies, while the small dots indicate harmonic and combination frequencies. The solid curves indicate calculated frequencies for α and δ with band energy adjustments. The dashed line for γ is a fit to the angle dependence expected for an ellipsoidal Fermi surface pocket. (b) Determined Fermi surface composed of α hole and γ and δ electron pockets.

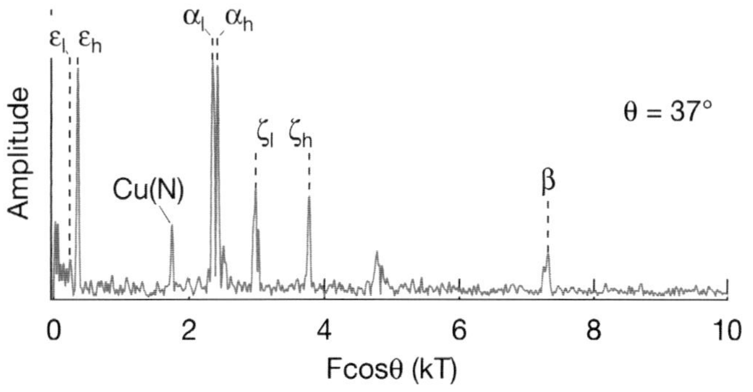

Figure 4. Fourier transform of dHvA oscillations in KFe_2As_2.

appropriate band energy adjustments, the α and δ frequencies are accounted for by the hole and electron pockets, respectively. The effective masses associated with these frequencies are only moderately enhanced over the band masses by a factor of 2–3. The angle dependence of γ indicates that it arises from an ellipsoidal pocket. Although the tiny pockets predicted by the original calculation disappear after the energy adjustments, it is reasonable to place the γ pocket somewhere along the original tiny-pockets lines. Taking account of laser ARPES data [12] we identify it as an electron pocket and place it in the Brillouin zone as shown in Fig. 3(b). It is to be noted here that, although the observed Fermi surface can be accounted for by the LSDA calculation reasonably well, it overestimates the antiferromagnetic moment (1.6 μ_B compared to the experimental value of 0.87 μ_B [13]).

Carrier numbers estimated from the determined Fermi surface are 0.0235 holes and 0.0246 electrons per primitive cell, satisfying the carrier compensation within experimental accuracy. The Sommerfeld coefficient estimated from the determined Fermi surface and effective masses is 5.0 mJ/K^{2}mol, which agrees well with the measured value of 5.1 mJ/K^{2}mol [14]. These indicate that the determined Fermi surface is complete and that there is no other pocket with a comparable volume to the observed ones.

The determined Fermi surface is composed of three-dimensional closed pockets [Fig. 3(b)]. This is in contrast to k_z-resolved ARPES data [15, 16, 17], which show quasi-two-dimensional Fermi surface cylinders even in the antiferromagnetic phase. This discrepancy arises from the limited k_z resolution of ARPES. The determined Fermi surface is not very anisotropic in the *ab* plane. Analyses of the measured effective masses and Dingle temperatures also suggest fairly isotropic *ab* plane electrical transport [6]. This is consistent with recent resistivity and optical measurements on detwinned crystals [8, 9].

3. KFe_2As_2

Single crystals grown by a self-flux method show residual resistivity ratios of ~600 [18]. de Haas-van Alphen oscillations were detected using the field-modulation technique. Figure 4 shows an example of a Fourier transform of observed dHvA oscillations. The field angle θ is measured from the c axis. We identify four fundamental frequencies ϵ, α, ζ, and β. Their angle dependences indicate that they correspond to quasi-two-dimensional Fermi surface cylinders with different degrees of warping. The former three appear in paris as expected, but the other pair to β is unclear at present. An LDA band structure calculation predicts three hole cylinders at the zone center, small hole cylinders near the zone boundary, and a tiny hole pocket. Using this as a guide, we place the α, ζ, and β cylinders at the center and the ϵ cylinders near the corner

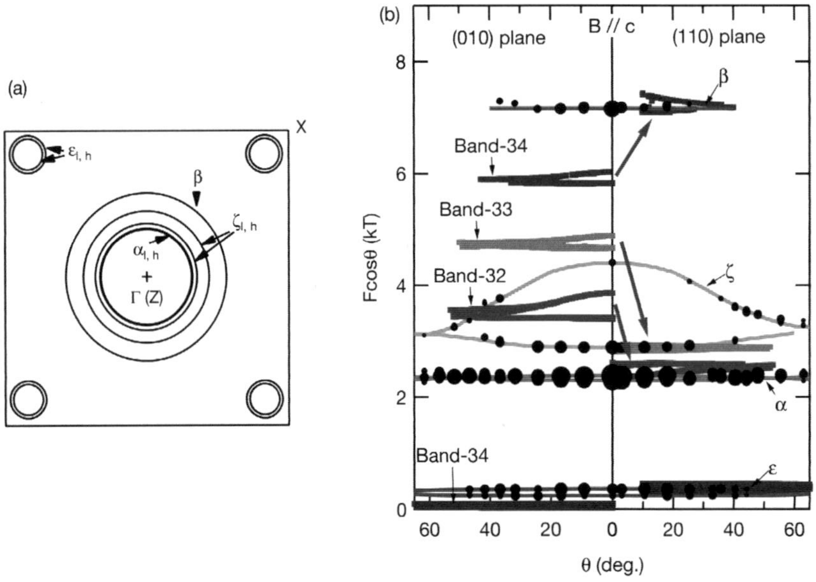

Figure 5. (a) Observed orbits projected along the c axis. In-plane anisotropy is neglected. (b) Comparison between the experimental (ϵ, α, ζ, and β) and calculated frequencies (Bands-32, 33, and 34). The right panel shows results of band-energy adjustments.

[Fig. 5(a)]. The Fermi surface volume is estimated to be 51.4% or 50.5% from the frequencies observed at $\theta = 0$ or 37 degrees, respectively. These estimates are in satisfactory agreement with the stoichiometry (50%). The Sommerfeld coefficient is estimated to be 94 or 82 mJ/K^{2}mol from the effective masses determined at $\theta = 0$ or 37 degrees, respectively. These values are consistent with the measured value of 93 mJ/K^{2}mol [19]. These indicate that the determined Fermi surface is complete. It is to be emphasized that the estimated Sommerfeld coefficients are 8–9 times larger than the band value, indicating strong electronic correlations in KFe_2As_2.

Quantitative agreement between the measured and calculated frequencies is poor [Fig. 5(b)]. In contrast to the case of $BaFe_2As_2$, band energy adjustments do not work as shown on the right side of the figure: the three dimensional shape of the ζ cylinder can not be reproduced, for example. Band structure calculations predict that the xz and yz bands are above the xy band at the Γ point. The former two are degenerate at Γ without spin-orbit coupling. On the other hand, experimentally, we have two similarly-sized cylinders, α and ζ, and a large cylinder, β. To explain this, it seems necessary to invert the calculated level scheme. Namely, the xz and yz bands are lower than the xy band and produce the two similarly-sized cylinders, α and ζ. Recent AREPS measurements have confirmed this conjecture [20].

4. Summary

We have determined the Fermi surface in $BaFe_2As_2$ and KFe_2As_2 based on quantum oscillation measurements. The Fermi surface in $BaFe_2As_2$ is composed of small three-dimensional electron and hole pockets and is explained by an LSDA calculation reasonably well despite the overestimated antiferromagnetic moment. The mass enhancements are only moderate, approximately 2–3. The Fermi surface in KFe_2As_2 is composed of quasi-two-dimensional hole cylinders. The Fermi surface calculated with LDA is not very consistent with the observed one.

It seems that LDA calculations yield wrong crystal field splitting of the Fe $3d$ states. Effective masses are heavy, up to $20m_e$, indicating strong electronic correlations in this compound.

Acknowledgments
TT thanks Teppei Yoshida (Univ. Tokyo) for valuable discussions.

References

[1] Kamihara Y, Watanabe T, Hirano M and Hosono H 2008 *J. Am. Chem. Soc.* **130** 3296
[2] Ishida K, Nakai Y and Hosono H 2009 *J. Phys. Soc. Jpn.* **78** 062001
[3] Johnston D C 2010 *Adv. Phys.* **59** 803–1061
[4] Hirschfeld P J, Korshunov M M and Mazin I I 2011 *Rep. Prog. Phys.* **74** 124508
[5] Rotter M, Tegel M and Johrendt D 2008 *Phys. Rev. Lett.* **101** 107006
[6] Terashima T, Kurita N, Tomita M, Kihou K, Lee C H, Tomioka Y, Ito T, Iyo A, Eisaki H, Liang T, Nakajima M, Ishida S, Uchida S, Harima H and Uji S 2011 *Phys. Rev. Lett.* **107** 176402
[7] Terashima T, Kimata M, Kurita N, Satsukawa H, Harada A, Hazama K, Imai M, Sato A, Kihou K, Lee C H, Kito H, Eisaki H, Iyo A, Saito T, Fukazawa H, Kohori Y, Harima H and Uji S 2010 *J. Phys. Soc. Jpn.* **79** 053702
[8] Nakajima M, Liang T, Ishida S, Tomioka Y, Kihou K, Lee C H, Iyo A, Eisaki H, Kakeshita T, Ito T and Uchida S 2011 *PNAS* **108** 12238
[9] Ishida S, Liang T, Nakajima M, Kihou K, Lee C H, Iyo A, Eisaki H, Kakeshita T, Kida T, Hagiwara M, Tomioka Y, Ito T and Uchida S 2011 *Phys. Rev. B* **84** 184514
[10] Chu J H, Analytis J G, De Greve K, McMahon P L, Islam Z, Yamamoto Y and Fisher I R 2010 *Science* **329** 824
[11] Tanatar M A, Blomberg E C, Kreyssig A, Kim M G, Ni N, Thaler A, Bud'ko S L, Canfield P C, Goldman A I, Mazin I I and Prozorov R 2010 *Phys. Rev. B* **81** 184508
[12] Shimojima T, Ishizaka K, Ishida Y, Katayama N, Ohgushi K, Kiss T, Okawa M, Togashi T, Wang X Y, Chen C T, Watanabe S, Kadota R, Oguchi T, Chainani A and Shin S 2010 *Phys. Rev. Lett.* **104** 057002
[13] Huang Q, Qiu Y, Bao W, Green M A, Lynn J W, Gasparovic Y C, Wu T, Wu G and Chen X H 2008 *Phys. Rev. Lett.* **101** 257003
[14] Rotundu C R, Freelon B, Forrest T R, Wilson S D, Valdivia P N, Pinuellas G, Kim A, Kim J W, Islam Z, Bourret-Courchesne E, Phillips N E and Birgeneau R J 2010 *Phys. Rev. B* **82** 144525
[15] Liu C, Kondo T, Ni N, Palczewski A D, Bostwick A, Samolyuk G D, Khasanov R, Shi M, Rotenberg E, Bud'ko S L, Canfield P C and Kaminski A 2009 *Phys. Rev. Lett.* **102** 167004
[16] Malaeb W, Yoshida T, Fujimori A, Kubota M, Ono K, Kihou K, Shirage P M, Kito H, Iyo A, Eisaki H, Nakajima Y, Tamegai T and Arita R 2009 *J. Phys. Soc. Jpn.* **78** 123706
[17] Yi M, Lu D H, Chu J H, Analytis J G, Sorini A P, Kemper A F, Mo S K, Moore R G, Hashimoto M, Lee W S, Hussain Z, Devereaux T P, Fisher I R and Shen Z X 2011 *PNAS* **108** 6878
[18] Kihou K, Saito T, Ishida S, Nakajima M, Tomioka Y, Fukazawa H, Kohori Y, Ito T, Uchida S, Iyo A, Lee C H and Eisaki H 2010 *J. Phys. Soc. Jpn.* **79** 124713
[19] Fukazawa H, Saito T, Yamada Y, Kondo K, Hirano M, Kohori Y, Kuga K, Sakai A, Matsumoto Y, Nakatsuji S, Kihou K, Iyo A, Lee C H and Eisaki H 2011 *Journal of the Physical Society of Japan* **80SA** SA118
[20] Yoshida T, Ideta S, Nishi I, Fujimori A, Yi M, Moore R G, Mo S K, Lu D H, Shen Z X, Hussain Z, Kihou K, Shirage P M, Kito H, Lee C H, Iyo A, Eisaki H and Harima H 2012 *arXiv:1205.6911*

Electron irradiation of Co, Ni, and P-doped $BaFe_2As_2$–type iron-based superconductors

Cornelis Jacominus van der Beek[1], Sultan Demirdiş[1], Dorothée Colson[2], Florence Rullier-Albenque[2], Yanina Fasano[3], Takasada Shibauchi[4], Yuji Matsuda[4], Shigeru Kasahara[4,5], Piotr Gierlowski[6], Marcin Konczykowski[1]

[1]Laboratoire des Solides Irradiés, CNRS UMR 7642 & CEA-DSM-IRAMIS, Ecole Polytechnique, F-91128 Palaiseau cedex, France
[2]Service de Physique de l'Etat Condensé, CEA Saclay, CEA-DSM-IRAMIS, CNRS URA 2464, F-91191 Gif-sur-Yvette, France
[3]Laboratorio de Bajas Temperaturas, Centro Atómico Bariloche & Instituto Balseiro, Avenida Bustillo 9500, 8400 San Carlos de Bariloche, Argentina
[4]Department of Physics, Kyoto University, Sakyo-ku, Kyoto 606-8501, Japan
[5]Research Center for Low Temperature and Materials Sciences, Kyoto University,Sakyo-ku, Kyoto 606-8501, Japan
[6]Institute of Physics of the Polish Academy of Sciences, 32-46 Aleja Lotników, 02-668 Warsaw, Poland

E-mail: `keesvanderbeek@polytechnique.edu`

Abstract. High energy electron irradiation is used to controllably introduce atomic-scale point defects into single crystalline $Ba(Fe_{1-x}Co_x)_2As_2$, $Ba(Fe_{1-x}Ni_x)_2As_2$, and $BaFe_2(As_{1-x}P_x)_2$. The appearance of the collective pinning contribution to the critical current density in $BaFe_2(As_{1-x}P_x)_2$, and the magnitude of its enhancement in $Ba(Fe_{1-x}Co_x)_2As_2$, conform with the hypothesis of quasi-particle scattering by Fe vacancies created by the irradiation. Whereas the insignificant modification of the temperature dependence of the superfluid density in $Ba(Fe_{1-x}Co_x)_2As_2$ and $Ba(Fe_{1-x}Ni_x)_2As_2$ points to important native disorder present before the irradiation, the critical temperatures of these materials undergo a suppression equivalent to that observed in the much cleaner $BaFe_2(As_{1-x}P_x)_2$. This lends credence to the hypothesis of line nodes of the order parameter (at finite k_z) in the former two materials.

1. Introduction

The premise of $s_{\pm}$ superconductivity in the multiband iron-based superconductors [1, 2], with a sign-changing order parameter between the electron-like and hole-like Fermi-surface sheets [2, 3, 4], has raised the question of the effect of atomic-scale point-like disorder on superconductivity in these materials [5, 6]. In particular, interband scattering of quasiparticles by point-like disorder was suggested to be pair-breaking, and, therefore, to lead to a strong suppression of the critical temperature, the appearance of sub-gap states [7], and a peculiar T^2 dependence of the London penetration depth at low temperature [8, 9, 10, 11, 12, 13, 14].

More recently, the question of the role of point-like disorder has become only more relevant [15]. This is because the anti-ferromagnetic coupling between Cooper pairs on different Fermi surface sheets [2, 3, 4], and the subtle changes of the Fermi surface with doping [3, 4], isovalent chemical substitution [17, 18, 19, 20, 21, 26, 27], and pressure [22, 23, 24, 25], allow for a wide variation of realizations of the order parameter [3, 4, 27]. In particular, the likelihood of line nodes of the order parameter in isovalently doped $BaFe_2(As_{1-x}P_x)_2$ [32], LaFePO [29], and KFe_2As_2 [28, 30, 31] are indicative of a nodal $s_\pm$– or a d-wave superconductivity [3, 4] in the first two and latter respective cases.

Nodal gap structure not imposed by symmetry, but arising from *e.g.*, the anisotropy of the Fermi surface and/or interband scattering [38], has also been evoked [16]. In the case of an $s_\pm$ state, the order parameter sign would not only change between bands but also within a given band [16]. One can also have an s–wave gap-structure in which nodes appear for non-zero k_z [15, 39, 40], a situation that might be relevant for $Ba(Fe_{1-x}Co_x)_2As_2$ [35], $Ba(Fe_{1-x}Ni_x)_2As_2$ [12], $Na_{1-\delta}(Fe_{1-x}Co_x)As$ [36], and overdoped $Ba_{1-x}K_xFe_2As_2$ [37].

1.1. Disorder effect on exotic superconductivity in iron pnictides

Given that the appearance of gap nodes in an s-wave superconductor does not imply symmetry breaking, indirect experiments yield, as yet, most insight into the order parameter structure. Such experiments include the low temperature behaviour of the penetration depth [8, 9, 10, 11, 12, 13, 14, 27, 29, 32] and the thermal conductivity [30, 32, 35, 37] as probes of the low-energy quasi-particle density of states (QPDOS), but also the sensitivity of the iron-based superconductors to quasi-particle scattering by homogeneous atomic-scale point-like disorder. This can be introduced by, *e.g.*, chemical substitution or energetic particle irradiation. Namely, the occurrence of sign changes of the order parameter implies that even non-magnetic point-like scatterers may act as pair-breakers [5, 6, 7, 15]. For $s_\pm$–superconductivity in particular, impurity scattering of quasi-particles between electron–like and hole–like bands (with opposite sign of the order parameter) would lead to a rapid suppression of the critical temperature T_c [5, 6, 7], and open the possibility for non-sign changing s_{++} superconductivity due to orbital fluctuations to manifest itself [5, 6, 41]. The vulnerability of the d–wave symmetry to both interband- and intraband impurity scattering would render this even more fragile than $s_\pm$. On the other hand, a less pronounced role of interband compared to interband scattering would reduce the sensitivity of $s_\pm$ superconductivity to disorder [15, 41]. Finally, the presence of "accidental" nodes would result in a rather high initial sensitivity of superconductivity to disorder, with a slower suppression of the critical temperature as function of disorder strength once the interband scattering is sufficiently strong to wash out, or "lift" the nodes on all Fermi surface sheets [16].

1.2. Effect of native disorder

Even if the problem at hand is characterized by a multitude of parameters, it is important to at least establish trends as controlled point–like disorder is added to iron-based superconductors. Early experiments have suggested a rather limited sensitivity of superconductivity in the iron-based materials to disorder. For example, the residual electronic term in the specific heat suggests a large low–temperature QPDOS [42, 43] away from optimal doping, that may be due to pair-breaking. The ubiquitous T^2–behaviour of the low–temperature ab–plane penetration depth $\lambda_{ab}(T)$ suggest strong pair-breaking [8, 9, 10, 11, 12, 13, 14]. Still, T_c's of the materials in question are high.

In our earlier work [44, 45], we have drawn attention to the fact that vortex pinning in the mixed state of the iron-based superconductors is indicative of the importance of native disorder. While the critical current density maximum j_c at low fields, and the subsequent $j_c \sim B^{-1/2}$ drop–off are due to nm-scale heterogeneities of the superconducting properties, leading to a so-called "strong–pinning" contribution j_c^s to the critical current (see Fig. 1 and Refs. [47, 48]), the so-called "weak collective pinning" [49, 50] contribution j_c^{coll} to the critical current density in the magnetic field range of several tenths of a T to several T is consistently interpreted in terms of quasi-particle scattering by the dopant atoms in charge–doped iron-based compounds. Estimates of scattering rates of the doping impurities, such as these are obtained from flux pinning, yield rather large values, which are at odds with a superconducting ground state that would be sensitive to point-like disorder should this be pair-breaking. Moreover, it may be noted that the occurrence of the quasi-particle scattering contribution to flux pinning consistently coincides with that of the T^2 behaviour of the penetration depth. Inversely, it is conspicuously absent in clean materials with a T-linear dependence of λ_{ab}, such as $BaFe_2(As_{1-x}P_x)_2$ [32, 45, 58].

The relative insensitivity of superconductivity in the iron-based materials may be due to these materials not having a sign-changing order parameter, but may also be attributed to limited interband scattering by the dopant impurities. Given the premise of Coulomb scattering (with a small change in crystal momentum) by the charged dopants [45], this is not unreasonable. Estimates from pinning situate the native point defects rather in the Born limit (see table 1).

1.3. Artificial disorder

An obvious manner to test the above ideas is through the artificial introduction of different kinds of point-like defects, either by chemical doping [59, 60, 61], or by irradiation [62, 63]. An example is the substitution of magnetic ions, which suppresses T_c rather more effectively than non-magnetic substitutions [59, 61]. A major drawback of chemical substitution in the iron-based compounds is, however, that this simultaneously leads to structural changes and/or doping. Energetic particle irradiation therefore seems preferable. Recent work by Tarantini *et al.* reported on T_c–suppression by α–particle irradiation of NdFeAs(O,F), but was prone to criticism, notably in that it induces magnetic impurities and defect clustering. The 3 MeV proton irradiation experiments by Nakajima *et al.* on $Ba(Fe_{1-x}Co_x)_2As_2$ reported monotonous T_c–depression as the residual resistivity increases; the authors provided estimates for the critical pair-breaking parameter as $\Gamma/2\pi T_c \sim 4-7$ and $\Gamma/2\pi T_c \sim 1.5-2.5$, as one goes from underdoped to overdoped [63]. However, apart from point defects, proton and neutron irradiation are known to induce, *in situ*, point defect cascades and clusters, which may play a different role than that of simple scatterers.

Here, we compare the effect of 2.5 MeV electron irradiation on differently substituted

Table 1. Contribution of dopant disorder to elastic scattering parameters of various iron pnictide superconductors, such as estimated from the weak collective pinning contribution to the (flux pinning) critical current density, j_c^{coll}. Here, k_F is the Fermi wavevector, ξ_0 is the Bardeen-Cooper-Schrieffer coherence length, n_d is the atomic point defect density, σ_{tr} is the transport scattering cross-section, δ_0 is the scattering phase angle, $\Gamma = n_d[\pi N(0)]^{-1}\sin^2\delta_0$ is the scattering rate, where $N(0) = mk_F/2\pi\hbar^2$ is the normal-state DOS.

Material	impurity	k_F Å^{-1}	ξ_0 nm	n_d nm^{-3}	σ_{tr} Å^2	$n_d\xi_0^3$	$\sin\delta_0$	Γ meV
$PrFeAsO_{1-y}$	O vacancy	0.33	2.4	1.5	6.7	21	0.3(2)	10
$NdFeAsO_{0.9}F_{0.1}$ [44]	F	0.33	3.3	1.5	2.5	54	0.2	4
$Ba_{0.72}K_{0.28}Fe_2As_2$ [55]	K	0.4	2.4	2.8	1.5	38	0.1(4)	3
$Ba_{0.6}K_{04}Fe_2As_2$ [56]	K	0.5	2.2	4	2.5 ± 1.3	43	0.2	8
$Ba_{0.45}K_{0.55}Fe_2As_2$ [45]	K	0.5	2.2	5.5	1.5	59	0.2	10
$Ba(Fe_{0.95}Co_{0.05})_2As_2$ [48]	Co	0.25	1.6	1	2.5	8	0.17	5
$Ba(Fe_{0.9}Co_{0.1})_2As_2$ [46]	Co	0.25	1.6	2	2.5	8	0.17	5
$Ba(Fe_{0.76}Ru_{0.24})_2As_2$ [54]	Ru	0.25	1.6	4.8	2.5	8	0.17	5
$BaFe_2(As_{0.67}P_{0.33})_2$ [45]	P	0.3 [57]	1.6	3.3	14	–	–	–

$BaFe_2As_2$. Such high–energy electron irradiation is known to produce homogeneously distributed vacancy–interstitial (Frenkel) pairs on all sublattices of a crystal, but, preferentially on that constituted by the lightest atomic species in the material – in the materials studied here, Fe. In the high-T_c cuprates, electron-irradiation defects are known to be strong unitary scatterers, comparable to Zn substitution. There, they are responsible for the suppression of the critical temperature [64, 65, 66], and the appearance of a T^2–temperature dependence of the penetration depth [66]. In semi-metals with extremely low carrier densities, such as Bi, the irradiation-induced defects are charged and lead to effective doping of the material [67]. The mobility of the created point defects at high temperature is a caveat, since it leads to partial annealing and clustering, albeit it to a lesser extent than that produced by proton or neutron irradiation.

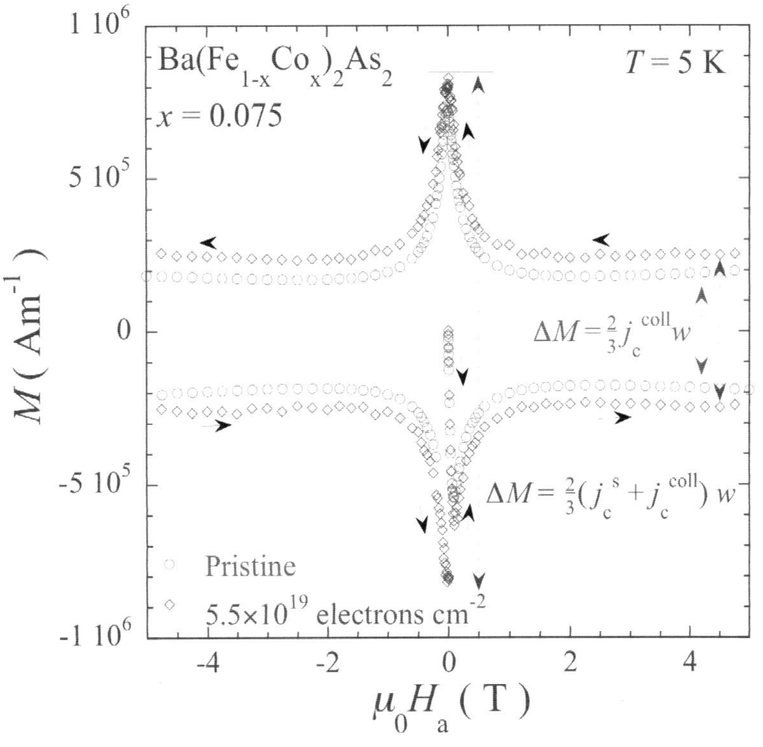

Figure 1. Single crystalline $Ba(Fe_{0.925}Co_{0.075})_2As_2$: Hysteresis loop of the irreversible magnetization, at 5 K, before (○) and after (◇) irradiation with 5.5×10^{19} cm^{-2} 2.5 MeV electrons. Closed arrows depict the direction in which the loop is traversed upon cycling the magnetic field. Dotted double arrows depict the width of the magnetization loop in the low–field, strong pinning regime, and in the higher field regime in which only the weak collective pinning contribution j_c^{coll} is relevant.

It turns out that electron irradiation leads to similar monotonic suppression of T_c in $Ba(Fe_{1-x}Co_x)_2As_2$, $Ba(Fe_{1-x}Ni_x)_2As_2$, and $BaFe_2(As_{1-x}P_x)_2$. In $Ba(Fe_{1-x}Co_x)_2As_2$, the suppression is weakest for the optimally doped material and strongest for underdoped crystals. In the range of intermediate electron fluences studied here, the resistance monotonically increases as function of fluence. In $Ba(Fe_{1-x}Co_x)_2As_2$ and $Ba(Fe_{1-x}Ni_x)_2As_2$, there is little to no change of the temperature dependence of the superfluid density. We have also measured the critical current density before and after the irradiation. In $Ba(Fe_{1-x}Co_x)_2As_2$ one has a clear increase of the weak collective pinning contribution to j_c, while in $BaFe_2(As_{1-x}P_x)_2$ the initially absent weak pinning contribution emerges after irradiation. This finding establishes that the field–independent contribution j_c^{coll} to the critical current density of iron-based superconductors is, indeed, due to atomic sized point pins. In both materials, this finding allows for an estimate of the density of produced defects per unit irradiation fluence. In $Ba(Fe_{1-x}Ni_x)_2As_2$, j_c is strongly suppressed by the irradiation.

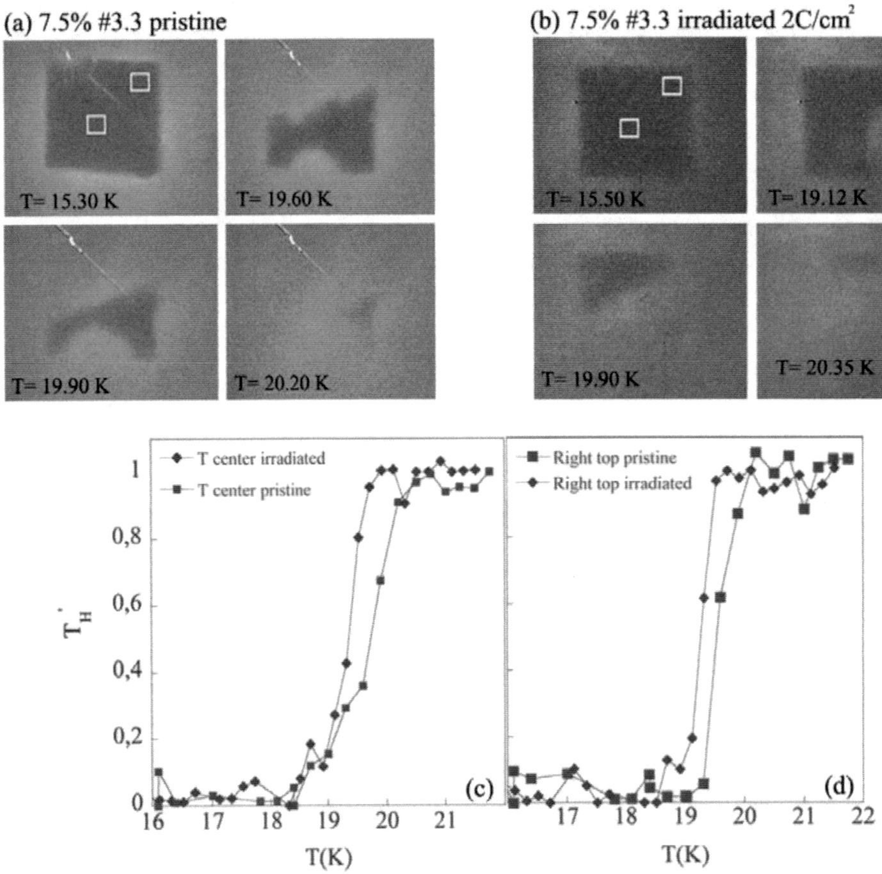

Figure 2. Transition from the superconducting to the normal state of a $Ba(Fe_{0.925}Co_{0.075})_2As_2$ single crystal before (a,c) and after irradiation with 2.1 Ccm^{-2} 2.5 MeV electrons (b,d), as imaged by the differential magneto-optical (DMO) technique [47]. The images (a,b) show the progressive admission of an ac magnetic field of magnitude 1 Oe, applied perpendicularly to the sample surface, as the temperature is raised. In these flux density maps, areas of high luminous intensity I correspond to the value of the applied field, while dark areas correspond to zero field, (*i.e.* full screening). The dark rectangles in the upper left hand panels of (a) and (b) correspond to the sample outline, *i.e.*, full Meissner expulsion of the magnetic field at the lowest temperature. The white squares indicate the positions at which the "transmittivity" (or "local ac susceptibility") defined as $[I(T) - I(T \ll T_c)]/[I(T \gg T_c) - I(T \ll T_c)]$ depicted in the lower panels (c,d) was determined.

2. Experimental details

2.1. Single crystalline samples

Single-crystals of $Ba(Fe_{1-x}Co_x)_2As_2$ [68] and $Ba(Fe_{1-x}Ni_x)_2As_2$ [69] were grown using the self-flux method. The high purity starting reagents Ba, FeAs, and CoAs were mixed in the molar ratio $1 : (4 - x) : x$, loaded in alumina crucibles and then sealed in evacuated quartz tubes. For each doping level, chemical analysis by an electron probe was performed on several crystals yielding the dopant content within 0.5% absolute accuracy. For this work we studied six Co–doping levels and two Ni–doping levels. The optimally doped $BaFe_2(As_{1-x}P_x)_2$ single crystals,

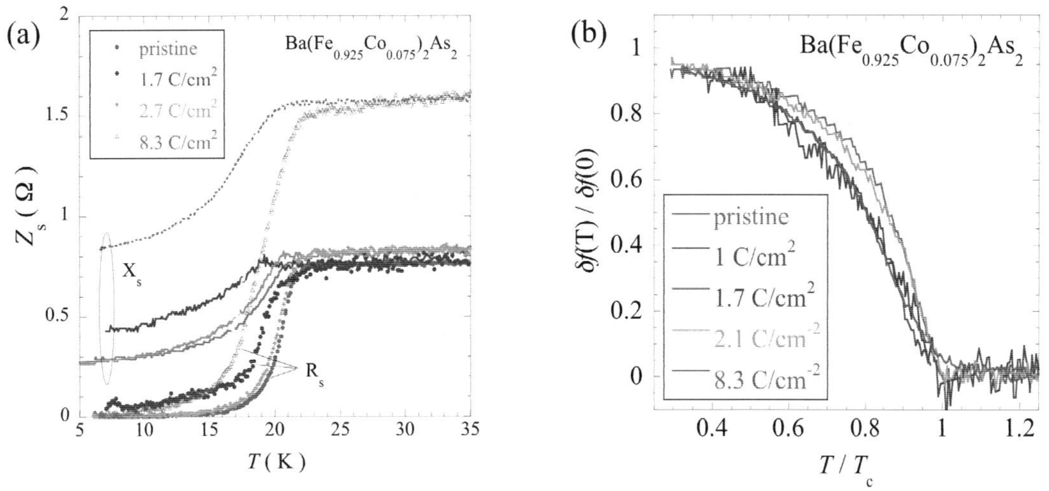

Figure 3. Surface impedance Z_s of $Ba(Fe_{0.925}Co_{0.075})_2As_2$ crystals before and after various low–temperature irradiation runs with 2.5 MeV electrons, and subsequent annealing at 300 K. Data points and thin lines show the surface resistance R_s and reactance X_s respectively, for various electron doses. (b) Shift $\delta f(T)$ of the resonant frequency of the superconducting Nb cavity (operating at 26.0 GHz) as function of temperature, normalized with respect to the low–temperature extrapolated $\delta f(0)$, for $Ba(Fe_{0.925}Co_{0.075})_2As_2$ crystals irradiated with various fluences of 2.5 MeV electrons.

with $x = 0.33$ and $x = 0.36$, were also grown by the self flux method [70], and characterized using x-ray diffraction and energy dispersive x-ray spectroscopy. No impurity phases were found within the experimental limits of $\sim \%1$.

2.2. Electron irradiation
The 2.5 MeV electron irradiation was performed at the SIRIUS Pelletron facility of the Laboratoire des Solides Irradiés (LSI) at the Ecole Polytechnique in Palaiseau, France [71]. Beam currents varied between 10 and 20 μA, depending on the irradiation run; the beam was swept over a 6×6 mm^2 area. Samples of different composition irradiated to the same dose were mounted together and irradiated simultaneously. In order to prevent *in-situ* defect migration, recombination, and clustering, the irradiations were carried out in a liquid H_2 bath ($T = 20$ K). In–situ resistance measurements were carried out on the $Ba(Fe_{1-x}Co_x)_2As_2$ material in order to assess the increase of the residual resistivity as a function of defect density at low temperature, as well as the effect of room-temperature annealing.

2.3. Measurements of T_c and the surface impedance
T_c–values before and after irradiation were determined using differential magneto-optical (DMO) imaging (see Fig. 2 and Ref. [47]). For the lower irradiation doses, in particular, the shift in T_c after irradiation is smaller than the transition width. DMO then provides an objective means to quantify the effect of the electron irradiation on different portions of a crystal. The change in T_c was further quantified using *ex-situ* measurements of the temperature dependent microwave surface impedance $Z_s = R_s + iX_s$. These were performed at the LSI using a superconducting Nb cavity, operating at 26.0 GHz, and cooled to 5 K using a 0.5 W cryocooler cold head. From

the temperature dependent resonance frequency f and unloaded quality factor Q_0, the surface resistance R_s and reactance X_s were determined as

$$R_s = \mu_0 f G \left[\frac{1}{Q_0(T)} - \frac{1}{\tilde{Q}_0} \right] \tag{1}$$

and

$$X_s(T) = \mu_0 G \left[f(T) - f(0) \right] + X_0 \tag{2}$$

respectively. Here $f(0)$ and $\tilde{Q}_0$ are the resonance frequency and the unloaded quality factor in the absence of the sample, and the geometrical factor $G = V/4\pi w\sqrt{wd}$, with V the inner volume of the cavity, d the thickness of the platelet-like single crystal sample, and w its smaller width. The additive constant X_0 was adjusted so that $R_s = X_s$ in the normal state.

2.4. Determination of the critical current density

Measurements of the magnetization M of the $Ba(Fe_{1-x}Co_x)_2As_2$ and $Ba(Fe_{1-x}Ni_x)_2As_2$ single crystals as function of the applied magnetic field H_a were performed using a commercial SQUID magnetometer. From the width $\Delta M(H)$ of the magnetic hysteresis loops, the field- and temperature dependence of the sustainable current density $j = \frac{3}{2}\Delta M/w$ was determined using the Bean model. The numeric prefactor $\frac{3}{2}$ is estimated from calculations of E.H. Brandt for rectangular bars of similar aspect ratio as the measured crystals [72]. Data on $BaFe_2(As_{1-x}P_x)_2$ were acquired by M. Konczykowski using the local Hall probe magnetometry technique [73, 74]. Briefly, the single crystalline superconducting crystal is centered on top of an array of 11 Hall sensors, of active area 3×3 μm^2, and spaced by 20 μm, in such a manner that the short er crystal edge is parallel to the array, and the array spans the longer crystal edge. The Hall sensors are used to measure the spatial gradient dB/dx of the local induction component perpendicular to the crystal surface. The gradient dB/dx (in G/μm) is directly proportional to the sustainable screening current j, with $dB/dx \sim \frac{1}{6}j$ for a crystal aspect ratio of ~ 0.1 [72]. Measurements of the flux density gradient were performed both as function of H_a after zero field–cooling, and as function of temperature T, after field cooling, and the subsequent reduction of H_a by twice the full penetration field.

In what follows, we shall take the screening current at the lowest measurement temperature (5 K) to be representative of the critical current density. At higher temperature, the measured current density can be significantly lower than the pinning critical current density because of flux creep [58, 74].

3. Analysis of the critical current density

The critical current density is further analyzed along the lines of Refs. [44, 45]. The hysteresis loops of all charge–doped iron-based superconductors, exemplified by Fig. 1, indicate the presence of two main pinning mechanisms in these materials. The central peak of the magnetization, at low applied fields, is due to the strong–pinning contribution j_c^s to the critical current, while the constant contribution at higher fields comes from "weak collective pinning" by the dopant atoms. In a critical current density - versus magnetic field plot (see Fig. 6) the strong pinning contribution is responsible for the low–field plateau and the subsequent $j_c \propto B^{-1/2}$ decrease, while the collective pinning contribution yields the high–field constant j_c. It can be written in terms of superconducting parameters, defect density n_d, and the elementary pinning force f_p of a single defect, as

$$j_c^{coll} = j_0 \left(\frac{n_d \langle f_p^2 \rangle \xi_{ab}^3}{\varepsilon_\lambda \varepsilon_0} \right)^{2/3}. \tag{3}$$

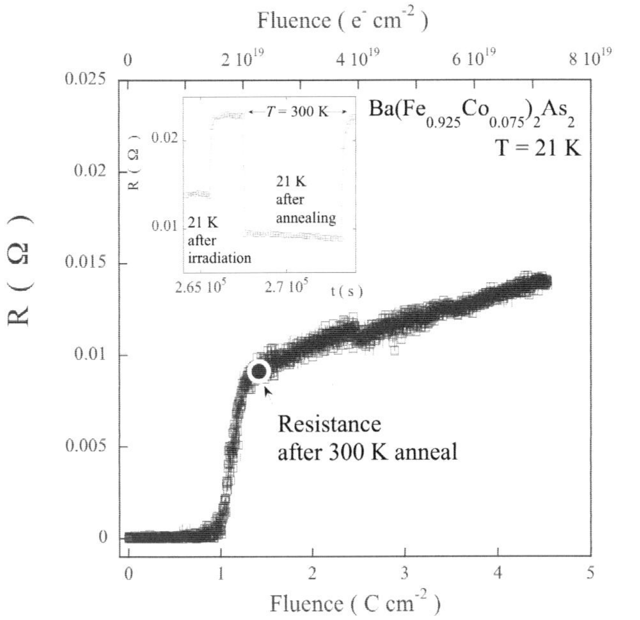

Figure 4. $Ba(Fe_{0.925}Co_{0.075})_2As_2$: Resistance as function of dose of 2.5 MeV electrons, measured at 21 K. The sample transits to the normal state after a dose of 1.2 Ccm^{-2}, after which the resistance increases at a rate of $\delta R/R = 0.14[Ccm^{-2}]^{-1}$. The Inset shows the effect of annealing. The resistance of the crystal is represented as function of time. The protocol comprises initial measurements at 21 K following irradiation with 4.6 Ccm^{-2} 2.5 MeV electrons. The crystal is then warmed to 300 K, cooled to 21 K, and heated once again to 300 K. After the first anneal at 300 K, the resistance drops back to the value reached after low-temperature irradiation with only 1.4 Ccm^{-2} (white circle on the curve in the main panel).

Here, j_0 is the depairing current density, ξ_{ab} and λ_{ab} are the ab–plane coherence length and penetration depth respectively, $\varepsilon_0 \equiv \Phi_0^2/4\pi\mu_0\lambda_{ab}^2$ is the vortex line energy, ε_λ is the penetration depth anisotropy [51], and $\Phi_0 = h/2e$ is the flux quantum. The averaging $\langle \ldots \rangle$ is performed over the vortex core.

The magnitude of j_c^{coll} is compatible with scattering of the quasi-particles in the vortex cores as being at the origin of the weak collective pinning contribution in pristine iron-based superconductors, provided that the dopant atoms are the scattering defects. This gives rise to [52, 53]

$$f_p = 0.3g(\rho_D)\varepsilon_0 \left(\frac{\sigma_{tr}}{\pi\xi_{ab}^2}\right)\left(\frac{\xi_0}{\xi_{ab}}\right) \quad (4)$$

which depends not only on the Gor'kov impurity parameter $\rho_D = \hbar v_F/2\pi T_c l = \xi_0/l$, but also on the transport cross-section $\sigma_{tr} = (2\pi/k_F^2)\sin^2\delta_0$ (δ_0 is the scattering phase angle, and k_F the Fermi wavevector). The analysis of the intermediate field critical current density of several common iron based superconductors allows one to estimate the scattering parameters compiled in Table 1. The weak collective pinning contribution is absent in isovalently doped $BaFe_2(As_{1-x}P_x)_2$ for all doping levels x [see Ref. [58] and Fig. 9(a)]; surprisingly, it is present in $Ba(Fe_{1-x}Ru_x)_2As_2$ [54] and LiFeAs [73].

4. Results and Discussion

4.1. $Ba(Fe_{1-x}Co_x)_2As_2$

Fig. 4 shows the resistance of an optimally doped $Ba(Fe_{0.925}Co_{0.075})_2As_2$ single crystal, measured *in-situ* during exposure to the 10 μA, 2.5 MeV beam, at T = 21 K. The sample transits to the normal state at a fluence of 1.2 Ccm^{-2}, after which the resistance increases at a rate of $\delta R/R = 0.14[Ccm^{-2}]^{-1}$. After a fluence of 4.6 Ccm^{-2} was reached, the electron beam was stopped, the sample heated to 300 K, and the resistance measured. The crystal was subsequently cooled down once again to 21 K, and the resistance measured again (Inset to Fig. 4). It appears that after the 300 K anneal, the resistance has dropped to the value first

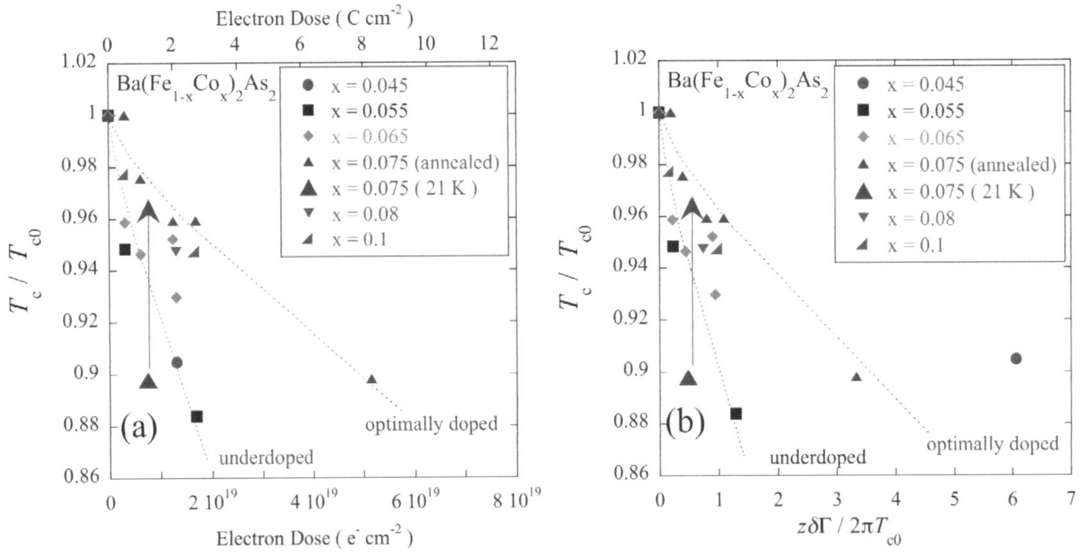

Figure 5. $Ba(Fe_{1-x}Co_x)_2As_2$: evolution of T_c (normalized by the initial critical temperature T_{c0}) as function of (a) the fluence of 2.5 MeV electrons (b) the estimated normal state scattering rate Γ (normalized by T_{c0}), for different doping levels x. The small data points show the T_c–values after annealing at 300 K. The large green triangle shows the drop of T_c following low-temperature irradiation of optimally doped $Ba(Fe_{0.925}Co_{0.075})_2As_2$, the arrow shows the effect of annealing at 300 K.

measured after irradiation with a fluence of 1.4 Ccm^{-2}. Therefore, annealing at 300 K leads to the annihilation and clustering of the point defects produced by the irradiation, with a resulting drop of the irradiation–induced resistance change of 65 %. Since all data presented below concern *ex-situ* experiments performed after heating the crystals to 300 K, the effect of annealing should be taken into account.

The evolution of the reduced critical temperature T_c/T_{c0} with electron fluence (after 300 K annealing) is shown for various Co-substitution levels x in Fig. 5(a). Here, T_{c0} is the critical temperature of the crystal before irradiation. It is to be stressed that, given the modest T_c changes involved, and the typical dispersion of T_c–values between similarly grown samples, even from the same batch, it is essential to compare the values of T_c and T_{c0} on the same crystal before and after irradiation. For the same reason, a plot of the absolute T_c versus dose would be meaningless in the regime of small doses. Fig. 5 also shows the effect of 300 K annealing, for one sample. It is seen that the change in T_c drops by a factor three after annealing.

Fig. 5(b) shows the same data, as function of the product of the induced change in the scattering rate $\delta\Gamma/2\pi T_c$ and the electronic effective mass enhancement $z = m^*/m_e$. Here m_e is the free electron mass, and m^* the effective mass. $z\delta\Gamma = e\delta\rho/R_H m_e$ (with e the electronic charge) is estimated from the change of the resistance due to the irradiation (after annealing), together with the published data for the resistivity ρ and the Hall coefficient R_H published in Ref. [68]. In spite of the contributions of multiple bands to the conductivity of $Ba(Fe_{1-x}Co_x)_2As_2$, this approach can be justified given that the conductivity contribution from the electron sheets dominates the transport properties [68]. Typically, $\delta R/R \sim 0.05$ [Ccm^{-2}]$^{-1}$ for the optimally doped sample (cf Fig. 3). Note, however, that the resistivity change with electron dose depends on x, and that, in order to produce the "simulated" $\delta\Gamma$ values of Fig. 5b, we use data measured in-situ on crystals with different x. Estimates of the scattering rate from

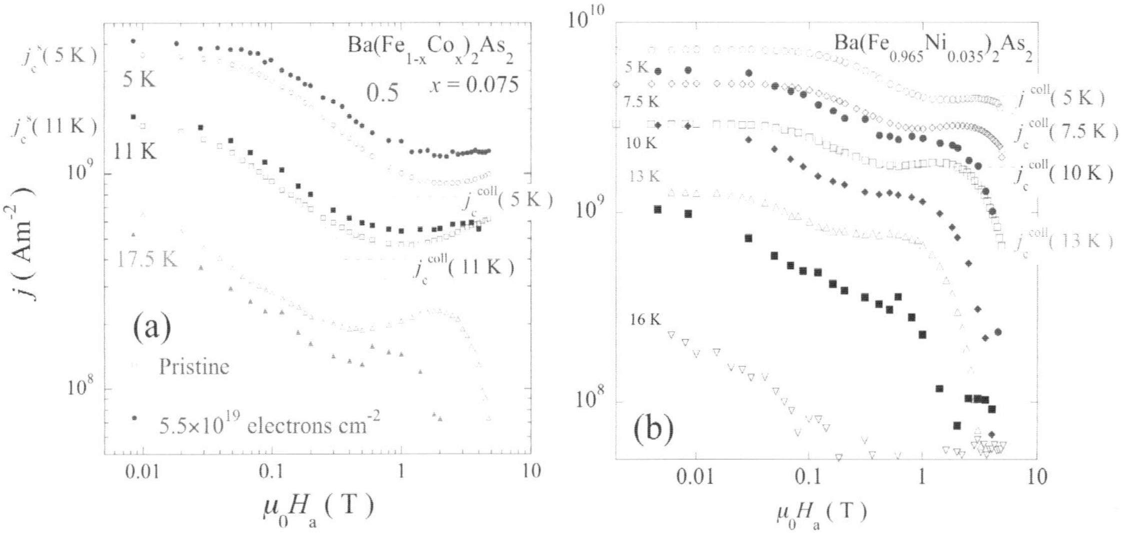

Figure 6. Sustainable current density j of (a) optimally doped $Ba(Fe_{1-x}Co_x)_2As_2$ ($x = 0.075$) and (b) optimally doped $Ba(Fe_{1-x}Ni_x)_2As_2$ (x = 0.035) before (open symbols) and after irradiation at 21 K with 5.5×10^{19} cm^{-2} 2.5 MeV electrons (closed symbols). The measurements were performed at 5 K (○ , ●), 7.5 K (◇ , ◇), 10 K (□ , ■), 11 K (□ , ■), 13 K (△), 16 K (▽), and 17.5 K (△ , △).

critical current measurements (see below) yield qualitatively similar results as those shown in the Figure..

The surface impedance of optimally doped $Ba(Fe_{1-x}Co_x)_2As_2$ before and after electron irradiation is depicted in Fig. 3. A monotonous increase of the surface resistance is observed after irradiation; even though in this series of experiments, the increase was not linear as function of fluence. In particular, the crystal irradiated with 2.1 Ccm^{-2} electrons did not conform to the general trend. The normalized frequency shift of the Nb cavity, which is proportional to the superfluid density $n_s \propto \lambda_{ab}^{-2}$ of the superconducting crystals, showed a slight change to a less marked temperature dependence, indicative of disorder having been added to the crystal. The trend is similar to that observed by Hashimoto *et al.* in $Ba_{1-x}K_xFe_2As_2$ crystals with varying degrees of disorder [75], but is much less marked here.

The low-temperature critical current density of single crystalline $Ba(Fe_{0.925}Co_{0.075})_2As_2$ increases by a constant, field-independent amount after electron irradiation (see Fig. 6a). This indicates that it is the weak collective pinning contribution j_c^{coll} that is enhanced, presumably by atomic-sized point defects introduced by the irradiation. If one assumes that the relevant defects are Fe vacancy-interstitial pairs, *i.e.* that $\sigma_{tr} \sim \pi D_{Fe}^2 \sim 2.6$ Å^2, with $D_{Fe} = 0.92$ Å the ionic radius of Fe^{2+}, the increase of the critical current density is consistent, through Eqs. (3) and (4), with 0.006 dpa Fe / Ccm^{-2}. The increase of the low-temperature critical current density and the concomitant decrease of T_c after irradiation implies a steeper decrease of j_c with temperature, and a crossing of j_c-values of the pristine and the irradiated crystal at an intermediate temperature.

4.2. $Ba(Fe_{1-x}Ni_x)_2As_2$

Surface impedance measurements and DMO imaging on the irradiated $Ba(Fe_{1-x}Ni_x)_2As_2$ crystals show a more rapid depression of T_c with dose than in $Ba(Fe_{1-x}Co_x)_2As_2$. If anything,

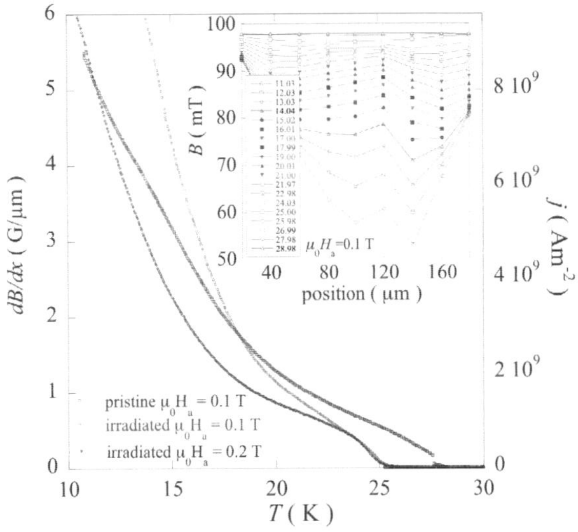

Figure 7. Temperature dependence of the flux density gradient dB/dx in a $BaFe_2(As_{0.64}P_{0.36})_2$ single crystal before and after irradiation with 0.8×10^{19} electrons cm^{-2}. Inset: Flux density profile across the crystal after zero-field cooling, application of the applied field $\mu_0 H_a = 100$ mT, and subsequent warming.

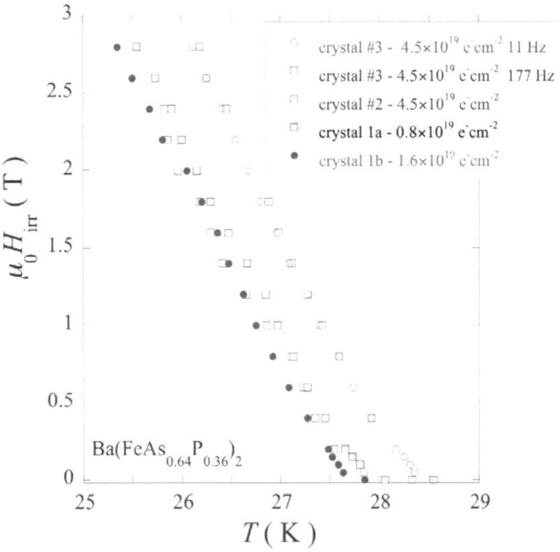

Figure 8. Irreversibility field $BH_{irr}(T)$ as function of temperature, for $BaFe_2(As_{0.64}P_{0.36})_2$ single crystals irradiated with 0.8×10^{19} (□), 1.6×10^{19} (•), and 4.5×10^{19} electrons cm^{-2} (○, □, □). $B_{irr}(T)$ is determined as the locus of temperatures T above which the nonlinearity of the crystal current-voltage characteristic vanishes at given induction H_a. T_c is determined from the extrapolation of $H_{irr}(T)$ to zero field.

T_c versus dose is comparable to that found in underdoped $Ba(Fe_{1-x}Co_x)_2As_2$ (see Fig. 11). As in $Ba(Fe_{1-x}Co_x)_2As_2$, the temperature dependence of the superfluid density of $Ba(Fe_{1-x}Ni_x)_2As_2$ changes little or not at all, even at the largest irradiation dose of 5.5×10^{19} electrons cm^{-2}.

The critical current density of the $Ba(Fe_{1-x}Ni_x)_2As_2$ crystals is strongly suppressed after the irradiation (see Fig. 6b). This is partially because the reduced measurement temperatures are higher in $Ba(Fe_{1-x}Ni_x)_2As_2$ than in its co-substituted counterpart, but mainly because the suppression of the prefactor $j_0 \propto n_s e(\Delta/hk_F)$ in Eq. (3) outweighs the increase in n_d (Δ is the superconducting gap amplitude).

4.3. $BaFe_2(As_{1-x}P_x)_2$

The Inset to Fig. 7 shows an example of flux density gradients measured on a $BaFe_2(As_{0.64}P_{0.36})_2$ single crystal irradiated with 0.9×10^{19} electrons cm^{-2}, after zero field cooling and the application of an external field of 100 mT. The temperature dependence of the flux density gradient obtained after field-cooling in 400 mT, reduction of the applied field to the indicated target value, and subsequent warming is shown in the main panel of the Figure. The flux density gradient is directly proportional to the sustainable current density j. As in $Ba(Fe_{1-x}Co_x)_2As_2$, the screening current in the irradiated crystal exceeds that of the pristine crystal at low temperature, but drops below it at higher temperature, due to the decrease of T_c.

The critical temperatures after electron-irradiation were determined as the extrapolation

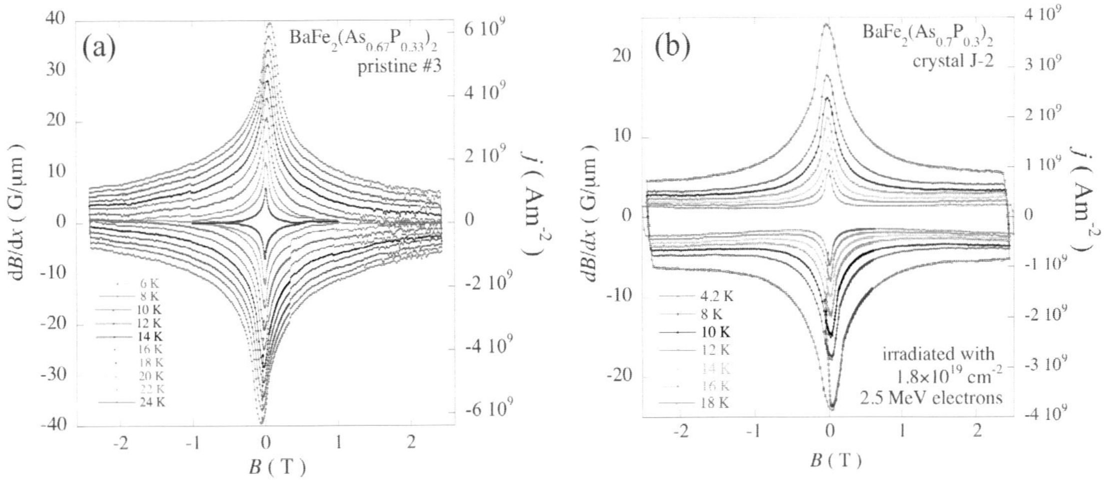

Figure 9. (a) Hysteretic loops of the local flux density gradient versus local induction B, measured on the surface of a pristine $BaFe_2(As_{0.67}P_{0.33})_2$ single crystal, at various temperatures (indicated). The astroid-shaped hysteresis loops are determined by the sole strong-pinning contribution to the critical current. (b) *ibid*, measured on the surface of a $BaFe_2(As_{0.7}P_{0.3})_2$ single crystal irradiated at 23 K with 1.8×10^{19} electrons cm^{-2}, at the indicated temperatures. The opening of the loops at higher flux densities reveal the emergence of a weak collective pinning contribution by the atomic–sized point defects introduced by the irradiation.

to zero of the "irreversibility field" $H_{irr}(T)$ above which the nonlinearity of the current-voltage characteristic – and therefore the critical current density – vanishes. Details on the determination of H_{irr}, can be found in, *e.g.*, Ref. [74]. In all cases, the $H_{irr}(T)$ values depend very little on the frequency of the ac field used to investigate the screening by the superconducting sample; the corresponding $T_{irr}(H)$ lies very close to the temperatures at which the dc screening current vanishes in Fig. 7. The T_c–values resulting from the extrapolation of $H_{irr}(T)$ to zero are gathered in Fig. 11. The T_c–depression as function of electron fluence is comparable to that in $Ba(Fe_{1-x}Ni_x)_2As_2$. Note that the resistivity increase of the irradiated $BaFe_2(As_{1-x}P_x)_2$ crystals corresponds to $\delta R/R \sim 0.16$ $[Ccm^{-2}]^{-1}$, three times higher than for $Ba(Fe_{1-x}Co_x)_2As_2$.

Figure 9 shows the hysteretic loops of the local flux density gradient on the surface of a pristine (a) and electron–irradiated $BaFe_2(As_{1-x}P_x)_2$ single crystal (b), for different temperatures. The pristine single crystal shows astroid–shaped magnetic hysteresis loops that are indicative of strong flux pinning by extended point-like defects only. After electron irradiation, the hysteresis loops open up due to the appearance of the field–independent collective pinning contribution from atomic-sized point defects. We interpret the appearance of a non-zero j_c^{coll} as being due to vortex pinning by vacancy-interstitial pairs introduced by the irradiation. Moreover, j_c^{coll} monotonically increases (at low temperature) as function of irradiation fluence (Fig. 10), which allows for a direct comparison with the theory for quasi-particle scattering mediated collective vortex pinning. Under the assumption that the only unknown parameter, σ_{tr}, is again approximated by the ionic cross–section of a Fe-vacancy, a fit to Eq. (3) yields a defect density of 0.0035 dpa / Ccm^{-2}. This number is taken as more precise than that obtained for $Ba(Fe_{1-x}Co_x)_2As_2$ in subsection 4.1, in which only two points were available. Note that the hypothesis of simple voids (non-scattering point defects) would necessitate an unphysical 1 dpa / Ccm^{-2} to explain the magnitude of the critical current density change.

4.4. Discussion; Comparison between differently substituted materials

The modification of the critical current density of $Ba(Fe_{1-x}Co_x)_2As_2$ and $BaFe_2(As_{1-x}P_x)_2$ by electron irradiation demonstrates that the field–independent contribution to j_c apparent in sub-T to T–fields is indeed due to collective pinning by atomic-sized point defects. The very reasonable number obtained for defect generation by the electron beam, 0.35% dpa / Ccm^{-2}, indicates that it is, in a first approximation, safe to assume that the most relevant produced defects are Fe vacancies, and that these vacancies are responsible for quasi-particle scattering. This interpretation is strengthened by the fact that scattering rates in $Ba(Fe_{1-x}Co_x)_2As_2$ estimated from the change of the resistivity are very much comparable to $\Gamma = n_d/[\pi N(0)]/\sin^2\delta_0$ estimated from the defect density. A comparison with Table 1, shows that the scattering cross-section of Fe vacancies should be comparable to that of Co and Ru impurities. Indeed, electron–irradiation of $Ba(Fe_{1-x}Co_x)_2As_2$ and $Ba(Fe_{1-x}Ni_x)_2As_2$ does not lead to appreciable changes of the temperature dependence of the superfluid density of those materials, which suggests that the effect of the dopant impurities overwhelms that of the defects added by the irradiation.

Irrespective of the precise changes in resistivity incurred by the irradiation, a striking feature of our results is the similar suppression of T_c by the electron irradiation of underdoped $Ba(Fe_{1-x}Co_x)_2As_2$, $Ba(Fe_{1-x}Ni_x)_2As_2$ and $BaFe_2(As_{1-x}P_x)_2$ for a given dose. The T_c–suppression in optimally doped $Ba(Fe_{1-x}Co_x)_2As_2$ is a factor 2 less. Given the clear evidence for line nodes in $BaFe_2(As_{1-x}P_x)_2$, this poses the question possible nodal structures of the order parameter in $Ba(Fe_{1-x}Ni_x)_2As_2$, and, possibly, $Ba(Fe_{1-x}Co_x)_2As_2$ as well. In the latter two materials, it was proposed that a nodal line might exist on the α–(hole–like) sheet, at finite k_z [35, 12].

It is tempting to compare our results to that obtained by other types of irradiation, as well as by doping. It turns out that the suppression of T_c for optimally doped $Ba(Fe_{1-x}Co_x)_2As_2$ is remarkably comparable to that reported for Co-doping of $Ba_{1-x}K_xFe_2As_2$ in Ref. [76], and much less than that obtained by Co-doping of KFe_2As_2 studied by the same authors.

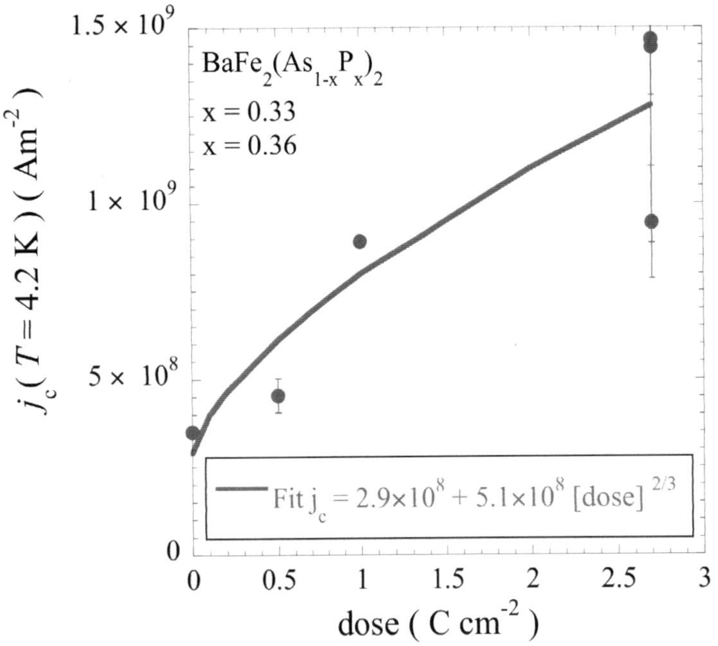

Figure 10. $BaFe_2(As_{1-x}P_x)_2$ (with $x = 0.33$, $x = 0.36$): Electron-fluence dependence of the weak collective pinning contribution j_c^{coll} to the critical current density. The drawn line denotes a fit to Eqs. (3,4). Assuming that the point defects most relevant for quasi-particle scattering are Fe vacancies induced by the irradiation, the parameter value 5.1×10^8 would correspond to 0.0035 dpa / Ccm^{-2}.

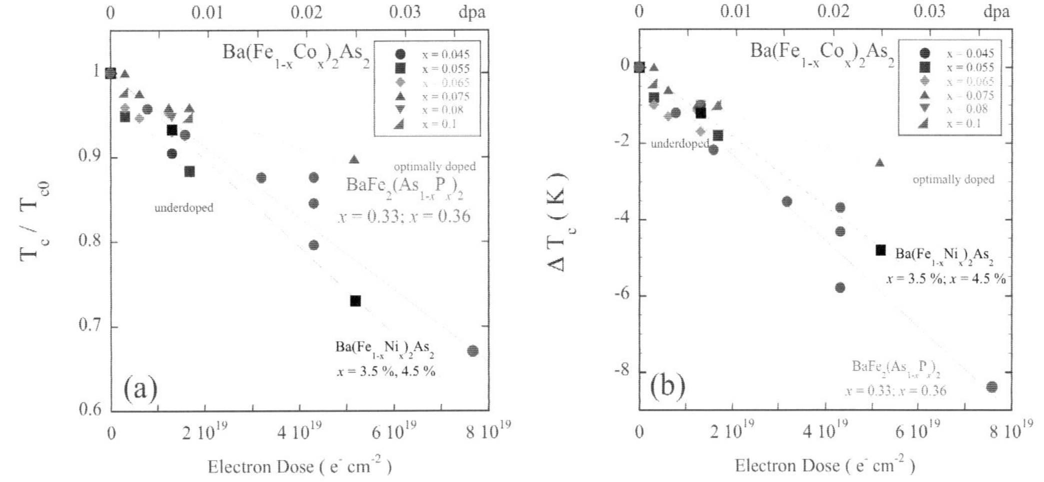

Figure 11. (a) Dose dependence of the T_c-change of $Ba(Fe_{1-x}Co_x)_2As_2$, $Ba(Fe_{1-x}Ni_x)_2As_2$, and $BaFe_2(As_{1-x}P_x)_2$, after irradiation with 2.5 MeV electrons and annealing at 300 K. The upper scale shows the density of point defects (presumably Fe vacancies) added by the irradiation, such as determined from the dose-dependence of the critical current density of $BaFe_2(As_{1-x}P_x)_2$ (see Fig.10). (b) Dose dependence of T_c/T_{c0}, where T_{c0} is the critical temperature of the pristine crystal.

The decrease of T_c observed after electron irradiation is also comparable, as far as optimally doped $Ba(Fe_{1-x}Co_x)_2As_2$ is concerned, to the Co-doping experiments on $Ba_{1-x}K_xFe_2As_2$ by Li *et al.*, while the T_c suppression in underdoped $Ba(Fe_{1-x}Co_x)_2As_2$, $Ba(Fe_{1-x}Ni_x)_2As_2$, and $BaFe_2(As_{1-x}P_x)_2$ is similar to the Cu and Zn doping by those authors. Even if the comparison with chemical doping is tenuous, the observed trend clearly indicates a much weaker sensitivity of the materials studied here than what is expected for the scenario of $s_\pm$ superconductivity with strong interband scattering, and sets the three materials even further from the *d*-wave scenario.

A comparison with the previous irradiation studies reveals that T_c-suppression after electron irradiation is somewhat weaker than the results obtained by Nakajima by 3 MeV proton irradiation of optimally doped $Ba(Fe_{1-x}Co_x)_2As_2$ [63], and much weaker than in the α-particle irradiation of NdFeAs(O,F) by Tarantini *et al.* [62].

5. Summary and Conclusions

$Ba(Fe_{1-x}Co_x)_2As_2$ crystals of different dopng levels x, as well as $Ba(Fe_{1-x}Ni_x)_2As_2$ and $BaFe_2(As_{1-x}P_x)_2$ crystals, have been subjected to high energy (2.5 MeV) electron irradiation at low temperature (21 K). In spite of substantial annealing and clustering of the defects, the enhancement of the collective pinning contribution to the critical current density due to quasi-particle scattering in the vortex cores in $Ba(Fe_{1-x}Co_x)_2As_2$, and its appearance in $BaFe_2(As_{1-x}P_x)_2$ in which it is absent in the pristine material, demonstrate that the irradiation produces atomic-sized point defects. The increase in magnitude of the low-temperature critical current density is consistent with the pinning by Fe vacancies, created by the irradiation. In $Ba(Fe_{1-x}Co_x)_2As_2$ and $Ba(Fe_{1-x}Ni_x)_2As_2$, the introduction of these supplementary point defects does not lead to a significant modification of the temperature dependence of the superfluid density, which suggests that the defects are overwhelmed by pre-existing disorder (notably, the presence of the dopant atoms). Surprisingly, the critical temperature is similarly suppressed in all three materials. This, in spite of the fact that the order parameter in $BaFe_2(As_{1-x}P_x)_2$ is thought to have line nodes, while this possibility is much less certain in the other two materials. The results therefore lend credence to evidence for line nodes obtained from c-axis penetration depth [12] and thermal conductivity measurements [35].

Acknowledgements
We acknowledge support from the the grant MagCorPnic of the Réseau Thématique de Recherche Avancée "Triangle de la Physique du Plateau de Saclay", and by the Agence Nationale de la Recherche grant "PNICTIDES". Part of the work was also possible thanks to the support of the ECOS-Sud-MINCyt France-Argentina bilateral program, Grant No.A09E03. Work in Japan was supported by KAKENHI from JSPS, and by Grant-in-Aid for the Global COE program The Next Generation of Physics, Spun from Universality and Emergence from MEXT, Japan.

References

[1] Singh D J and Du M H 2008, *Phys. Rev. Lett.* **100** 237003
[2] Mazin I I , DJ Singh D J, Johannes M D, and Du M H 2008, *Phys. Rev. Lett.* **101** 057003
[3] Kuroki K , Onari S, Arita R, Usui H, Tanaka Y, Kontani H, and Aoki H, 2008, *Phys. Rev. Lett.* **101** 087004; *ibid 2009* , *Phys. Rev. Lett.* **102** 109902
[4] Kuroki K, Usui H, Onari S, Arita R, and Aoki H 2009, *Phys Rev B* **79** 224511
[5] Onari S and Kontani H 2009, *Phys. Rev. Lett.* **103** 177001
[6] Kontani H and Onari S 2010, *Phys. Rev. Lett.* **104** 157001
[7] Glatz A and Koshelev A E 2010, *Phys Rev* B **82** 012507
[8] Martin C, Tillman M E, H Kim, Tanatar M A, Kim S K, Kreyssig A, Gordon R T, Vannette M D , Nandi S, Kogan V G , Budko S L, Canfield P C, Goldman A I, and Prozorov R 2009, *Phys. Rev. Lett.* **102** 247002
[9] Kogan V G 2009, *Phys Rev* B **80** 214532
[10] Gordon R T, Martin C, Kim H, Ni N, Tanatar M A, Schmalian J, Mazin I I, Budko S L, Canfield P C, and Prozorov R 2009, *Phys. Rev.* B **79** 100506
[11] Gordon R T , Ni N , Martin C, Tanatar M A, Vannette M D, Kim H, Samolyuk G D, Schmalian J, Nandi S, Kreyssig A, Goldman A I, Yan Q J, Budko S L, Canfield P C, and Prozorov R 2009, *Phys. Rev. Lett.* **102** 127004
[12] Martin C, Gordon R T , Tanatar M A, Kim H, Ni N, Budko S L, Canfield P C, Luo H, Wen H H , Wang Z, Vorontsov A B, Kogan V G, and Prozorov R 2009, *Phys. Rev.* B **80** 020501
[13] Gordon R T, Kim H, Tanatar M A, Prozorov R, Kogan V G 2010, *Phys. Rev.* B **81** 180501 (*Preprint* arXiv:09125346)
[14] Prozorov R and Kogan V G 2011, *Rep. Prog. Phys.* **74** 124505
[15] Hirschfeld P J, Korshunov M M, and Mazin I I 2011, *Rep. Prog. Phys.* **74** 124508.
[16] Mishra V, Boyd G, Graser S, Maier T, Hirschfeld P J, and Scalapino D J 2009, *Phys. Rev.* B **79** 094512
[17] Jiang S, Xing H, Xuan G, Wang C, Ren Z, Feng C, Dai J, Xu Z, Cao G 2009, *J. Phys.: Condens. Matter* **21** 382203
[18] Wang C, Jiang S, Tao Q, Ren Z, Li Y, Li L, Feng C, Dai J, Cao G, Xu Z 2009, *EPL* **86** 47002
[19] Thirupathaiah S, Rienks E D L, Jeevan H S, Ovsyannikov R, Slooten E, Kaas J, van Heumen E, de Jong S, Duerr H A, Siemensmeyer K, Follath R, Gegenwart P, Golden M S, Fink J 2011, *Phys. Rev. B* **84** 014531
[20] Yoshida T, Nishi I , Ideta S, Fujimori A, Kubota M, Ono K, Kasahara S, Shibauchi T, Terashima T, Matsuda Y, Ikeda H, and Arita R 2011, *Phys. Rev. Lett.* **106** 117001
[21] Suzuki K, Usui H, and Kuroki K 2011, *J. Phys. Soc. Jpn.* **80** 013710
[22] Andersen O K and Lilia Boeri L 2011, *Annalen der Physik* **1** 8-50
[23] Ahilan K, F. L. Ning F L, T. Imai T, Sefat A S, McGuire M A, Sales B C, Mandrus D 2011, *Phys. Rev. B* **79** 214520
[24] Colombier E, Bud'ko S L, Ni N, and Canfield P C 2009, *Phys. Rev.* B **79** 224518
[25] Hassinger E, Valade F, Rene de Cotret S, Juneau-Fecteau A, Reid J-Ph, Kim H, Tanatar M A, Prozorov R, Shen B, Wen H H, Doiron-Leyraud N, and Taillefer L 2012, (*Preprint* arXiv:1205.6730)
[26] Putzke C, A. I. Coldea A I, Guillamón I, Vignolles D, McCollam A, LeBoeuf D, Watson M D, Mazin I I, Kasahara S, Terashima T, Shibauchi T, Matsuda Y, and Carrington A 2012, *Phys. Rev. Lett.* **108** 047002
[27] Hashimoto K, Kasahara S, Katsumata R, Mizukami Y, Yamashita M, Ikeda H, Terashima T, Carrington A, Matsuda Y, and T. Shibauchi T 2012, *Phys. Rev. Lett.* **108** 047003
[28] Hashimoto K, Serafin A, Tonegawa S, Katsumata R, Okazaki R, Saito T, Fukazawa H, Kohori Y, K. Kihou K, Lee C H, Iyo A, Eisaki H, Ikeda H, Matsuda Y, Carrington A, and Shibauchi T 2010, *Phys. Rev.* B **82** 014526

[29] Fletcher J D, Seran A, Malone L, Analytis J, Chu J-H, Erickson A S, Fisher I R, and Carrington A 2009, *Phys. Rev. Lett.* **102** 147001
[30] Reid J Ph, Tanatar M A, Juneau-Fecteau A, Gordon R T, Rene de Cotret S, Doiron-Leyraud N, Saito T, Fukazawa H, Kohori Y, Kihou K, Lee C H, Iyo A, Eisaki H, Prozorov R , and Taillefer L 2012, *Phys. Rev. Lett.* **109** 087001
[31] Maier T A , Graser S, Hirschfeld P J, and Scalapino D J 2011, Phys. Rev. B **83** 100515
[32] Hashimoto K, Yamashita M, Kasahara S, Senshu Y, Nakata N, Tonegawa S , Ikada K, Serafin A, Carrington A, Terashima T , Ikeda H, Shibauchi T, and Matsuda Y 2010, *Phys. Rev.* B **81**, 220501 (R) (*Preprint* arXiv:09074399)
[33] Yamashita M, Senshu Y, Shibauchi T, S. Kasahara, K. Hashimoto, D. Watanabe, H. Ikeda, T. Terashima, I. Vekhter, A. B. Vorontsov, and Y. Matsuda Y 2011, *Phys. Rev.* B **84** 0605075
[34] C Martin, H Kim, Gordon RT, Ni N , Kogan V G, Budko S L, Canfield P C , Tanatar M A, and Prozorov R 2010, *Phys Rev.* B **81** 060505
[35] Reid J-Ph, Tanatar M A, Luo X G, Shakeripour H, Doiron-Leyraud N, Ni N, Budko S L, Canfield P C, Prozorov R, and Louis Taillefer L, *Phys. Rev.* B **82** 064501
[36] Cho K, Tanatar M A, Spyrison N, Kim H, Song Y, Dai P, Zhang C L, and Prozorov R 2012 *Phys. Rev.* B **86** 02008
[37] Reid J Ph, Tanatar M A, Luo X G, Shakeripour H, S. René de Cotret S, Doiron-Leyraud N, Chang J, Shen B, Wen H H, Kim H, Prozorov R, and Taillefer L 2012, *Phys. Rev. Lett.* **109** 087001 (*Preprint* arXiv:1105.2232)
[38] Maier T A, Graser S, Scalapino D J, and Hirschfeld P J 2009, *Phys. Rev. B* **79** 224510
[39] Graser S, Kemper A F, Maier T A, Cheng H-P, Hirschfeld P J, and Scalapino D J 2010, *Phys. Rev. B* **81** 214503
[40] Hirschfeld P J and Scalapino D J 2010, *Physics* **3** 64
[41] Efremov D V, Korshunov M M, Dolgov O V, Golubov A A, and Hirschfeld P J 2011, *Phys. Rev. B* **84**, 180512(R)
[42] Hardy F, Burger P, Wolf T, Fisher R A, Schweiss P, Adelmann P, Heid R, Fromknecht R, Eder R, D. Ernst1, v. Löhneysen H, and C. Meingast C 2010, *EPL (Europhysics Letters)* **91** 47008
[43] Mu G, Zeng B, Cheng P, Wang Z, Fang L, Shen B, Shan L, Ren C, Wen HH 2010, *Chin Phys Lett.* **27** 037402
[44] van der Beek C J, Rizza G, Konczykowski M, Fertey P, Monnet I, Klein T, Okazaki R, Ishikado M, Kito H, Iyo A, Eisaki H, Shamoto S, Tillman M E, Bud'ko S L, Canfield P C, Shibauchi T, and Matsuda Y 2010, *Phys Rev B* **81** 174517
[45] van der Beek C J, Konczykowski M, Kasahara S, Terashima T, Okazaki R, I, Shibauchi T, and Matsuda Y 2010, *Phys. Rev. Lett.* **105** 267002
[46] Yamamoto A, Jaroszynski J, Tarantini C, Balicas L, Jiang J, Gurevich A, Larbalestier D C, Jin R, Sefat A S, McGuire M A, Sales B C, Christen D K, and Mandrus D 2009, *Appl. Phys. Lett.* **94**, 062511
[47] Demirdiş S, van der Beek C J, Fasano Y, Cejas-Bolecek N R, Pastoriza H, Colson D, and Rullier-Albenque F 2011, *Phys. Rev. B* **84** 094517
[48] van der Beek C J, Demirdis S, Konczykowski M, Fasano Y, Cejas Bolecek N R, Pastoriza H, Colson D, Rullier-Albenque F 2012 *Physica B* **407** 1746
[49] Larkin A I and Ovchinnikov Yu 1979, *J. Low Temp. Phys.* **21** 409
[50] Blatter G, Feigel'man M V, Geshkenbein V B, Larkin A I, and Vinokur V M 1994, *Rev Mod Phys* **66** 1125
[51] van der Beek C J, Konczykowski M, and Prozorov R 2012, *Supercond. Sci. Technol.* **25** 084010
[52] Thuneberg E V, Kurkijärvi J, and Rainer D 1982, *Phys. Rev. Lett.* **48** 1853
[53] Thuneberg E V, Kurkijärvi J, and Rainer D 1984, *Phys Rev B* **29** 3913
[54] Konczykowski M, Demirdiş S, van der Beek C J, Prozorov R, Tanatar M A, Canfield P C, Kasahara S, Shibauchi T , and Matsuda Y 2011, *Bulletin of the American Physical Society* **56** 1
[55] Wang X-L, Ghorbani S R, Lee S-I, Dou S X, Lin C T, Johansen T H, Cheng Z X, Peleckis G, Muller K, Shabazi M, Sun G L, and Sun D L 2010, *Phys. Rev. B* **82** 024525
[56] Yang H, Luo H, Wang Z, and Wen H H 2008, *Appl. Phys. Lett.* **93** 142506
[57] Shishido H, Bangura A F, Coldea A I, Tonegawa S, Hashimoto K, Kasahara S, Rourke P M C, Ikeda H, Terashima T, Settai R, Onuki Y, Vignolles D, Proust C, Vignolle B, McCollam A, Matsuda Y, Shibauchi T, and Carrington A 2010, *Phys. Rev. Lett.* **104** 057008
[58] Demirdiş S, Fasano Y, van der Beek C J, Pastoriza H, Konczykowski M, Kasahara S, Shibauchi T, and Matsuda Y 2012 (*Preprint*)
[59] Cheng P, Shen B, Hu J, and Wen H H 2010, *Phys. Rev. B* **81** 174529
[60] Li J, Guo Y, Zhang S, Yu S, Tsujimoto Y, Kontani H, Yamaura K, Eiji Takayama-Muromachi E 2011, *Phys. Rev. B* **84** 020513
[61] Li J, Guo Y F, Zhang S B, Yuan J, Tsujimoto Y, Wang X, Sathish C I, Sun Y, Yu S, Yi W , Yamaura K,

E. Takayama-Muromachiu E, Shirako Y, Akaogi M, and H. Kontani 2012, (*Preprint* ArXiv:1206.0811)
[62] Tarantini C, Putti M, Gurevich A, Shen Y, Singh R K , Rowell J M, Newman N, Larbalestier D C, Cheng P, Jia Y, and Wen H H 2010 , *Phys. Rev. Lett.* **104** 087002
[63] Nakajima Y, Taen T, Tsuchiya Y, Tamegai T, Kitamura H, and Murakami T 2011, *Phys. Rev B* **82** 220504
[64] Rullier-Albenque F, Vieillefond P A, Alloul H, Tyler A W, Lejay P, and Marucco J F 2000, *Europhys. Lett.* **50** 81
[65] Rullier-Albenque F, Alloul H, and Tourbot R 2003, *Phys. Rev. Lett.* **91** 047001
[66] Spathis P, Colson S, Yang F, van der Beek C J, Gierlowski P, Shibauchi T, Matsuda Y, Gaifullin M, Li M, and Kes P 2008, *Phys. Rev B* **77** 104503
[67] Beuneu F and Bois P 1988, *J. Phys. F, Metal Phys.* **37** 6041
[68] Rullier-Albenque F, Colson D, Forget A, and Alloul H 2009, *Phys. Rev. Lett.* **103**, 057001
[69] Olariu A, Rullier-Albenque F, Colson D, and Forget A 2011, *Phys. Rev. B* **83** 054518
[70] Kasahara S, Shibauchi T, Hashimoto K, Ikada K, Tonegawa S, Okazaki R, Sishido H, Ikeda H, Takeya H, Hirata H, Terashima T, and Matsuda Y 2010, *Phys. Rev. B* **81** 184519
[71] See http://www.lsi.polytechnique.fr/accueil/equipements/accelerateur-sirius/
[72] Brandt E H 1996, *Phys. Rev. B* **54** 4246
[73] Konczykowski M,van der Beek C J, Tanatar M A, Mosser V, Song Y J, Kwon YY S, and Prozorov R 2012, *Phys. Rev. B* **84**180514(R)
[74] Konczykowski M,van der Beek C J, Tanatar M A, Luo H Q, Wang Z S, Shen B, Wen H H, and Prozorov R 2012, *Phys. Rev. B* **86** 024515
[75] Hashimoto K, Shibauchi T, Kasahara S, Ikada K, Tonegawa S, Kato T, Okazaki R, van der Beek C J, Konczykowski M, Takeya H, Hirata K, Terashima T, and Matsuda Y 2009, *Phys. Rev. Lett.* **102** 207001
[76] Wang A F, Zhou S Y, Luo X G, Hong X C, Yan Y J, Ying J J, Cheng P, Ye G J, Xiang Z J, Li S Y, and Chen X H 2012 (*Preprint* arXiv:1206.2030)

Journal of Physics: Conference Series **449** (2013) 012024 doi:10.1088/1742-6596/449/1/012024

Biquadratic magnetic interaction in parent ferropnictides

A L Wysocki[1,2], K D Belashchenko[1], L Ke[3], M van Schilfgaarde[4] and V P Antropov[3]

[1]Department of Physics and Astronomy and Nebraska Center for Materials and Nanoscience, University of Nebraska-Lincoln, Lincoln, Nebraska 68588, USA
[2]School of Applied and Engineering Physics, Cornell University, Ithaca, New York 14853, USA
[3]Ames Laboratory, Ames, Iowa, 50011, USA
[4]Department of Physics, King's College London, Strand, London WC2R 2LS, United Kingdom

Abstract. We discuss an effective spin Hamiltonian with biquadratic interaction for ferropnictide superconductors from the point of view of band structure theory and available experimental data. This model is consistent with electronic structure calculations and captures many observed magnetic properties, including the anisotropy of the exchange coupling, thin domain walls, and the crossover from first to second-order phase transition under doping. The parameters of the model are analyzed as a function of the local spin moment using first-principles calculations. Calculations show the biquadratic coupling is negative in stoichiometric KFe_2Se_2, and the phase diagram is extended into this region. We also consider magnetic short-range order and discuss the limitations of this model in comparison with experiment.

1. Introduction

Strong evidence points toward spin fluctuations as the dominant pairing mechanisms in ferropnictide superconductors [1]. Therefore, intense attention has been focused on the magnetic properties of these materials, and in particular on understanding the character of spin fluctuations in both parent and doped compounds. An intermediate magnitude of the local moments ($\mu \sim 1\mu_B$) observed in neutron scattering experiments, as well as the fact that parent compounds are metallic, suggest that strong itinerant effects may be expected, but, on the other hand, that an effective model based on localized spin variables may provide a reasonable starting point. It has also been suggested that spin dynamics should be described based on significantly larger local moments, which are strongly reduced as a result of quantum spin fluctuations.

Application of the Heisenberg model to ferropnictides meets with problems. Although the collinear antiferromagnetic (AFM) ground state ("stripe phase") may be obtained in the widely studied J_1-J_2 Heisenberg model with $J_2 > J_1/2$, the magnetic properties are poorly described by this model. First, it is well established both experimentally [2–5] and theoretically [6, 7] that the nearest-neighbor exchange constant (obtained from spin wave analysis or from linear response calculations) is strongly anisotropic in the stripe phase. This anisotropy strongly depends on the local moment and can even change sign at large values of μ [7, 8]. Second, the Heisenberg model of any range fails to describe the energies of noncollinear structures connecting the degenerate AFM domains. While in this model the $(\pi, 0)$ and $(0, \pi)$ AFM domains are connected by a continuously degenerate set of noncollinear states, band structure calculations find a rather high energy barrier between them [7, 9]. This is consistent with the experimental observation of microscopically thin twin domain walls between such stripe states [10]. Within the Heisenberg model such domain walls would be very thick due to the continuous degeneracy (in fact, infinitely thick in the classical description at zero temperature).

These essential features missing in the Heisenberg model can be reintroduced in different ways. It has been argued that anisotropic spin fluctuations are linked to orbital ordering [11–14]. Orbital degrees of freedom can be included in the model either as electronic degrees of freedom coupled to local spins [15, 16], or as Ising variables in a Kugel-Khomski-type model [13, 14]. However, a simpler description of magnetic properties may be achieved by adding non-Heisenberg terms in the effective spin-only Hamiltonian. Biquadratic coupling naturally appears when the total energy of an itinerant electron system is mapped to an effective spin Hamiltonian. This coupling comes from the complicated dependence of the electronic energy on the spin configuration, and a specific mechanism for ferropnictides was proposed [17]. The importance of biquadratic interaction in ferropnictides and related materials was discussed in Refs. [7, 9, 18], and the thermodynamics of this model was studied in Ref. [19]. The viewpoint based on the fully itinerant picture may be found in Refs. [17, 20].

In this paper we discuss the magnetic properties of ferropnictides within the effective spin model with biquadratic interaction [19]. After briefly reviewing the model in Section 2, we discuss the numerical values of the parameters based on first-principles calculations in Section 3. In this section we analyze the dependence of the interaction parameters on the local moment in $CaFe_2As_2$, as well as the frequency dependence of the biquadratic coupling. Using the stoichiometric KFe_2Se_2 as an example, we also show that the biquadratic coupling can change sign depending on the band filling, while remaining quite sizeable. In Section 4 we describe the phase diagram of the model, extending prior results [19] to the case of negative biquadratic coupling. Magnetic short-range order is discussed in Section 5, and the overall properties of the effective spin model are summarized in Section 6.

2. Model spin Hamiltonian

Consider the following effective classical spin Hamiltonian:

$$H = \sum_{i<j} \left[J_{ij} \mathbf{S}_i \cdot \mathbf{S}_j - \tilde{K}_{ij} \left(\mathbf{S}_i \cdot \mathbf{S}_j \right)^2 \right] \tag{1}$$

The first term is the conventional (isotropic) exchange interaction, and the second term is the pairwise biquadratic interaction. Ferropnictides with the stripe ground state can be described by the J_1-K-J_2-J_c model, including two in-plane exchange parameters J_1 and J_2, an interplane coupling J_c, and a nearest-neighbor biquadratic coupling $K = \tilde{K}S^2$. Based on the electronic structure calculations, it is known that in ferropnictides the biquadratic coupling of electronic origin is much larger than the one generated by the "order-from-disorder" mechanism [21–23] or by magnetostructural coupling [24].

The linear-response exchange parameters J_{ij}^{LR}, which determine the spin-wave spectra, are defined as second derivatives of the total energy with respect to spin rotations. From the Hamiltonian (1) we find

$$J_{ij}^{\mathrm{LR}} = J_{ij} - 2\tilde{K}_{ij} S^2 \mathbf{e}_i \cdot \mathbf{e}_j, \tag{2}$$

where $\mathbf{e}_i = \mathbf{S}_i / S$. For the stripe ground state, we find $J_{1a} = J_1 + 2\tilde{K}S^2$ and $J_{1b} = J_1 - 2\tilde{K}S^2$. Thus, the anisotropy of the nearest-neighbor linear-response exchange parameter in the stripe phase is captured by the biquadratic term in the Hamiltonian (1). The biquadratic term also eliminates the continuous degeneracy for noncollinear states connecting the $(\pi, 0)$ and $(0, \pi)$ stripe states [7, 9]. This barrier also explains why twin domain walls in ferropnictides are microscopically thin [10].

For example, the spin wave spectrum of $CaFe_2As_2$ was measured in Refs. [3, 4] and fitted to the anisotropic J_1-J_2-J_c Heisenberg model. Mapping these anisotropic linear-response exchange parameters to model (1), we find $SJ_1 = 22$, $SK = 14$, $SJ_2 = 19$, and $SJ_c = 5.3$ meV, where $K = \tilde{K}S^2$.

3. First-principles analysis of interaction parameters

Band structure theory can provide unique information about the spin Hamiltonian of the system. Apart from identifying the ground magnetic state, it is able to describe the adiabatic energy surface for a wide range of spin configurations with different magnitudes of the spin moments and angles between them. The energies of such configurations can be evaluated using constrained density functional theory [25] and then mapped onto a model spin Hamiltonian.

Band structure calculations based on density functional theory (DFT) correctly produce the stripe antiferromagnetic configuration as the ground state for ferropnictides [26]. As explained above, contrary to the Heisenberg model, in DFT the ground state does not have an internal degeneracy [7, 9]. Let us consider the total energy as a function of the rotation angle θ between the two interpenetrating Néel sublattices for stoichiometric $CaFe_2As_2$ and KFe_2Se_2 systems (Fig. 1). Here we use local density approximation (LDA) and the full-potential linear muffin-tin orbital (LMTO) method, and the distance R_{Fe-As} between Fe and As (Se) atoms is varied to obtain different magnitudes of the spin moments. In $CaFe_2As_2$ the ground state is the collinear stripe AFM configuration (Fig. 2a). The total energy depends strongly on the angle θ and reaches a maximum of 20-40 meV at $\theta = 90°$ (depending on the local spin moment). The simplest way to map this dependence to an effective spin Hamiltonian is to use Eq. 1 with a nearest-neighbor biquadratic coupling. For $CaFe_2As_2$, as well as for all compounds with the collinear stripe ground state, K is positive.

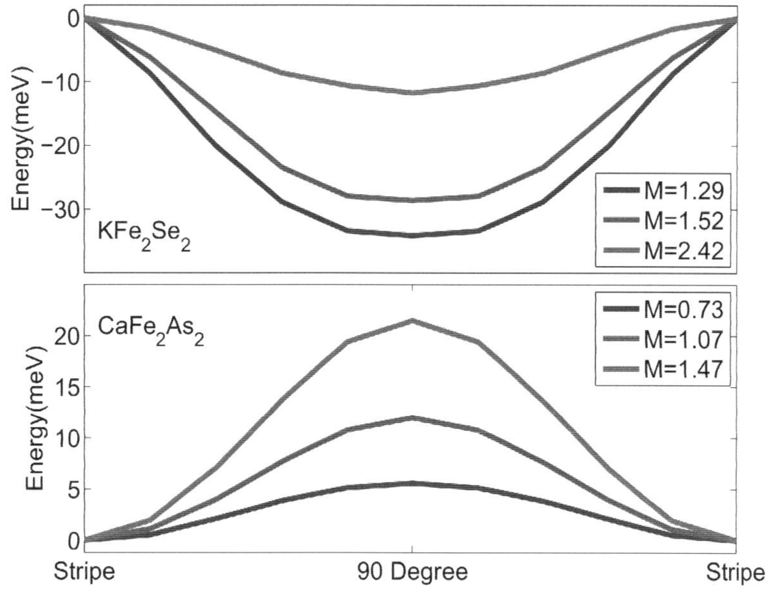

Figure 1. Calculations of the total energy in $CaFe_2As_2$ (bottom panel) and KFe_2Se_2 (top panel) as a function of the angle between the two interpenetrating Néel sublattices for different values of the Fe-As(Se) bond length. The curves are labeled by the value of the magnetic moment μ on Fe atoms in the stripe structure. The end points (0 and 180°) represent two degenerate stripe configurations.

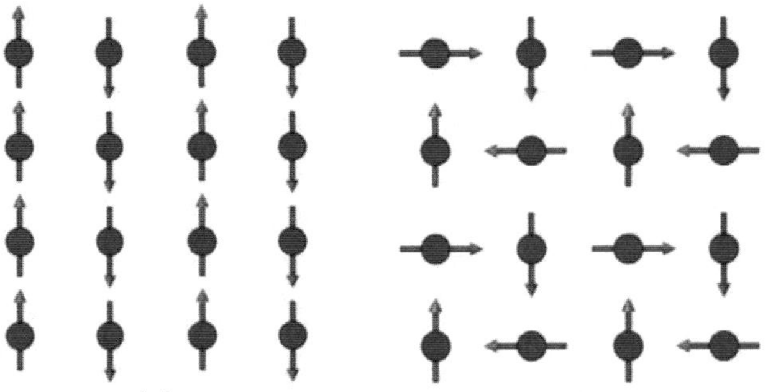

Figure 2. Ground state spin orderings for the J_1-K-J_2-J_c model. (a) Stripe ordering for $K > 0$. (b) Noncollinear ordering with a 90° angle between two Néel sublattices for $K < 0$.

The situation in the case of KFe_2Se_2 is opposite: the total energy reaches its minimum at $\theta = 90°$ (Fig. 2b), corresponding to negative K in Eq. (1). This result shows that depending on the electronic structure (band filling) the biquadratic coupling can change sign, stabilizing either collinear or noncollinear magnetic orderings. Thus, in materials like stoichiometric KFe_2Se_2 the intrinsic non-Heisenberg interaction can lift the degeneracy in an opposite way compared to the order-from-disorder mechanism.

To further study the character of magnetic interactions in ferropnictides, we evaluate the effective exchange coupling for $CaFe_2As_2$ using the linear response technique. Table 1 and Fig. 3 show the effective exchange parameters for several nearest neighbors as a function of the local spin moment, which is manipulated by varying R_{Fe-As}. Table 1 also includes the values of SK^* obtained from the energy barrier determined as shown in Fig. 1.

Large values of SK at large spin moments μ show that in this regime the AFM stripe order is very stable, while the isotropic J_1-J_2 Heisenberg model is decidedly inappropriate for the description of the low-lying magnetic states. Comparing the values of SK (obtained from the linear-response calculation for the nearest neighbors) with SK^* (obtained from the energy barrier), we find that at small μ they differ by about a factor of 3. This means that non-Heisenberg coupling at small μ extends over at least a few coordination spheres. A significant anisotropy of the J_3 parameter (Table 1) confirms this interpretation. Note that the J_3 interaction is between the sites on the same Néel sublattice. Therefore, in order to include its anisotropy in the effective spin model, it would be necessary to add non-pairwise quartic terms in (1).

For larger μ the values of SK and SK^* come fairly close to each other, demonstrating a gradual transition from itinerant (long-ranged) to localized magnetic interaction. Note that the dependence of K on μ (Fig. 3) is much stronger than that of J_{ij}, consistent with the fact that it may be interpreted as a higher-order term in the expansion of the total energy with respect to the non-magnetic state.

The anisotropy of J_1 was obtained by linear response technique [8] even before it was found experimentally. Once mapped to the J_1-K-J_2-J_c model, the experimental anisotropy [4] appears to be larger compared to calculations by nearly a factor of 3 (J_{1b} even changes sign in this model). Anisotropy of J_1 of this magnitude can only be obtained in band theory at very large values of μ [8]. This discrepancy may, at least partially, be due to the anisotropic contribution from non-nearest neighbors (as mentioned above for J_3). Another possibility is that it is necessary to consider *dynamical* magnetic interaction, which cannot be directly incorporated in a localized

μ	SJ_1	SK	SJ_2	SJ_c	SJ_{3a}	SJ_{3b}	SK^*	T_N
0.76	37	2.1	24	10	0.64	-8.6	6.6	99
1.07	50	4.6	40	9.4	9.7	1.1	11	182
1.39	49	8.8	44	8.6	14	5.3	13	245
1.57	43	11.2	42	7.9	16	6.6	14	253

Table 1. The exchange coupling parameters (in meV) and the Neel temperature (in K) in the stripe AFM phase of $CaFe_2As_2$ as a function of the magnetic moment μ (in μ_B) of Fe. Indices a and b correspond to antiparallel and parallel spin pairs, respectively. SK^* is obtained from the height of the energy barrier (Fig. 1). T_N is estimated using Monte Carlo results from Section 4.

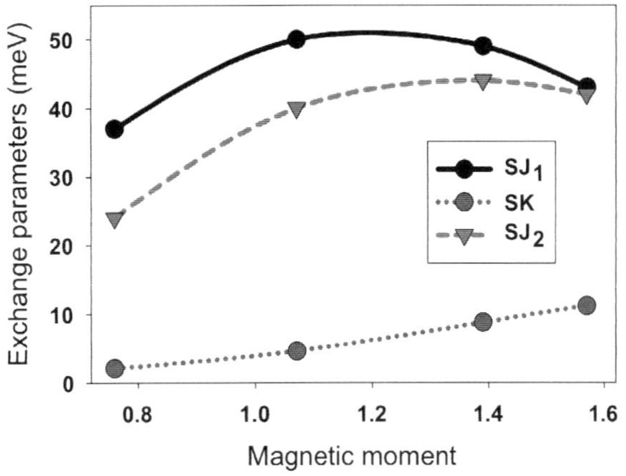

Figure 3. Exchange and biquadratic coupling parameters as a function of the local moment.

spin model. Consider the *frequency-dependent* exchange coupling defined as

$$J_{ij}(\omega) = \mathrm{Re}\,\chi_{ij}^{-1}(\omega) \tag{3}$$

where $\chi_{ij}^{-1}(\omega)$ is the inverse Fourier transform of the dynamical transverse spin susceptibility. The dynamic biquadratic coupling can be obtained from the anisotropy of J_1 in the stripe AFM phase. To calculate the transverse dynamic spin susceptibility $\chi(\mathbf{q},\omega)$, we use an all-electron linear response technique developed in Refs. [27, 28]. In this method, the susceptibility $\hat{\chi}(\mathbf{q},\omega)$ is first calculated in the full product basis set representation. Then it is projected onto the functions representing local spin densities on each magnetic site, which gives a matrix in basis site indices. This projection corresponds to the rigid spin approximation. Inversion of this matrix with a subsequent Fourier transform provides the real-space representation of the inverse susceptibility. Note that the local exchange enhancement does not affect the intersite exchange parameters.

The results for the anisotropy of the dynamic nearest-neighbor coupling in $CaFe_2As_2$ are shown in Fig. 4. (At $\omega = 0$ it corresponds to the static SK in Table 1.) It is seen that $K(\omega)$ strongly depends on frequency, changing by a factor of 2 or 3 over the range of the spin-wave spectrum. This frequency dependence can not be incorporated in an effective spin Hamiltonian and complicates the interpretation of the measured spin excitation spectra. The relative importance of the longer-range interactions contributing to SK^* and of the dynamical effects is currently unclear, but it is quite likely that the high-frequency part of the spectrum is more appropriately described in the itinerant picture. In the following we assume that the thermodynamic properties are not sensitive to these high-frequency excitations and may be described by a spin model (1) with effective interaction parameters that may already include some reasonable averaging over the thermal frequency range.

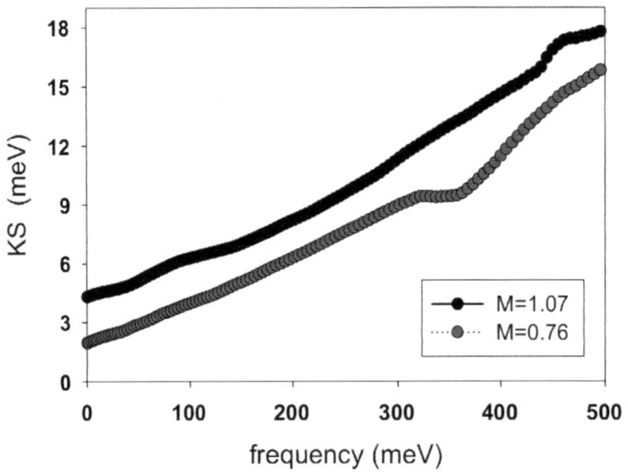

Figure 4. Dynamical biquadratic exchange coupling in the spin wave region for $CaFe_2As_2$ in the antiferromagnetic (stripe) magnetic state, for two values of the local moment obtained by modifying the Fe-As bond length.

4. Phase diagram

The spin ground state of the J_1-K-J_2-J_c model depends on the sign of K. At $K = 0$ the ground state has an internal O(3) degeneracy in addition to the global spin rotation. It is convenient to divide the original square lattice into two equivalent sublattices I and II, which are connected by next-nearest-neighbor bonds. The antiferromagnetic J_2 coupling enforces the Neél order within each sublattice, but the classical energy does not depend on the relative angle between them. The biquadratic interaction lifts this degeneracy. At $K > 0$ the collinear stripe ordering is stabilized (Fig. 2a), while at $K < 0$ the energy is minimized in a noncollinear (2Q) configuration with nearest-neighbor spins aligned perpendicular to each other (Fig. 2b). While the internal symmetry of the magnetic phase becomes discrete (Z_2) at $K > 0$, it is only lowered from O(3) to O(2) at $K < 0$. Here we consider the magnetic phase diagram of this model at both positive and negative K.

4.1. Mean-field approximation

At $K > 0$ the Hamiltonian takes the following form in the mean-field approximation (MFA):

$$H_{\mathrm{MFA}}^{K>0} = -mJ_s \sum_i \eta_i S_{iz} - qK_0 \sum_i S_{iz}^2 \tag{4}$$

where $J_s = 4J_2 + 2J_c$, $K_0 = 4K$ and $\eta_i = \pm 1$ for $i \in \mathrm{I}$ and $i \in \mathrm{II}$. The MFA equations contain dipole and quadrupole effective fields [29], demanding self-consistency for both the staggered magnetization $m = \eta_i \langle S_i^z \rangle$ and the quadrupole order parameter $q = \left(3\langle S_{iz}^2 \rangle - 1\right)/2$. The lowest free energy may be obtained in the paramagnetic phase ($m = q = 0$), the antiferromagnetic phase ($m \neq 0$, $q \neq 0$), or in the quadrupolar phase($m = 0$, $q \neq 0$).

For $K < 0$ we assume that the magnetization on the sublattice I (II) lies along the x (y) axis. The MFA Hamiltonian can be then written as

$$H_{\mathrm{MFA}}^{K<0} = -mJ_s \sum_i \eta_i S_{i\parallel} - K_0 \sum_i \left(q_\perp S_{i\parallel}^2 + q_\parallel S_{i\perp}^2\right) \tag{5}$$

where $S_{i\parallel}$ ($S_{i\perp}$) is the component of $\mathbf{S}_i$ parallel (perpendicular) to the magnetization of site i, while $q_\perp = \langle S^2_{\mathrm{I}y} - S^2_{\mathrm{I}z}\rangle = \langle S^2_{\mathrm{II}x} - S^2_{\mathrm{II}z}\rangle$ and $q_\parallel = \langle S^2_{\mathrm{I}x} - S^2_{\mathrm{I}z}\rangle = \langle S^2_{\mathrm{II}y} - S^2_{\mathrm{II}z}\rangle$ are the two parameters characterizing the quadrupole ordering tensor. Minimization of the free energy with respect to parameters m, $q_\perp$, and $q_\parallel$ leads to a set of three self-consistent equations. Similarly to the case of positive K, there are three possible types of solutions: paramagnetic ($m = q_\parallel = q_\perp = 0$), antiferromagnetic ($m \neq 0$, $q_\parallel \neq 0$, $q_\perp \neq 0$, and quadrupolar ($m = 0$, $q_\parallel \neq 0$, $q_\perp \neq 0$).

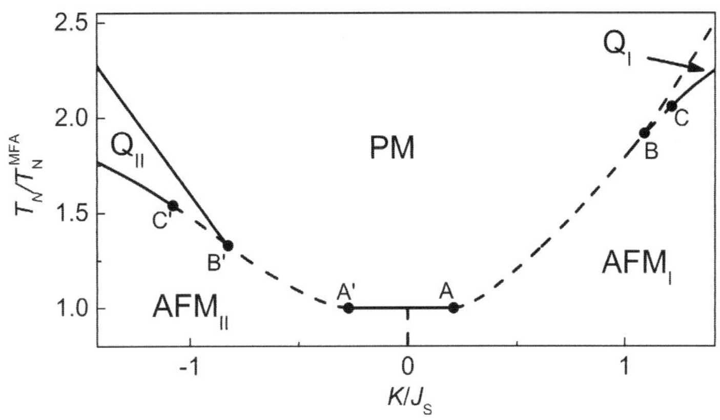

Figure 5. MFA phase diagram for the J_1-K-J_2-J_c model. Temperature is measured in units of $T_N^{\mathrm{MFA}} = J_s/3$, which is the second-order T_N in MFA. AFM$_\mathrm{I}$ and AFM$_\mathrm{II}$ denote stripe and noncollinear ($\theta = 90°$) orderings, respectively, while Q$_\mathrm{I}$ and Q$_\mathrm{II}$ denote the corresponding quadrupolar phases. Solid and dashed lines denote second and first-order transitions, respectively. Points A, A$'$, C and C$'$ are tricritical points, point B is a triple point, and B$'$ is a critical endpoint.

The MFA phase diagram is shown in Fig. 5. For small positive K there is a second order transition from the AFM stripe phase to the paramagnetic state with $T_N^{MFA} = J_s/3$ being independent of K. At the tricritical point A ($K_A = 5J_s/24$) the transition changes to first order, and at $K > K_A$ the transition temperature grows approximately linearly in K. Note that first-order transitions are common in systems with biquadratic coupling [29, 30]. The tricritical point A is within the range of realistic parameters for ferropnictides.

The quadrupolar phase appears at a much larger value $K \approx J_s$ (point B). This phase is separated from the paramagnetic phase by a first-order transition, but the transition from the AFM to quadrupolar phase is first order at $K < K_C$ and second-order at $K > K_C$.

For $K < 0$ the phase diagram is quite similar. The noncollinear AFM phase (labeled AFM$_\mathrm{II}$ in Fig. 5) disorders through a second-order transition up to the tricritical point A$'$ at $K_{\mathrm{A}'} \approx -0.27J_s$. At $|K| < |K_{\mathrm{A}'}|$ the transition temperature is constant and equal to $J_s/3$, while the first-order transition temperature at larger $|K|$ grows approximately linearly in K. The quadrupolar phase appears at a critical endpoint at $K \approx -0.9J_s$. The transition from the noncollinear phase to the quadrupolar phase changes from first to second order at a slightly higher $|K|$ (point C$'$). The quadrupolar phase disorders through a second-order transition, contrary to the quadrupolar phase appearing at large $K > 0$. Both quadrupolar phases appear at very large values of $|K|$ and are unlikely to be realized in ferropnictides and related materials.

4.2. Monte Carlo simulations
The phase diagram at $K > 0$ was previously evaluated in Ref. [19]; here we extend it to the region of $K < 0$. This extension allows one to better understand the effects of biquadratic coupling, and, as shown in Section 3, negative K may be realized in appropriately doped stoichiometric iron chalcogenides. We used cubic lattices $D \times D \times D$ with periodic boundary conditions and $D = 14$, 16, 18, and 20. Monte Carlo simulations were performed using the Metropolis algorithm, each step consisting in a trial random change of the spin direction for one site. The lengths of the equilibration and averaging runs were adjusted to achieve sufficient accuracy. We restrict ourselves to the realistically small range of $|K|/J_s < 0.4$, where the quadrupolar phase does not appear. For the noncollinear ordering we used two alternative definitions of the order parameter: $\sqrt{(L_{\rm I}^2 + L_{\rm II}^2)/2}$ and $|\mathbf{L}_{\rm I} \times \mathbf{L}_{\rm II}|$, where $\mathbf{L}_J$ is the Néel order parameter for sublattice J. The results for both choices of the order parameters agree within the computational error bar.

The order of the phase transition was determined by analyzing the behavior of the fourth-order energy cumulant [31]. If this cumulant converges to 2/3 for all temperatures with increasing D, the transition is second-order. If the energy cumulant develops a minimum at some temperature, which sharpens with increasing D, the transition is of first order [31]. Second-order transition temperatures were determined using finite-size scaling of the Binder cumulant [32]. First-order transition temperatures were found from the peaks of the appropriate susceptibilities.

The magnetic phase diagram is shown in Fig. 6. To illustrate the dependence of the transition temperature on the model parameters, we include the dependence of the Néel temperature on K for a few values of the J_2/J_1 and J_c/J_1 ratios. At $K > 0$ and $T = 0$ the stripe phase is stable for $J_2/J_1 > 0.5$. At $K < 0$ the noncollinear phase is stable with respect to the simple Néel order at $J_2/J_1 > 0.5(1 - |K|)$. Close to these points, as well as at smaller J_c/J_1, the transition temperature is significantly reduced compared to MFA due to strong fluctuations.

Using the parameters taken from the fitting of experimental spin wave dispersions for $CaFe_2As_2$ and $BaFe_2As_2$, the Néel temperatures are estimated at 90 K and 65 K, respectively (in both cases the transition is first-order in agreement with experiment [33]). These values are notably smaller compared to the experimental T_N of 170 K and 140 K. We have also estimated T_N using the calculated exchange parameters for $CaFe_2As_2$ (see the last column of Table 1). In this estimate, we first calculated J_s taking J_2, symmetrized J_3, and J_c into account, found the mean-field estimate $T_N^{\rm MFA}$, and then appropriately reduced it using Monte Carlo results (Fig. 6) for the corresponding values of the parameter ratios. (Specifically, the mean-field T_N were multiplied by 0.51, 0.56, 0.59, and 0.6 for the four lines in Table 1.) We see that the experimental T_N for $CaFe_2As_2$ is reproduced for $\mu \sim 1$, in reasonable agreement with the measured values of the local spin moment. We also see that the contribution of J_3 is not negligible; for example, at $\mu = 1.39\mu_B$ it reduces the T_N estimate by 20%. More distant intrasublattice couplings (not listed) are at least a few times smaller than J_3.

At $K > 0$ the phase transition changes from second to first order at increasing K. This is similar to MFA results, but the tricritical value of K is significantly reduced by fluctuations. On the other hand, for $K < 0$ the transition remains second-order even beyond the MFA tricritical point, although the incipient tendency toward first-order transition increases with the magnitude of K. It is possible that the tricritical point at $K > 0$ is driven down by interaction with the nematic order parameter [34].

5. Magnetic short-range order
Spin excitations in the paramagnetic phase AFe_2As_2 materials (A=Ca,Sr,Ba) were investigated using inelastic neutron scattering [35–37]. In general, these studies show the existence of strong short-ranged correlations peaked at the wave vector $\mathbf{Q}_{AFM}$ of the low-temperature AFM structure. These correlations exhibit in-plane anisotropy [36, 37], whose magnitude is similar to

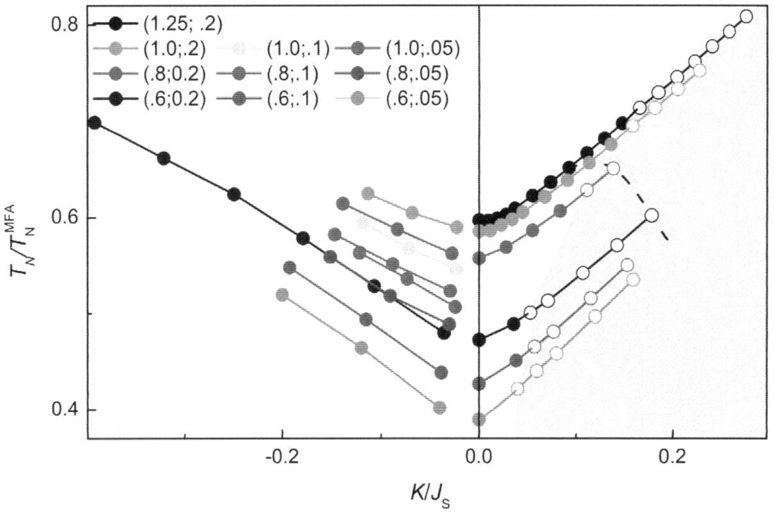

Figure 6. Magnetic phase diagram of the J_1-K-J_2-J_c model obtained by MC simulations. The plot shows the transition temperatures for stripe-to-paramagnetic (positive K) and noncollinear-to-paramagnetic (negative K) phase transitions as well as the order of these transition as a function of biquadratic interaction K. Temperature is measured in units of $T_N^{\rm MFA} = J_s/3$ which is the second-order T_N in MFA. Each line type is labeled by a set of two parameters (J_2/J_1; J_c/J_1). The region of first-order transitions (empty symbols) is schematically highlighted by shading and thicker lines. The dashed line shows the point of inversion of J_{1b}.

the anisotropy of spin waves at low temperatures. The interpretation of this anisotropy is an unresolved question.

The excitation spectrum of the two-dimensional J_1-J_2-K model deep below the nematic phase transition has properties consistent with experimental measurements for the paramagnetic phase [38]. Although evidence of electronic nematicity in a wide range of temperatures above the Néel point has been reported in ferropnictides [39], it is currently unclear whether the underlying nature of the corresponding phase is captured by a strongly two-dimensional spin model with biquadratic interaction.

Here we show that in the paramagnetic phase of model (1) with a moderate interlayer coupling (which does not have a broad nematic phase) the anisotropy of the excitation spectrum is insensitive to biquadratic coupling K; this implies that in this model the anisotropy of the excitation spectrum increases as the temperature is raised into the paramagnetic phase.

Consider the static structure factor $C(\mathbf{q}) = \langle|\mathbf{S}(\mathbf{q})|^2\rangle$, where $\mathbf{S}(\mathbf{q})$ is the Fourier transform of $\mathbf{S}_i$. This quantity is equal to the dynamical structure factor integrated over frequency, and its anisotropy is therefore representative of the anisotropy of the dynamical excitation spectrum. We have calculated $C(\mathbf{q}_\parallel)$ (at $q_z = 0$) in Monte Carlo simulations by collecting a number of spin configurations (each taken after an equilibration run), evaluating $|\mathbf{S}(q_x, q_y, z_k)|^2$ using the fast Fourier transform for each atomic plane z_k, and averaging over the atomic planes and the spin configurations. For comparison, we also evaluated $C(\mathbf{q})$ using the Onsager reaction field approximation, in which the fluctuation-dissipation theorem is enforced by construction [40]. Similar to the random phase approximation (RPA), this method gives the transition temperature

$$\frac{T_N^{\rm RPA}}{T_N^{\rm MFA}} = \left(\sum_{\mathbf{q}} \frac{1}{1 - J_{\mathbf{q}}}\right)^{-1} \tag{6}$$

where $J_{\mathbf{q}}$ is the Fourier transform of the normalized exchange parameters J_{ij}/J_s. Note that the biquadratic term has no effect on Eq. (6), because it does not contribute to the paramagnetic susceptibility; the phase transition is always second-order in this approximation. The structure factor above T_N^{RPA} is given by

$$C_{\mathrm{RPA}}(\mathbf{q}) = \frac{S^2}{1-(J_{\mathbf{q}}-\lambda)T_N^{\mathrm{MFA}}/T} \tag{7}$$

where the parameter λ satisfies the equation

$$\lambda = \sum_{\mathbf{q}} \frac{J_{\mathbf{q}}}{1-(J_{\mathbf{q}}-\lambda)T_N^{\mathrm{MFA}}/T} \tag{8}$$

The results for $C(\mathbf{q}_{\|})$ at $J_2/J_1 = 0.8$, $J_c/J_1 = 0.2$, and $T = 1.1T_N$ are shown in Fig. 7. First, we see that Monte Carlo results at $K = 0$ (panel (b)) agree very well with the RPA formula (panel (a)). As expected, the structure factor has elliptical peaks centered around the AFM wavevectors $\{1, 0\}$, with the major axis directed toward the corners of the Brillouin zone. This feature agrees with experimental excitation spectra. The peaks have elliptical shape reminiscent of the anisotropic spin wave excitations at low temperatures, even though the system is in the paramagnetic state, and the structure factor retains the full four-fold symmetry of the square lattice. (In Ref. [38] this symmetry is obtained in the nematic phase only after symmetrization, assuming the existence of macroscopic AFM domains.) The ellipticity of the peaks is a generic property of the Heisenberg model [35]. Further, Fig. 7c shows the structure factor for $K/J_1 = 0.4$

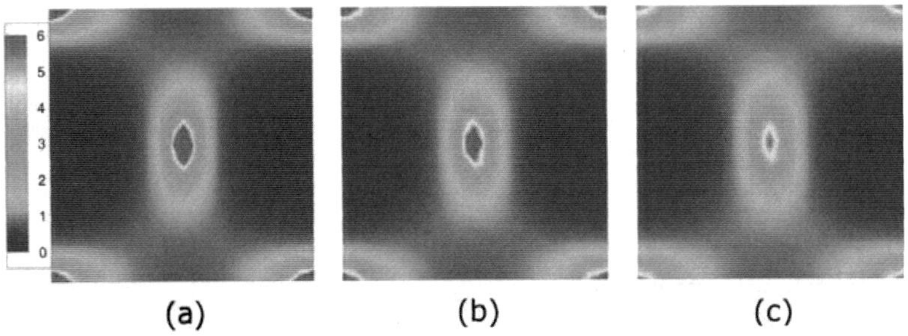

Figure 7. Static structure factor $C(q_x, q_y, 0)$ within the first Brillouin zone at $T = 1.1T_N$. (a) $C_{\mathrm{RPA}}(q_x, q_y, 0)$ for $J_2/J_1 = 0.8$, $J_c/J_1 = 0.2$ and $K = 0$. (b) Monte Carlo results for the same parameters. (c) Monte Carlo results for $J_2/J_1 = 0.8$, $J_c/J_1 = 0.2$ and $K/J_1 = 0.4$. The X point is placed in the center of the frame; the Γ point is in the middle of the vertical edges. Normalization corresponds to $S = 1$. The functions in all panels have full four-fold symmetry.

(other parameters are the same as in panel (b)). We see that the biquadratic coupling of this sizeable magnitude does not affect the eccentricity of the elliptical zone-boundary peaks.

Using the RPA structure factor (7), it is easy to see that in the Heisenberg model (at $K = 0$) the anisotropy of the paramagnetic excitation spectrum is identical to the anisotropy of the spin wave velocity at $T = 0$. Positive K decreases the anisotropy of the spin-wave velocity at $T = 0$ [19]. Therefore, the fact that K has little effect on the anisotropy of the paramagnetic spectrum implies that in model (1) with large K the anisotropy of the excitation spectrum should notably increase as the temperature is raised into the paramagnetic phase.

Note that the anisotropy of the spin excitation spectrum in 122 compounds increases with Co doping [41, 42]. According to first-principles calculations (Table 1), the J_2/J_1 ratio decreases with decreasing μ. As long as the effect of doping is mainly to decrease μ, these two trends are consistent with each other.

6. Discussion and conclusions

The addition of biquadratic terms to the Heisenberg model allows one to reconcile it with numerous experimental observations, including the anisotropy of the spin wave spectrum, microscopic thickness of the twin domain walls between different variants of the stripe phase, as well as the existence of a tricritical point as a function of doping [33, 43–45]. (Doping is expected to reduce the spin moments, and thereby the importance of biquadratic coupling, see Section 3.) Biquadratic terms eliminate the continuous degeneracy of the ground state, in agreement with electronic structure calculations. With the inclusion of the third-neighbor exchange coupling J_3, this model was also used to describe the magnetic interaction in other families of iron-based superconductors [46].

The localized spin model should be viewed as an effective mapping of complicated itinerant interactions. The significant frequency dependence of the dynamic analog of the biquadratic term (Section 3) indicates a limitation of the model, which may also be related to the high-frequency behavior of the spin fluctuation spectrum. As we have seen in Section 5, in a moderately three-dimensional effective spin model with strong biquadratic interaction the anisotropy of the spin fluctuation spectrum should increase as the system is heated above the Néel temperature. On the other hand, experiments show that this anisotropy is essentially independent of temperature [35–37]. One possible interpretation is that the effective biquadratic interaction is not very strong in ferropnictides. It was argued [42] that the anisotropy of J_1 extracted from the spin-wave spectrum [4] reflects the high-energy part of the spectrum dominated by itinerant effects, while the low-energy part can be described by an isotropic J_1-J_2 model [42]. This interpretation is consistent with first-principles calculations of the frequency-dependent exchange parameters (Section 3). On the other hand, pure Heisenberg model has a continuous ground-state degeneracy, which is inconsistent with first-principles calculations and with the observation of microscopically thin domain walls. Another possible interpretation is the existence of a strong magnetic short-range order above the Néel temperature [47]. Although strong short-range order is unattainable in models with moderate interplane coupling (as assumed in Section 5), it can be reached in the symmetry-broken nematic phase in the strongly two-dimensional limit [38]. The situation remains unresolved, particularly in view of recent experiments suggesting broken symmetry in a broad temperature range above T_N [39].

References

[1] Hirschfeld P J, Korshunov M M and Mazin I I 2011 *Rep. Progr. Phys.* **74** 124508
[2] Lynn J W and Dai P 2009 *Physica* C **469** 469
[3] Diallo S O, Antropov V P, Perring T G, Broholm C, Pulikkotil J J, Ni N, Bud'ko S L, Canfield P C, Kreyssig A, Goldman A I and McQueeney R J 2009 *Phys. Rev. Lett.* **102** 187206
[4] Zhao J, Adroja D T, Yao D X, Bewley R, Li S, Wang X F, Wu G, Chen X H, Hu J and Dai P 2009 *Nature Phys.* **5** 555
[5] Johnston D C 2010 *Adv. Phys.* **59** 803
[6] Yin Z P, Lebègue S, Han M J, Neal B P, Savrasov S Y and Pickett W E 2008 *Phys. Rev. Lett.* **101** 047001
[7] Pulikkotil J, Ke L, van Schilfgaarde M, Kotani T and Antropov V 2010 *Supercond. Sci. Technol.* **23** 054012
[8] Han M J, Yin Q, Pickett W E and Savrasov S Y 2009 *Phys. Rev. Lett.* **102** 107003
[9] Yaresko A N, Liu G Q, Antonov V N and Andersen O K 2009 *Phys. Rev.* B **79** 144421
[10] Chuang T M, Allan M P, Lee J, Xie Y, Ni N, Budko S L, Boebinger G S, Canfield P C and Davis J C 2010 *Science* **327** 181
[11] Krüger F, Kumar S, Zaanen J and van den Brink J 2009 *Phys. Rev.* B **79** 054504
[12] Lee C C, Yin W G and Ku W 2009 *Phys. Rev. Lett.* **103** 267001
[13] Chen C C, Moritz B, van den Brink J, Devereaux T P and Singh R R P 2009 *Phys. Rev.* B **80** 180418
[14] Chen C C, Maciejko J, Sorini A P, Moritz B, Singh R R P and Devereaux T P 2010 *Phys. Rev.* B **82** 100504
[15] Lv W, Krüger F and Phillips P 2010 *Phys. Rev.* B **82** 045125
[16] Yin W G, Lee C C and Ku W 2010 *Phys. Rev. Lett.* **105** 107004
[17] Knolle J, Eremin I, Chubukov A V and Moessner R 2010 *Phys. Rev.* B **81** 140506
[18] Turner A M, Wang F and Vishwanath A 2009 *Phys. Rev.* B **80** 224504

[19] Wysocki A L, Belashchenko K D and Antropov V P 2011 *Nature Phys.* **7** 485
[20] Fernandes R M, Chubukov A V, Knolle J, Eremin I and Schmalian J 2012 *Phys. Rev.* B **85** 024534
[21] Chandra P, Coleman P and Larkin A I 1990 *Phys. Rev. Lett.* **64** 88
[22] Fang C, Yao H, Tsai W F, Hu J and Kivelson S A 2008 *Phys. Rev.* B **77** 224509
[23] Xu C, Müller M and Sachdev S 2008 *Phys. Rev.* B **78** 020501
[24] Ke L, van Schilfgaarde M, Pulikkotil J, Kotani T and Antropov V 2011 *Phys. Rev.* B **83** 060404
[25] Dederichs P H, Blügel S, Zeller R and Akai H 1984 *Phys. Rev. Lett.* **53** 2512
[26] Mazin I I, Singh D J, Johannes M D and Du M H 2008 *Phys. Rev. Lett.* **101** 057003
[27] Kotani T, van Schilfgaarde M and Faleev S V 2007 *Phys. Rev.* B **76** 165106
[28] Kotani T and van Schilfgaarde M 2008 *Journal of Physics: Condensed Matter* **20** 295214
[29] Matveev V M and Nagaev E L 1972 *Sov. Phys. Solid State* **14** 408
[30] Nagaev E L 1982 *Sov. Phys. Usp.* **25** 31
[31] Challa M S S, Landau D P and Binder K 1986 *Phys. Rev.* B **34** 1841
[32] Binder K 1981 *Z. Phys.* B **43** 119
[33] Ni N, Nandi S, Kreyssig A, Goldman A I, Mun E D, Bud'ko S L and Canfield P C 2008 *Phys. Rev.* B **78** 014523
[34] Hu J and Xu C 2012 *Physica* C **481** 215
[35] Diallo S O, Pratt D K, Fernandes R M, Tian W, Zarestky J L, Lumsden M, Perring T G, Broholm C L, Ni N, Bud'ko S L, Canfield P C, Li H F, Vaknin D, Kreyssig A, Goldman A I and McQueeney R J 2010 *Phys. Rev.* B **81** 214407
[36] Harriger L W, Luo H Q, Liu M S, Frost C, Hu J P, Norman M R and Dai P 2011 *Phys. Rev.* B **84** 054544
[37] Ewings R A, Perring T G, Gillett J, Das S D, Sebastian S E, Taylor A E, Guidi T and Boothroyd A T 2011 *Phys. Rev.* B **83** 214519
[38] Yu R, Wang Z, Goswami P, Nevidomskyy A H, Si Q and Abrahams E 2012 *Phys. Rev.* B **86** 085148
[39] Kasahara S, Shi H J, Hashimoto K, Tonegawa S, Mizukami Y, Shibauchi T, Sugimoto K, Fukuda T, Terashima T, Nevidomskyy A H and Matsuda Y 2012 *Nature* **486** 382
[40] Eastwood M P and Logan D E 1995 *Phys. Rev.* B **52** 9455
[41] Park J T, Inosov D S, Yaresko A, Graser S, Sun D L, Bourges P, Sidis Y, Li Y, Kim J H, Haug D, Ivanov A, Hradil K, Schneidewind A, Link P, Faulhaber E, Glavatskyy I, Lin C T, Keimer B and Hinkov V 2010 *Phys. Rev.* B **82** 134503
[42] Tucker G S, Fernandes R M, Li H F, Thampy V, Ni N, Abernathy D L, Bud'ko S L, Canfield P C, Vaknin D, Schmalian J and McQueeney R J 2012 *Phys. Rev.* B **86** 024505
[43] Ni N, Thaler A, Kracher A, Yan J Q, Bud'ko S L and Canfield P C 2009 *Phys. Rev.* B **80** 024511
[44] Mun E D, Bud'ko S L, Ni N, Thaler A N and Canfield P C 2009 *Phys. Rev.* B **80** 054517
[45] Fernandes R M, Pratt D K, Tian W, Zarestky J, Kreyssig A, Nandi S, Kim M G, Thaler A, Ni N, Canfield P C, McQueeney R J, Schmalian J and Goldman A I 2010 *Phys. Rev.* B **81** 140501
[46] Hu J, Xu B, Liu W, Hao N N and Wang Y 2012 *Phys. Rev.* B **85** 144403
[47] Mazin I I and Johannes M D 2009 *Nature Phys.* **5** 141

Acknowledgments

We are grateful to P. Dai and I. Eremin for useful discussions. Work at UNL was supported by NSF DMR-1005642 and EPS-1010674. Work at Ames Laboratory was supported by Department of Energy-Basic Energy Sciences, under Contract No. DE-AC02-07CH11358. K. B. is a Cottrell Scholar of Research Corporation.

Electron Correlation and Spin Dynamics in Iron Pnictides and Chalcogenides

Rong Yu[1], Qimiao Si[1], Pallab Goswami[2], Elihu Abrahams[3]

[1] Department of Physics and Astronomy, Rice University, Houston, Texas 77005, USA
[2] National High Magnetic Field Laboratory and Department of Physics, Florida State University, Tallahassee, Florida 32306, USA
[3] Department of Physics and Astronomy, University of California Los Angeles, Los Angeles, CA 90095, USA

E-mail: qmsi@rice.edu

Abstract. Superconductivity in the iron pnictides and chalcogenides is closely connected to a bad-metal normal state and a nearby antiferromagnetic order. Therefore, considerable attention has been focused on the role of electron correlations and spin dynamics. In this article, we summarize some key experiments that quite directly imply strong electron correlations in these materials, and discuss aspects of the recent theoretical studies on these issues. In particular, we outline a w-expansion, which treats the correlation effects using the Mott transition as the reference point. For the parent systems, it gives rise to an effective J_1-J_2 model that is coupled to the itinerant electrons in the vicinity of the Fermi energy; this model yields an isoelectronically-tuned quantum critical point, and allows a study of the distribution of the spin spectral weight in the energy and momentum space in the paramagnetic phase. Within the same framework, we demonstrate the Mott insulating phase in the iron oxychalcogenides as well as the alkaline iron selenides; for the latter system, we also consider the role of an orbital-selective Mott phase. Finally, we discuss the singlet superconducting pairing driven by the short-range J_1-J_2 interactions. Our considerations highlight the iron pnictides and chalcogenides as exemplifying strongly-correlated electron systems at the boundary of electronic localization and itinerancy.

1. Introduction

Superconductivity in iron pnictides and chalcogenides [1] occurs by chemical doping of their antiferromagnetic parents, compounds in which the Fe valence is +2. The relationship between superconductivity and parent antiferromagnetism has been discussed from the beginning of the field [2]. The materials have a layered structure, with each layer containing a square lattice of Fe ions. Their electronic properties are primarily associated with the Fe $3d$ orbitals. The appropriate microscopic Hamiltonian contains a kinetic-energy part involving both intra- and inter-orbital hopping, as well as an interaction component featuring Coulomb repulsions and Hund's coupling.

The study of spin dynamics has played a central role in the developing understanding of the microscopic physics and the phases of these compounds. Based on the fact that the parent compounds are "bad" metals, a strong-coupling approach was advanced early on [3]. This approach expands around the Mott transition in terms of a parameter w, which measures the coherent part of electronic excitation spectrum as is depicted in Fig. 1; this parameter is small when the ratio of effective electron-electron interaction to bandwidth is large.

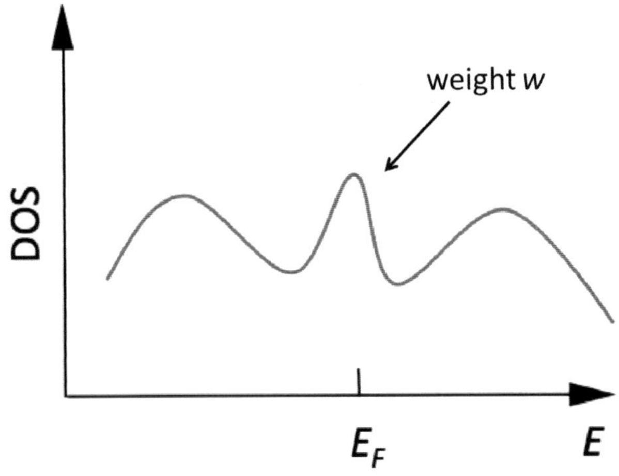

Figure 1. Single-electron spectral function as the sum of coherent and incoherent parts. The single-electron density of states (DOS) is plotted against energy (E); E_F is the Fermi energy. Each peak may contain additional structure due to the multi-orbital nature of the iron pnictides.

To zeroth order in w, the entire spectrum is incoherent and represents quasi-localized local moments and a J_1-J_2 coupling of these local moments was proposed [3, 4]; this provided the understanding of the collinear antiferromagnetic order of the parent compounds. At higher order in w, the local moments become coupled to the itinerant carriers [3, 5, 6]. In the doped systems, the approach leads to a multiband t-J_1-J_2 model that has been used to study the superconductivity. This is a strong-coupling approach, and is in sharp contrast to a weak-coupling description that invokes nesting of hole and electron Fermi surface pockets to drive both the spin fluctuations and superconductivity. Related strong-coupling approaches have been taken from a number of perspectives [7, 8, 9, 10, 11, 12, 13, 14, 15, 16, 17, 18, 19, 20, 21].

Here we provide a brief status report on this line of approach to the iron pnictides. We begin by expounding, in Sec. 2, on the experimental evidences for the importance of electronic correlations.

In Sec. 3, we summarize the w-expansion with respect to the Mott transition point. Sec. 4 is devoted to the magnetic quantum critical point (QCP) that arises by increasing w, which is associated with the reduction of U/W, the ratio of the direct Coulomb and Hund's exchange interactions to the characteristic bandwidth. In addition, we summarize the theoretical proposal for realizing the QCP by P-substitution for As in the parent iron pnictides, and the subsequent experimental realizations.

In Sec. 5, we consider the quantum fluctuations in the magnetic sector. For the parent iron pnictides, we show that our J_1-J_2 based results compare well with the inelastic neutron scattering results in the parent iron pnictides. For the parent alkaline iron selenides, the predictions based on an extended J_1-J_2 model have been confirmed by experiments.

In Sec. 6, we consider the effects of tuning the parent systems in the opposite direction, by enhancing U/W. This occurs in $La_2O_2Fe_2O(Se,S)_2$, in which the expansion of the Fe square lattice leads to a Mott insulator.

In Sec. 7, we address the correlation effects in the alkaline iron selenides. The enhanced electron correlations by the ordered iron vacancies leads to Mott insulating behavior in the parent compounds. We consider how the vacancy-ordered Mott insulating parent compounds are connected to the vacancy-disordered metallic phase via a novel orbital-selective Mott phase.

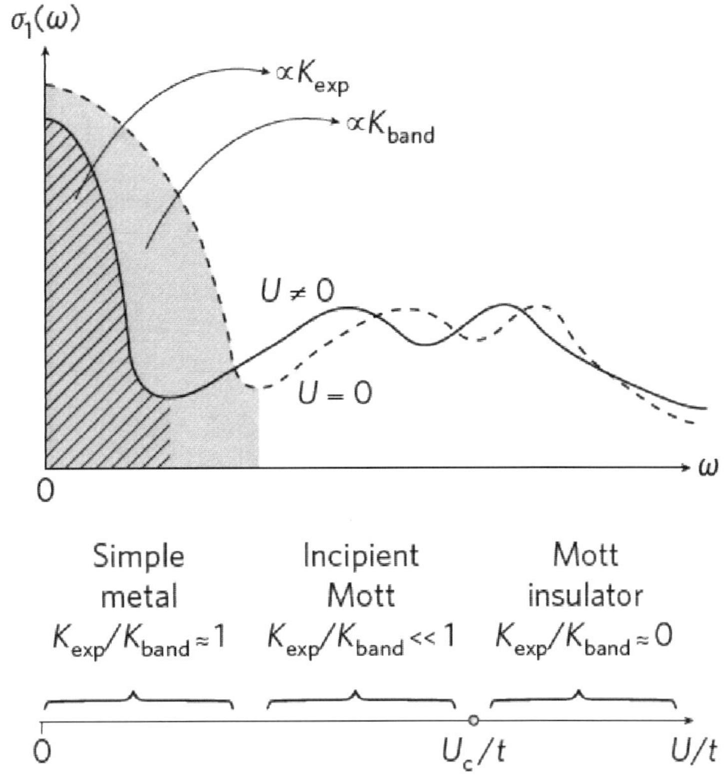

Figure 2. Electronic correlations in the undoped iron pnictides. The optical conductivity as a function of frequency. The area K under the $\omega = 0$ Drude peak is proportional to the kinetic energy of the coherent electrons near the Fermi energy. K_{exp} is measured and K_{band}, is the expected value from (non-interacting) band theory. The degree to which K_{exp}/K_{band} is smaller than one is a measure of the ratio of the electron-electron repulsion to bandwidth, U/t, or a measure of the coherent quasiparticle weight w. The interaction transfers spectral weight from low energies to high energies, up to order U. U_c is the threshold interaction for a Mott localization transition. That K_{exp}/K_{band} is substantially smaller than one means that U/t is smaller than but close to U_c/t, where the electrons are on the verge of losing their itinerancy.

Finally, we discuss the effect of the reduced/suppressed vacancy order on the spin spectral weight at temperatures above the Néel/structural transitions in the insulating alkaline iron selenides.

In Sec. 8, we discuss the superconducting pairing for iron pnictides and alkaline iron selenides based on the strong-coupling model. Sec. 9 includes some summary remarks.

2. Experimental manifestations of strong-coupling physics

The strong-coupling approach has its roots in a number of experimental observations. One is the fact that the room temperature resistivity is so high (~ 0.4 mΩ-cm) that the mean free path is only of order the Fermi wavelength, as is typical for bad metals near a Mott transition. Because this is the case even in compounds that are quite pure as inferred from their small residual resistivity, it signifies the existence of strong inelastic scatterings that are attributable to the sizable electron-electron interactions [3, 22]. This arises if the metallic parent iron pnictides are

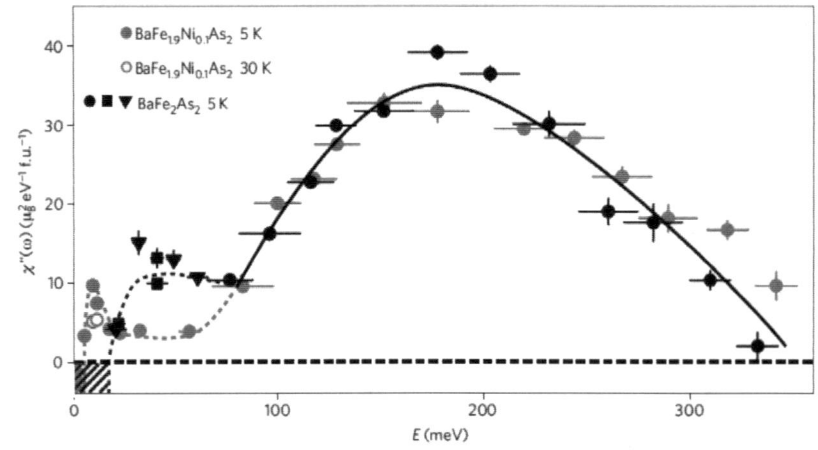

Figure 3. Spin spectral weight for several iron arsenides. Adapted with permission from Ref. [28].

close to a Mott transition, when only a small portion of the single-electron excitation spectrum lies in the coherent low-energy part, of weight w (Fig. 1).

This bad-metal interpretation has been confirmed by optical conductivity measurements [23, 24, 25, 26, 27]. As illustrated in Fig. 2, the experimentally measured spectral weight for the Drude peak normalized by the non-interacting value extracted from band-structure calculations, K_{exp}/K_{band}, provides a measure of the correlation strength: it would be close to 1 for weakly interacting systems, and close to 0 for Mott insulators. The experimentally determined value [23, 24] is close to about 1/3, falling into the regime of bad metals in proximity to Mott localization. This implies that a large portion of the single-electron spectral weight lies in the incoherent part, which would be associated with the large Coulomb (direct plus Hund's) interaction energy scale. At the same time, this reduction of Drude weight is accompanied by the transfer of optical spectral weight between low and high frequencies.

The spin dynamics, as revealed in inelastic neutron scattering experiments [28, 29, 30], also gives strong support to the picture of interacting local moments, whose sizeable spectral weight arises from the *incoherent* part (weight $\sim 1 - w$) of the electronic excitation spectrum. This is illustrated in Fig. 3, adapted from Liu et al [28], where the spin spectral distribution is shown over a range of energies which includes spin waves throughout the Brillouin zone. The integrated spectral weight is very large ($\int d\omega\chi'' \sim 3\mu_B^2$ per Fe), consistent with local moments of spin ~ 1. Such a large spectral weight readily arises from the incoherent part of the single-electron spectral weight, as DMFT calculations have shown [28, 31, 32], but is very difficult to obtain from the single-electron excitations in the vicinity of the small Fermi pockets in the weak-coupling approaches. Indeed, an RPA calculation fails to represent the data [28].

The observation of the Mott insulating behavior in the iron oxychalcogenides [33] and vacancy-ordered alkaline iron selenides [34, 35] has provided further evidence for the strong electron correlations. These materials should be metallic according to the LDA bandstructure calculations in the paramagnetic state (even in the presence of the ordered vacancies in the alkaline iron pnictides), but are observed to be insulating with a very large ordered moment. As discussed in Secs. 6 and 7, the Mott insulating nature of these materials - as opposed to the metallic behavior of the parent iron pnictides - can be understood in terms of a reduced bandwidth but a comparable strength of the Coulomb interaction (along with the Hund's coupling). Since the reduction of the LDA bandwidth is rather modest, the observed Mott

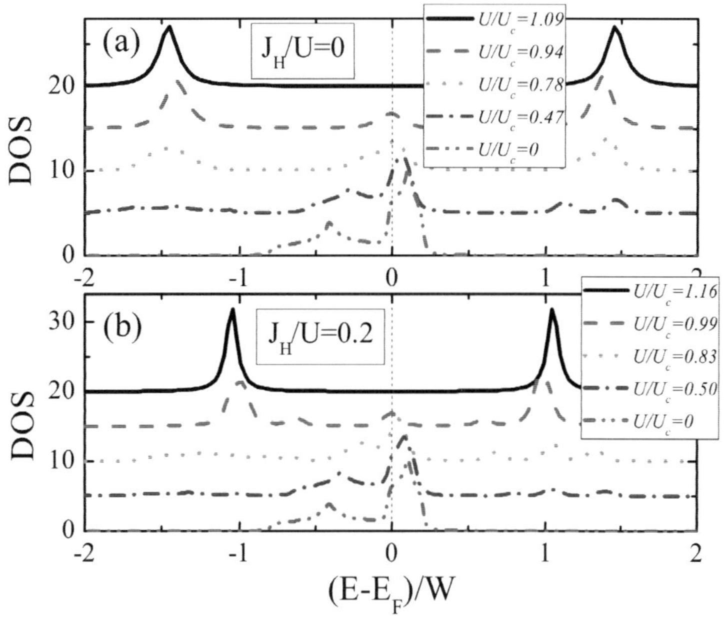

Figure 4. Density of states (DOS) of a two-orbital model derived in a slave-spin study. Results for both the zero Hund's coupling (a) and a non-zero Hund's coupling (b) show incoherent electronic excitations across the Mott transition, $U = U_c$. In both (a) and (b), the curves for $U/U_c > 0$ have been shifted upward for clarity. Adapted from Ref. [36].

insulating behavior provides evidence for the strong electron correlations in the iron pnictides as well. More generally, the evolution in the degree of itinerancy across the various materials families in the iron-based materials can be understood in a picture in which the strength of the local interactions (U) is fixed but the kinetic energy($\propto W$) is variable, as can be seen through calculations of multi-orbital models [36, 37] and *ab initio* LDA+DMFT studies [7].

3. The w-expansion of a multi-orbital Hubbard model with respect to the Mott transition

For bad metals near a Mott transition, the incoherent electronic excitations play a dominant role in the spin dynamics. This effect is difficult to describe in a direct expansion of U/W, with respect to the non-interacting reference point. It can instead be more readily captured by an expansion with respect to the Mott transition point, with $(U_c - U)/W$ being the small parameter. In practice, this is equivalent to expanding in terms of w, the percentage of the single-electron spectral weight in the coherent part [3, 5, 38].

When $w = 0$, at the Mott transition, we can integrate out all the gapped single-electron excitations. The vanishing w implies the absence of the coherent spectral weight, as is already inferred in Fig. 1 and is further illustrated in Fig. 4, which shows the spectral function across the Mott transition U_c in a multi-orbital model with or without the Hund's coupling J_H as determined using a slave-spin technique [36, 37]. This leads to localized magnetic moments, with the exchange interactions on the Fe-square lattice containing both the nearest-neighbor component J_1 and the next-nearest-neighbor one J_2 [3, 5]. General considerations suggest that $J_2 \gtrsim J_1/2$. In this parameter regime, the ground state of the $J_1 - J_2$ Heisenberg model on the square lattice is expected to have the collinear $(\pi, 0)$ antiferromagnetic order [39], as is observed experimentally in the iron pnictides. Furthermore, the antiferromagnetic order in this model is accompanied by an Ising order, whose linear coupling to structural degrees of freedom induces a structural distortion which is also seen experimentally in the pnictides.

Because the charge gap is relatively small, we may also expect significant higher-spin couplings. Of particular interest is the four-spin biquadratic interaction K, whose effect we will discuss in Sec. 5.

At higher order in w, there are itinerant coherent carriers which are coupled to the local moments. This is summarized in the following equations for the couplings: At $w = 0$,

$$H_J = \sum_{\langle ij\rangle} J_1^{\alpha\beta} \mathbf{s}_{i,\alpha} \cdot \mathbf{s}_{j,\beta} + \sum_{\langle\langle ij\rangle\rangle} J_2^{\alpha\beta} \mathbf{s}_{i,\alpha} \cdot \mathbf{s}_{j,\beta} + \sum_{i,\alpha\neq\beta} J_H^{\alpha\beta} \mathbf{s}_{i,\alpha} \cdot \mathbf{s}_{i,\beta}, \tag{1}$$

where $\langle ij\rangle$ and $\langle\langle ij\rangle\rangle$ label nearest-neighbor (n.n.) and next-nearest-neighbor (n.n.n.) Fe sites on its square lattice. The greek indices label the orbitals and J_H is the on-site Hund's coupling. Because of the multiple orbitals, both J_1 and J_2 are matrices. At linear order in w, the kinetic energy H_c of the itinerant carriers and their coupling H_m to the quasi local moments appear when integrating out the high energy degrees of freedom:

$$H_c = w \sum_{\mathbf{k},\alpha,\sigma} E_{\mathbf{k}\alpha\sigma} c^\dagger_{\mathbf{k}\alpha\sigma} c_{\mathbf{k}\alpha\sigma} \tag{2}$$

$$H_m = w \sum_{\mathbf{k}\mathbf{q}\alpha\beta\gamma} G_{\mathbf{k},\mathbf{q}\alpha\beta\gamma}\, c^\dagger_{\mathbf{k}+\mathbf{q}\alpha\sigma} \frac{\boldsymbol{\tau}_{\sigma\sigma'}}{2} c_{\mathbf{k}\beta\sigma'} \cdot \mathbf{s}_{\mathbf{q}\gamma}, \tag{3}$$

where $\boldsymbol{\tau}$ are the Pauli matrices. The final form for the low-energy effective Hamiltonian is

$$H_{eff} = H_J + H_c + H_m. \tag{4}$$

4. Quantum criticality in the iron pnictides

An important consequence of the strong-coupling approach is that by tuning the strong-coupling parameter w, one can access a QCP, separating a paramagnetic metallic phase from an antiferromagnetic or possibly insulating phase. Indeed, it was predicted [6] and subsequently experimentally confirmed that doping phosphorus into a magnetic Fe-As parent compound would reveal a magnetic quantum critical point, as shown in Fig. 4. The reason is that, since P is smaller than As, the lattice contracts a bit upon P doping, with a consequence that the kinetic energy (bandwidth) increases so that w increases to a critical value w_c at which the magnetism disappears. This can be understood by examining the action that describes the quantum fluctuations of the J_1-J_2 model, which is appropriate when $w = 0$: $\mathcal{S} = \mathcal{S}_2 + \mathcal{S}_4 + \cdots$, with

$$\mathcal{S} = \int d\mathbf{q}d\omega[r_0 + c_1 q_x^2 + c_2 q_y^2 + \omega^2[|\mathbf{m}(\mathbf{q},\omega)|^2 + |\mathbf{m}'(\mathbf{q},\omega)|^2]. \tag{5}$$

Here, $r_0 < 0$ and $\mathbf{m}$ and $\mathbf{m}'$ are vector order parameters describing the Néel vectors of two interpenetrating sublattices on the square lattice [with $\mathbf{m} \pm \mathbf{m}'$ respectively corresponding to the $(\pi, 0)$ and $(0, \pi)$ order]. The couplings to the itinerant fermions at finite order in w bring about added terms to this effective theory. To the linear order in w, there are two added terms:

$$\Delta\mathcal{S} = \int d\mathbf{q}d\omega[wA_{\mathbf{Q}} + \gamma|\omega|[|\mathbf{m}(\mathbf{q},\omega)|^2 + |\mathbf{m}'(\mathbf{q},\omega)|^2] \tag{6}$$

The wA_Q term is positive, providing a mass shift that weakens the $(\pi, 0)$ order; when $r_0 + wA_Q = 0$, the magnetic order disappears at a QCP. The $\gamma|\omega|$ term describes the Landau

Journal of Physics: Conference Series **449** (2013) 012025 doi:10.1088/1742-6596/449/1/012025

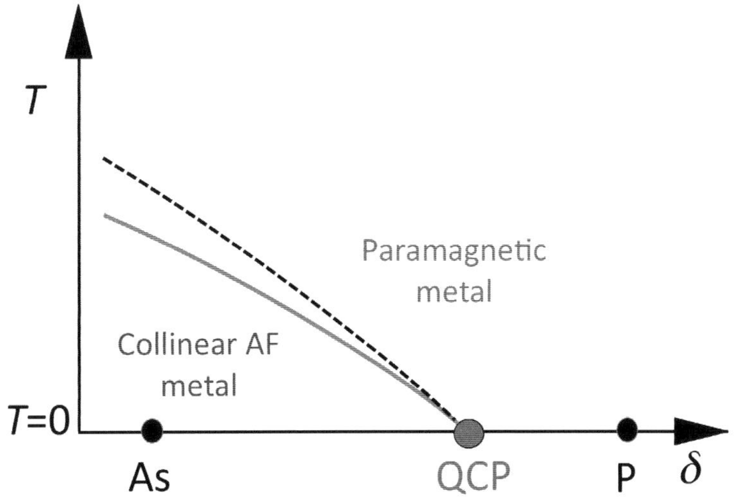

Figure 5. Pnictide phase diagram near a magnetic quantum critical point. Here δ is a non-thermal control parameter: increasing δ enhances the spectral weight w in the coherent part of the single-electron excitations. A specific example for δ is the concentration of P doping for As. The red dot denotes the quantum critical point that separates a two-sublattice collinear AF ground state from a paramagnetic one. The blue solid/black dashed lines respectively represent the magnetic/structural transitions at non-zero temperatures. Here the structural transition reflects the transition associated with the Ising order parameter, $\mathbf{m}\cdot\mathbf{m}'$. Adapted from Ref. [6].

damping; here the prefactor γ is order w^0, but the linear-in-ω dependence has an upper cutoff frequency that is linear in w. Thus, the total $\mathcal{S}+\Delta\mathcal{S}$ gives rise to a magnetic QCP, as illustrated in Fig. 4, where at a critical value of the P-doping, $r_0 + wA_Q = 0$. An important ingredient of the analysis given in Ref. [6] is that the quartic coupling, $\tilde{u}(\mathbf{m}\cdot\mathbf{m}')^2$, is only marginally relevant with respect to the underlying fixed point associated with the O(3) transition of $\mathbf{m}$ (and $\mathbf{m}'$). This makes the Ising and antiferromagnetic transitions to be essentially coinciding and second order. (By contrast, for the thermally driven transition, $\tilde{u}$ is strongly relevant with respect to the classical O(3) fixed point, and the Ising and antiferromagnetic transitions become either significantly split or first order.)

This theoretical proposal for a magnetic QCP in P-doped parent iron arsenides [6] has now been extensively confirmed in experiments carried out in P-doped CeFeAsO [40, 41, 42] and BaFe2As2 [43, 44, 45].

5. Spin dynamics in the paramagnetic phase of parent iron pnictides

Quantum criticality represents one way to study the quantum fluctuations, through the tuning of a control parameter. Alternatively, these fluctuations can be directly probed by studying the frequency and wave-vector dependences of the spin structure factor. A particularly instructive case to consider is the parent arsenides in their paramagnetic phases, at $T > T_N$ [47, 48, 49]; the absence of order fascilitates the study of the underlying interactions. With that in mind, we have studied the $J_1 - J_2 - K$ model:

$$H = J_1 \sum_{i,\delta} \mathbf{S}_i \cdot \mathbf{S}_{i,\delta} + J_2 \sum_{i,\delta'} \mathbf{S}_i \cdot \mathbf{S}_{i+\delta'} - K \sum_{i,\delta} (\mathbf{S}_i \cdot \mathbf{S}_{i,\delta})^2 \qquad (7)$$

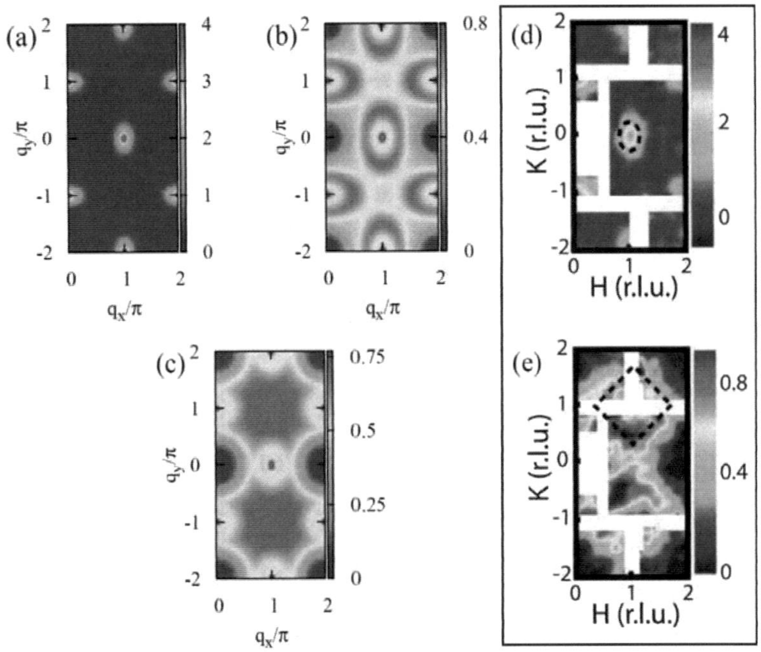

Figure 6. Evolution of the spin structure factor $S(q,\omega)$ in the paramagnetic phase of the $J_1 - J_2 - K$ model and the comparison with inelastic neutron scattering results of $BaFe_2As_2$. The calculated $S(q,\omega)$ shows elliptical features near $(\pi, 0)$ at low energies (in panels (a) and (b)). They are split into features that are centered around (π,π), as the energy is increased towards the zone-boundary spin-excitation energy (panel (c)). This trend is consistent with the inelastic neutron scattering experiments, shown in the box at the right for two energies measured in the paramagnetic phase of BaFe2As2 (data taken from Ref. [48]). Adapted from Ref. [50].

which is combined with the damping term specified in Eq. 6. Our approach represents one way to describe the spin dynamics by incorporating the contributions of the incoherent single-electron excitations in bad metals close to a Mott transition. An alternative approach is to study the spin dynamics in this regime using the DMFT method [31]. The results from the two approaches are complementary.

To calculate the dynamical structure factor, in Ref. [50] we have decoupled the biquadratic coupling using Hubbard-Stratonovich fields and applied a modified spin-wave method [51] to treat the resulting spin Hamiltonian. At low energies, $S(\mathbf{q},\omega)$ as a function of $\mathbf{q}$ displays elliptic features centered around $(\pi, 0)$ and its symmetry-equivalent points. Such elliptic features reflect the existence of two correlation lengths whose ratio is controlled by J_2/J_1, and are not sensitive to the magnitude of the biquadratic coupling K. [Indeed, the results at low energies are similar to those for the $K = 0$ case [46].] At high energies, $S(\mathbf{q},\omega)$ vs. $\mathbf{q}$ is sensitive to the K coupling. Fig. 6 shows the results of the energy evolution of the momentum-distribution of the dynamical structural factor, which compares well with the experimental observations [48].

For the insulating iron alkaline iron selenides in the presence of the $\sqrt{5} \times \sqrt{5}$ vacancy order, both the block spin order and spin-wave excitations can be studied in an extended $J_1 - J_2$ model [52]. Spin-wave measurements have fully confirmed the calculated spectrum [53].

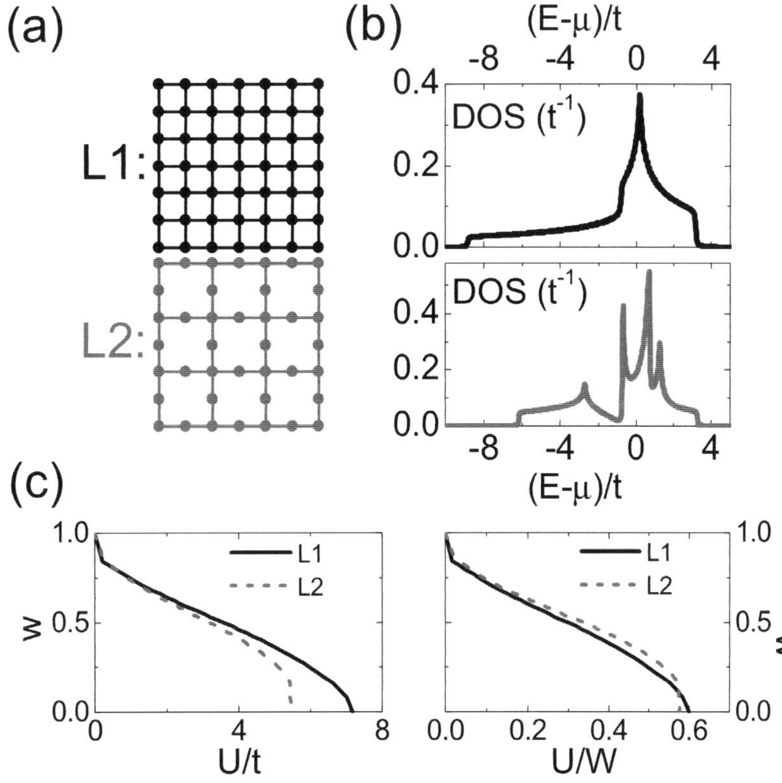

Figure 7. Kinetic energy reduction and Mott localization by ordered Fe vacancies. (a) The regular square lattice and its counterpart with 2×2 ordered vacancies (the pattern is chosen for an illustrative purpose, such that the model calculation is simplified, and similar results arise for other patterns including $\sqrt{5} \times \sqrt{5}$); (b) DOS of the two lattices; (c) Left panel: w vs. U/t. The Mott threshold U_c is reduced by the ordered vacancies. Here, t is the nearest-neighbor hopping parameter. Right panel: w plotted against U/W, showing that U_c tracks W, the bandwidth. Adapted from Ref. [54].

6. Parent Mott insulator arising from kinetic energy reduction

In Sec. 4, we discussed how to induce a QCP by increasing the kinetic energy, which increases w and weakens the magnetic order. In the same spirit, one would expect to be able to tune the system in the opposite way, by reducing the kinetic energy and thereby increasing the correlation effect. This corresponds to moving to the left on the U/t axis shown in Fig. 2.

The iron oxychalcogenides $La_2O_2Fe_2O(Se,S)_2$ provide a case study [33]. These are also parent compounds, *i.e.* they have a composition such that the nominal valence of Fe is 2+. They also contain an Fe square lattice, which is expanded compared to that of the iron pnictides; this leads to narrower Fe $3d$-electron bands. According to LDA calculations, the paramagnetic ground state should be metallic. However, the band narrowing enhances correlation effects and promotes the Mott insulating state. Measurements of the transport and magnetic properties provide the experimental evidence that the system is indeed a Mott insulator. Because the bandwidth of $La_2O_2Fe_2O(Se,S)_2$ is narrower than that of the parent iron pnictides by only a moderate amount (on the order of 25%), this result provides evidence that the parent pnictides are on the verge of Mott localization. Furthermore, elastic neutron scattering measurements have determined the magnetic order [55]. and found a large ordered moment, about 2.8 μ_B/Fe,

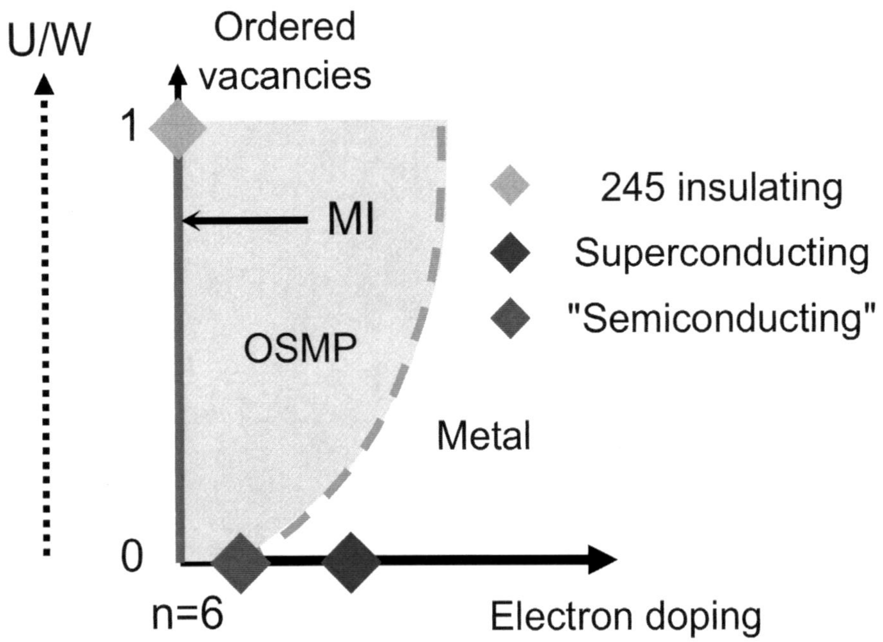

Figure 8. Overall phase diagram for $K_{1-x}Fe_{2-y}Se_2$**.** The vertical axis represents the degree of iron vacancy order, with 1 corresponding to fully $\sqrt{5} \times \sqrt{5}$ vacancy ordering and 0 marking complete vacancy disorder or absence of vacancies. The degree of the iron vacancy order tunes U/W. MI and OSMP stand for the Mott insulator and the orbital-selective Mott phase, respectively, and the orange dashed line describes the transition from OSMP to the delocalized metallic phase. The diamond symbols mark the approximate positions for the various phases that have been observed experimentally.

consistent with the discussion in Sec. 2.

Mott insulating behavior also exists in $R_2O_2Fe_2OSe_2$ (R=Ce,Pr,Nd, and Sm) which possess a similarly large ordered moment [56] and in several other iron oxychalcogenides whose ordered moments are yet to be determined: $(Sr,Ba)_2F_2Fe_2O(Se,S)_2$ (Ref. [57]), $Na_2Fe_2OSe_2$ (Ref. [58]), and BaFe2OSe2 (Refs. [59, 60]).

7. Alkaline iron selenides: Mott insulator and orbital-selective Mott insulator

Recently, superconductivity with T_c comparable to the iron pnictides was discovered in $K_{1-y}Fe_{2-x}Se_2$ ("KFS") [61] and related iron selenides with K replaced by Rb, Cs, or Tl [34, 62, 63]. Here too, the superconductivity occurs near antiferromagnetic order [34]. Angle-resolved photoemission (ARPES) experiments [64, 65, 66] have shown that the Fermi surface has only electron pockets, in contrast to the iron pnictides , which have both electron and hole pockets.

The parent selenide compounds, with Fe valence at 2+, are insulating [34, 67]. These compounds have Fe vacancies forming ordered patterns [34, 68]. LDA band-structure calculations show that the paramagnetic ground state of such a vacancy-ordered structure is metallic [69, 70], therefore the insulating ground state must result from electron interactions. How strong are these interactions? A clue is provided by the size of the ordered moment, which is about 3.3 μ_B/Fe [68, 53]. Recognizing that the maximal possible moment is 4 μ_B/Fe, the observed ordered moment is indeed very large. Therefore, the interaction energy must be sizable compared to the kinetic energy.

Microscopic considerations have suggested that the Mott insulating nature of the vacancy-ordered system is very natural [54]. The ordered vacancies lead to a band-narrowing, which enhances the ratio U/W, and pushes the system to the Mott insulator side. This is shown in Fig. 7, with an illustrative model having a 2×2 vacancy ordering pattern [54].

In these iron-based compounds, there are five $3d$ orbitals which are non-degenerate; a slave-spin study [71] has identified an orbital-selective Mott phase ("OSMP") in which the xy orbitals are localized and the remaining $3d$ orbitals are delocalized. ARPES measurements in the superconducting compound have shown that raising the temperature causes a suppression of the spectral weight of the xy orbitals near the Fermi energy, while keeping the weight of its xz and yz counterparts non-zero [72], Combining the ARPES results and the theoretical phase diagram yields an overall phase diagram shown in Fig. 8.

The OSMP provides the link between the vacancy-ordered/Mott insulating phase and the vacancy disordered/free superconducting phase. Experimental evidence has come from transport measurements in the insulating alkaline iron selenides under pressure [73].

We close this section by discussing the expected paramagnetic spin excitations in the insulating $Rb_{0.8}Fe_{1.6}Se_2$. Because T_N is very large (on the order of 500 K), and is only slightly (a few 10's K) below the vacancy ordering temperature T_s, the $\sqrt{5} \times \sqrt{5}$ iron vacancy order becomes fragile above T_N [68]. As we have just discussed in this section, the reduction of the iron vacancy order leads to an effective reduction of the electron correlation strength U/W from that of the fully vacancy-ordered case, destabilizing the Mott insulating state. Correspondingly, the spin spectral weight will be reduced from the very large value in the fully vacancy-ordered state. This has recently been observed: Above the Néel temperature, Wang et al. [74] reported spin excitations that are more strongly damped than those of the fully vacancy-ordered state as well as a significant reduction of the total moment. In other words, the experimental observations in the paramagnetic regime, both below and above the vacancy ordering temperature, are consistent with the Mott-insulating behavior of the fully vacancy-ordered state. The fact that the magnetic excitations at low temperatures are well described by spin waves with a very large stiffness constant reinforces the Mott-insulating behavior of the fully vacancy-ordered state.

8. Superconductivity in iron pnictides and alkaline iron selenides

In the carrier doped case, the w-expansion discussed earlier leads to a 5-band $t - J_1 - J_2$ model [5, 3]. Spin-singlet superconducting pairing from this model was studied in Ref. [75] in the case of the iron pnictides. Magnetic frustration by itself leads to a large degeneracy in the pairing states. The kinetic energy breaks this into a quasi-degeneracy among a reduced set of pairing states, as shown in Fig. 9. For small electron and hole Fermi pockets, an A_{1g} $s_{x^2y^2}$ state dominates over the phase diagram but a B_{1g} $d_{x^2-y^2}$ state is close by in energy; an $A_{1g} + iB_{1g}$ state, which breaks time-reversal symmetry, occurs at low temperatures in part of the phase diagram. Compared to the two-band case, there is an enhanced amplitude for the A_{1g} $s_{x^2+y^2}$ channel, which causes anisotropy of the gap on the electron pockets or even an accidental gap near M.

An important advantage of this strong-coupling approach is that it gives rise to pairing amplitudes for the alkaline iron selenides that are comparable to those for the iron pnictides, in spite of the very different Fermi surfaces. This is consistent with the fact that the superconducting transition temperatures are very similar.

9. Summary and outlook

We have summarized the phenomenological basis for strong electron correlations in the iron pnictides and chalcogenides. Chief among these are the bad-metal behavior seen in the electrical transport and charge dynamics properties in both classes of materials; the large spectral weight

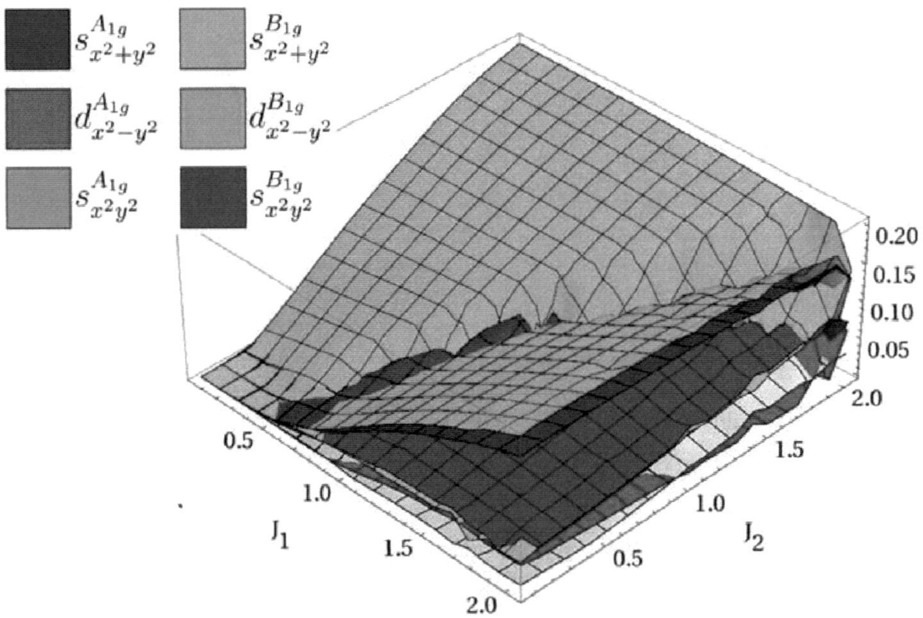

Figure 9. The pairing amplitudes (for d_{xz}, d_{yz} orbitals) in the five orbital model for electron doping $\delta = 0.14$. Adapted from Ref. [75].

observed through spin dynamics measurements, also seen in both cases; and the observation of Mott insulating states in the iron oxychalcogenides and alkaline iron selenides.

We have also outlined a w-expansion, which uses the Mott transition as a reference point. This expansion treats the bad-metal behavior through a proximity to the Mott transition, which separates the single-electron excitations into coherent and incoherent ones whose weights are w and $1 - w$ respectively. In this way, it naturally incorporates the large spin spectral weight in the bad-metal (*i.e.*, small w) regime.

For the parent arsenides, this expansion gives rise to an effective spin Hamiltonian in the form of a J_1-J_2 model that is coupled to the itinerant electrons in the vicinity of the Fermi energy. Varying w amounts to tuning the relative degree of itinerancy vs. localization and provides a means to vary the strength of magnetic ordering. We have discussed an early theoretical proposal for a quantum critical point resulting from w-tuning and for its realization through a P-for-As isoelectronic substitution in the parent iron arsenides. This prediction has since been experimentally confirmed by a number of groups using transport, neutron, NMR and quantum oscillation measurements. Using the same approach, theoretical studies have been carried out for the spin dynamics in the paramagnetic phase of the parent arsenides. In the presence of a biquadratic interactions, the theoretical results compare well with the inelastic neutron scattering results above T_N, both at low energies near the magnetic-zone centers and at high energies close to the magnetic-zone boundaries.

We have also demonstrated how a Mott insulating phase may arise from a kinetic-energy reduction, and its experimental evidence in both the iron oxychalcogenide compounds and the vacancy-ordered alkaline iron selenides. In the latter families, we show how an orbital-selective Mott phase appears in the phase diagram.

Finally, we have discussed the singlet superconducting pairing driven by the short-range J_1-J_2 interactions. For the carrier-doped cases, the w-expansion gives rise to a five-orbital matrix

t-J_1-J_2 model. In this model, the dominant pairing channels are the nearly degenerate A_{1g} $s_{x^2y^2}$ and B_{1g} $d_{x^2-y^2}$ states, with A_{1g} $s_{x^2+y^2}$ channel also playing a significant role in the iron pnictides.

More generally, we have highlighted the phenomenological and theoretical evidences that the iron pnictides and chalcogenides are bad metals at the boundary between electronic localization and itinerancy. As such, these materials not only are important in their own right, but also provide a setting to study the physics in this important parameter regime of strongly correlated electrons in general.

Acknowledgments

We are grateful to J. Dai, A. H. Nevidomskyy, P. Nikolic, Z. Wang and J.-X. Zhu for collaborations, and to them as well as many other colleagues for useful discussions. This work has been in part supported by the NSF Grant No. DMR-1006985 and the Robert A. Welch Foundation Grant No. C-1411. P. G. was supported at the National High Magnetic Field Laboratory by NSF Cooperative Agreement No. DMR-0654118, the State of Florida, and the U. S. Department of Energy. Part of this work was carried out at the Aspen Center for Physics (NSF grant 1066293).

References

[1] Kamihara Y, Watanabe T, Hirano M, and Hosono H 2008 *J. Am. Chem. Soc.* **130** 3296
[2] de la Cruz C et al 2008 *Nature* **453** 899
[3] Si Q and Abrahams E 2008 *Phys. Rev. Lett.* **101** 076401
[4] Yildirim T 2008 *Phys. Rev. Lett.* **101** 057010
[5] Si Q, Abrahams E, Dai J, and Zhu J-X 2009 *New J. Phys.* **11** 045001
[6] Dai J, Si Q, Zhu J-X, and Abrahams E 2009 *Proc. Natl. Acad. Sci.* **106** 4118
[7] Yin Z P, Haule K and Kotliar G 2011 *Nat. Mater.* **10**, 932
[8] Seo K, Bernevig B A and Hu J 2008 *Phys. Rev. Lett.* **101**, 206404
[9] Chen, W-Q, Yang, K-Y, Zhou, Y and Zhang, F-C 2009 *Phys. Rev. Lett.* **102**, 047006
[10] Moreo A, Daghofer M, Riera J A and Dagotto E 2009 *Phys. Rev. B* **79**, 134502
[11] Berg E, Kivelson S A and Scalapino D J 2009 *New J. Phys.* **11**, 085007
[12] Lv W, Krüger F and Phillips P 2010 *Phys. Rev. B* **82**, 045125
[13] Ma F, Lu Z-Y and Xiang T 2008 *Phys. Rev. B* **78** 224517
[14] Han M J, Yin Q, Pickett W E and Savrasov S Y 2009 *Phys. Rev. Lett.* **102** 107003
[15] Wysocki A L, Belashchenko K D and Antropov V P 2011 *Nat. Phys.* **7** 485
[16] Fang C, Yao H, Tsai W-F, Hu J and Kivelson S A 2008 *Phys. Rev. B* **77** 224509
[17] Xu C, Muller M and Sachdev S 2008 *Phys. Rev. B* **78**, 020501(R)
[18] Uhrig G S, Holt M, Oitmaa J, Sushkov O P and Singh R R P 2009 *Phys. Rev. B* **79**, 092416
[19] Ishida H and Liebsch A 2010 *Phys. Rev. B* **81** 054513
[20] Giavannetti G, Ortix C, Marsman M, Capone M, van den Brink J and Lorenzana J 2011 *Nat. Commun.* **2** 398
[21] Calderón M J, Le'on G, Valenzuela B and Bascones E 2012 *Phys. Rev. B* **86** 104514
[22] Abrahams E, Si Q 2011 *J. Phys.: Condens. Matter* **23** 223201
[23] Qazilbash M et al 2009 *Nat. Phys.* **5** 647
[24] Hu W Z et al 2008 *Phys. Rev. Lett.* **101** 257005
[25] Yang J et al 2009 *Phys. Rev. Lett.* **102** 187003
[26] Boris A V et al 2009 *Phys. Rev. Lett.* **102** 027001
[27] Degiorgi L 2011 *New J. of Phys.* **13** 023011
[28] Liu M et al 2012 *Nat. Phys.* **8** 376
[29] Zaliznyak I A et al 2011 *Phys. Rev. Lett.* **107** 216403
[30] Zhao J et al 2009 *Nat. Phys.* **5** 555
[31] Park H, Haule K, and Kotliar G 2011 *Phys. Rev. Lett.* **107** 137007
[32] Toschi A, Arita R, Hansmann P, Sangiovanni G, and Held K 2012 *Phys. Rev. B* **86** 064411
[33] Zhu J-X et al 2010 *Phys. Rev. Lett.* **104** 216405
[34] Fang M et al 2011 *Euro. Phys. Lett.* **94** 27009
[35] Wang D M, He J B, Xia T-L and Chen G F 2011 *Phys. Rev. B* **83** 132502
[36] Yu R and Si Q 2011 *Phys. Rev. B* **84** 235115

[37] Yu R and Si Q 2012 *Phys. Rev. B* **86** 085104
[38] Moeller G, Si Q, Kotliar G, Rozenberg M, and Fisher D S 1995 *Phys. Rev. Lett.* **74** 2082
[39] Chandra P, Coleman P, and Larkin A I 1990 *Phys. Rev. Lett.* **64** 88
[40] de la Cruz C et al 2010 *Phys. Rev. Lett.* **104** 017204
[41] Luo Y et al 2010 *Phys. Rev. B* **81** 134422
[42] Jesche A, Krellner C, de Souza M, Lang M, and Geibel C 2009 *New J. Phys.* **11** 103050
[43] Jiang S et al 2009 *J. Phys. Condens. Matter* **21** 382203
[44] Kasahara S et al 2010 *Phys. Rev. B* **81** 184519
[45] Hashimoto K et al 2012 *Science* **336** 1554
[46] Goswami P, Yu R, Si Q and Abrahams E 2011 *Phys. Rev. B* **84** 155108
[47] Diallo S O et al 2010 *Phys. Rev. B* **81** 214407
[48] Harriger L W et al 2011 *Phys. Rev. B* **84** 054544
[49] Ewings R A et al 2011 *Phys. Rev. B* **83** 214519
[50] Yu R et al 2012 *Phys. Rev. B* **86** 085148
[51] Takahashi M 1989 *Phys. Rev. B* **40** 2494
[52] Yu R, Goswami P, Si Q 2011 *Phys. Rev.* B **84** 094451
[53] Wang M et al 2011 *Nat. Commun.* **2** 580
[54] Yu R, Zhu J-X, and Si Q 2011 *Phys. Rev. Lett.* **106** 186401
[55] Free D G and Evans J S O 2010 *Phys. Rev. B* **81** 214433
[56] Ni N, Jia S, Huang Q, Climent-Pascual E and Cava R J 2011 *Phys. Rev. B* **83** 224403
[57] Kabbour H et al 2008 *J. Am. Chem. Soc.* **130** 8261
[58] He J B, Wang D M, Shi H L, Yang H X, Li J Q and Chen G F 2011 *Phys. Rev. B* **84** 205212
[59] Han F, Wan X, Shen B and Wen H-H 2012 *Phys. Rev. B* **86** 014411
[60] Lei H, Ryu H, Warren J, Frenkel A I, Ivanovski V, Cekic B and Petrovic C *arXiv:1206.5788*
[61] Guo J et al 2010 *Phys. Rev. B* **82** 180520
[62] Krzton-Maziopa A et al 2011 *J. Phys.: Condens. Matter* **23** 052203
[63] Mizuguchi Y et al 2011 *Appl. Phys. Lett.* **98** 042511
[64] Zhang Y et al 2011 *Nat. Mater.* **10** 273
[65] Qian T et al 2011 *Phys. Rev. Lett.* **106** 187001
[66] Mou D et al 2011 *Phys. Rev. Lett.* **106** 107001
[67] Wang D M, He J B, Xia T-L, and Chen G F, 2011 *Phys. Rev. B* **83** 132502
[68] Bao W et al 2011 *Chin. Phys. Lett.* **28** 086104
[69] Cao C and Dai J 2011 *Phys. Rev. Lett.* **107** 056401
[70] Yan X-W et al 2011 *Phys. Rev. B* **83** 233205
[71] Yu R and Si Q 2012 arXiv:1208.5547
[72] Yi M et al 2012 arXiv:1208.5192
[73] Gao P et al 2012 arXiv:1209.1340
[74] Wang M et al 2012 arXiv:1211.7328.
[75] Goswami P, Nikolic P, and Si Q 2010 *EuroPhys. Lett.* **91** 37006

Journal of Physics: Conference Series **449** (2013) 012026 doi:10.1088/1742-6596/449/1/012026

Origin of the Large Anisotropy in the χ_3 Anomaly in URu_2Si_2

P. Chandra[1], P. Coleman[1,2] and R. Flint[3]

[1] Center for Materials Theory, Department of Physics and Astronomy, Rutgers University, Piscataway, NJ 08854

[2]Department of Physics, Royal Holloway, University of London, Egham, Surrey TW20 0EX, UK.

[3] Department of Physics, Massachusetts Institute for Technology, 77 Massachusetts Avenue, Cambridge, MA 02139-4307

E-mail: pchandra@physics.rutgers.edu

Abstract. Motivated by recent quantum oscillations experiments on URu_2Si_2, we discuss the microscopic origin of the large anisotropy observed many years ago in the anomaly of the nonlinear susceptibility in this same material. We show that the magnitude of this anomaly emerges naturally from hastatic order, a proposal for hidden order that is a two-component spinor arising from the hybridization of a non-Kramers Γ_5 doublet with Kramers conduction electrons. A prediction is made for the angular anisotropy of the nonlinear susceptibility anomaly as a test of this proposed order parameter for URu_2Si_2.

The character of the long-range "hidden order" (HO) in the heavy fermion material URu_2Si_2 continues to fascinate and elude theorists and experimentalists [1]. Despite a large entropy released at the HO transition temperature ($T_0 = 17.5K$) accompanied by many sharp thermodynamic anomalies there[1, 2, 3, 4], there is no consensus on the important local degrees of freedom and more importantly how they form the hidden order state.

A revisit to the nonlinear susceptibility (χ_3) and its sharp anomaly at T_0 [3, 4] may provide insight into the mysterious nature of the hidden order. Originally the behavior of this mean-field-like anomaly inspired a density-wave description. A simple Landau theory, $f = f(T - T_c(B^2))$, was constructed that yielded the relation

$$\Delta\left(\frac{c_v}{T}\right)\Delta\chi_3 = 12\left[\Delta\left(\frac{\partial\chi}{\partial T}\right)\right]^2 \tag{1}$$

that was tested successfully experimentally[5], suggesting underlying itinerant order; here we recall that $M = \chi_1 B + \frac{\chi_3}{3!}B^3$ where M is the magnetization and χ_1 and χ_3 are the linear and nonlinear susceptibilities respectively. However the large c-axis anisotropy of this χ_3 anomaly was then emphasized and ascribed to local Ising f-moment behavior [6, 7]. The appropriate Landau theory is then

$$f(T, B_z) = [\alpha(T_c - T) - \eta_z B_z^2]|\Psi|^2 + \beta|\Psi|^4 \tag{2}$$

where Ψ is the hidden order parameter; here a key question arises: what is the microscopic origin of the large anisotropy coefficient η_z?

The anisotropic field-response of the U-ions in URu_2Si_2 is most naturally understood if their low-energy configurations are $5f^2$, and this point has been noted by several authors[6, 7, 8, 9, 10]. Such strong Ising anisotropy was also observed in dilute $U_xTh_{1-x}Ru_2Si_2$, thus indicating that it is a key property of the U ions [8]. It was originally described by quadrupolar ordering of the local f-moments[6, 7, 9], but unfortunately recent resonant X-ray measurements see no signature for such multipolar order[11, 12]. Furthermore there is experimental evidence from scanning tunnelling microscopy[17, 18], and de Haas-van Alphen (dHvA) and Shubnikov-de Haas (SdH)[13, 14, 15, 16] measurements for itinerant behavior below T_o, so a local-moment scenario cannot capture the full picture at the transition.

Recent quantum oscillations provide a clue towards resolving this dilemma, as they reveal that the quasiparticles deep in the HO phase exhibit a large Ising anisotropy[16]. More specifically their Zeeman splitting depends only on the c-axis component of the magnetic field, $\Delta E = g(\theta)\mu_B B$ with a g-factor $g(\theta) = g\cos\theta$. The size of the measured g-factor anisotropy is resolution limited and exceeds thirty, corresponding to an anisotropy of the Pauli susceptibility in excess of 900 that is too large to be attributed simply to spin-orbit coupling. These experiments suggest that the anisotropy of the U $5f^2$ moments is transferred to the mobile quasiparticles in the HO phase via hybridization. Angle-dependence of the Pauli-limited critical field of the superconducting state[19, 20] confirms the g-factor anisotropy; furthermore the presence of Ising quasiparticles in a superconductor with $T_c \sim 1.5K$ indicate that the U $5f^2$ configuration is degenerate to within an energy resolution of $g\mu_B B \sim 5K$. The tetragonal symmetry of URu_2Si_2 protects the non-Kramers doublet Γ_5 that is quadrupolar in the basal plane and magnetic along the c-axis, and it has been proposed as the origin of the magnetic anisotropy in both dilute and dense URu_2Si_2[8, 9, 16, 21].

Conventionally in heavy fermion materials, hybridization involves valence fluctuations between a ground-state Kramers doublet and an excited singlet; it develops via a crossover, leading to mobile heavy quasiparticles. However if the ground-state is a non-Kramers doublet, the Kondo effect will involve an excited Kramers doublet[8]. Hybridization now carries spin and coherence; its development breaks time-reversal and spin rotational invariance at a true phase transition. Furthermore because it mixes states of half-integer (non-Kramers doublets) and integer (conduction electrons) it breaks Kramers parity and thus also breaks double-time reversal[21]. In URu_2Si_2 optical[22] and tunnelling probes[17, 18] indicate that hybridization develops abruptly at the HO transition, and thus we have proposed that the hybridization is a two-component order parameter ("hastatic" order) that transforms as a spinor[21].

Let us briefly digress to discuss the microscopic valence fluctuations of a non-Kramers doublet. The valence fluctuation physics of a non-Kramers doublet are described by an Bolech-Andrei model[23] (ABM) $H = \sum_{\mathbf{k}} \epsilon_{\mathbf{k}} c^\dagger_{\mathbf{k}\sigma} c_{\mathbf{k}\sigma} + \sum_j H_{ABM}(j)$, where the valence fluctuations and atomic physics of each U ion are described by

$$H_{ABM}(j) = V\left[c^\dagger_{\sigma\alpha}(j)X_{\sigma\alpha}(j) + \text{H.c}\right] + \epsilon X_{\sigma\sigma}(j) \tag{3}$$

where the $X_{\sigma\alpha} = |5f^3, \sigma\rangle\langle 5f^2, \alpha|$ is the Hubbard operator between the Kramers and non-Kramers state, $X_{\sigma\sigma} = |5f^3, \sigma\rangle\langle 5f^3, \sigma|$ is the projection onto the excited Kramer's doublet, and $\epsilon = E[5f^3] - E[5f^2]$ is the energy difference between the $5f^3$ excited Kramers doublet and the $5f^2$ non-Kramers ground-state. The operator $c^\dagger_{\sigma\alpha}(j)$ creates a conduction *hole* with channel and spin index α and σ at site j. Valence fluctuations out of a non-Kramer's doublet involve at least two different crystal symmetry channels, the Γ_7 and Γ_6 channels. The ABM results from the projection of an Anderson model into the low-lying $5f^2$, $5f^3$ subspace. $H_{ABM}(j) = \mathcal{P}H_{Anderson}(j)\mathcal{P}$. Typically the $c^\dagger_{\sigma\alpha}$ is expanded in terms of the two valence

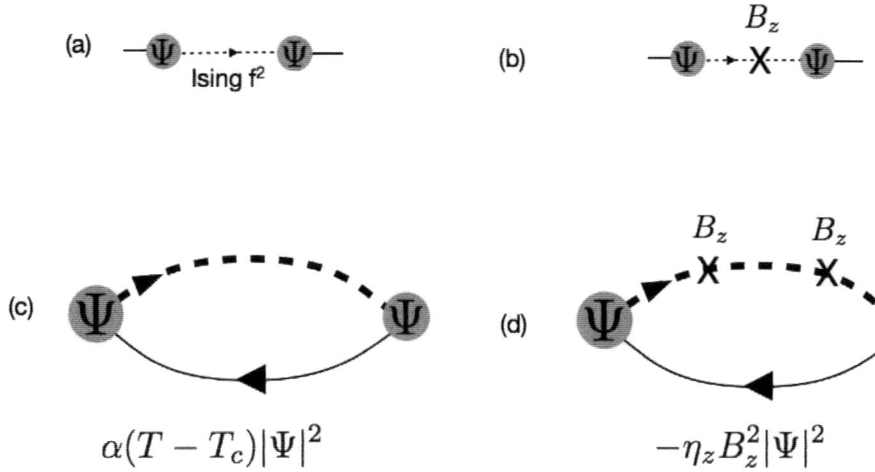

Figure 1. Showing (a) resonant Kondo scattering via hybridization off non-Kramers $5f^2$ doublet, (b) coupling to magnetic field via f-state is purely Ising like (c) zero field quadratic contribution to the Landau energy (d) Ising field-dependent contribution $-B_z^2\Psi^2$ term in the Landau theory.

fluctuation channels $\lambda = 1, 2 \equiv \Gamma_{6,7}$ as follows

$$c^\dagger_{\sigma\alpha} = \sum_{\lambda=1,2} c^\dagger_{\lambda\sigma'}\Gamma^{\sigma'}_{\sigma\alpha}(\lambda), \qquad \Gamma^{\sigma'}_{\sigma\alpha}(\lambda) = \langle 5f^2\alpha|f_{\lambda\sigma'}|5f^3\sigma\rangle. \tag{4}$$

where $c^\dagger_{\lambda\sigma} = \sum_{\mathbf{k},\sigma'} \gamma^\lambda_{\sigma\sigma'}(\mathbf{k})c^\dagger_{\mathbf{k}\sigma}e^{-i\mathbf{k}\cdot\mathbf{R}_j}$ defines the spin-orbit coupled Wannier states of the conduction holes in terms of the crystal-field form-factors $\gamma^\lambda_{\sigma\sigma'}(\mathbf{k})$ and plane wave conduction states.

If we represent $X_{\sigma\alpha} = \hat{\Psi}^\dagger_\sigma f_\alpha$ using a slave boson decomposition, where $\Psi^\dagger_\sigma$ is a slave boson describing the excited Kramers' doublet $\hat{\Psi}^\dagger_\sigma|0\rangle \equiv |5f^3, \sigma\rangle$ while $f^\dagger_\alpha$ is an Abrikosov pseudo-fermion describing the low-lying non-Kramer's doublet $f^\dagger_\alpha|0\rangle \equiv |5f^2, \alpha\rangle$. Hastatic order corresponds to the condensation of the slave boson,

$$X_{\sigma\alpha} \to \langle\hat{\Psi}^\dagger_\sigma\rangle f_\alpha \equiv \Psi^*_\sigma f_\alpha \tag{5}$$

The condensed slave boson defines the two-component Hastatic order parameter, giving rise to hybridization terms of the form shown in Fig. 1 (a). Although this term looks like a conventional hybridization in a heavy fermion system, it breaks time reversal symmetry in a new way, by mixing an integer spin non-Kramers doublet with a half-integer spin conduction electron. The coherent, broken symmetry physics of the HO phase is is determined by the hybridization Hamiltonian

$$H^*(j) = V\left[\Psi^*_\sigma c^\dagger_{\sigma\alpha}(j)f_\alpha(j) + \text{H.c}\right] + \epsilon|\Psi|^2 \tag{6}$$

Let us now return to the anomaly in χ_3. In the field-dependent Landau theory of hastatic order, we get the first temperature-dependent term (Fig. 1 a) from the hybridization of the conduction electrons and the Ising f-moments. We can obtain the Landau theory from a Feynman diagram expansion of the Free energy in powers of $\langle\Psi\rangle$, as illustrated in Fig. 1. Applying a magnetic field, the conduction electrons couple to it isotropically, whereas the non-Kramers doublet couples linearly only to B_z(Fig. 1b). Thus the resulting field-dependent Landau theory, depicted schematically in Fig1c and Fig1d, is

$$f[\psi] = [\alpha(T - T_c) - \eta_z B_z^2 - \eta_\perp B_\perp^2]\Psi^2 + \beta\Psi^4 \tag{7}$$

where $B_z = B\cos\theta$ and $B_\perp = B\sin\theta$. Minimizing the free energy with respect to Ψ, we obtain

$$f = -\frac{[\alpha(T_c - T) + B^2(\eta_z\cos^2\theta + \eta_\perp\sin^2\theta)]^2}{4\beta} \tag{8}$$

from which

$$\Delta\chi_3 = -\frac{\partial^4 f}{\partial B^4} = \frac{6}{\beta}(\eta_z\cos^2\theta + \eta_\perp\sin^2\theta)^2$$

We can calculate η_z and $\eta_\perp$ within a two-channel Anderson lattice model of hastatic order [21], and we find that $\frac{\eta_\perp}{\eta_z} = \frac{T_0^2}{D^2}$ where T_o and D are the hidden order transition temperature and the bandwidth respectively. Taking a conservative $\frac{T_0}{D} \sim \frac{1}{30}$, leads to an anisotropy of 10^3 in $\frac{d\chi_1}{dT}$ and thus a factor of 10^6 in $\Delta\chi_3$. Of course our model is simplified and there will be many material features (e.g. f-electron contributions to η_z arising from fluctuations to excited states) that will reduce the anisotropy. The crucial point here is that the anisotropy in the χ_3 anomaly will be several orders of magnitude greater than the single-ion anisotropy in χ_1, and that $\chi_3 \propto \cos^4\theta$ where θ is the angle from the c-axis. This enormous Ising anisotropy in the anomaly of χ_3 is a natural consequence of hastatic order and is one of its many testable consequences. Careful new measurements to determine the angular dependence of $\Delta\chi_3$ would verify the Ising nature of the quasiparticles.

In conclusion we have shown that hybridization between half-integer conduction electrons and spin-half non-Kramers doublets provides a natural microscopic explanation for the large anomaly observed in the nonlinear susceptibility anomaly[3, 4] in URu_2Si_2. Here the f-moments couple only to the c-axis component of the applied field and, via hybridization, transfer their Ising anisotropy to the conduction electrons. These mobile Ising quasiparticles form Cooper pairs at lower temperatures, and the observable consequences of their hastatic nature in the superconducting state remain to be explored.

Acknowledgments. We are grateful to helpful discussions with N. Andrei, N. Harrison, K. Haule, G. Kotliar, Y Matsuda and J. Mydosh. We would like to thank the Aspen Center for Physics, where this project was started and then completed during two separate visits. This work was supported under grants DMR 0907179 (P. Coleman), NSF grant 1066293 (P. Coleman, P. Chandra and R. Flint) and a Simons Foundation fellowship (R. Flint).

References

[1] J.A. Mydosh and P.M. Oppeneer, *Rev. Mod. Phys.* **83**, 1301 (2011).
[2] T.T.M. Palstra et al, *Phys. Rev. Lett.* **55** 2727 (1085).
[3] Y. Miyako et al, *J. Appl. Phys.* **70** 5791 (1991).
[4] A.P. Ramirez et al. *Phys. Rev. Let.* **68**, 2680 (1992).
[5] A.P. Ramirez et al. *Physica B* **199 & 200**, 426 (1994).
[6] P. Santini and G. Amoretti, *Phys. Rev. Lett.* **73**, 1027 (1994).
[7] P. Santini, *Phys. Rev. B* **57**, 5191 (1998).
[8] H. Amitsuka and T. Sakakibara, *J. Phys. Soc. Japan* **63**, 736 (1994).
[9] F.J. Ohkawa and H. Simizu, *J. Phys. Cond. Mat.* **11**, L519 (1999).
[10] R. Flint, P. Chandra and P. Coleman, arXiv:1207.2433 (2012).

[11] H. Amitsuka et al, *J. Phys.: Conf. Series* **200**, 012007 (2010).
[12] H. Walker et al. *Phys. Rev. B* **Phys. Rev. B 83**, 193102 (2011).
[13] H. Ohkuni et al, *Phil. Mag* **79** 1045 (1999).
[14] Y. Jo et al., *Phys. Rev. Lett.* **98** 166404 (2007).
[15] E. Hassinger et al.,*Phys. Rev. Lett.* **105** 216409 (2010).
[16] M. M. Altarawneh et al, *Phys. Rev. Lett.* **106** 146403 (2010).
[17] A.R. Schmidt et al, *Nature* **465**, 570 (2010).
[18] P. Aynajian et al., *Proc. Natl. Acad. USA* **107** 10383 (2010).
[19] J.P. Brison et al., *Physica C* **250**, 128 (1995).
[20] M.M. Altarawneh et al., *Phys. Rev. Lett.***108** 066407 (2012).
[21] P. Chandra, P. Coleman and R. Flint, arXiv:1207.4828 (2012).
[22] D.A. Bonn et al., *Phys. Rev. Lett.* **61**, 1305 (1988).
[23] C.Bolech and N. Andrei, *Phys. Rev. Lett* **88**, 237206 (2002).

Journal of Physics: Conference Series **449** (2013) 012027 doi:10.1088/1742-6596/449/1/012027

Heavy Fermion Superconductor Ce_2PdIn_8 studied by ^{115}In Nuclear Quadrupole Resonance

H. Fukazawa[1], R. Nagashima[1], S. Shimatani[1], Y. Kohori[1], D. Kaczorowski[2]

[1]Department of Physics, Graduate School of Science, Chiba University, Chiba 263-8522, Japan
[2]Institute of Low Temperature and Structure Research, Polish Academy of Sciences, P.O. Box 1410, PL-50-950 Wroclaw, Poland

E-mail: hideto@nmr.s.chiba-u.ac.jp

Abstract. We have performed nuclear quadrupole resonance (NQR) measurements on a recently-discovered heavy-fermion superconductor Ce_2PdIn_8 with superconducting critical temperature $T_c = 0.64$ K. Below coherent temperature $T_{coh} \sim 30$ K, the spin-lattice relaxation rate $1/T_1$ decreases with decreasing temperature T and is proportional to $T^{1/2}$ between T_c and T_{coh}. This is clearly different from the Fermi-liquid behavior in which the T dependence is proportional to T, and indicates that Ce_2PdIn_8 is located on the verge of antiferromagnetic quantum critical point from the view point of the NQR. Below T_c, $1/T_1$ shows no coherence peak and is proportional to T^3. This is clear evidence for the realization of unconventional superconductivity with line nodes in this compound.

1. Introduction

The relationship between superconductivity and magnetic spin fluctuations is one of the central issues in strongly correlated electron systems [1]. Unconventional superconductivity has been discovered in pressurized $CeIn_3$ [2], pressurized $CeRhIn_5$ [3], and $CeCoIn_5$ [4] near the magnetic instability point. In these systems, magnetic correlation arises from the competition and/or cooperation between the Ruderman-Kittel-Kasuya-Yosida interaction and the Kondo effect which are based on the hybridization effect of localized f-electrons and itinerant conduction electrons. It has been recognized that the dimensionality of a system also plays a vital role in unconventional superconductivity near a quantum critical point (QCP) [5]. Such importance has recently been recognized by experimentally investigating artificial superlattices of the antiferromagnetic (AF) heavy-fermion (HF) compound $CeIn_3$ and the conventional metal $LaIn_3$ [6]. Ce-based HF compounds with the formula Ce_nMIn_{3n+2} ($n = 1, 2, \infty$, M = Co, Rh, Pd, Ir) are also good candidates for the investigation of dimensionality, quantum criticality, and superconductivity [2, 3, 4, 7, 8, 9, 10, 11, 12], because their crystal structure can be viewed as successive layers of quasi-two-dimensional $CeMIn_5$ and/or three-dimensional $CeIn_3$.

Among the quasi-2D HF superconductors, $CeCoIn_5$ is one of the most intensively studied systems. This compound clearly exhibits non-Fermi-liquid (NFL) behavior in the normal state [4] and d-wave superconductivity below the superconducting (SC) transition temperature $T_c = 2.3$ K [10]. The outstanding phenomenon of $CeCoIn_5$ is a magnetic field-induced spin-density-wave state (so-called Q phase) that forms in the low-temperature and high-field (LTHF) corner of the $H - T$ phase diagram [13, 14]. Most importantly, it has been suggested that this

Journal of Physics: Conference Series **449** (2013) 012027 doi:10.1088/1742-6596/449/1/012027

new phase may be the long sought inhomogeneous Fulde-Ferrell-Larkin-Ovchinnikov (FFLO) SC state [15].

Recently superconductivity was discovered in Ce_2PdIn_8 with $T_c \simeq 0.7$ K [9]. This compound has a Ho_2CoGa_8-type tetragonal crystal structure (space group $P4/mmm$). The electronic specific heat coefficient is approximately 1.5 J/K^{2}mol-fu [9, 16, 17, 18], which indicates heavy-fermion superconductivity. Resistivity has clearly shown a logarithmic dependence on the temperature T between 40 and 100 K with a local maximum at approximately 30 K($\sim T_{\rm coh}$) corresponding to a cross-over temperature between the coherent and incoherent dense Kondo regimes and a linear dependence on T between T_c and approximately 10 K, which is evidence for NFL behavior [9, 16, 17]. Moreover, it has been clarified recently that the magnetic field H and/or pressure can tune the QCP [18, 19, 20, 21, 22]. As for the superconductivity, it has been shown that the most probable candidate for the SC gap symmetry is d-wave superconductivity. Thermal conductivity measurements under magnetic field have clearly shown the presence of residual density of states (R-DOS) at the Fermi level [20]. Penetration depth measurements using a tunnel diode resonator have also indicated a nodal-line structure in the SC gap [23]. All these properties found in Ce_2PdIn_8 are quite similar to those in the more two-dimensional $CeCoIn_5$. However, the T_c and upper critical field H_{c2} under zero and finite pressure for Ce_2PdIn_8 are considerably lower than those for $CeCoIn_5$. This is probably associated with the dimensionality of the systems.

These fascinating phenomena in Ce_2PdIn_8 have been revealed by bulk-property measurements such as electrical transport measurements. Therefore, it is vital to perform microscopic and site-selective measurements to clarify the magnetic properties in the normal state and the superconductivity of this compound. In this paper, we report ^{115}In nuclear quadrupole resonance (NQR) measurements of Ce_2PdIn_8.

2. Experiments

Polycrystalline samples of Ce_2PdIn_8 were synthesized by the arc melting method. The detailed procedure is described in Ref. [17]. The ^{115}In NQR measurements were performed using a phase-coherent pulsed NQR spectrometer in the frequency range of 7-60 MHz. The crystals were powdered in order to reduce heating up of the samples at low temperature and improve signal intensity. Measurements above 1.5 K were performed using a ^{4}He cryostat and below 1.5 K with a ^{3}He-^{4}He dilution refrigerator. The spin-lattice relaxation time T_1 was obtained from the recovery of the nuclear magnetization after a saturation pulse. T_c was estimated to be 0.64 K through ac susceptibility by measuring the temperature dependence of the characteristic frequency of the NQR sample-coil $f_{\rm coil}$.

3. Results and Discussion

There are three inequivalent crystalline In sites in Ce_2PdIn_8: the most symmetric site In(1) ($4mm$), the less symmetric site In(2) ($mmm.$), and the least symmetric site In(3) ($2mm.$). By diagonalizing the electric quadrupole $\mathcal{H}_Q$ for the In nuclei ($I = 9/2$) and considering the three inequivalent In sites, we can assign all the In sites in the NQR spectrum of Ce_2PdIn_8. All the feature of the NQR spectrum is shown elsewhere [24]. For each In site, we denote the resonance lines $1\nu_Q$, $2\nu_Q$, $3\nu_Q$ and $4\nu_Q$ according to increasing frequency. The $1\nu_Q$ line is from the transition between $|I_z = \pm 1/2\rangle$ and $|I_z = \pm 3/2\rangle$; the $2\nu_Q$ line is from the transition between $|I_z = \pm 3/2\rangle$ and $|I_z = \pm 5/2\rangle$, etc.

As clearly shown in Fig. 1, we did not observe any signals other than those from the ^{113}In and ^{115}In nuclei from Ce_2PdIn_8. Because natural abundance of ^{113}In is 4.3% and that of ^{115}In is 95.7%, the ^{113}In line intensities are quite small. The lines due to ^{113}In overlap with those of ^{115}In because the ratio of quadrupole moments is $^{113}Q/^{115}Q = 0.82/0.83$. This confirms the quality of the samples that were free of any clear parasitic phases which were always found

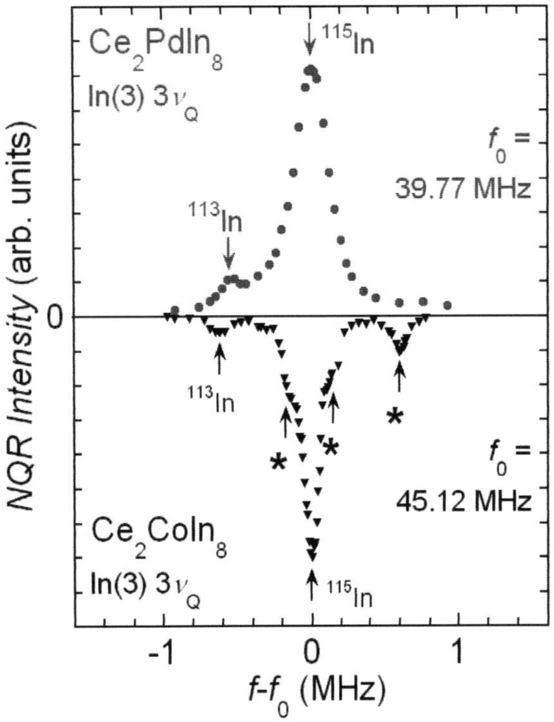

Figure 1. (Color online) In-NQR spectra of Ce_2PdIn_8 and Ce_2CoIn_8. The data of Ce_2CoIn_8 was cited from ref. [12].

in Ce_2CoIn_8 (* marks in Fig. 1) and Ce_2RhIn_8 [12]. The line-width is about 1.8 times wider than the corresponding line-widths of single-crystalline $CeCoIn_5$ [10] and Ce_2CoIn_8 [12]. This line-width broadening may be due to some structural disorder in the sample studied or/and may arise from the averaged electric-field-gradient in the powdered samples.

We obtained T_1 at the frequency of the spectral center at each In site over the range of 0.075-70 K. The nuclear magnetization recovery curve was fitted by the quadruple exponential function as expected for each line of the spectrum of the nuclear spin $I = 9/2$ of the ^{115}In nuclei [25]. For In sites with finite asymmetry parameter η, we adopted the modification of the recovery curves given by Chepin and Ross [26]. All the data were well fitted to the corresponding curve with a single T_1 component even at the lowest measurement temperature.

Obtained temperature dependence of $1/T_1$ is displayed elsewhere [24]. Instead, we show temperature dependence of the spin-lattice relaxation rate $1/T_1$ at the In(1) site of Ce_2PdIn_8, $CeCoIn_5$, and $CeIrIn_5$, normalized at the respective T_c in Fig. 2. In the normal state of Ce_2PdIn_8, the temperature dependence of $1/T_1$ remains constant at higher temperatures above about 30 K($\sim T_{coh}$). This indicates that the $4f$ electrons of the Ce ions are nearly localized above this temperature. This is consistent with the behavior that the resistivity possesses a broad peak around T_{coh} and exhibits a $-\log T$ temperature dependence between 40 and 100 K [9, 16, 17]. $1/T_1$ varies nearly as $T^{1/2}$ between 0.7($\simeq T_c$) and 30 K. This is clearly different from the Fermi-liquid behavior in which the T dependence is proportional to T (namely, the Korringa relation) and is consistent with the NFL behavior confirmed in various previous measurements [9, 16, 17, 18, 19, 20, 21, 22]. However, in order to discuss the difference of the normal state properties of these compounds shown here, we need Knight shift measurements related with transferred hyperfine field from Ce moments and more detailed data at higher

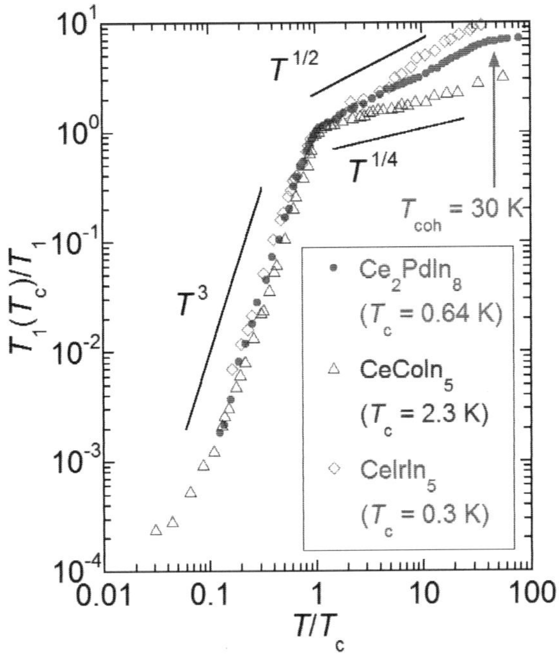

Figure 2. (Color online) Temperature dependence of the spin-lattice relaxation rate $1/T_1$ at the In(1) site of Ce_2PdIn_8, $CeCoIn_5$ and $CeIrIn_5$, normalized at the respective T_c. The $1/T_1$ data of $CeCoIn_5$ and $CeIrIn_5$ were taken from Ref. [10].

temperature above approximately 100 K related with contribution from localized f electrons and conduction electrons.

In the SC state, $1/T_1$ decreases rapidly just below T_c and the Hebel-Slichter coherence peak was not observed. $1/T_1$ varies nearly proportional to T^3 at low temperatures. We did not observe any saturation or T linear dependence of $1/T_1$ at lowest measurement temperatures, which would arise from residual density of states often caused by a small amount of paramagnetic impurities in nodal superconductors. This is because the present lowest measurement temperature of 75 mK is approximately $T_c/9$ and the temperature is not sufficiently low to observe such an extrinsic contribution. As apparent from Fig. 2, there is a good agreement of temperature dependence of Ce_2PdIn_8, $CeCoIn_5$ and $CeIrIn_5$ in the SC state. This coincidence is attributable to the similar energy gap amplitudes in these compounds. The similar gap sizes for both compounds suggest that the SC states of Ce_2PdIn_8 and $CeCoIn_5$ are quite similar to each other. Hence, we may speculate that the SC symmetry of this compound is *d*-wave. However, in order to conclude on this issue, we must clarify the SC symmetry of the spin part of this compound by measuring the Knight shift with a single crystal.

4. Summary

We have performed NQR measurements on the HF superconductor Ce_2PdIn_8 with $T_c = 0.64$ K. Below $T_{coh} \sim 30$ K, $1/T_1$ decreases with decreasing temperature and varies nearly as $T^{1/2}$ between T_c and T_{coh}. This is clearly different from the Fermi-liquid behavior in which the T dependence is proportional to T, and indicates that Ce_2PdIn_8 is located close to the AF QCP from the viewpoint of NQR. Below T_c, $1/T_1$ shows no coherence peak and is proportional to T^3. This is clear evidence of unconventional superconductivity with line nodes. Further studies under finite magnetic field and/or pressure are also needed to understand the properties of the field-/pressure-induced QCP and the LTHF SC phase of the compound.

Acknowledgments

The authors thank S. Ohara for fruitful discussion. This work was supported by Grants-in-Aid for Scientific Research (Nos. 21540351 & 22684016) from MEXT and JSPS, Innovative Areas "Heavy Electrons" (No. 21102505) from MEXT, and Global COE and AGSST financial support program of Chiba University. The work in Poland was supported by the National Science Centre (Poland) under research grant No. 2011/01/B/ST3/04482.

References

[1] S. Sachdev and B. Keimer, Phys. Today **64**, 29 (2011).
[2] N. D. Mathur, F. M. Grosche, S. R. Julian, I. R. Walker, D. M. Freye, R. K. W. Haselwimmer, and G. G. Lonzarich, Nature **394**, 39 (1998).
[3] H. Hegger, C. Petrovic, E. G. Moshopoulou, M. F. Hundley, J. L. Sarrao, Z. Fisk, and J. D. Thompson, Phys. Rev. Lett. **84**, 4986 (2000).
[4] C. Petrovic, P. G. Pagliuso, M. F. Hundley, R. Movshovich, J. L. Sarrao, J. D. Thompson, Z. Fisk, and P. Monthoux, J. Phys.: Condens. Matter **13**, L337 (2001).
[5] T. Takimoto and T. Moriya, Phys. Rev. B **66**, 134516 (2002).
[6] H. Shishido, T. Shibauchi, K. Yasu, T. Kato, H. Kontani, T. Terashima, and Y. Matsuda, Science **327**, 980 (2010).
[7] C. Petrovic, R. Movshovich, M. Jaime, P. G. Pagliuso, M. F. Hundley, J. L. Sarrao, Z. Fisk, and J. D. Thompson, Europhys. Lett. **53**, 354 (2001).
[8] G. Chen, S. Ohara, M. Hedo, Y. Uwatoko, K. Saito, M. Sorai, and I. Sakamoto, J. Phys. Soc. Jpn. **71**, 2836 (2002).
[9] D. Kaczorowski, A. P. Pikul, D. Gnida, and V. H. Tran, Phys. Rev. Lett. **103**, 027003 (2009):*ibid* **104**, 059702 (2010).
[10] Y. Kohori, Y. Yamato, Y. Iwamoto, T. Kohara, E. D. Bauer, M. B. Maple, and J. L. Sarrao: Phys. Rev. B **64** (2001) 134526.
[11] G.-Q. Zheng, K. Tanabe, T. Mito, S. Kawasaki, Y. Kitaoka, D. Aoki, Y. Haga, and Y. Onuki, Phys. Rev. Lett. **86**, 4664 (2001).
[12] H. Fukazawa, T. Okazaki, K. Hirayama, Y. Kohori, G. Chen, S. Ohara, I. Sakamoto, and T. Matusmoto, J. Phys. Soc. Jpn. **76** (2007) 124703.
[13] B. L. Young, R. R. Urbano, N. J. Curro, J. D. Thompson, J. L. Sarrao, A. B. Vorontsov, and M. J. Graf, Phys. Rev. Lett. **98**, 036402 (2007).
[14] M. Kenzelmann, Th. Strässle, C. Niedermayer, M. Sigrist, B. Padmanabhan, M. Zolliker, A.D. Bianchi, R. Movshovich, E. D. Bauer, J. L. Sarrao, and J. D. Thompson, Science **321**, 1652 (2008).
[15] Y. Matsuda and H. Shimahara, J. Phys. Soc. Jpn. **76**, 051005 (2007), and references therein.
[16] D. Kaczorowski, A. P. Pikul, B. Belan, L. Sojka, and Ya. Kalychak, Physica B **404**, 2975 (2009).
[17] D. Kaczorowski, D. Gnida, A. P. Pikul, and V. H. Tran, Solid State Commun. **150**, 411 (2010).
[18] Y. Tokiwa, P. Gegenwart, D. Gnida, and D. Kaczorowski, Phys. Rev. B **84**, 140507(R) (2011).
[19] M. Matusiak, D. Gnida, and D. Kaczorowski, Phys. Rev. B **84**, 115110 (2011).
[20] J. K. Dong, H. Zhang, X. Qiu, B. Y. Pan, Y. F. Dai, T. Y. Guan, S. Y. Zhou, D. Gnida, D. Kaczorowski, and S.Y. Li, Phys. Rev. X **1**, 011010 (2011).
[21] V. H. Tran, D. Kaczorowski, R. T. Khan, and E. Bauer, Phys. Rev. B **83**, 064504 (2011).
[22] D. Gnida, M. Matusiak, and D. Kaczorowski, Phys. Rev. B **85**, 060508(R) (2012).
[23] K. Hashimoto *et al.* (unpublished).
[24] H. Fukazawa, R. Nagashima, S. Shimatani, Y. Kohori, D. Kaczorowski, Phys. Rev. B **86**, 094508 (2012).
[25] D. E. MacLaughlin, J. D. Williamson, and J. Butterworth, Rhys. Rev. B **4**, 60 (1971).
[26] J. Chepin and J. H. Ross, J. Phys.: Condens. Matter, **3**, 8103 (1991).

Journal of Physics: Conference Series **449** (2013) 012028 doi:10.1088/1742-6596/449/1/012028

Routes to heavy-fermion superconductivity

F Steglich[1], O Stockert[1], S Wirth[1], C Geibel[1], H Q Yuan[2], S Kirchner[1,3] and Q Si[4]

[1]Max Planck Institute for Chemical Physics of Solids, Nöthnitzer Str. 40, 01187 Dresden, Germany

[2]Department of Physics and Center for Correlated Matter, Zhejiang University, Hangzhou, Zhejiang 310027, China

[3]Max Planck Institute for Physics of Complex Systems, Nöthnitzer Str. 38, 01187 Dresden, Germany

[4]Department of Physics and Astronomy, Rice University, Houston, TX 77005, USA

steglich@cpfs.mpg.de

Abstract. Superconductivity in lanthanide- and actinide-based heavy-fermion (HF) metals can have different microscopic origins. Among others, Cooper pair formation based on fluctuations of the valence, of the quadrupole moment or of the spin of the localized 4*f*/5*f* shell have been proposed. Spin-fluctuation mediated superconductivity in $CeCu_2Si_2$ was demonstrated by inelastic neutron scattering to exist in the vicinity of a spin-density-wave (SDW) quantum critical point (QCP). The isostructural HF compound $YbRh_2Si_2$ which is prototypical for a Kondo-breakdown QCP has so far not shown any sign of superconductivity down to $T \approx 10$ mK. In contrast, results of de-Haas-van-Alphen experiments by Shishido et al. (J. Phys. Soc. Jpn. 74, 1103 (2005)) suggest superconductivity in $CeRhIn_5$ close to an antiferromagnetic QCP beyond the SDW type, at which the Kondo effect breaks down. For the related compound $CeCoIn_5$ however, a field-induced QCP of SDW type is extrapolated to exist inside the superconducting phase.

1. Heavy-fermion metals

Heavy-fermion (HF) metals are intermetallics containing certain lanthanide (e.g., Ce and Yb) or actinide (e.g., U and Pu) elements. The lanthanide-based HF metals are commonly regarded as prototypes of Kondo-lattice systems, for which the Doniach ($T-J$) phase diagram (figure 1) holds [2]. Here, T is the absolute temperature and J describes the strength of the antiferromagnetic exchange interaction between the effective spin of the localized 4*f* shell and the spins of the conduction electrons. J determines both of the competing fundamental interactions: the inter-site Ruderman-Kittel-Kasuya-Yoshida (RKKY) interaction which tends to stabilize the local 4*f*-electron moments as well as the on-site Kondo interaction which tends to screen them. At small J ("weak coupling"), the RKKY interaction dominates, and long-range magnetic [mostly: antiferromagnetic (AF)] order forms at low temperatures. At sufficiently large J, where the Kondo effect prevails, the latter eventually quenches the local moments. Well below the Kondo temperature T_K, local Kondo singlets (i.e., 4*f*-electron states entangled with conduction electron states) form. In a periodic Kondo lattice they

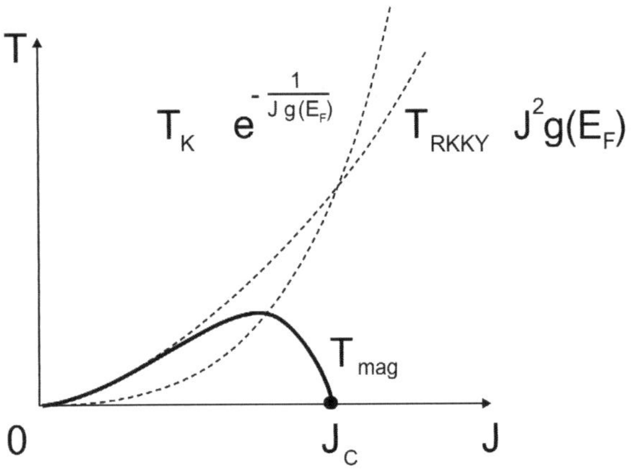

Figure 1. Doniach phase diagram. Binding energies for the Kondo effect, k_BT_K, and for the RKKY interaction, k_BT_{RKKY}, as a function of the 4*f*-conduction electron exchange integral *J* ($J > 0$ for antiferromagnetic spin exchange). T_{mag}: magnetic ordering temperature ($T_{mag} \rightarrow 0$ at $J = J_c$), cf. Ref. 1.

propagate, due to the Bloch theorem, and act as charge carriers with the same quantum numbers as the non-interacting conduction electrons. Because of the extremely large on–site Coulomb repulsion, these "composite fermions" exhibit a very small Fermi velocity, which is only of the order of the velocity of sound. The effective mass of these composite charge carriers, as determined from the very large Sommerfeld coefficient of the electronic specific heat *C*(*T*), $\gamma = C/T$, is correspondingly large ("heavy electrons" or "heavy fermions").

Close to where the RKKY and Kondo interactions cancel each other ($J = J_c$, see figure 1), magnetic order can be suppressed by means of (external or chemical) pressure which influences *J*. Magnetic field, *B*, which typically causes a larger reduction of the antiferromagnetic order than that of the Kondo effect, is another non-thermal control parameter by which AF order can be suppressed. If the AF order terminates at the critical value of the control parameter in a continuous fashion, this disappearance marks a quantum critical point (QCP) [3].

Two variants of QCPs have been established for HF metals [4], depending on the behavior of the HFs upon approaching the magnetic instability at $J = J_c$ (figure 1). In both cases, these composite charge carriers are defined on the strong coupling side, where the 4*f*-states - being entangled with the conduction-electron states and thus constituent parts of the propagating Kondo singlets – are delocalized and contribute to a large Fermi surface.

In the first scenario, the composite charge carriers also exist on the weak-coupling side of the QCP and, therefore, AF order cannot be due to local 4*f*-derived magnetic moments. In this case, the magnetic order is of an itinerant nature, i.e., HF spin-density-wave (SDW) order. The QCP is then called a *SDW QCP* which denotes a continuous classical phase transition in higher dimensions [5-7]. The Fermi-surface volume at a SDW QCP stays large and undergoes a smooth change due to the gradual opening of the SDW gap. In this case, the fluctuations of the AF order parameter present the only relevant degrees of freedom. $CeCu_2Si_2$ [8] and $CeNi_2Ge_2$ [9] have been considered exemplary materials exhibiting a SDW QCP rather early.

On the other hand, if the composite fermions exist only on the paramagnetic side of the phase diagram (figure 1), the onset of AF order at the QCP is accompanied by a break-up of the Kondo singlets. On the weak-coupling side of this *Kondo-breakdown QCP* [10, 11] (strongly, incompletely yet, Kondo screened) local 4*f*-derived magnetic moments undergo AF order, in the presence of a small Fermi surface, made up by the conduction electrons exclusively. Consequently, the continuous AF quantum phase transition was predicted to concur with an abrupt change of the Fermi-surface volume [10, 11]. This was indeed inferred from isothermal measurements of the Hall coefficient as a function of the control parameter magnetic field, *B*, for the quantum critical material $YbRh_2Si_2$ [12, 13]. Further evidence for this unique type of QCP to be present in $YbRh_2Si_2$ is derived from the vanishing of a quantum critical energy scale $k_BT^*(B)$ (which denotes the Fermi surface crossover at finite *T*) [14] and a violation of the Wiedemann-Franz law [15]. Contrarily, $k_BT^*(B)$ remains finite at a QCP of the SDW type [4].

2. Variants of heavy-fermion superconductors

The advent of heavy-fermion (HF) superconductivity (SC) in $CeCu_2Si_2$ [16] followed the discoveries of superfluidity in ^3He [17] and HF phenomena in $CeAl_3$ [18]. Given the antagonistic nature of SC and magnetism, the observation of SC in $CeCu_2Si_2$ came as big surprise: All superconductors known at that time lose their SC when doped with a tiny amount (~1 at%) of magnetic impurities – whereas in $CeCu_2Si_2$ a dense, periodic lattice of (100 at%) magnetic Ce^{3+} ions is necessary to generate the superconducting state. This was inferred from the observation that the non-magnetic reference compound $LaCu_2Si_2$ is not a superconductor [16] and that SC in $CeCu_2Si_2$ is fully suppressed by doping with a tiny amount of non-magnetic impurities [19]. For $CeCu_2Si_2$, the ($T \rightarrow 0$) Sommerfeld coefficient $\gamma \approx 1$ J/K^{2}mole exceeds the γ value of a simple metal like Cu by about three orders of magnitude. From the observation that the jump $\Delta C(T)/T$ at $T_c \approx 0.6$ K is of the same gigantic order as the normal-state value of C/T at T_c, it was concluded [16] that the heavy-mass charge carriers make up the Cooper pairs. Similar conclusions were drawn for a few U-based HF superconductors discovered in the mid 1980ies, i.e., UBe_{13} [20], UPt_3 [21] and URu_2Si_2 [22].

SC in HF metals involves pairing order parameters which are distinct from the BCS s-wave type. Strong support for this is lent by the existence of multiple superconducting phases, similar to what was observed for superfluid ^3He [23]. Multiphase superconductivity was found for UPt_3 [24], $U_{1-x}Th_xBe_{13}$ [25] and $PrOs_4Sb_{12}$ [26]; the latter compound is unique here, since electric quadrupole fluctuations rather than magnetic dipole fluctuations, as commonly assumed for HF superconductors, are believed to mediate the Cooper pairing. Quite generally, the microscopic pairing mechanism in HF superconductors is driven by electronic interactions, contrasting SC in classical (BCS) superconductors mediated by electron-phonon coupling. In most HF superconductors, pairing is intimately related to magnetism. This was first inferred for the weak antiferromagnet UPt_3 ($T_N = 5$ K) from inelastic neutron-scattering (INS) spectra, which showed a drop of the magnetic intensity below $T_c = 0.5$ K [27]. For UPd_2Al_3, an analysis of tunnelling [28] and INS [29] results revealed that SC ($T_c \approx$ 2K), which microscopically coexists with local-moment AF order ($T_N = 14.3$ K, $\mu_{ord} = 0.85\ \mu_B$/U), is mediated by the acoustic magnon at the AF ordering wave vector [29]. Beyond these examples of SC inside AF order, SC often also occurs in the immediate vicinity of AF order. In $CeCu_2Ge_2$ [30] and $CeRh_2Si_2$ [31], SC appears near a threshold pressure p_c, where antiferromagnetism disappears abruptly. In many other HF metals, AF order is suppressed continuously by pressure, and SC develops near a QCP. For example, $CePd_2Si_2$ exhibits a narrow dome of SC ($T_{c,max} \approx 0.4$ K), centered around the inferred critical pressure $p_c \approx 2.8$ GPa at which AF order seems to go away smoothly [32]. Apart from an AF instability, other types of phase transitions may exist in the vicinity of SC in HF metals: UGe_2 [33], like URhGe [34] and UCoGe [35], shows SC within a regime of ferromagnetic (FM) order. Recent NMR results highlight FM fluctuations as glue for superconductivity in UCoGe [36]. In $CeCu_2Si_2$ [37] and $CeCu_2Ge_2$ [38], SC is observed up to high pressure, where a weak valence instability occurs. Here, quantum critical valence fluctuations [39] are believed to mediate the formation of the Cooper pairs. In URu_2Si_2, SC ($T_c = 1.5$ K) coexists with some symmetry-broken state of yet unidentified origin, so-called "hidden order" ($T_{HO} = 17.5$ K) [40].

Within the last decade, the number of HF superconductors has significantly increased to about 40. The vast majority of these recently discovered HF superconductors belongs to two distinct groups of intermetallics: (i) the $Ce_nT_mIn_{3n+2m}$ family of compounds (T: transition metal), like $CeCoIn_5$ [41] and (ii) systems lacking a center of inversion symmetry, like $CePt_3Si$ [42].

The former systems are quasi-two-dimensional (2D) variants of the cubic superconductor $CeIn_3$ ($T_c \approx 0.2$ K) [32]. They are formed by stacking alternating n layers of $CeIn_3$ and m layers of TIn_2 sequently along the tetragonal c-axis. As predicted in Ref. 43, the so reduced dimensionality causes an increase of T_c by more than one order of magnitude ($T_c = 2.3$ K for $CeCoIn_5$). A further substantial T_c-enhancement was achieved upon replacing Ce (with localized 4f shell) by Pu (whose 5f shell is spatially more extended). Among HF superconductors, $PuCoGa_5$ with $T_c = 18.5$ K is presently the record holder [44]; its Rh homologue [45] as well as $NpPd_2Al_5$ [46] also show enhanced T_c values.

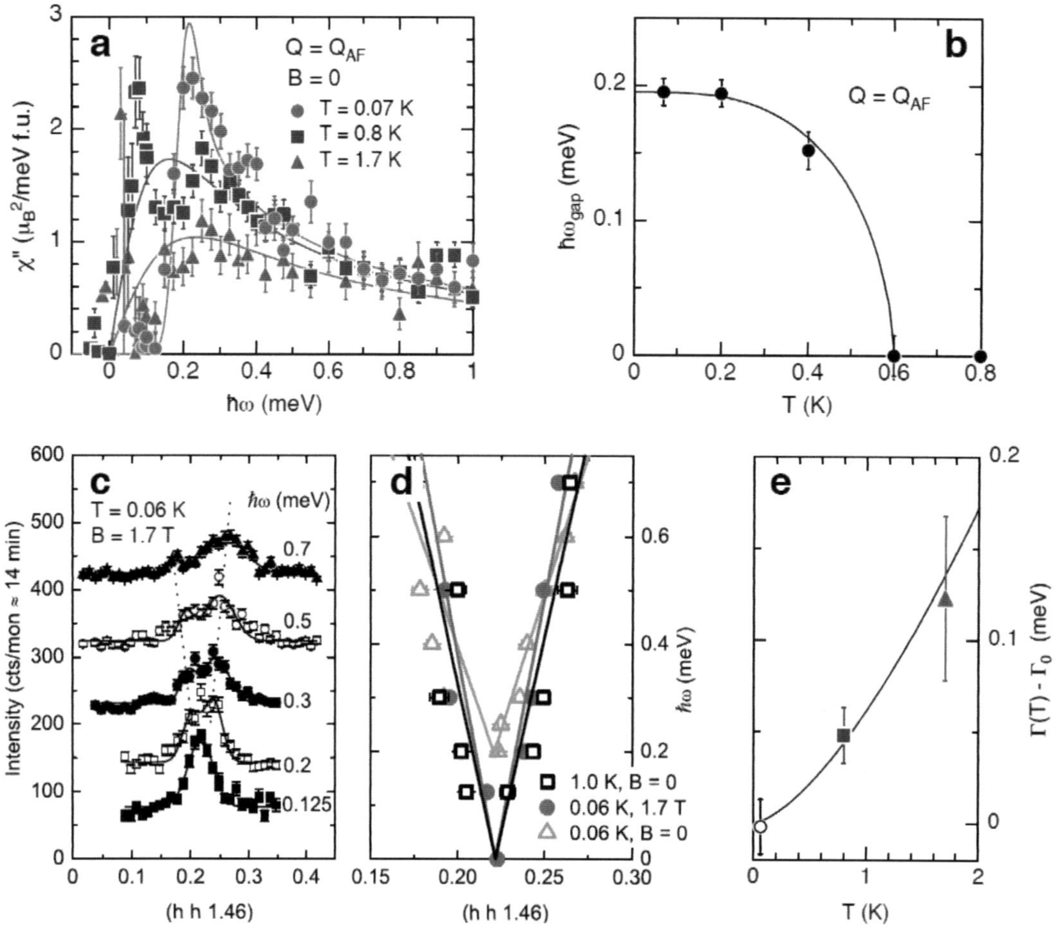

Figure 2. Magnetic response, relaxation rate and spin gap at the SDW ordering wave vector $\mathbf{Q}_{AF}$ in S-type $CeCu_2Si_2$. **a** χ'' vs $\hbar\omega$ for different temperatures. From Ref. 53. **b** Temperature dependence of the spin-excitation gap in the superconducting state, together with the scaled gap function for a weak-coupling d-wave superconductor (solid line). From Ref. 53. **c** Momentum dependence in the normal state for different energy transfers $\hbar\omega$ at $T = 0.06$ K, $B = 1.7$ T. Scans are shifted by 100 counts with respect to each other. Solid lines indicate fits with two Gaussian peaks; dashed lines are guides to the eye. From Ref. 54. **d** Spin fluctuation dispersion as derived from fits to $\boldsymbol{Q}$ scans in the normal state at $B = 0$, $T = 1.0$ K or at $T = 0.06$ K and $B = 1.7$ T as well as in the superconducting state at $T = 0.06$ K and $B = 0$. Solid lines indicate linear fits. From Ref. 54. **e** Linewidth of the quasielastic magnetic response in the normal state, as obtained by fitting the data in **a** (lines therein). The solid line represents a $T^{1.5}$ dependence. The $T = 0$ offset is 0.112 meV. From Ref. 53.

In the past few years, the group of "non-centrosymmetric" superconductors has attracted much theoretical interest. This derives from the fact that the lack of inversion symmetry allows for a mixing of even-parity ($S = 0$) and odd-parity ($S = 1$) pair states, the degree of mixing depending on the strength of the antisymmetric spin-orbit coupling [47, 48].

The majority of HF superconductors *with* inversion symmetry show a highly anisotropic, even-parity order parameter, while a small number of them are prime candidates for odd-parity pairing [33-35, 49, 50]. Interestingly, in each of these latter cases SC coexists with either AF or FM order, cf. Refs. [27] and [33], respectively.

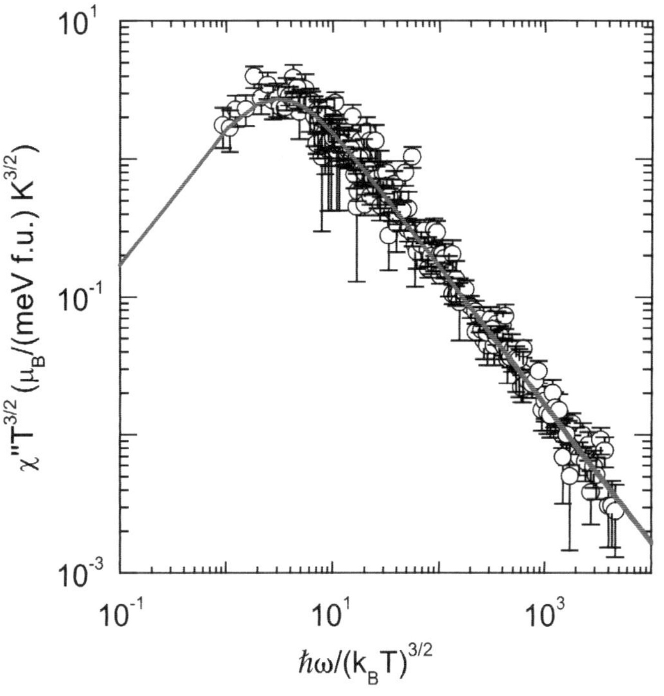

Figure 3.Universal scaling of the imaginary part of the dynamical susceptibility χ'' ($\boldsymbol{Q}$, ω) as a function of $\hbar\omega/(k_BT)^{3/2}$ for S-type $CeCu_2Si_2$. From Ref. 54.

3. Superconductivity near antiferromagnetic quantum critical points

Many of the HF superconductors exhibit a non-Fermi-liquid (nFL) low-*T* normal state, highlighting a nearby AF QCP. For most of these nFL superconductors the QCP occurs at high pressure, $CePd_2Si_2$ being a well-known example [32]. This makes it challenging to unravel the magnetic structure as well the spectrum of critical fluctuations near the QCP by neutron scattering. $CeCu_2Si_2$ is well suited for such an investigation as here, the QCP is accessible already at ambient pressure. It is located very closely to the true stoichiometry point inside the narrow homogeneity range of the 122 phase in the chemical phase diagram [51]. Within this range, homogeneous single crystals can be grown with, e.g., a tiny deficiency or a tiny excess of Cu. The former crystals show weak AF order ("A-type"), the latter ones are superconducting ("S-type").

Neutron-diffraction measurements on A-type $CeCu_2Si_2$ revealed HF-SDW order below $T_N \approx 0.8$ K, with a tiny staggered moment of ≈ 0.1 μ_B/Ce and an incommensurate ordering wave vector $\mathbf{Q}_{AF}$ = (0.215 0.215 0.53) [52].

Recently, INS experiments have been performed on an S-type $CeCu_2Si_2$ single crystal [53]. As shown in figure 2a, this sample exhibits a broad quasielastic magnetic response in its low-temperature normal state. SC induces a gap below 0.2 meV in the quasielastic spectrum. This gap cannot be related to the opening of a SDW gap, as it is absent in A-type $CeCu_2Si_2$. It rather refers to the opening of a superconducting gap in the HF density of states at the Fermi level (figure 2b). The quasielastic response is extremely localized in $\boldsymbol{Q}$-space, i.e., only around the incommensurate ordering wave vector $\mathbf{Q}_{AF}$ where SDW order occurs nearby in the phase diagram (figures 2c,d). This clearly shows that the spectra in figure 2a represent a broad distribution of dynamical SDW correlations. Their lifetime strongly increases upon lowering the temperature ("slowing down"), cf. figure 2e, indicating the close proximity of AF order. As shown in figure 3, these "almost quantum critical spin fluctuations" obey a universal scaling over wide ranges of energy transfer and temperature, as expected for 3D critical modes associated with a SDW QCP [54].

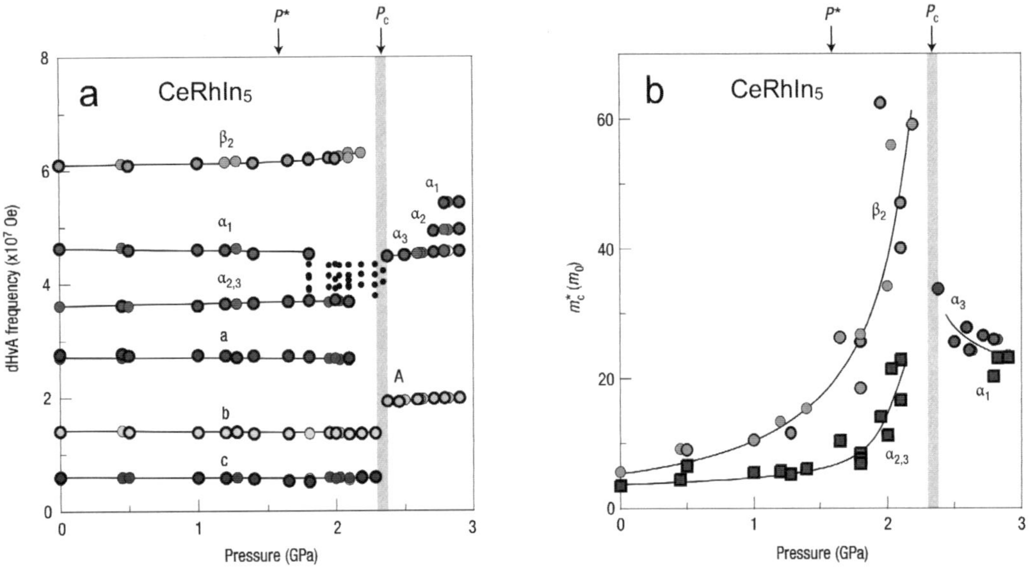

Figure 4. Changes of Fermi-surface properties across a pressure-induced Kondo-breakdown quantum critical point in $CeRhIn_5$. Pressure dependence of de Haas-van Alphen frequencies (**a**) and cyclotron masses (**b**). From Ref. 61.

In ***Q***-space, the magnetic response of S-type $CeCu_2Si_2$ discussed before appears as an overdamped, dispersive mode ("AF paramagnon") [53]. From the slope of the dispersion relation in the normal state (figure 2d), the paramagnon velocity v_{pm} was found to be smaller by almost one order of magnitude than the Fermi velocity v_F* of the composite charge carriers [53]. As mentioned in Sect. 1, v_F* is of the order of the velocity of sound only [55], which prevents retardation of the electron-phonon coupling. On the other hand, because $v_F^* >> v_{pm}$, the coupling between the heavy fermions and the SDW fluctuations is well retarded. It implies that in this interaction the direct Coulomb repulsion between the charge carriers is avoided and the magnetic excitations can provide the glue for SC. This is also concluded from the large difference between the spectral weight in the normal and superconducting states, which highlights a huge saving in magnetic exchange energy, when compared to the superconducting condensation energy [53]. The latter is reliably determined from the specific heat both in the superconducting and field-driven normal state. The exchange energy saving exceeds the condensation energy by a factor larger than 20. Correspondingly, the loss of kinetic energy exceeds the condensation energy by also a factor of order 20 – compared to a factor of only 2- 3 in the case of classical (BCS) superconductors. This observation is understood if there is a break-up of the Kondo effect at energies above a relatively small Kondo-breakdown (Fermi-surface crossover) energy scale k_BT^*, which causes a shift of spectral weight in the electron spectrum from below k_BT^* to above it and up to several eV, i.e., the order of the local Coulomb interaction. I.o.w., even for this canonical example of SC near an SDW QCP, the f-electron localization comes into play dynamically.

For a HF metal like $CeCu_2Si_2$, this Fermi-surface crossover temperature T^*, which generically is below T_K = 15-20 K, must be finite [56]. With respect to the magnetic response in the INS spectra shown in figure 2a, which extends to about 2 meV, corresponding to k_BT_K, this implies that only the low-frequency spinfluctuations involved in the Cooper pairing are of the collective paramagnon type, while those with frequencies larger than k_BT^* must be related to the Kondo effect.

In the future, searching for a Kondo-breakdown transition at which $T^* \rightarrow 0$ (at a critical value of the control parameter) inside the magnetically ordered phase of HF antiferromagnets will be of high timely interest. Measurements under both chemical [57] and hydrostatic [58] pressure have suggested

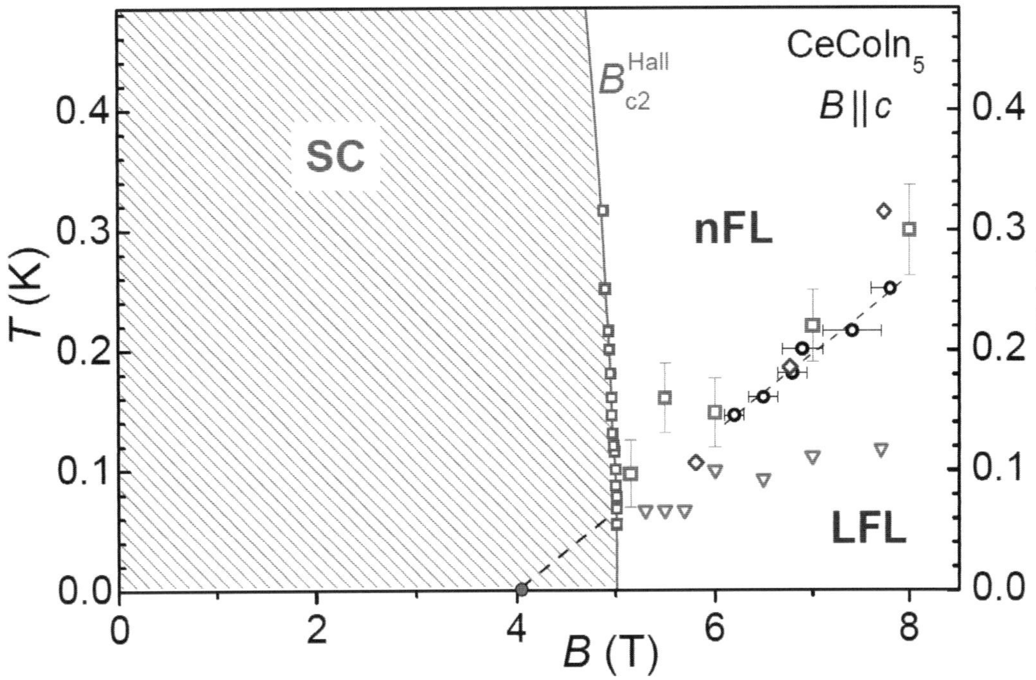

Figure 5. Temperature-magnetic field (T-B) phase diagram of $CeCoIn_5$ for $B \parallel [001]$. The black circles mark results from Hall measurements [64] indicating the cross-over from Landau Fermi-liquid (LFL) to non-Fermi-liquid (nFL) behavior. An extrapolation hints at a putative QCP at $B_{QCP} \sim 4.1$ T(red dot), i.e,. inside the superconducting phase (SC). Results from thermal expansion (red squares, [65]) and early magnetoresistance (purple diamonds, [66]) measurements point toward a similar value. the deviation from T^2-behavior in the longitudinal magnetoresistance (triangles, [67]) was also interpreted as $B_{QCP} < B_{c2,0}$.

the existence of such a Kondo-breakdown transition in volume compressed $YbRh_2Si_2$. Very recent Fermi-surface studies via measurements of the de-Haas-van Alphen effect under high magnetic fields have identified a Kondo-breakdown transition inside the AF phase in $CeRhIn_5$: It occurs near $B^* \approx 40$ T, i.e., well below the critical field, where AF order is smoothly suppressed at absolute zero temperature, $B_N \approx 50$ T [59]. For magnetic fields $B < B^*$ and at $p = 0$, $CeRhIn_5$ is, therefore, a local-moment HF antiferromagnet. This holds true also at lower fields and finite pressure. In the field range 10 – 17 T and near the critical pressure $p_N = 2.3$ GPa where $T_N \rightarrow 0$ [60], de Haas-van Alphen oscillations indicate an abrupt reconstruction of the Fermi surface (figure 4a) that is accompanied by an incipient divergence of the cyclotron mass (figure 4b) [61]. Therefore, also at finite pressure, AF order at $B = 0$ is very likely of the local-moment type. Further on, the putative AF QCP masked by the pressure-induced superconducting dome in $CeRhIn_5$ is most probably of the Kondo-breakdown variety. This suggests that HF SC not only arises in the vicinity of SDW QCPs, like in $CeCu_2Si_2$ [53, 54], but can also be driven by the purely electronic fluctuations of the Kondo-breakdown QCP.

It is interesting to compare $CeRhIn_5$ with its Co homologue, when the field is applied along the c-axis. Early reports on the magneto-resistance [62] and specific heat [63] of $CeCoIn_5$ claimed the AF QCP to coincide with the upper critical field at $T = 0$, i.e., $B_{QCP} \approx B_{c2,0}$, while later reports point toward a putative QCP inside the superconducting phase [64-66], with $B_{QCP} \approx 4.1$ T (see figure 5). In subsequent current-voltage measurements inside the Shubnikov phase of superconducting $CeCoIn_5$, Hu et al. observed a sharp increase in the flux-flow resistivity upon decreasing either the temperature or the magnetic field [68]. This increase was ascribed to quasiparticle scattering off of critical AF fluctuations and, consequently, provides a strong indication for the existence of an AF phase boundary in the *B-T* phase diagram, i.e., below $T = 1.6$ K and $B = B_{QCP} = 4.1$ T, in excellent agreement with the results of [64, 65, 67]. Also, a transition very recently observed at roughly 4 T in the isothermal field-dependent entropy was suggested to be related to a QCP hidden by the superconducting phase [69]. The latter findings [69] as well as thermal expansion studies [70] are compatible with SDW order – in contrast to $CeRhIn_5$ where AF order is of the local-moment type, except for magnetic fields $B > B^* \approx$ 40 T [59].

4. Perspective

HF superconductors provide a multitude of unconventional SC scenarios. Most frequently, SC in close vicinity of AF QCPs has been established. As proposed in, e.g., Ref. 71 and 43, HF SC near a SDW QCP, mediated by almost quantum critical SDW fluctuations, could indeed be verified for $CeCu_2Si_2$ [53, 54]. It will be important to experimentally check whether this SDW QCP scenario can be applied to other (pressure-induced) nFL superconductors, such as $CePd_2Si_2$ [32]. Pressure-induced SC in $CeRhIn_5$ [72] appears to occur [60, 61] near a Kondo-breakdown QCP, which is often labeled a zero-temperature 4*f*-selective Mott transition. This offers a link between HF SC and unconventional SC in other families of strongly correlated electron systems, including the newly discovered Fe-based pnictides/chalcogenides and the doped Mott insulators of the cuprates and organic charge transfer salts, cf. various contributions to this conference.

Acknowledgements

Part of the work done in Dresden was supported by the DFG Research Unit 960, "Quantum Phase Transitions". HQY is partially supported by NSFC, the National Basic Research Program of China (973 Program) and the Zhejiang Provincial Natural Science Foundation of China. The Work at Rice University has in part been supported by NSF Grant No. DMR-1006985 and the Robert A. Welch Foundation Grant No. C-1411.

References

[1] Doniach S 1977 *Physica B* **91** 231
[2] Hewson A C 1993 *The Kondo Problem to Heavy Fermions* (Cambridge: Univ. Press)
[3] Sachdev S 1999 *Quantum Phase Transitions* (Cambridge: Univ. Press)
[4] Gegenwart P, Si Q and Steglich F 2008 *Nature Phys.* **4** 186
[5] Hertz J A 1976 *Phys. Rev. B* **14** 1165
[6] Moriya T 1985 *Spin Fluctuations in Itinerant Electron Magnetism* (Berlin: Springer)
[7] Millis A J *1993 Phys. Rev. B* **48** 7183
[8] Gegenwart P et al. *1998 Phys. Rev. Lett.* **81** 1501.
[9] Gegenwart P, Kromer F, Lang M, Sparn G, Geibel C and Steglich F 1999 *Phys. Rev. Lett.* **82** 4548
[10] Si Q, Rabello S, Ingersent K and Smith J L 2001 *Nature* **413** 804
[11] Coleman P, Pépin C, Si Q and Ramazashvili R 2001 *J. Phys.: Condens. Matter* **13** R723
[12] Paschen S, Lühmann T, Wirth S, Gegenwart P, Trovarelli O, Geibel C, Steglich F, Coleman P and Si Q 2004 *Nature* **432** 881
[13] Friedemann S, Oeschler N, Wirth S, Krellner C, Geibel C, Steglich F, Paschen S, Kirchner S and Si Q 2010 *Proc. Natl. Acad. Sci.* USA **107** 14547
[14] Gegenwart P, Westerkamp T, Krellner C., Tokiwa Y, Paschen S, Geibel C, Steglich F, Abrahams E and Si Q. 2007 *Science* **315** 969
[15] Pfau H et al. 2012 *Nature* **484** 493
[16] Steglich F, Aarts J, Bredl C D, Lieke W, Meschede D, Franz W and Schäfer H 1979 *Phys. Rev. Lett.* **43** 1892
[17] Osheroff D D, Richardson R C, Lee D M 1972 *Phys. Rev. Lett.* **28** 885
[18] Andres K, Graebner J E and Ott H R 1975 *Phys. Rev. Lett.* **35** 1779
[19] Spille H, Rauchschwalbe U and Steglich F 1983 *Helv. Phys. Acta* **56** 165

[20] Ott H R, Rudigier H, Fisk Z and Smith J L 1983 *Phys. Rev. Lett.* **50** 1595
[21] Stewart G R, Fisk Z, Willis J O and Smith J L 1984 *Phys. Rev. Lett.* **52** 679
[22] Schlabitz W, Baumann J, Pollit B, Rauchschwalbe U, Mayer H M, Ahlheim U and Bredl C D 1984 *Abstr. 4th Int. Conf. Valence Fluctuations* (Cologne) unpublished; 1986 *Z. Phys. B* **62** 171
[23] See, e.g., Vollhardt D and Wölfle P 1990 *The superfluid Phases of ^{3}He* (London: Taylor and Francis)
[24] See, e.g., Sauls J 1994 *Adv. Phys.* **43** 113
[25] Ott H R, Rudigier H, Fisk Z and Smith J L 1985 *Phys. Rev. B* **31** 1651
[26] Maple M B, Ho C-P, Zapf V S, Frederick N A, Bauer E D, Yuhasz W M, Woodward F M and Lynn J W 2002 *J. Phys. Soc. Jpn.* **71** (Supplement) 23
[27] Aeppli G, Bishop D, Broholm C, Bucher E, Siemensmeyer K, Steiner M and Stüsser N 1989 *Phys. Rev. Lett.* **63** 676
[28] Jourdan M, Huth M and Adrian H 1999 *Nature* **398** 47
[29] Sato N K, Aso N, Miyake K, Shiina R, Thalmeier P, Varelogiannis G, Geibel, C, Steglich F, Fulde P and Komatsubara T 2001 *Nature* **410** 340
[30] Jaccard D, Behnia K and Sierro J 1992 *Phys. Lett. A* **163** 475
[31] Movshovich R, Graf T, Mandrus D, Thompson J D, Smith J L and Fisk Z 1996 *Phys. Rev. B* **53** 8241
[32] Mathur N D, Grosche F M, Julian S R, Walker I R, Freye D M, Haselwimmer R K W and Lonzarich G G 1998 *Nature* **394** 39
[33] Saxena S S et al. 2000 *Nature* **406** 587
[34] Lévy F, Sheikin I, Grenier B and Huxley A D 2005 *Science* **309** 1343
[35] Huy N T, Gasparini A, de Nijs D E, Huang Y, Klaasse J C P, Gortenmulder T, de Visser A, Hamann A, Görlach T and v. Löhneysen H 2007 *Phys. Rev. Lett.* **99** 067006
[36] Hattori T et al. 2012 *Phys. Rev. Lett.* **108** 066403
[37] Yuan H Q, Grosche F M, Deppe M, Geibel C, Sparn G and Steglich F 2003 *Science* **302** 2104
[38] Holmes A, Jaccard D and Miyake K 2004 *Phys. Rev. B* **69** 024508
[39] Onishi Y and Miyake K 2001 *J. Phys. Soc. Jpn.* **69** 3955
[40] See, e.g., Wiebe C R, Janik J A, Mac Dougall G J, Luke G M, Garrett J D, Zhou H D, Jo Y J, Balicas L, Qiu Y, Copley J R D, Yamani Z, Buyers W J L 2007 *Nature Phys.* **3** 96
[41] Petrovic C, Pagliuso P G, Hundley M F, Movshovich R, Sarrao J L, Thompson J D, Fisk Z and Monthoux P 2001 *J. Phys.: Condens. Matter* **13** L337
[42] Bauer E, Hilscher G, Michor H, Paul Ch, Scheidt E W, Gribanov A, Seropegin Yu, Noël H, Sigrist M and Rogl P 2004 *Phys. Rev. B* **92** 027003
[43] Monthoux P and Lonzarich G G 1999 *Phys. Rev. B* **59** 14598
[44] Sarrao J L, Morales L A, Thompson J D, Scott B L, Stewart G R, Wastin F, Rebizant J, Boulet P. Colineau E and Lander G H 2002 *Nature* **420** 297
[45] Wastin F, Boulet P, Rebizant J, Colineau E and Lander G H 2003 *J. Phys.: Condens. Matter* **15** 52279
[46] Aoki D et al. 2007 *J. Phys. Soc. Jpn.* **76** 063701
[47] Gor'kov L P and Rashba R I 2001 *Phys. Rev. Lett.* **87** 037004
[48] Agterberg D F, Frigeri P A, Kaur R P, Koga A and Sigrist M 2006 *Physica B* **378-380** 351
[49] Tou H, Kitaoka Y, Ishida K, Asayama K, Kimura N, Ōnuki Y, Yamomoto E, Haga Y and Maezawa K 1998 *Phys. Rev. Lett.* **80** 3129
[50] Ishida K et al. 2001 *Phys. Rev. Lett.* **89** 037002
[51] Steglich F et al. 1996 *Physica B* **223-224** 1
[52] Stockert O et al. 2004 *Phys. Rev. Lett.* **92** 136401
[53] Stockert O et al. 2011 *Nature Phys.* **7** 119
[54] Arndt J, Stockert O, Schmalzl K, Faulhaber E, Jeevan H S, Geibel C, Schmidt W, Loewenhaupt M and Steglich F 2011 *Phys. Rev. Lett.* **106** 246401

[55] Rauchschwalbe U, Lieke W, Bredl C D, Steglich F, Aarts J, Martini K M and Mota A C 1982 *Phys. Rev. Lett.* **49** 1448
[56] Stockert O, Kirchner S, Steglich F and Si Q 2012 *J. Phys. Soc. Jpn.* **81** 011001
[57] Friedemann S, Westerkamp T, Brando M, Oeschler N, Wirth S, Gegenwart P, Krellner C, Geibel C and Steglich F 2009 *Nature Phys.* **5** 465
[58] Tokiwa Y, Gegenwart P, Geibel C and Steglich F 2009 *J. Phys. Soc. Jpn.* **78** 123708
[59] Yuan H Q et al. (to be published)
[60] Park T, Ronning F, Yuan H Q, Salamon M B, Movshovich R, Sarrao J L and Thompson J D 2006 *Nature* **440** 65
[61] Shishido H, Settai R, Harima H, Ōnuki Y 2005 *J. Phys. Soc. Jpn.* **74** 1103
[62] Paglione J, Tanatar M A, Hawthorn D G, Boaknin E, Hill R W, Ronning F, Sutherland M, Taillefer L, Petrovic C and Canfield P C 2003 *Phys. Rev. Lett.* **91** 246405
[63] Bianchi A, Movshovich R, Vekhter I, Pagliuso P G and Sarrao J L, 2003 *Phys. Rev. Lett.* **91** 257001
[64] Singh S, Capan C, Nicklas M, Rams M, Gladun A, Lee H, DiTusa J F, Fisk Z, Steglich F and Wirth S *2007 Phys. Rev. Lett.* **98** 057001
[65] Zaum S, Grube K, Schäfer R, Bauer E D, Thompson J D and von Löhneysen H 2011 *Phys. Rev. Lett.* **106** 087003
[66] Ronning P, Capan C, Bianchi A, Movshovich R, Lacerda A, Hundley M F, Thompson J D, Pagliuso P G and Sarrao J L 2005 *Phys. Rev. B* **71** 104528
[67] Howald L, Seyfarth G, Knebel G, Lapertot G, Aoki D and Brison J-P 2011 *J. Phys. Soc. Jpn.* **80** 024710
[68] Hu T, Xiao H, Sayles T A, Dzero M, Maple M B and Almasan C C 2012 *Phys. Rev. Lett.* **108** 056401
[69] Tokiwa Y, Bauer E D and Gegenwart P *arXiv*: 1203.5567
[70] Donath J G, Steglich F, Bauer E D, Sarrao J L and Gegenwart P 2008 *Phys. Rev. Lett.* **100** 136401
[71] Scalapino D J, Loh Jr. E and Hirsch J E 1986 *Phys. Rev. B* **34** 8190
[72] Hegger H, Petrovic C, Moshopoulou E G, Hundley MF, Sarrao J L, Fisk Z and Thompson J D 2000 *Phys. Rev. Lett.* **84** 4986

10th International Conference on Materials and Mechanisms of Superconductivity (M2S-X) IOP Publishing
Journal of Physics: Conference Series **449** (2013) 012029 doi:10.1088/1742-6596/449/1/012029

Spin-Triplet Superconductivity Induced by Longitudinal Ferromagnetic Fluctuations in UCoGe: Theoretical Aspect

Y. Tada,[1]* S. Fujimoto,[2], N. Kawakami,[2] T. Hattori,[2] Y. Ihara,[3] K. Ishida,[2] K. Deguchi,[4] N. K. Sato,[4] and I. Satoh[5]

[1] Institute for Solid State Physics, the University of Tokyo, Kashiwa, Chiba 277-8581, Japan
[2]Department of Physics, Graduate School of Science, Kyoto University, Kyoto 606-8502, Japan
[3]Department of Physics, Graduate School of Science, Hokkaido University,Sapporo 060-0810, Japan
[4]Department of Physics, Graduate School of Science, Nagoya University, Nagoya 464-8602, Japan
[5]Institute for Materials Research, Tohoku University, Sendai 980-8577 Japan

E-mail: tada@issp.u-tokyo.ac.jp

Abstract. Identification of pairing mechanisms leading to the unconventional superconductivity realized in copper-oxide, heavy-fermions, and organic compounds is one of the most challenging issues in condensed-matter physics. Clear evidence for an electron-phonon mechanism in conventional superconductors is seen by the isotope effect on the superconducting transition temperatures T_{SC}, since isotopic substitution varies the phonon frequency without affecting the electronic states. In unconventional superconductors, magnetic fluctuations have been proposed to mediate superconductivity, and considerable efforts have been made to unravel relationships between normal-state magnetic fluctuations and superconductivity. Here, we show that characteristic experimental results on the ferromagnetic (FM) superconductor UCoGe ($T_{\rm Curie} \sim 2.5$ K and $T_{SC} \sim 0.6$ K) can be understood consistently within a scenario of the spin-triplet superconductivity induced by FM spin fluctuations. Temperature and angle dependencies of the upper critical magnetic field of the superconductivity (H_{c2}) are calculated on the basis of the above scenario by solving the Eliashberg equation. Calculated H_{c2} well agrees with the characteristic experimental results observed in UCoGe. This is a first example that FM fluctuations are shown to be a pairing glue of superconductivity.

1. Introduction

The discovery of superconductivity in ferromagnetic (FM) UGe_2 under pressure gave great impact for condensed-matter community [1], since ferromagnetism and superconductivity have been considered to be mutually exclusive phenomena. After the discovery, superconductivity has been found in several U-based FM compounds [2, 3]. UCoGe is one of the compounds among the FM superconductors discovered so far which can be readily explored in experiments, because of its high superconducting (SC) transition temperature (T_{SC}) and low Curie temperature ($T_{\rm Curie}$) at ambient pressure. Actually, important and intriguing experimental results have been reported for UCoGe [4, 5], and some of them are summarized as follows.

Journal of Physics: Conference Series **449** (2013) 012029 doi:10.1088/1742-6596/449/1/012029

(i) Microscopic measurements of μSR [Fig.1(a)] [6] and Co-NQR [Fig.1(b)] [7] have shown that the superconductivity occurs within the FM region, resulting in homogeneous microscopic coexistence of the ferromagnetism and superconductivity. The superconductivity is also found in the paramagnetic state and T_{SC} shows a broad maximum around a critical pressure of ferromagnetism [Fig.1(c)] [3], which is in sharp contrast with UGe_2 where the SC is seen only in the FM phase [1].

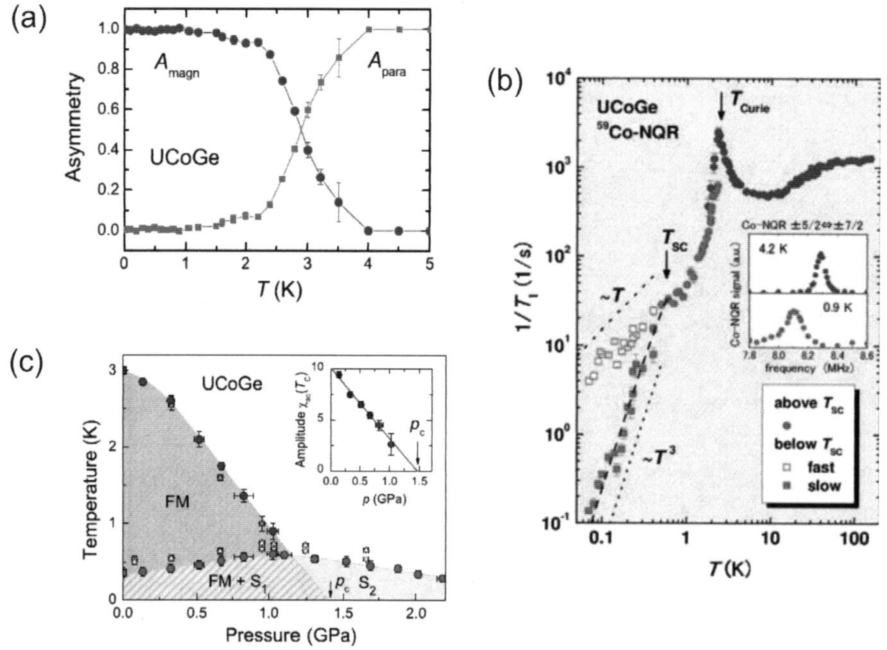

Figure 1. (a)Temperature variation of the normalized asymmtries ($A_{\rm magn} + A_{\rm para} = 1$) of the FM and paramagnetic phases measured with zero-field muon-spin rotation on UCoGe [6]. Below ~ 1.5 K, magnetism is observed in the whole sample volume although superconductivity occurs at ~ 0.8 K, indicative of coexistence of ferromagnetism and superconductivity on a microscopic scale. (b) Temperature dependence of ^{57}Co $1/T_1$ in the single-crystal sample [7]. $1/T_1$ above 2.3 K, shown by the blue dots, was measured at the paramagnetic (PM)-frequency (8.3 MHz) peak arising from the $\pm5/2 \leftrightarrow \pm7/2$ transitions, and $1/T_1$ below 2.3 K, shown by the red dots, was measured at the FM-frequency (8.1 MHz) peak shifted from the PM peak due to the appearance of the internal field at the Co site. The inset shows the NQR signal arising from the $\pm5/2 \leftrightarrow \pm7/2$ transitions obtained at 4.2 K (PM) and 0.9 K (FM). Two $1/T_1$ components were observed in the SC state, the faster (slower) component denoted by red open (closed) squares; the red broken curve below $T_{\rm SC}$ represents the temperature dependence calculated assuming a line-node gap with $\Delta_0/k_{\rm B}T_{\rm SC} = 2.3$. The SC anomaly observed from the FM signal is clear evidence that superconductivity occurs in the FM region. (c) Pressure-temperature phase diagram of UCoGe [11]. Ferromagnetism is denoted as blue area, superconductivity is as yellow area. $T_{\rm SC}(p)$ extrapolates to a FM quantum critical point at the critical pressure $p_{\rm c}$. Superconductivity coexists with ferromagnetism below $p_{\rm c}$ as blue-yellow hatched area. Inset: Amplitude of $\chi_{\rm ac}(T)$ at $T_{\rm C}$ as a function of pressure. The data follow a linear p dependence and extrapolate to $p_{\rm c} = 1.46 \pm 0.10$ GPa.

(ii) The temperature variation of the ^{59}Co-NQR spectra shown in Fig. 2 indicates that the FM region grows continuously around T_{Curie} coexisting with the PM region, and that the FM moment in the FM region appears discontinuously, whereas the hysteresis behavior was not observed in the temperature variation of the spectra and $1/T_1$ in the both states below T_{Curie} is almost identical [8]. Since the divergence of $1/T_1$ and the clear anomaly in the specific heat were observed at T_{Curie} [7, 5] and the FM ordered moment is small ($\mu_s \sim 0.05\mu_B$) at low temperatures [5] , the FM phase transition in UCoGe is considered to possess weak first-order character with critical FM fluctuations, which is close to second-order one. Although we need further investigation of sample and pressure dependence on the FM transition to understand the transition character thoroughly, the presence of the critical fluctuations implies the validity of the scenarios of spin fluctuation-induced superconductivity.

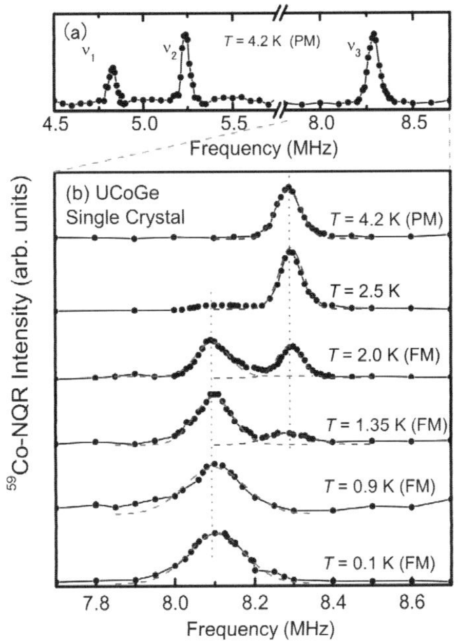

Figure 2. (a) ^{59}Co-NQR spectra (ν_1, ν_2, ν_3) at 4.2 K in the PM state[8]. (b) Temperature dependence of the NQR spectrum from the $\pm 5/2 \leftrightarrow \pm 7/2$ transitions (ν_3)[8]. The PM and FM signals coexist between 1 and 2.7 K. The blue (red) broken lines represent Gaussian fits to the PM (FM) peaks; the solid lines are guides to the eye.

(iii) Although transport properties show rather three dimensional isotropic characters [9], studies of the SC upper critical magnetic field (H_{c2}) [Fig. 2 (a)] and angle dependence of H_{c2} [Fig. 2 (b)] have shown that SC properties are highly anisotropic [10, 11, 12]; the superconductivity survives in fields exceeding 15 T along the a and b axes, whereas H_{c2} for fields along the c direction (H_{c2}^c) is as small as 0.8 T.

(iv) The static magnetic susceptibility is strongly anisotropic [10]; the magnetic anisotropy is Ising like with the c axis as a magnetic easy axis [Fig. 4 (a)]. In addition, direction-dependent nuclear-spin lattice relaxation rate ($1/T_1$) measurements on a single crystalline sample have revealed the magnetic fluctuations in UCoGe to be Ising-type FM ones along the c axis (longitudinal-mode fluctuations) as shown in Fig. 4 (b) [13]. The Ising-type fluctuations are also suggested from the anisotropic correlation length measured by the inelastic neutron measurements [23].

(v) From detailed angle-resolved NMR and Meissner measurements [9], it was found that magnetic fields along the c axis (H^c) is a tuning parameter of the longitudinal FM fluctuations and strongly suppresses the FM fluctuations, as shown in Fig. 5 (a) and 5 (b). This sharp angle dependence of $1/T_1$ was also confirmed at 600 mK just above T_{SC} [14]. It

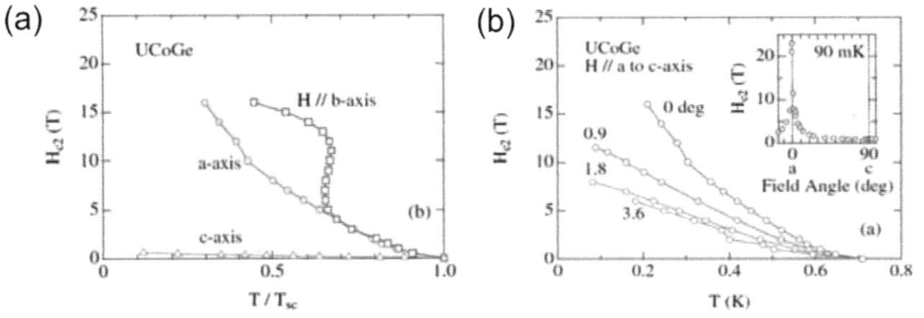

Figure 3. (a) Temperature dependence of the upper critical fields for $H \parallel a, b$ and c-axis [12]. The temperature is normalized by the SC critical temperature T_{SC} at zero field. (b) Temperature dependence of the upper critical field H_{c2} close to $H \parallel a$-axis [12]. The field direction is tilted from the a to c-axis. The inset shows the angular dependence of H_{c2} at 90 mK. The values of H_{c2} greater than 16 T are the results of linear extrapolations to 90 mK.

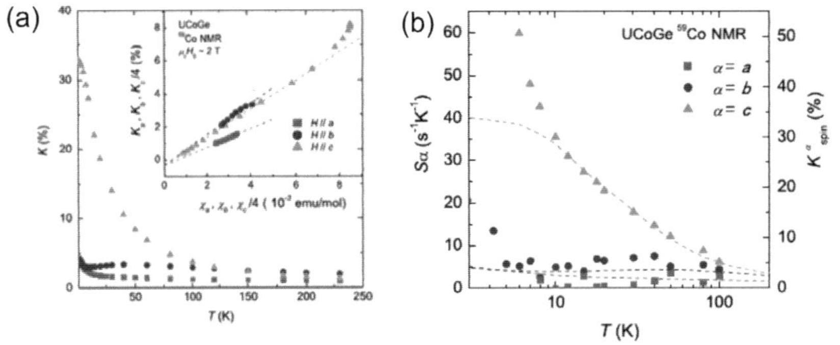

Figure 4. NMR shift K measured in field along the a, b and c directions [13]. Remarkable anisotropy was observed. In the inset, K is plotted against the bulk susceptibility χ measured in 2 T. The nearly identified slopes for all the directions indicate that the hyperfine coupling constants are isotropic. (b) Direction-decomposed dynamic susceptibility S_α and static susceptibility $K_{\rm spin}$ along each direction [13]. Identical Ising anisotropy for both quantities found above 8 K suggests that the longitudinal mode dominates these fluctuations. It is noteworthy that in this T range, the Knight shift scales with S_α, which is predicted by the self-consistent renormalization theory for a 3-dimensional nearly FM metals [15].

was also shown that the superconductivity is observed in the limited magnetic-field region where the longitudinal FM spin fluctuations are active [Fig. 4 (c)]. These results suggest that the superconductivity in UCoGe is tightly coupled with the longitudinal FM spin fluctuations along the c-axis.

From a theoretical point of view, when an itinerant FM superconductor possesses the large energy splitting between the majority and minority spin Fermi surfaces, exotic spin-triplet superconductivity is anticipated, in which pairing is between parallel spins within each spin Fermi surface. In addition, it has been argued that critical FM fluctuations near a quantum phase transition could mediate spin-triplet superconductivity [16, 17, 18, 19, 20, 21], and actually, (iv) in the above indicates that the superconductivity in UCoGe is intimately coupled with the Ising FM spin fluctuations. However, as far as we know, there have been no reports identifying

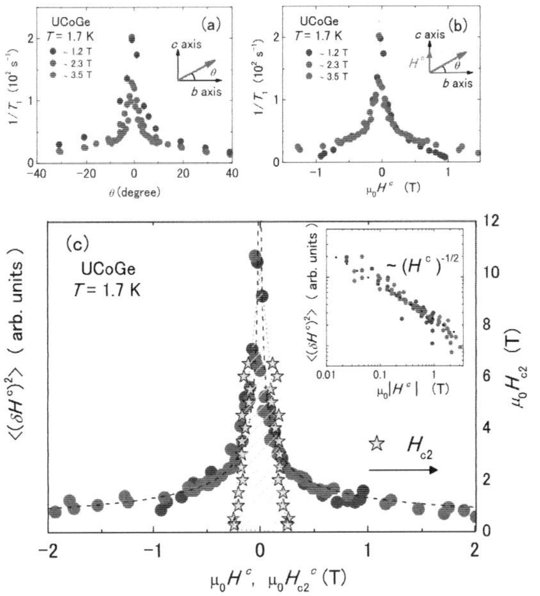

Figure 5. (a) Angle dependence of $1/T_1$ in the bc plane measured in three different magnetic fields at 1.7 K [9]. (b) Plot of the $1/T_1$ against H^c. The $1/T_1$ data collapse onto a single curve when plotted against H^c. (c) H^c dependence of magnetic fluctuations along the c axis $\langle(\delta H^c)^2\rangle$ at 1.7 K, extracted using $\langle(\delta H^c)^2\rangle \propto \frac{1}{\cos^2\theta}\left(\frac{1}{T_1}(\theta) - \frac{(1+\sin^2\theta)}{2}\frac{1}{T_1^c}\right)$. Angle dependence of H_{c2}, determined by the onset of the Meissner signal, is plotted against $H_{c2}^c \equiv H_{c2}\sin\theta$. It is found that the superconductivity is observed in the limited field region (yellow slashed area), where the longitudinal fluctuations are active. Inset: plot of $\langle(\delta H^c)^2\rangle$ against $|H^c|$. The relation of $\langle(\delta Hc)^2\rangle \propto 1/\sqrt{H^c}$ is shown by dashed lines in (c) and the inset.

magnetic fluctuations inducing unconventional superconductivity. Here, we show, from model calculations on the basis of the scenario of FM spin fluctuation-mediated superconductivity in UCoGe, that unconventional temperature and angle dependence of H_{c2} is consistently understood by adopting the characteristic longitudinal FM fluctuations revealed experimentally. Besides, by comparing calculations for possible pairing states, we also discuss candidates of SC gap symmetries in UCoGe.

2. Calculations and Results

2.1. Model

In this section, we introduce a simple model to study H_{c2} in UCoGe at ambient pressure where the ferromagnetism and superconductivity coexist. Although UCoGe has a complicated multi-band structure, since the spin fluctuations dominate the low energy physics in UCoGe, we can use a simplified model to study its superconductivity with appropriately taking into account the critical FM spin fluctuations observed in the experiments [9, 13, 23]. The model relevant to UCoGe in the FM phase with an applied magnetic field $\boldsymbol{H}$ is

$$S = S_0 + S_{\rm ex} + S_{\rm int}, \tag{1}$$

$$S_0 = \sum_k \int_0^{1/T} d\tau c_k^\dagger(\tau)[\partial_\tau + \varepsilon_k]c_k(\tau), \tag{2}$$

$$S_{\rm ex} = -(\boldsymbol{h}_{ex} + \mu_B \boldsymbol{H}) \cdot \sum_k \int_0^{1/T} d\tau c_k^\dagger(\tau)\boldsymbol{\sigma} c_k(\tau), \tag{3}$$

$$S_{\rm int} = -\frac{2g^2}{3}\sum_q \int_0^{1/T} d\tau \int_0^{1/T} d\tau' S_q^c(\tau)\chi^c(\boldsymbol{q}, \tau - \tau')S_{-q}^c(\tau'), \tag{4}$$

and effects of the vector potential on electron motions are taken into account in a semiclassical way. $c_k = (c_{k\uparrow}, c_{k\downarrow})^T$ is an annihilation operator of low energy quasiparticles which are formed

Journal of Physics: Conference Series **449** (2013) 012029 doi:10.1088/1742-6596/449/1/012029

through hybridizations between conduction electrons and f-electrons. A typical energy scale of the quasiparticles is $40 \sim 50$ (K) estimated from the specific heat [3] and the coherence temperature [13]. Since the anisotropy in the resistivity in UCoGe is small as discussed in the introduction, we use an isotropic dispersion $\varepsilon_k = -2t\sum_{i=a,b,c}\cos k_i a_0 - \mu$ where a_0 is the lattice constant and μ is the chemical potential. We fix the electron density per site as $n = 0.15$ which leads to a small but nearly spherical Fermi surface. As discussed in Ref.[9], the anisotropic mass model cannot explain the anomalous angle dependence of H_{c2}. In our model, the anisotropy in H_{c2} arises from the exchange term and the pairing interaction term in Eq.(1). We use an energy unit $t = 1$ and a length unit $a_0 = 1$.

The second term of Eq.(1) includes the exchange splitting of the Fermi surface in the FM phase which is assumed to be large compared to the SC transition temperature at zero-field T_{SC0}. The exchange field is parallel to the c-axis and fixed to be $\boldsymbol{h}_{ex} = (0, 0, h_{ex}) = (0, 0, 0.5t)$, which would lead to suppression of the Pauli depairing effect against in-plane magnetic fields in the equal spin pairing states of the superconductivity [22]. This is a key assumption to understand the large anisotropy in H_{c2} in the present study, because the equal spin pairing states are not robust against the in-plane Zeeman field while the observed $H_{c2}^{a,b}$ are huge in UCoGe. However, we put a remark on this assumption. UCoGe has a multi-band structure with a strong spin-orbit interaction and $5f_{5/2}$ states are located near the Fermi level. In such a system, it is possible that, even if the magnetization in the FM state is small, the exchange splitting for each Fermi surface is large. Actually, in the band calculations, the orbital magnetizations and the spin polarizations cancel to result in a small total magnetic moment. The present assumption can be checked by NMR Knight shift measurements in the SC states.

Finally, the last term in Eq. (1) describes the interaction between the quasiparticles through the Ising spin fluctuations, and $S_q^c(\tau) = \sum_k c_{k-q}^{\dagger}(\tau)(\sigma^c/2)c_k(\tau)$. The dynamical susceptibility $\chi^c(\boldsymbol{q}, i\Omega_n)$ strongly depends on the c-axis magnetic field as revealed in the NMR experiments [9].

$$\chi^c(\boldsymbol{q}, i\Omega_n) = \frac{\chi_0}{\delta + q^2 + |\Omega_n|/\gamma_q}, \tag{5}$$

$$\delta(H^c) = h_{ex}^2 + c'\sqrt{H^c}, \tag{6}$$

where $\gamma_q = vq$ and $v = 4$ is approximately the Fermi velocity. Here, we have assumed a simple Landau damping in $\chi^c(\boldsymbol{q}, i\Omega_n)$, although existence of non-Landau damping was reported [23, 24]. The coefficient c' is determined by the NMR experiments, and a normalization factor $H_0^c \sim 1$ T is introduced as $c' \equiv ch_{ex}^2/\sqrt{H_0^c}$ so that the dimensionless parameter c is $c \sim \mathcal{O}(1)$. Although the origin of this magnetic field dependence in the susceptibility is not so clear, we have a possible explanation for it. According to the band calculation, the $5f_{5/2}$ bands in UCoGe have some semi-metallic like structures around several k-points [25]. For such structures, the density of states would be $\sim \sqrt{H^c}$, which results in $(\chi^c)^{-1} \sim$ const. $+ \sqrt{H^c}$ near the criticality. In this scenario, detailed structures in the bands are important, and therefore, the spin fluctuations in UCoGe under a magnetic field can depend on sample qualities. This problem will be studied elsewhere. We note that the coupling constant g should be regarded as a renormalized one including the vertex corrections which enhance the SC transition temperatures. Although we have postulated a simple analytic q-dependence in χ^c, there might exist non-analytic corrections $\sim q^2 \log q$ in 3-dimensional isotropic systems [26, 27, 28, 29]. The non-analytic corrections are important in connection with fluctuation-induced first order phase transitions. However, since the first order character of the FM transition is weak, and critical FM fluctuations are well developed in UCoGe as discussed in Sec.1, it would be legitimate to neglect the first order character, and assume that χ^c is analytic in q as long as calculations of the superconducting properties are concerned. In addition, if exist in the Ising-like system, the non-analytic corrections would not change the present study qualitatively as discussed in [30].

2.2. Eliashberg equation

In this study, we focus on the anisotropy between the a and c-axes upper critical fields (H^a_{c2} and H^c_{c2}). However, it was reported that the b-axis upper critical field H^b_{c2} shows a "S-shaped" curve at high fields, and relations to the reentrant superconductivity in URhGe were discussed [12]. In URhGe, the magnetization flips at a critical $H^b = H^{b*}$ and the spin fluctuations at H^{b*} are considered to be different from those of $H^b = 0$ [31]. Similar scenarios are expected to hold also in UCoGe where the S-shaped H^b_{c2} would indicate a crossover of the superconductivities of different mechanisms under magnetic field along the b axis. The problem of H^b_{c2} in UCoGe is an another issue and will be studied elsewhere.

To determine the upper critical field H_{c2} in the ac-plane, we solve the linearized Eliashberg equation in the presence of a vector potential [32]. The Eliashberg equation for an equal-spin pairing state in the FM phase is

$$\Delta_{\sigma\sigma}(k) = -\frac{T}{2N}\sum_{k'} V(k,k')[G_{\sigma\sigma'}(k'+\Pi)G_{\sigma\sigma'}(-k') + G_{\sigma\sigma'}(k')G_{\sigma\sigma'}(-k'+\Pi)]\Delta_{\sigma'\sigma'}(k'), \quad (7)$$

where $\Pi = (-i\nabla_R - 2e\boldsymbol{A}(\boldsymbol{R}), 0)$ and $\nabla \times \boldsymbol{A} = \boldsymbol{H} = H(\cos\theta, 0, \sin\theta)$. A short notation $k = (\boldsymbol{k}, i\omega_n)$ is used hereafter. Here we have neglected the effect of the magnetization in the orbital motion, because the observed moment is very small in UCoGe [33]. Note that the Green's function $\hat{G}(k)$ has off-diagonal elements when $\theta \neq 90°$. The gap function depends both on the momentum in the relative coordinate and the center-of-mass coordinate, and is expanded as $\Delta_{\sigma\sigma}(k, \boldsymbol{R}) = \Delta_{\sigma\sigma}(k)\cdot\phi_0(\boldsymbol{R})$, where ϕ_0 is the lowest level Landau function.

We solve the linearized Eliashberg equation for two cases corresponding to the classification by the magnetic point group $D_2(C_2^z)$ for UCoGe which has an orthorhombic crystal structure [33, 34]. One is the so-called A state which corresponds to point node symmetry with a d-vector $\boldsymbol{d} \sim (a_1k_a + ia_2k_b, a_3k_b + ia_4k_a, 0)$ near the Γ point. The other one is the so-called B state corresponding to horizontal line node symmetry with $\boldsymbol{d} \sim (b_1k_c + ib_2k_ak_bk_c, ib_3k_c + b_4k_ak_bk_c, 0)$. Here, $\{a_i\}$ and $\{b_i\}$ are real coefficients. Both of the gap functions are determined by self consistent calculations, and the resulting SC states are non-unitary because of the presence of the exchange field $\boldsymbol{h}_{ex}$. We note that, under the applied magnetic field $\boldsymbol{H}$, the calculated gap functions are distorted and has lower symmetry than that of the cubic lattice in the present model.

The pairing interaction is evaluated at the first order in $g^2\chi_0$,

$$V(k,k') = -\frac{g^2}{6}\chi^c(k-k') + \frac{g^2}{6}\chi^c(k+k'). \quad (8)$$

For the selfenergy, we include only the term

$$\begin{aligned}\Sigma_0(k) &\equiv (\Sigma_{\uparrow\uparrow}(k) + \Sigma_{\downarrow\downarrow}(k))/2 \\ &= \frac{T}{6N}\sum_q g^2\chi^c(q)[G^0_{\uparrow\uparrow}(k+q) + G^0_{\downarrow\downarrow}(k+q)], \end{aligned} \quad (9)$$

where $G^0_{\sigma\sigma}$ is the non-interacting Green's function. Other neglected terms in the self energy are much smaller than Σ_0 in magnitude. As noted in the previous section, the coupling constant g is a renormalized one.

In the present study, the coupling constant is fixed as $g^2\chi_0 = 125t$ which is rather large to give a high transition temperature so that the numerical costs are reduced. However, qualitatively, our main results do not depend on the value of $g^2\chi_0$. For this value of $g^2\chi_0$, the mass enhancement factor $1/z$ is ~ 1.3, which is not so large because the density of states at the Fermi surface is small in the present parameters. The transition temperature for the A state without magnetic field is $T_{SC0} = 0.020t$ which we regard as 1.0 K according to the experiments for UCoGe. Then, the hopping integral is fixed as $t = 50$ K throughout in our study, and the order of this value is consistent with the coefficient of the linear specific heat $\gamma \sim 50$ mJ/K^{2}mol [3] and the coherence temperature $T^* \sim 40$ K [13]. We also fix the lattice constant, $a_0 = 4.0$ Åwhich leads to an effective mass for the orbital motion of the Cooper pairs $m_{\rm eff} \equiv \hbar/ta_0^2 \sim 440m_0$ where m_0 is the bare electron mass. Correspondingly, we fix $H_0^c = 0.02t \sim 1$ T, which is consistent with the NMR experiments [9].

2.3. Results for A state

Before discussing numerical results of H_{c2} of the present model, we note that the orbital limit $H_{\rm orb}$ depends on positions of gap nodes in general [30]. A simple explanation is as follows. Let us consider a superconductor with isotropic Fermi surface which has nodes in a gap function. The effective velocity for the cyclotron motion of the orbital depairing effect can be of the form of $\tilde{\boldsymbol{v}}_k = \varphi_k \boldsymbol{v}_k$ for the basis function φ_k corresponding to the SC gap symmetry, where $\boldsymbol{v}_k$ is a velocity of electrons. For φ_k with horizontal line nodes, when the magnetic field is parallel to the c-axis, the cyclotron motion is suppressed at the nodes where in-plane $\tilde{\boldsymbol{v}}_k$ is zero, resulting in a large orbital limiting field. The cyclotron motion is not suppressed for $H \perp c$ because $\tilde{\boldsymbol{v}}_k \perp H$ is non-zero on the Fermi surface, which leads to the anisotropy of the orbital limiting field $H_{\rm orb}^{\parallel c} > H_{\rm orb}^{\perp c}$. On the other hand, for the point node gap function, $\tilde{\boldsymbol{v}}_k$ cannot be zero except for the poles of the Fermi surface $k_a = k_b = 0$ which are realized as cross points of two line nodes $k_a = 0$ and $k_b = 0$. Therefore, for point nodes at the poles of the Fermi surface, the order is turned over, $H_{\rm orb}^{\parallel c} < H_{\rm orb}^{\perp c}$. Following this consideration, if there is no suppression of the pairing interaction and the Pauli depairing effect is negligible, $H_{c2}^c < (>) H_{c2}^a$ is expected for the A (B) state.

We now turn to discussions of calculation results. Temperature dependence of $H_{c2}^{a,c}$ with $c = 1, 3$ for the A state is shown in Fig. 6. It is found that the qualitative behaviors of H_{c2}

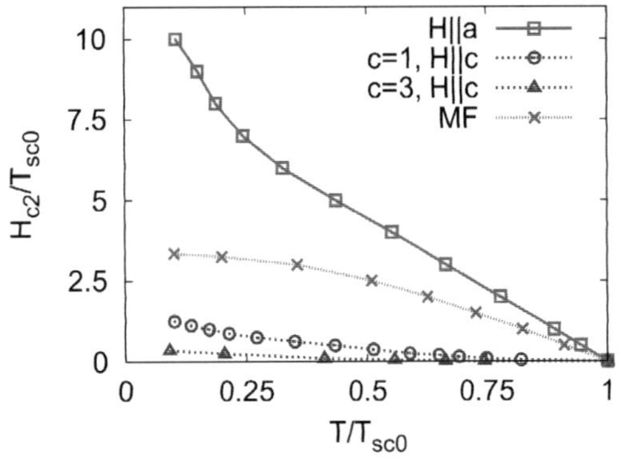

Figure 6. Temperature dependence of H_{c2} for the a and c axes in the A state. The purple curve "MF" is H_{c2}^c calculated with the mean field $\delta(H^c)$.

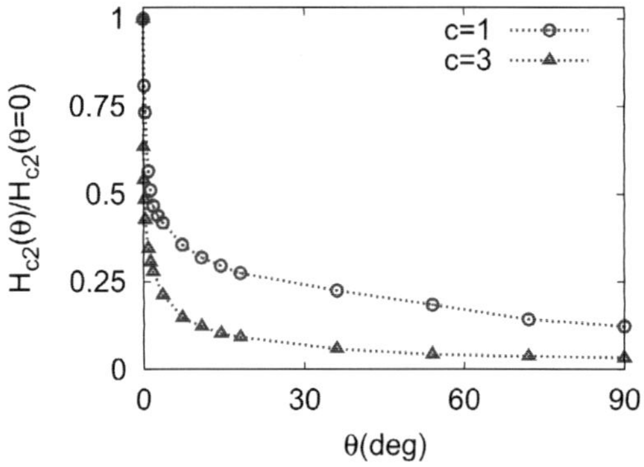

Figure 7. Field angle dependence of H_{c2} at $T = 0.1T_{SC0}$ in the A state.

are independent of c. H_{c2}^a exhibits a strong coupling behavior and reaches $H_{c2}^a/T_{SC0} \sim 10$. Although the precise value of H_{c2}^a/T_{SC0} depends on the choice of h_{ex} and $g^2\chi_0$, strong coupling calculations of H_{c2} naturally explain the upward curvature in H_{c2}^a and its large value observed in the experiments, which is a key to understand the anomalous anisotropic behavior in H_{c2}.

The calculated anisotropy ratio H_{c2}^a/H_{c2}^c becomes $10 \sim 20$ in consistent with the experiments, although the Fermi surface of our model is merely spherical. We emphasize that, for H_{c2} to have a large anisotropy between the a and c-axes, the experimentally observed $\sqrt{H^c}$-dependence in the susceptibility is crucially important. For a comparison, we calculate H_{c2}^c in the case that χ^c has weaker H-dependence, mean-field like dependence of $\delta(H^c) = (h_{ex} + \mu_B H^c)^2$, for which results are shown with purple symbols in Fig. 6. In this case, suppression of H_{c2} for the c-axis is only moderate, and we cannot obtain anomalously large anisotropy in H_{c2}. In Fig. 7, we show field angle dependence of $H_{c2}(\theta)$ at $T = 0.1T_{\rm SC0}$, where θ is the angle from the a-axis to the c-axis. It is seen that $H_{c2}(\theta)$ is strongly suppressed when the field direction is slightly tilted from the a-axis to the c axis. This strong angle dependence is again due to the $\sqrt{H^c}$ suppression of the pairing interaction. The calculation results well explain the unusual angle dependence of H_{c2} observed experimentally. This good agreement is strong evidence that the superconductivity in UCoGe is actually mediated by the critical Ising spin fluctuations.

We note that the relation between H^c and H_{c2} is considered to be an analog of the isotope effect in conventional superconductors where ion mass M controls transition temperature $T_{\rm SC}$. In both cases, suppression of the pairing interaction by increasing H^c or M leads to reduction of $T_{\rm SC}$. One important reason why we can establish such a relation between H^c and H_{c2} in the f-electron system UCoGe is that the critical spin fluctuations dominate the low energy physics. Details of the system and complexity of multi-band are sub-dominant factors for the superconductivity, and therefore, the SC properties are well understood once the spin fluctuations are appropriately taken into account.

2.4. Results for B state

The transition temperature at zero-field for the B state is $T_{\rm SC0} = 0.0213t$ which differs from that for the A state, because of the presence of the exchange field along the c-axis. However, the difference is not important in our study. Temperature dependence of $H_{c2}^{a,c}$ for the B state is shown in Fig. 8. For the B state, strong-coupling behaviors for the a-axis are not so significant,

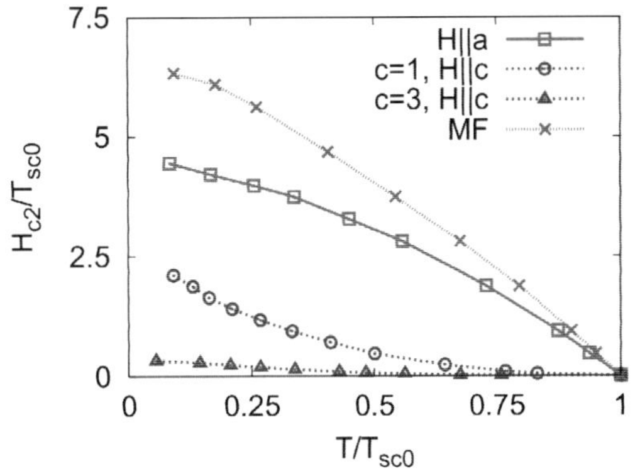

Figure 8. Temperature dependence of H_{c2} for the a and c axes in the B state. The purple curve "MF" is H_{c2}^c calculated with the mean field $\delta(H^c)$.

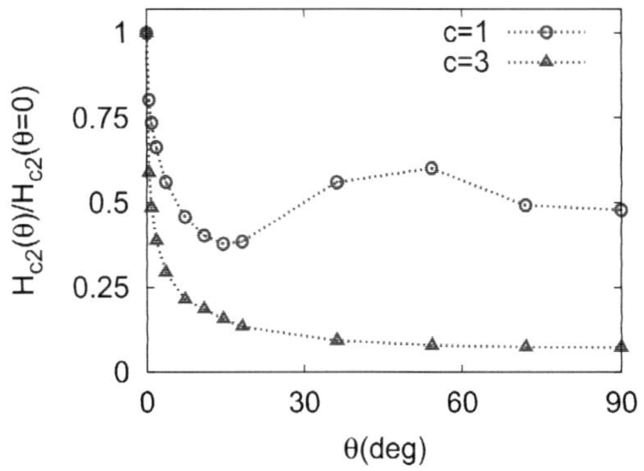

Figure 9. Field angle dependence of H_{c2} at $T = 0.1T_{\rm SC0}$ in the B state.

because the orbital limiting fields depend on the positions of gap nodes as discussed in the previous section, and the enhancement of H_{c2}^a for the B state is weaker than that for the A state. For the present value of $m_{\rm eff}$, the Pauli depairing effect cannot be negligibly small at high magnetic fields, which leads to the suppression of H_{c2}^a in that region. For the c-axis, detailed behaviors depend on the value of the parameter c. In the case of rather weak suppression of the Ising fluctuations, i.e. $c = 1$, H_{c2} is enhanced at low temperatures. If the suppression is sufficiently strong, for example $c = 3$, such enhancement cannot be seen. We also show H^c_{c2} calculated with the mean field $\delta(H^c)$ by the purple curve as in the A state. In the B state, H^c_{c2} is not strongly suppressed and H^c_{c2} is larger than H^a_{c2} in contrast to the experiments. The relation $H^c_{c2} > H^a_{c2}$ with the mean-field $\delta(H^c)$ is directly understood from the general relation $H^c_{\mathrm{orb}} > H^a_{\mathrm{orb}}$ for horizontal line node gap functions in simple systems.

Behaviors of H_{c2} as a function of the field angle also depend on the value of c as shown in Fig. 9. For $c = 1$, $H_{c2}(\theta)$ shows a non-monotonic behavior. This is because, for the B state, the relation $H_{c2}(\theta = 0°) < H_{c2}(\theta = 90°)$ holds in the orbital limit if there is no suppression of the Ising fluctuations, while in fact the suppression in $\chi^c(q)$ gets stronger as $\theta \to 90°$. On the other hand, for $c = 3$, H_{c2} monotonically decreases. Comparing these results with those for the A state which are robust against the parameter value, one may conclude that the A state is a promising candidate for the SC realized in the FM state in UCoGe, although the B state is not ruled out.

3. Summary

In this paper, we have studied the upper critical field H_{c2} for the FM superconductor UCoGe in the FM phase. In the introduction, we discussed several experimental observations which are key clues to understand relations between the FM and SC. To clarify the relationship, we introduced the simple model which includes the critical FM spin fluctuations with anomalous $\sqrt{H^c}$ dependence revealed by the NMR experiments. The linearized Eliashberg equations are solved within the first order in the spin fluctuations. In the A state with point nodes, temperature dependence of H^a_{c2} shows an upward curvature while H^c_{c2} is suppressed. When the field angle is tilted from the a to the c axis, H_{c2} is strongly suppressed. These behaviors are totally due to the characteristic suppression of the FM spin fluctuations by H^c, and in nice agreement with the experimentally observed H_{c2} in UCoGe. This is strong evidence of the scenario that the spin-triplet superconductivity is actually mediated by the critical Ising FM spin fluctuations in UCoGe. The relation between H^c and H_{c2} is an analog of the isotope effect in conventional superconductors. In the B state with a horizontal line node, both enhancement of H^a_{c2} and suppression of H^c_{c2} are moderate compared to those in the A state. Therefore, the A state is a promising candidate for the pairing symmetry in UCoGe, although the B state is not ruled out.

Finally, we put remarks on some remaining problems related to the present study. Firstly, in the present study, we have simply adopted the experimentally observed $\sqrt{H^c}$ dependence in $\chi^c(q)$. Although we suggested a possible explanation, the origin of this anomalous dependence is not clear.

Secondly, we have assumed that the exchange splitting is large enough to suppress the Pauli depairing effect. However, this assumption should be checked experimentally by e.g. NMR Knight shift measurements. There is also a related problem. We have focused on the coexistence region of the superconductivity and the ferromagnetism where the exchange splitting can be large. Under pressure [11, 12], T_{Curie} is suppressed and the exchange splitting becomes small. In this case and in the paramagnetic phase, the Pauli depairing effect would not be suppressed. Actually, it was reported that, around the critical pressure where T_{Curie} is extrapolated to zero, $H^{a,b}_{c2}(T \to 0)$ is suppressed while $H^c_{c2}(T \to 0)$ is increasing [35]. These behaviors could be understood as a result of two competing effects; although the Pauli depairing effect would not be suppressed by the exchange field near the criticality, the FM spin fluctuations are enhanced.

Investigations of these issues are needed for a comprehensive understanding of UCoGe, and they would be studied in the future.

Acknowledgments
The authors thank S. Yonezawa, and Y. Maeno for experimental support and valuable discussions, and D. Aoki, J. Flouquet, A. de Visser, A. Huxley, H. Harima, and H. Ikeda for valuable discussions. This work was partially supported by Kyoto Univ. LTM centre, Yukawa Institute, the "Heavy Electrons" Grant-in-Aid for Scientific Research on Innovative Areas (No. 20102006, No. 21102510, No. 20102008, No. 23102714, and No. 23840009) from The Ministry of Education, Culture, Sports, Science, and Technology (MEXT) of Japan, a Grant-in-Aid for the Global COE Program "The Next Generation of Physics, Spun from Universality and Emergence" from MEXT of Japan, a grant-in-aid for Scientific Research from Japan Society for Promotion of Science (JSPS), KAKENHI (S) (No. 20224015), KAKENHI (C) (No. 23540406) and FIRST program from JSPS .

References

[1] Saxena S S *et al.* 2000 *Nature* **406** 587
[2] Aoki D, Huxley A, Ressouche E, Braithwaite D, Flouquet J, Brison J-P, Lhotel E, and Paulsen C 2001 *Nature* **413** 613
[3] Huy N T *et al.* 2007 *Phys. Rev. Lett.* **99** 067006
[4] Gasparini A, Huang Y K, Huy N T, Klaasse J C P, Naka T, Slooten E and de Visser A 2010 *J. Low. Temp. Phys.* **161** 134-147
[5] Aoki D, Taupin M, Paulsen C, Hardy F, Taufour V, Kotegawa H, Hassinger E, Malone L, Matsuda T D, Miyake A, Sheikin I, Knafo W, Knebel G, Howald L, Brison J-P, and Flouquet J arXiv:1201.3030
[6] de Visser A, Huy N T, Gasparini A, de Nijs D E, Andreica D, Baines C, and Amato A 2009 *Phys. Rev. Lett.* **102** 167003
[7] Ohta T, Hattori T, Ishida K, Nakai Y, Osaki E, Deguchi K, Sato N K, and Satoh I 2010 *J. Phys. Soc. Jpn.* **79** 023707
[8] Hattori T, Ishida K, Nakai Y, Ishida K, Ohta T, Deguchi K, Sato N K, and Satoh I 2010 *Physica C* **470** S561
[9] Hattori T, Ihara Y, Nakai Y, Ishida K, Tada Y, Fujimoto S, Kawakami N, Osaki E, Deguchi K, Sato N K and Satoh I 2012 *Phys. Rev. Lett.* **108** 066403
[10] Huy N T, de Nijs D E, Huang Y K and de Visser A 2008 *Phys. Rev. Lett.* **100** 077002
[11] Slooten E, Naka T, Gasparini A, Huang Y K and de Visser A Huy N T, de Nijs D E, Huang Y K and de Visser A 2009 *Phys. Rev. Lett.* **103** 097003
[12] Aoki D, Matsuda T D, Taufour V, Hassinger E, Knebel G and Flouquet J 2009 *J. Phys. Soc. Jpn.* **78** 113709
[13] Ihara Y, Hattori T, Ishida K, Nakai Y, Osaki E, Deguchi K, Sato N K and Satoh I 2011 *Phys. Rev. Lett.* **105** 206403
[14] Hattori T, *et al. unpublished.*
[15] Moriya T 1991 *J. Magn. Magn. Matter* **100** 261
[16] Monthoux P and Lonzarich G G 1999 *Phys. Rev. B* **59** 14598
[17] Kirkpatrick T R, Belitz D, Vojta T and Narayanan R 2001 *Phys. Rev. Lett.* **87** 127003
[18] Wang Z, Mao W and Bedell K 2001 *Phys. Rev. Lett.* **87** 257001
[19] Roussev R and Millis A J 2001 *Phys. Rev. B* **63** 140504
[20] Chubukov A V, Finkel'stein A M, Haslinger R and Morr D K 2003 *Phys. Rev. Lett.* **90** 077002
[21] Fujimoto S 2004 *J. Phys. Soc. Jpn.* **73** 2061
[22] Mineev V P 2010 *Phys. Rev. B* **81** 180504(R)
[23] Stock C, Sokolov D A, Bourges P, Tobash P H, Gofryk K, Ronning F, Bauer E D, Rule K C and Huxley A D 2011 *Phys. Rev. Lett.* **107** 187202
[24] Mineev V P and Michal V P *arXiv*:1206.3468
[25] Samsel-Czekala M, Elgazzar S, Oppeneer P M, Tallik E, Walerczyk W and Troc R 2010 *J. Phys.:Condens. Matter* **22** 015503
[26] Belitz D, Kirkpatrick T R and Vojta T 1997 *Phys. Rev. B* **55** 9452; Belitz D, Kirkpatrick T R and Vojta T1999 *Phys. Rev. Lett.* **82** 4707; Belitz D, Kirkpatrick T R and Rollbühler J 2005 *Phys. Rev. Lett.* **94** 247205; Kirkpatrick T R and Belitz D 2012 *Phys. Rev. B* **85** 134451.
[27] Chubukov A V, Pépin C and Rech J 2004 *Phys. Rev. Lett.* **92** 147003; Efremov D V, Betouras J J and A. V. Chubukov 2008 *Phys. Rev. B* **77** 220401(R); Maslov D L, Chubukov A V and Saha R 2006 *Phys. Rev. B* **74** 220402(R).
[28] Duine R A and MacDonald A H 2005 *Phys. Rev. Lett.* **95** 230403
[29] Conduit G J, Green A G and Simons B D 2009 *Phys. Rev. Lett.* **103** 207201

[30] Tada Y, Kawakami N and Fujimoto S 2011 *J. Phys. Soc. Jpn. suppl.* **81** SA006
[31] Lévy F, Sheikin I, Grenier B and Huxley A D 2005 *Science* **309** 1343
[32] Tada Y, Kawakami N and Fujimoto S 2010 *Phys. Rev. B* **80** 104506
[33] Mineev V P 2002 *Phys. Rev. B* **66** 134504
[34] Mineev V P 2008 *J. Phys. Soc. Jpn.* **66** 103702; Mineev V P *arXiv*:0812.2171
[35] Aoki D and Flouquet J 2012 *J. Phys. Soc. Jpn.* **81** 011003

Journal of Physics: Conference Series **449** (2013) 012030 doi:10.1088/1742-6596/449/1/012030

NMR investigation of the pressure induced Mott transition to superconductivity in Cs_3C_{60} isomeric compounds

H Alloul[1], Y Ihara[1], T Mito[1], P Wzietek[1], M Aramini[2], D Pontiroli[2] and M Ricco[2]

[1] Laboratoire de Physique des Solides, CNRS UMR 8502, Université Paris-Sud 11 91405 Orsay (France)

[2]Dipartimento di Fisica, Università di Parma Via U.P. Usberti 7/a, 43100 Parma (Italy)

Email: alloul@lps.u-psud.fr

Abstract. The discovery in 1991 of high temperature superconductivity (SC) in A_3C_{60} compounds, where A is an alkali ion, has been initially ascribed to a BCS mechanism, with a weak incidence of electron correlations. However various experimental evidences taken for compounds with distinct alkali content established the interplay of strong correlations and Jahn Teller distortions of the C_{60} ball. The importance of electronic correlations even in A_3C_{60} has been highlighted by the recent discovery of two expanded fulleride Cs_3C_{60} isomeric phases that are Mott insulators at ambient pressure. Both phases undergo a pressure induced first order Mott transition to SC with a (p, T) phase diagram displaying a dome shaped SC, a common situation encountered nowadays in correlated electron systems. NMR experiments allowed us to establish that the bipartite A15 phase displays Néel order at 47K, while magnetic freezing only occurs at lower temperature in the fcc phase. NMR data do permit us to conclude that well above the critical pressure, the singlet superconductivity found for light alkalis is recovered. However deviations from BCS expectations linked with electronic correlations are found near the Mott transition. So, although SC involves an electron-phonon mechanism, correlations have a significant incidence on the electronic properties, as had been anticipated from DMFT calculations.

1. Introduction

In fullerides A_nC_{60}, the charge transferred from the alkali atoms (A) to the fullerene molecules is nearly complete, and is expected to participate in delocalized metallic states. However the weak transfer integrals between C_{60} molecules imply electronic structures with small bandwidths [1], so that electronic correlations should be important [2]. Furthermore adding electrons on the C_{60} balls cannot be done without taking into consideration the coupling of the molecular electronic states with the inner vibration modes of the molecule, which induce Jahn-Teller effects [3]. This yields the observation of a series of effects associated with molecular physics and strong correlations in the A_nC_{60} compounds.

However, from the normal state and superconducting properties of A_3C_{60} compounds one considered initially that those were weakly correlated metals displaying superconducting BCS behaviour [4]. The latter was still thought somewhat original as the pairing appears mediated by local on ball optical phonon modes.

While this convinced many researchers that electronic correlations were not important in the A_3C_{60}, further studies of various A_nC_{60} allowed to evidence insulating states with large optical gaps (~0.5eV) in the even electron compounds A_4C_{60} [5-6] and Na_2C_{60} [7] which cannot be explained by a simple progressive band filling of the lowest unoccupied C_{60} six-fold degenerate t_{1u} molecular level.

The absence of magnetism in the ground state and the small spin gap observed by NMR spin lattice relaxation data [7] could only be ascribed to the simultaneous influence of strong electron correlations and Jahn-Teller Distortions (JTD) of the C_{60} ball, which localizes the electrons on the balls in a low spin state [8-9].Those energetically favour evenly charged C_{60} molecules, hence their qualification as Mott-Jahn-Teller insulators [10].

As for the A_3C_{60}, recent efforts to study them in expanded states confirm that they are often magnetic [11]. Furthermore, as will be recalled here, the large ionic radius of Cs permits to stabilize two expanded Cs_3C_{60} phases which are magnetic and insulating at ambient pressure and exhibit a Mott transition to a metallic and SC state when the distance between C_{60} is decreased by an applied pressure[12-13]. Those results confirm anticipations from calculations of the ground state energies assuming strong correlations and Jahn -Teller effects [14].

We shall recall first in section 2 the usual attempts to explain SC in A_3C_{60} within the BCS scenario and shall discuss then in section 3 the expected incidence of correlations on this picture. In section 4 the NMR data on the two Mott phases of Cs_3C_{60} will be reported. The phase diagram and the electronic properties of the SC state will be described in section 5. The NMR data on the electronic properties of the metallic state near the Mott transition will be displayed in section 6, and will allow us to evidence the recovery of a metallic state upon thermal expansion of the lattice at high T.

2. BCS Superconductivity in A_3C_{60}?

Contrary to the evenly charged fullerides, the A_3C_{60} compounds are usually metallic and present a regular Drude peak in their optical conductivity [15]. Also the normal state electronic susceptibility, as measured by the electron spin resonance (ESR) intensity is practically T independent [16], that is Pauli like, as confirmed as well by NMR Knight shift and T_1T data [17]. Quantitatively the effective mass enhancement is $m^*/m_0 \sim 3$, so that one is tempted to conclude that these compounds are not far from regular Fermi liquids [4, 18].

In the superconducting state, a jump of the specific heat is detected and increases with T_c as expected for a BCS transition [19]. Similarly, the SC gap 2Δ scales with T_c, with a $2\Delta/k_BT_c$ ratio not far from 3.5, value expected in the theoretical BCS scenario [4]. NMR in the superconducting state shows up the decrease of Knight shift, that is the loss of spin susceptibility, expected for singlet pairing [20-21]. Finally Hebel-Slichter coherence peaks in both the NMR $(T_1)^{-1}$ [20-21] and the muon spin relaxation rate [22] have been observed below T_c in weak applied fields, which also points towards *conventional s-wave symmetry of the order parameter.*

But, among the experimental observations, the most emphasized one has been the variation of T_c with the chemical nature of the inserted alkali, which permits to change the inter-ball distance [4]. One could monitor, as shown in figure 1, a continuous decrease of T_c with lattice compression in fcc A_3C_{60}.

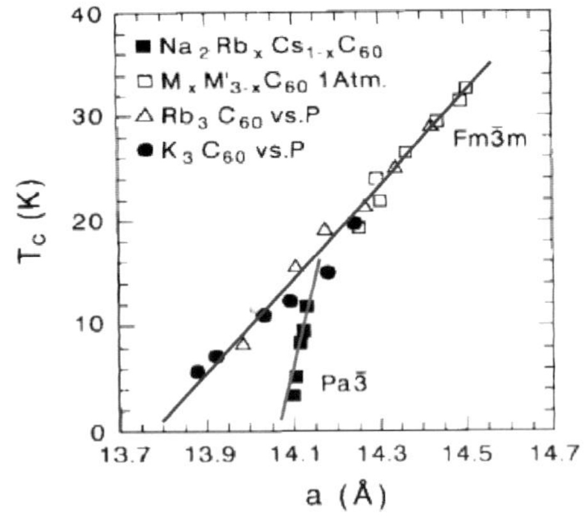

Fig. 1. T_c variation versus fcc lattice parameter obtained by chemical exchange of the alkali (M /M') or by applying a pressure which leads to a decrease of DOS and of T_c. The samples with cubic *Pa*-3 structure do not follow the same trend as those with *Fm*-3*m* structure (adapted from [23]).

There, the detected T_c variation is the same for chemical [24] and pressure changes [25] of the lattice parameter. Moreover, the absence of alkali mass isotope effect has been initially assigned to a BCS weak coupling relation for T_c

$$k_B\, T_c = \hbar\omega_D \exp\,[\,-1/VN(E_F)], \qquad (1)$$

in which the density of states $N(E_F)$ appears to be the single driving parameter. This has led to consider an on ball electron-phonon coupling V, the active phonons in ω_D being then the high energy optical phonons (Hg), which are responsible for the JT effects [4] [26-28].

However, it has been pointed out that, as displayed in figure 1, this relation does not extend to samples of Na_2AC_{60}, which order in the cubic *Pa*3 structure [29-30], though a similar law might hold with a larger slope [23]. This difference has raised questions about the possible incidence of the structure, of the nature of the alkali and of the merohedral disorder on the density of states [31] which has to be considered in equ.(1). Actually, it has been shown by ARPES that the arrangement of the molecules on deposited single layer fulleride films at the half filled composition does have a dominant influence on the electronic properties of the films [32].

3. Incidence of JTD and electronic correlations

Considering these results one is led to question whether the correlation effects and JTD which induce the Mott insulating state of the evenly charged compounds do play any role in the electronic properties of these A_3C_{60} compounds. We have seen that the Mott JT states for even n are already not so far from a metallic state. Does that mean that odd parity of n would indirectly favor hopping rather than localization?

To answer that question, one might consider the energy cost of transferring one electron on a nearby site, that is transforming $2C_{60}^{-n}$ into $C_{60}^{-(n+1)}$ and $C_{60}^{-(n-1)}$. If $E(n)$ is the molecular energy of a ball with n electrons, this energy cost is given by

$$U_{\text{eff}} = E(n+1) + E(n-1) - 2E(n). \qquad (2)$$

One has been led to conclude that the contribution of the JTD to U_{eff} will be positive for even n, and negative for odd n. The JTD stabilisation of evenly charged C_{60} therefore adds to the on ball coulomb repulsion for even n and reduces it for odd n [33-34]. This explains the occurrence of large correlation effects and of the JTD Mott insulating state in the evenly charged compounds. But conversely this can be taken as an argument for smaller overall electronic correlations in the A_3C_{60} compounds. However the magnitude of U_{eff}<0.5 eV, as compared to $U\sim$ 1 eV, would hardly justify a total loss of electronic correlations [4].

Various efforts have then been undertaken to estimate the incidence of electronic correlations on electron phonon superconductivity in the case of these A_3C_{60}. Usually, the Coulomb electronic interactions are taken into account within the McMillan extension of BCS through an empirical parameter μ^*

$$k_B\, T_c = \hbar\omega_D \exp\,[\,-1/(\lambda-\mu^*)], \qquad (3)$$

where $\lambda = VN(E_F)$. This equation applies when $\hbar\omega_D \ll E_F$ while here the phonons, the JTD and the Fermi energies have the same order of magnitude, which does not allow a standard use of this approach[1]. More recent calculations done within Dynamical Mean Field Theory (DMFT), suggest that in this specific case of Jahn-Teller on ball phonons, the electronic correlations would not be very detrimental to phonon driven superconductivity [35].

[1] Using the more accurate Migdal-Eliashberg equation does not change this conclusion, see [4].

In any case a different school of thought has been suggesting for long that electron-electron correlations might even drive superconductivity in such compounds [2-3]. It has even been proposed that A_nC_{60} should be Mott insulators and are only metallic due to a non stoichiometry of the alkalis [36]. Although our result for the charge segregation in the cubic quenched phase of CsC_{60} [37] could be supportive of this possibility, no experimental evidence along this line could be found so far in A_nC_{60}. It has rather been seen that T_c peaks at n=3 if one varies the alkali doping [38].

To better qualify the proximity to a Mott insulating state, attempts to produce expanded fulleride compounds have been done for long. A successful route has been to insert NH_3 neutral molecules to expand the A_3C_{60} lattice. It was found that $(NH_3)K_3C_{60}$ is insulating at ambient pressure [39], and becomes SC with T_c up to 28K under pressure [40]. Furthermore at 1bar it was shown to be an AF [11]. However this insertion of NH_3 induces an orthorhombic distortion of the lattice which is retained, though modified, under applied pressure [41]. So, although this system undergoes a Mott transition to a metallic state, this occurs in an electronic structure where the t_{1u} level degeneracy has been lifted by the specific spatial order of the C_{60} balls. Let us recall then that, for a single orbital case the Mott transition occurs for U/W=1, while for N degenerate orbitals the critical value U_c is expected to be larger [33], typically U_c=$N^{1/2}W$. Therefore U_c is reduced in the case of $(NH_3)K_3C_{60}$ with respect to that of A_3C_{60}, for which the t_{1u} orbital degeneracy is preserved. Though this complicates any direct comparison between these two cases, these results establish that A_3C_{60} are not far from a Mott transition, and hence that electronic correlations cannot be neglected.

4. Mott insulating state in Cs_3C_{60} isomeric compounds

A much simpler way to expand the C_{60} fcc lattice has been of course to attempt the insertion of Cs, the largest alkali ion, by the synthesis procedure used for the other A_3C_{60}. But, the bct Cs_4C_{60} and the polymer phase Cs_1C_{60} being stable at room T, the Cs_3C_{60} composition is found metastable. In a mixed composition sample it was found that a Cs_3C_{60} phase becomes SC at high pressure with $T_c \sim 40$ K [42]. A distinct chemical route finally recently permitted to produce samples containing the fcc-Cs_3C_{60}, mixed with an A15 isomeric phase. The latter has been found insulating at ambient pressure, and becomes SC under pressure, exhibiting a SC dome with a maximum T_c of 38K at about 10 kbar [12].

In similar mixed phase samples, NMR experiments allowed us to separate the specific ^{133}Cs spectra of the two isomers [13]. As displayed in figure 2, the A15 phase has a single Cs site with non cubic local symmetry, hence its seven lines quadrupole split spectrum (^{133}Cs has a nuclear spin I=7/2). The fcc-Cs_3C_{60} NMR spectra exhibit the usual features found for all other fcc-A_3C_{60}, displaying the signals of the orthorhombic (O) site and of two tetrahedral sites (T, T') [43]. Those are associated with the merohedral disorder [44] of the orientation of the C_{60} balls, which differentiates two local orientation orderings of the C_{60} around the alkali [45].

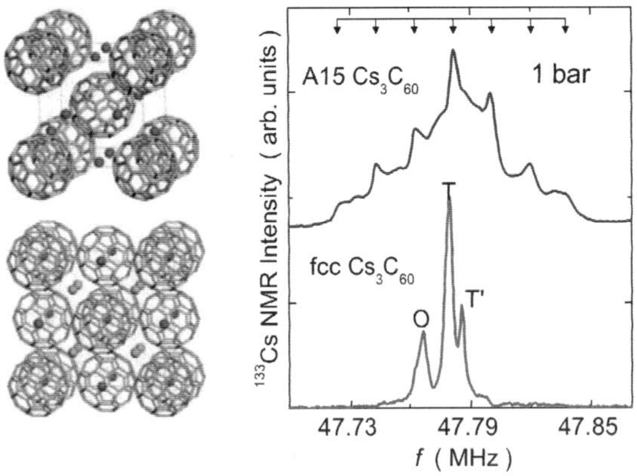

Fig. 2. The unit cells of the A15 and fcc isomeric phases of Cs_3C_{60} are displayed on the left [12]. Their ^{133}Cs NMR spectra [13] taken at 1bar and T=300 K are reported on the right. The specific features of the spectra are discussed in the text.

This possibility to select the signals of the two phases allowed us to demonstrate that both are in a paramagnetic insulating state at p=1 bar, and display a Mott transition at distinct applied pressures[13]. We could further study their magnetic behaviour at ambient pressure using data on the ^{13}C nuclear spins, which are directly coupled to the electron spins of the t_{1u} molecular orbitals. The coupling being dipolar [21], it induces a characteristic anisotropic broadening of the ^{13}C NMR which allowed us to monitor the C_{60}^{3-} magnetism.

As can be seen in figure 3 this anisotropic shift K_{ax} evidences a Curie–Weiss paramagnetism of the C_{60}^{3-}, which does not differ for the two isomers above 100K [13]. These data are compatible with a low spin S=1/2 effective moment as expected from figure 3 for a JT stabilized state. Furthermore the structural difference between the two isomers results in distinct magnetic ground states at p=1 bar. The bipartite A15 structure displays a Néel order below T_N=47 K [12,13,46], while the frustration of the fcc lattice results in a spin freezing only below 10 K for the standard fcc- Cs_3C_{60} [13,47].

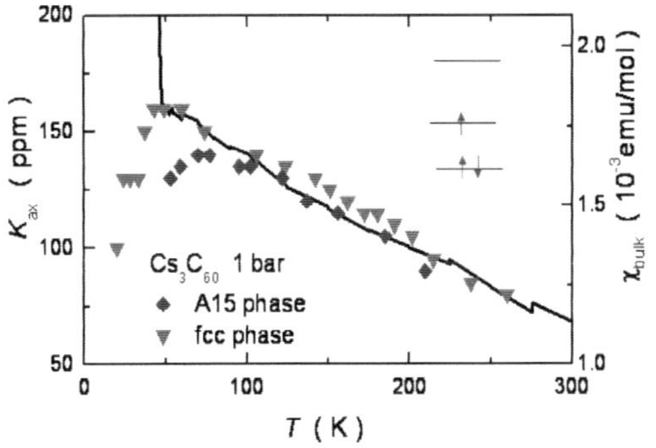

Fig. 3. Paramagnetic behaviour detected by SQUID data (full line, right scale) and from the anisotropic shift K_{ax} of the ^{13}C NMR in the two isomeric Cs_3C_{60} phases [13]. No significant difference can be detected above 100 K and the effective moment corresponds to the low spin JTD state sketched in insert. The magnetization increase below 47 K is that of the A15 phase Néel state.

5. SC state induced under applied pressure and low *T* phase diagram

Under an applied pressure, macroscopic diamagnetism is found to appear abruptly in both phases though at a slightly lower critical pressure p_c for the fcc phase than for the A15. From NMR data in the A15 phase we could evidence that this appearance of SC is concomitant with an abrupt loss of magnetism, which definitely points toward a first order transition. The phase diagrams slightly differ for the two Cs_3C_{60} isomers, but as shown in figure 4, they roughly merge together [13,47] if plotted versus V_{C60}, the unit volume per C_{60} ball. There, the T_c data can be scaled as well with that for the other known fcc-A_3C_{60}, and it has been evidenced that a similar maximum of T_c versus V_{C60} applies for the two structures [13,47,48].

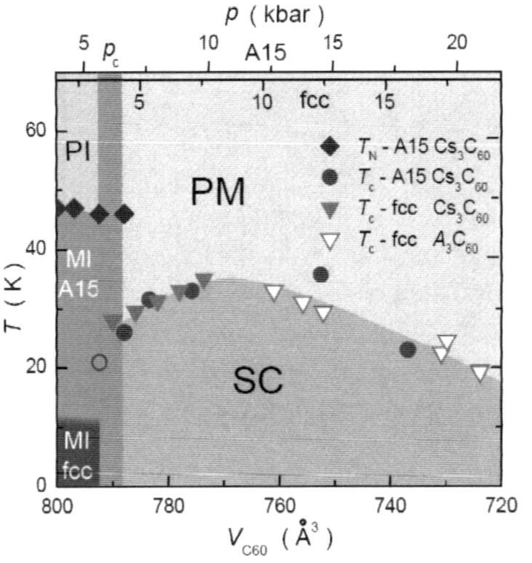

Fig. 4. Phase diagram representing T_c versus the volume V_{C60} per C_{60} ball, and the Mott transition at p_c (hatched bar) for the two isomer phases [47]. The corresponding pressure scales are shown on the upper scale. Data for the other known fcc-A_3C_{60} are reported as empty symbols. The transition to a magnetic insulating (MI) phase occurs only below 10 K in fcc-Cs_3C_{60}.

Journal of Physics: Conference Series **449** (2013) 012030 doi:10.1088/1742-6596/449/1/012030

We could probe the electronic excitations by taking ^{133}Cs NMR spin lattice relaxation data. Well above p_c, drops of $(T_1T)^{-1}$, which evidence the opening of the superconducting gap below T_c, were detected for the two phases [13]. So the electronic properties are found to behave as for the non expanded fcc-phases, with a singlet state s-wave SC state. This ensemble of results reveal that the SC state does not depend significantly on the actual structure, except through the modification of the inter-ball distance and transfer integrals between neighbouring C_{60}.

As seen in figure 5, we could evidence in the A15 phase that such a drop of $(T_1T)^{-1}$ already appears just above p_c, together with a diamagnetic shift of the NMR signal [49] which confirms that *bulk SC* occurs immediately after the first order Mott transition.

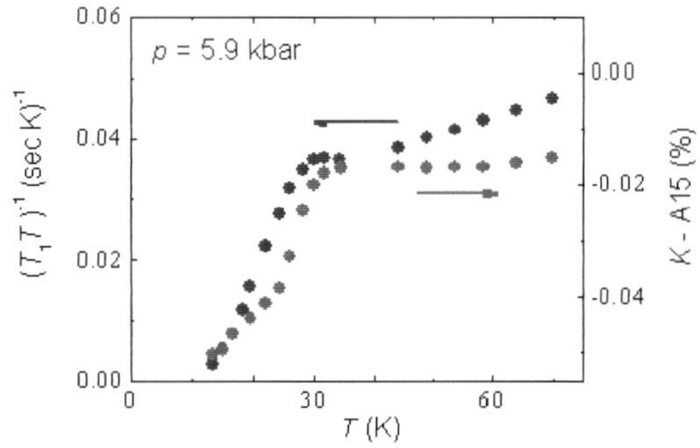

Fig. 5. ^{133}Cs NMR $(T_1T)^{-1}$ (left scale) and shift data (right scale) taken at a pressure just above p_c ~ 5.1 kbar for a nearly pure A15 sample. Bulk SC is evidenced to occur below about 30K at this pressure.

Above the magnetic and SC ground states, the x-ray and NMR spectra did not display any significant change with increasing pressure through the paramagnetic insulator to metal transition, for both isomers. This absence of significant structural modifications at p_c establishes then that the evolution with p is dominated by electronic degrees of freedom.

6. Normal state electronic properties and high *T* insulating properties

A great advantage of NMR $(T_1T)^{-1}$ data is that it permits us to probe the evolution of the electronic properties from the paramagnetic insulator to the paramagnetic metallic state. This is illustrated in figure 6 by a series of measurements performed in the fcc phase on the ^{133}Cs NMR of the T' site [47].

One can see that, in the Mott state, below the pressure p_c of the Mott transition, $(T_1T)^{-1}$ increases with decreasing T as expected for a dense paramagnet, this being true for both phases. In fact in this regime $(T_1)^{-1}$ is nearly T independent well above the magnetically ordered state, its value being mainly determined by the exchange coupling J between the moments localized on the C_{60} balls [13,46,47].

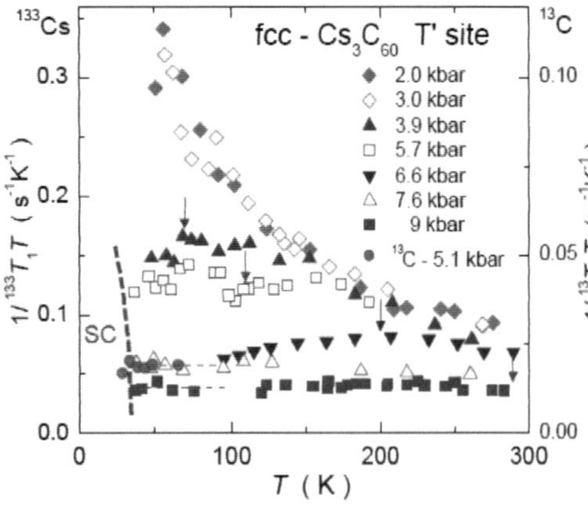

Fig. 6. ^{133}Cs $(T_1T)^{-1}$ data taken at increasing pressures on the T' site of fcc-Cs_3C_{60}. The low T Mott transition occurs at p_c~3.9 kbar. Notice the sharp decrease of $(T_1T)^{-1}$ at the MIT, and then its regular decrease with increasing pressure in the metallic state. The recovery of an insulating state upon expansion of the lattice at high T can be seen from the progressive evolution of $(T_1T)^{-1}$ toward that of the low pressure paramagnetic insulator. From [47]

At high pressures above 7 kbar a nearly T independent $(T_1T)^{-1}$ is recovered, as indeed seen in most other non expanded fcc phases, and as expected for the Korringa nuclear relaxation of a metallic state

$$(T_1T)^{-1} = (4\pi k_B / \hbar)\ A_{hf}^2\ N(E_F)^2, \qquad (4)$$

where A_{hf} is the hyperfine coupling of the ^{133}Cs nuclear spin with the electronic density of state $N(E_F)$ taken at the Fermi level.

However for intermediate pressures the variation of $(T_1T)^{-1}$ is quite unexpected as it goes through a maximum, pointed out by arrows in figure 6. One can notice that it shifts towards higher T at increasing pressures, while $(T_1T)^{-1}$ progressively approaches the value found in the insulating state.

Let us consider first the behaviour found just above the SC state at about 50 K. We can see in figure 6 the regular increase of $(T_1T)^{-1}$ with decreasing pressure, which would suggest from equ (4) a regular increase of the density of state $N(E_F)$ until the Mott transition at p_c. But such an increase of $N(E_F)$, would correspond in the BCS formalism of Equ(1) to an increase of T_c, while in this range of pressures, T_c decreases when p_c is approached from above (this corresponds to the decreasing part on the left of the dome in figure 4). Consequently equations (4) or (1), or both, do not apply [50]. In other words the magnetic excitations detected above the Mott transition have some specific incidence on the SC condensation energy.

The second important observation illustrated in figure 6 is the recovery of the magnetic excitations of the paramagnetic insulating state which occurs at high temperature for any fixed pressure. The fulleride compounds being highly compressible, they also of course expand with increasing T, so that the Mott transition is crossed as well at high T and the insulating state is recovered in this expanded state. In other words the critical pressure p_c increases with increasing T. Let us point out that such a transition toward an insulating state had been suggested about a decade ago to explain the constant value of $(T_1)^{-1}$ found above room T in the cubic Na_2CsC_{60} compounds [51]. Finally, as discussed in [47] the transition line $p_c(T)$ seems to reach around room T a critical point similar to that found for the liquid gas transition. At higher temperatures a continuous evolution from the paramagnetic metallic state to the paramagnetic insulating state should occur. This behaviour appears quite analogous to that observed for the 2D Mott transition in organic compounds [52].

7. Discussion

We have recalled here that electronic correlations and on ball Jahn-Teller effects have significant incidence on the physical properties of A_nC_{60} compounds. Those help to establish a robust insulating state for evenly doped compounds. While electronic correlations were not expected to be dominant in the n=3 cubic compounds, we have definitely shown here that those remain quite close to a Mott transition.

The expanded Cs_3C_{60} compounds are unique inasmuch as they are Mott insulators at p=1 bar and can be driven through the Mott transition by an applied pressure. The Mott state is found to correspond to low spin as could be expected for a Jahn-Teller distorted ground state. This has been confirmed recently by infrared experiments [53] which do evidence a dynamic loss of symmetry of the C_{60} balls. The phase diagram displays a SC dome, reminiscent of that found in many other correlated electron systems, with a T_c maximum at a pressure somewhat above that of the first order Mott insulator state boundary.

In the SC state, the NMR experiments have established that the pure BCS equation or its Migdal-Eliashberg extension do not yield even a qualitative explanation of the deviations found near the Mott transition, where T_c decreases with decreasing pressure. There the normal state electronic properties deviate from Fermi liquid behaviour as spin fluctuations in the metallic state become prominent before the system switches into the Mott state [47]. Also, the orientational disorder of the C_{60} in fcc-Cs_3C_{60} does not appear detrimental to the s-wave SC, even near the Mott transition, as the T_c dome does not differ from that of the A15 ordered isomer. This opposes the case of cuprates for which both normal state and d-wave SC are very sensitive to disorder [54-55].

On the theory side, serious efforts to include JT effects and correlations using DMFT have allowed to evidence that the Coulomb repulsion is at least not that detrimental to this on ball pairing [35]. It is important to recall here that ground state energy calculations done with DMFT have even allowed the anticipation [56] of the phase diagram with a SC dome near the Mott transition [57]. This is a strong indication that detailed theoretical understanding of the properties of this series of materials is within reach. Further developments should then help to decide whether the large T_c values in A_3C_{60} compounds do result from a fundamental cooperation between correlations and electron-phonon interactions, and to determine above which pressure or bandwidth the BCS limit would be recovered.

Finally, we propose here that Cs_3C_{60} is a rather good model system to study a 3D Mott transition. Indeed the latter is not controlled by doping, but is only driven by the inter-ball distance, that is the t_{1u} bandwidth. In the *(p,T)* range probed in [47] the qualitative behaviour detected appears quite similar to that expected for a single orbital Mott transition [58] although the S=1/2 ground sate has orbital degeneracy. Extensive measurements of other spectral and thermodynamic properties, possibly on larger T ranges, are required to complete the experimental insight on this Mott transition.

Acknowledgments
The authors acknowledges particularly V. Brouet who has been the major actor in the work on the non SC compounds and shows a continuing interest on the matter. We thank as well L. Forró and his group, with whom collaborations have been extending over many years.

We also warmly thank E. Tosatti, M. Fabrizio and M Capone with whom we have had over the years many fruitful discussions about the theoretical aspects of the electronic structure of the fulleride compounds.

References

[1] Erwin SC and Pickett WE 1991 *Science* **254** 842
[2] Chakraverty S, Gelfand MP and Kivelson S 1991 *Science* **254** 970
[3] Auerbach A, Manini N and Tosatti E 1994 *Phys. Rev. B* **49** 12998
[4] For a review on SC properties of A_3C_{60} : Gunnarsson O 1997 *Rev. of Modern Physics* **69**, 575
[5] Iwasa Y and Kaneyasu T 1995 *Phys. Rev. B* **51** 3678
[6] Knupfer M and Fink J 1997 *Phys. Rev. Lett.* **79** 2714
[7] Brouet V, Alloul H, Thien-Nga L, Garaj S and Forró L 2001 *Phys. Rev. Lett.* **86**, 4680
[8] For a review on Jahn-Teller effects in fullerenes: Brouet V, Alloul H, Garaj S. and Forró L 2004 in "Fullerene-based materials", Structure and Bonding series, series editor Mingos DMP, volume editor Prassides K, *Springer-Verlag (Berlin- Heidelberg),* 125
[9] Alloul H 2012 *EPJ Web of conferences* **23** 0015
[10] Fabrizio M and Tosatti E 1997 *Phys. Rev. B* **55** 13 465
[11] Prassides K, Margadonna S, Arcon D, Lappas A, Shimoda H and Iwasa Y 1999 *J. Am. Chem. Soc.* **121** 11227
[12] Ganin AY, Takabayashi Y, Khimyak YZ, Margadonna S, Tamai A, Rosseinsky MJ and Prassides K 2008 *Nature Mater.* **7** 367 ; Takabayashi Y, Ganin AY, Jeglič P, Arčon D, Takano T, Iwasa Y, Ohishi Y, Takata M, Takeshita N, Prassides K and Rosseinsky MJ 2009 *Science* **323** 1585
[13] Ihara Y, Alloul H, Wzietek P, Pontiroli D, Mazzani M and Riccò M 2010 *Phys. Rev. Lett.* **104** 256402
[14] Capone M, Fabrizio M, Castellani C, and Tosatti E 2002 *Science* 296 2364
[15] Degiorgi L, Nicol EJ, Klein O, Grüner G, Wachter P, Huang SM, Wiley J and Kaner R.B 1994 *Phys. Rev. B* **49** 7012
[16] Robert J, Petit P, Yldirim T and Fischer JE 1998 *Phys. Rev. B* **57** 1226
[17] Tycko R, Dabbagh G, Rosseinsky MJ, Murphy DW, Ramirez AP and Fleming RM 1992 *Phys. Rev. Lett.* **68** 1912
[18] Ramirez AP, Rosseinsky MJ, Murphy DW and Haddon RC 1992 *Phys. Rev. Lett.* **69** 1687
[19] Burkhart G.J and Meingast C. 1996 *Phys. Rev. B* **54**, R6865
[20] Stenger VA, Pennington CH, Buffinger DR and Ziebarth RP 1995 *Phys. Rev. Lett.* **74** 1649
[21] Pennington CH and Stenger V A 1996 *Rev. Mod. Phys.* **68** 855
[22] Kiefl RF, MacFarlane WA, Chow KH, Dunsiger S, Duty TL, Johnston TMS, Schneider JW, Sonier J, Brard L, Strongin RM, Fischer JE, and Smith AB III 1993 *Phys. Rev. Lett.* **70** 3987

[23] Yildirim T, Fischer JE, Dinnebier R, Stephens PW and Lin CL 1995 *Solid State Commun.* **93** 269
[24] Fleming RM, Ramirez AP, Rosseinsky MJ, Murphy DW, Haddon RC, Zakurak SM and Makhija AV 1991 Nature **352** 787
[25] Zhou O, Vaughan GBM, Zhu Q, Fischer JE, Heiney PA, Coustel N, McCauley JP Jr and Smith AB III 1992 Science **255** 833
[26] Varma CM, Zaanen J and Raghavachari K 1991 *Science* **254** 989
[27] Schluter M, Lannoo M, Needels M, Baraff GA and Tomanek D 1992 *Phys. Rev. Lett.* **68** 526
[28] Mazin I I, Rashkeev SN, Antropov VP, Jepsen O, Liechtenstein AI and Andersen OK 1992 *Phys. Rev. B* **45** 5114
[29] Kniaz K, Fischer JE, Zhu Q, Rosseinsky MJ, Zhou O and Murphy DW 1993 *Solid State Commun.* **88** 47
[30] Prassides K, Christides C., Thomas IM, Mizuki J, Tanigaki K, Hirosawa I and Ebbesen TW 1994 *Science* **263** 950
[31] Gelfand M.P and Lu J P 1992 *Phys. Rev. Lett.* **68** 1050; *Phys. Rev. B* **46** 4367
[32] Brouet V, Yang WL, Zhou XJ, Choi HJ, Louie SG, Cohen ML, Goldoni A, Parmigiani F, Hussain Z and Shen ZX 2004 *Phys. Rev. Lett.* **93** 197601
[33] Gunnarsson O, Koch E and Martin RM 1996 *Phys. Rev. B* **54** R11026
[34] Victoroff W and Heritier M 1996 *J. Phys. I (France)* **6** 2175
[35] Han JE, Gunarsson O and Crespi VH 2003 *Phys. Rev. Lett.* **90** 167006
[36] Lof RW, van Veenendaal MA, Koopmans B, Jonkman HT and Sawatzky GA 1992, *Phys. Rev. Lett.* **68** 3924
[37] Brouet V, Alloul H, Quéré F, Baumgartner G, and Forró L 1999 *Phys. Rev. Lett.* **82** 2131
[38] Yildirim T, Barbedette L, Fischer JE, Lin CL, Robert J, Petit P and Palstra TTM 1996 *Phys. Rev. Lett.* **77** 167
[39] Rosseinsky MJ, Murphy DW, Fleming RM and Zhou O 1993 *Nature* **364** 425
[40] Zhou O, Palstra TTM, Iwasa Y, Fleming RM, Hebard AF, Sulewski PE, Murphy DW and Zegarski BR 1995 *Phys. Rev. B* **52** 483
[41] Margadonna S, Prassides K, Shimoda H, Iwasa Y and Mezouar M. 2001 *EPL,* **56** 61
[42] Palstra TTM, Zhou O, Iwasa Y,. Sulewski PE, Fleming RM and Zegarski BR 1195 *Solid State Commun.* **93** 327
[43] Walstedt RE, Murphy DW and Rosseinsky M 1993 *Nature (London)* **362** 611
[44] Stephens PW, Mihály L, Lee P, Whetten R, Huang SM, Kaner R, Diederich F and Holczer K 1991 *Nature (London)* **351** 632;
[45] Matus P, Alloul H, Kriza G, Brouet V, Singer PM, Garaj S and Forró L 2006 *Phys. Rev. B* **74** 214509
[46] Jeglič P, Arčon D, Potočnik A, Ganin AY, Takabayashi Y, Rosseinsky MJ and Prassides K 2009 *Phys. Rev.* B **80** 195424
[47] Ihara Y, Alloul H, Wzietek P, Pontiroli D, Mazzani M and Riccò M 2011 *EPL* **94** 37007
[48] Ganin AY, Takabayashi Y, Jeglič P, Arčon D, Potočnik A, Baker PJ, Ohishi Y, McDonald MT, Tzirakis M, McLennan A, Darling GR, Takata M, Rosseinsky MJ and Prassides K 2010 *Nature* **466** 221
[49] T. Mito *et al* (in preparation)
[50] Similar discrepancies have been found in $(NH_3)_x$ NaK_2C_{60} see Riccò M, Shiroka T, Sartori A, Bolzoni F and Tomaselli M 2001 *EPL* **53**, 762
[51] Brouet V, Alloul H, Garaj S and Forro L 2002 *Phys. Rev .B* **66** 155124
[52] Kagawa F, Miyagawa K and Kanoda K, 2009, *Nature Physics* **5** 880
[53] Klupp G, Matus P, Kamarás K, Ganin AY, McLennan A, Rosseinsky MJ, Takabayashi Y, McDonald MT and Prassides K 2012, *Nature Communications* **3** 912
[54] Rullier-Albenque F, Alloul H, Balakirev F and Proust C 2008 *EPL* **81** 37008
[55] Alloul H, Bobroff J, Gabay M and Hirschfeld P 2009 *Rev. Mod. Phys.* **81** 45
[56] Capone M, Fabrizio M, Castellani C and Tosatti E 2009 *Rev. Modern Physics* **81** 943
[57] See also Capone M, this conference.
[58] Georges A, Kotliar G, Krauth W and Rozenberg MJ 1996 *Rev. Mod. Phys.* **68** 13

Theory of 'hidden' quasi-1D superconductivity in Sr_2RuO_4

S Raghu, Suk Bum Chung and Samuel Lederer
Department of Physics, Stanford University, Stanford, California 94305, USA
E-mail: sraghu@stanford.edu

Abstract. Is the mechanism of unconventional superconductivity in Sr_2RuO_4 closer in spirit to superfluid ^{3}He, or to the cuprates, pnictides, and organic superconductors? We challenge prevailing assumptions in this field and using well-controlled perturbative renormalization group calculations, we suggest that superconductivity in Sr_2RuO_4 resembles more closely the quasi-one dimensional organic superconductors. Our theory has certain phenomenological consequences that are consistent with the experimentally observed phenomena.

1. Introduction

Strontium Ruthenate, Sr_2RuO_4, is in many ways an archetypal unconventional superconductor. We say this for several reasons. Firstly, Knight shift measurements[1] convincingly show that the ground state is a spin-triplet (i.e. odd-parity[2, 3]) superconductor. In a metal with time-reversal and inversion symmetry, such states cannot arise from the electron-phonon interactions and must occur as a consequence of the bare *repulsive* electron-electron interactions. Secondly, its layered perovskite structure can be synthesized to a high degree of perfection, permitting us to neglect the effects of disorder. Thirdly, its electronic structure is relatively simple compared to related unconventional superconductors such as heavy fermion systems. Most importantly, its normal state is a well-behaved Fermi liquid, as is known from quantum oscillations[4, 5], heat capacity[6], and photoemission experiments[7](For a comprehensive review, see Ref. [8]). Therefore, it should be possible to construct a microscopic theory of unconventional superconductivity that is based on well-controlled effective field theoretic descriptions of a Landau Fermi liquid theory and its instabilities. Such a theory would be an anchor point for understanding other unconventional superconductors such as the organic, and perhaps even some aspects of the Iron pnictide and cuprate superconductors.

This conference proceeding is based on the following works[9, 10, 11]. However, in addition to elaborating and clarifying some aspects of this work, we will introduce some new points of perspective that we hope may be of some use to those working in this fascinating field.

2. Electronic structure considerations

The quasiparticle bands at the Fermi energy of this system are derived primarily from the Ruthenium d_{xz} d_{yz}, d_{xy} orbitals. The electrons in these orbitals tunnel to neighboring sites by making use of intervening oxygen atoms; however, for a low energy description that focuses on states near the Fermi level and preserves all the symmetries of the system, it is sufficient to consider solely the d-orbitals. We discuss the electronic structure at a qualitative level in

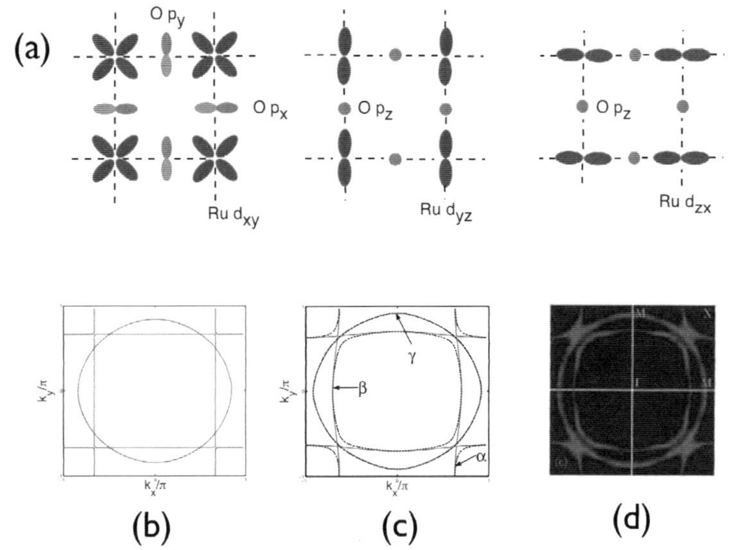

Figure 1. Electronic structure of Sr_2RuO_4. The Fermi surfaces are derived from the Ru t_{2g} electrons. Electrons in the d_{xy} orbital tunnel equally well in both directions, and their dispersion produces the circular Fermi surface shown in the photoemission data (d). By contrast, electrons in $d_{xz}(d_{yz})$ orbitals tunnel mainly along the x(y) direction. Their quasi-one dimensional character produce the straight portions of the Fermi surface. When orbital mixings are excluded, the Fermi surface of a simple tight-binding model has the form in (b); including the lowest order inter-orbital mixings produce the Fermi surface in (c), which is an excellent approximation to the experimentally observed Fermi surface (d).

this section; further details will be found in Sec. 4. Due to the tetragonal symmetry of the system, the d_{xz}, d_{yz} orbitals remain degenerate and transform as a doublet under point group operations whereas the d_{xy} orbital is non-degenerate. The electron kinetic energies are strongly dependent on the orbital character of the wave functions (Fig. 1). Electrons in a d_{xy} orbital tunnel with equal amplitude along the x,y nearest neighbor displacements. This gives rise to a fairly isotropic Fermi surface, shown in Fig. 1. By contrast, electrons in the d_{xz} d_{yz} orbitals have a quasi-one dimensional character: an electron in a $d_{xz}(d_{yx})$ orbital has a substantially larger nearest-neighbor tunneling amplitude along the $\hat{x}(\hat{y})$ direction. This produces the highly anisotropic Fermi sheets in Fig. 1. In the artificial limit where there is no mixing among the different orbitals, the Fermi surfaces are shown in Fig. 1b. However, when spin-orbit coupling and longer range hoppings between the d_{xz} , d_{yz} orbitals are included, the Fermi surfaces take the form shown in Fig. 1c, which is an excellent approximation to the experimentally observed Fermi surface, reproduced from Ref.[7, 12] in Fig. 1d. It is important to stress that the bandstructure here incorporates renormalization effects due to Fermi liquid parameters, such as mass enhancements, etc. Below, when we present a weak-coupling theory, it is to be understood that while the bare electron-electron interactions are always large, the residual interactions between dressed quasiparticles are assumed to be weak compared to their kinetic energy.

In the experimentally observed Fermi surface, there are three distinct sheets, denoted α, β, γ.

The γ sheet represents the circular Fermi surface and consists primarily of electrons in the d_{xy} orbital. The $\{\alpha, \beta\}$ sheets are quasi-one dimensional and are primarily built from d_{xz}, d_{yz} orbitals. Due to the tetragonal symmetry and the distinct orbital character of the three Fermi surfaces, the superconducting pairing is believed to be predominantly derived from either the γ or the $\{\alpha, \beta\}$ bands; a superconducting state that develops equally in all 3 sheets requires some fine-tuning and is therefore non-generic[13]. One of the open problems in the microscopic theory is to identify the so-called "active" bands. Below T_c, superconductivity will be induced in the remaining "passive" degrees of freedom via a proximity effect. Thus, while at zero temperature, the system behaves as a p-wave superconductor with gaps on all 3 Fermi sheets, there are a range of circumstances in which the system behaves as though only the active bands exhibit appreciable pairing.

3. Orbital dependent superconductivity

The prevailing assumption in the field has been that γ is the active band. In addition to making our lives easier, there are good reasons for supposing this to be the case. Firstly, from quantum oscillation studies, it is known that the γ band has the largest mass enhancement relative to the values based on first principles calculations. Secondly, it is close to a van Hove singularity, which makes it natural to believe that ferromagnetic fluctuations could play a key role in mediating spin-triplet pairing among these electrons. Such a scenario would be reminiscent of the physics of ^{3}He, where ferromagnetic fluctuations (paramagnons) are thought to play a key role. However, as can be seen from the experimental Fermi surface in Fig. 1(d), γ is nearly perfectly circular; this makes the particle-hole susceptibility nearly a constant, as we discuss below, and paramagnon mediated spin-triplet pairing is much weaker in such a system.

More importantly, inelastic neutron scattering studies have revealed that the in-plane magnetic excitations are predominantly antiferromagnetic in character[14]; these are derived from the reasonably well-nested $\{\alpha, \beta\}$ Fermi surfaces. If these degrees of freedom were the active bands, then spin-triplet superconductivity would arise mainly from incommensurate particle-hole fluctuations, and would be closer to what is found in the quasi-one dimensional organic superconductors. This dichotomy, namely whether the microscopic pairing mechanism in Sr_2RuO_4 is closer to ^{3}He or to the organics, cuprates and pnictides, was emphasized in a recent review by D. Scalapino[15]. Despite the observation of incommensurate antiferromagnetic fluctuations and the absence of clear ferromagnetic correlations, the idea that γ is the active band, and that superconductivity in this material is reminiscent of superfluid ^{3}He, has persisted.

The main experimental evidence that is cited in favor of γ as the active band is the heat capacity measurements of Ref.[6]. In this beautiful work, the specific heat was obtained in ultra-pure samples both as a function of temperature and pressure. Since the specific heat coefficient is directly proportional to the total density of states at the Fermi level, $C/T \sim k_B^2 \rho(E_F)$, the authors, using knowledge from dHvA oscillations, attempted to identify the degrees of freedom that become gapped below the superconducting transition. From the dHvA measurements, it is known that 57% of $\rho(E_F)$ is derived from the γ sheet and $\{\alpha, \beta\}$ contribute to the remaining 43%[4]. Indeed, the heat capacity data as a function of field shows features suggestive of a two-gap superconductor (Fig. 2a). There is a sharp anomaly near the upper critical field, below which there is a "shoulder" at lower fields, followed at even lower fields by another feature. The authors of ref. [6] claimed that the shoulder region matches well with the value of 43% of the normal-state density of states, from which they concluded that the $\{\alpha, \beta\}$ bands are the passive degrees of freedom.

However, there are several reasons to challenge this interpretation. Firstly, there is an underlying assumption that the pairing gap is fully isotropic on both the active and the passive bands so a sharp separation may be invoked in identifying the source of the shoulder in the heat

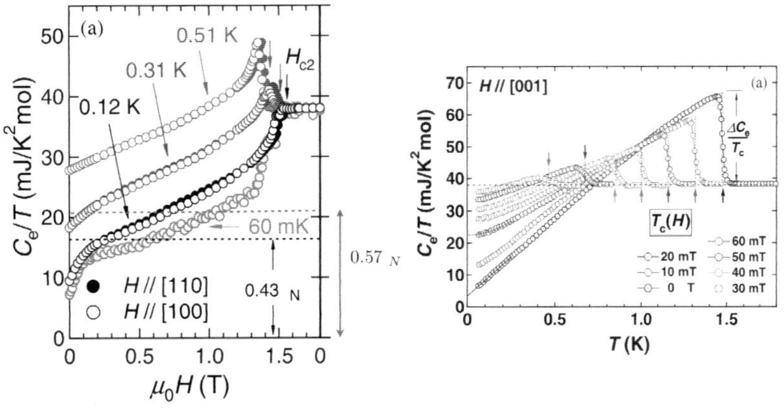

(a) (b)

Figure 2. Data from [6]. (a) Heat capacity in an in-plane magnetic field. The black dashed line corresponds to 43% of the normal state density of states and falls in the "shoulder" region. We have added the red dashed line, which corresponds to 57% of the normal state density of states. This line equally well falls into the shoulder region. Both these lines are estimates of the remaining density of states of the passive bands assuming that a full gap opens on the active band(s). The data is consistent with either scenario for the active bands. (b) Heat capacity as a function of temperature. Note that the "two gap" feature is absent at zero field.

capacity. Secondly, there is a disagreement of approximately 20% between the heat-capacity measurements and the dHvA measurements for the total normal state density of states. Thirdly, the "shoulder" region is sufficiently wide that even the opposite scenario in which the $\{\alpha, \beta\}$ bands are the active ones, could account for it. We have added a dashed red line in Fig. (2a) that would correspond to the case in which 43% of the normal-state specific heat derives from the active bands, which also lies in the shoulder region of the data. Lastly, the two-gap features are not seen in the temperature dependence of the heat-capacity in the absence of a field, which is perplexing. The data is undeniably important and possesses many clues as to the nature of the microscopic pairing structure. However taken at face value, it does not seem to rule out either scenario for the active degrees of freedom in this system.

4. Weak-Coupling theory

Having shown that the heat capacity measurements do *not* unambiguously favor either scenario for the active band(s), we now turn to the main topic and present a microscopic theory of superconductivity. We show that in the weak-coupling limit, there is a very sharp and inescapable conclusion: the active bands are $\{\alpha, \beta\}$, and triplet pairing in this system is reminiscent more of the quasi-one dimensional organics than of superfluid ^{3}He. Using a simple extension of a recently developed asymptotically exact weak-coupling analysis of the Hubbard model[9], we show that the dominant superconducting instability is in the triplet channel and occurs among the quasi-1D Fermi surfaces of Sr_2RuO_4.

We consider the dynamics of electrons in a tetragonal lattice of Ru t_{2g} orbitals governed by a Hamiltonian of the form

$$H = H_0 + U\sum_{i\alpha} n_{i\alpha\uparrow}n_{i\alpha\downarrow} + \frac{V}{2}\sum_{i,\alpha\neq\beta} n_{i\alpha}n_{i\beta} + \delta H \tag{1}$$

Here, $n_{i\alpha\sigma}$ is the density of electrons having spin σ at position i in orbital $\alpha = xy, yz, zx$, and $n_{i\alpha} = \sum_\sigma n_{i\alpha\sigma}$. The strength of the repulsive interaction between two electrons on like (distinct) orbitals at the same lattice site is given by U(V). $H_0 = \sum_\alpha \sum_{\vec{k}\sigma} \left(\epsilon^0_{\alpha\vec{k}} - \mu\right) c^\dagger_{\vec{k}\alpha\sigma} c_{\vec{k}\alpha\sigma}$ is the dominant intra-orbital kinetic energy and gives rise to three decoupled energy bands at the Fermi level as shown in Fig. 1. Here, we make use of the following tight-binding parametrization of these energies[16, 17]:

$$\begin{aligned} \epsilon^0_{xz}(\vec{k}) &= -2t\cos k_x - 2t^\perp \cos k_y \\ \epsilon^0_{yz}(\vec{k}) &= -2t^\perp \cos k_x - 2t\cos k_y \\ \epsilon^0_{xy}(\vec{k}) &= -2t'\left(\cos k_x + \cos k_y\right) - 4t''\cos k_x \cos k_y \end{aligned} \tag{2}$$

with $(t, t^\perp, t', t'', \mu) = (1.0, 0.1, 0.8, 0.3, 1.0)$. The quantity δH represents smaller terms such as longer range hopping and spin orbit coupling which mix the distinct orbitals. It plays a relatively minor role in determining the superconducting transition temperature, and can therefore be considered perturbatively. However, it plays a crucial role in selecting a superconducting state which breaks time-reversal symmetry, as we discuss below. When $\delta H = 0$, the non-interacting charge and spin susceptibilities of the normal state are separate functions for each orbital:

$$\chi_\alpha(\vec{q}) = -\int \frac{d^2k}{(2\pi)^2} \frac{f(\epsilon_{\alpha,\vec{k}+\vec{q}}) - f(\epsilon_{\alpha,\vec{k}})}{\epsilon_{\alpha,\vec{k}+\vec{q}} - \epsilon_{\alpha,\vec{k}}} \tag{3}$$

where $f(\epsilon)$ is the Fermi distrubition function, and the total susceptibility is obtained by summing the contributions from all three orbitals. Since the quasi-two dimensional band is almost circular with a radius k_f^{2d}, its susceptibility is nearly constant $\chi_{xy} \sim 1/4\pi t'$ for $q < 2k_f^{2d}$, whereas the quasi-1D bands have a susceptibility that is peaked at a wave-vector $\vec{q}_x = (2k_f^{1d}, \pi)$ for the xz orbital and $\vec{q}_y = (\pi, 2k_f^{1d})$ for the yz orbital. It is this latter feature which, in the weak-coupling framework, gives rise to the incommensurate spin fluctuations in the material[14].

Since the superconductivity in Sr_2RuO_4 evolves out of a Fermi liquid with $T_c \ll E_f$, we proceed with a weak coupling analysis and consider the limit $U, V \ll W$ where W is the bandwidth. In this limit, superconductivity is the only instability of the Fermi liquid, and it can be treated in an asymptotically exact manner via a two-stage renormalization group analysis[9]. In the first stage, high energy modes are perturbatively integrated out above an unphysical cutoff, and an effective particle-particle interaction in the Cooper channel is derived:

$$\begin{aligned} \Gamma_s(\hat{k}, \hat{q}, \alpha) &= U^2\chi_\alpha(\hat{k} + \hat{q}) - 2V^2 \sum_{\beta\neq\alpha} \chi_\beta(\hat{k} - \hat{q}) \\ \Gamma_t(\hat{k}, \hat{q}, \alpha) &= -U^2\chi_\alpha(\hat{k} - \hat{q}) - 2V^2 \sum_{\beta\neq\alpha} \chi_\beta(\hat{k} - \hat{q}) \end{aligned} \tag{4}$$

where $\Gamma_{s(t)}(\hat{k}, \hat{q}, \alpha)$ is the effective interaction in the singlet(triplet) channel. Figure 3 displays the Feynman diagrams which contribute to these leading terms in the perturbative expansion.

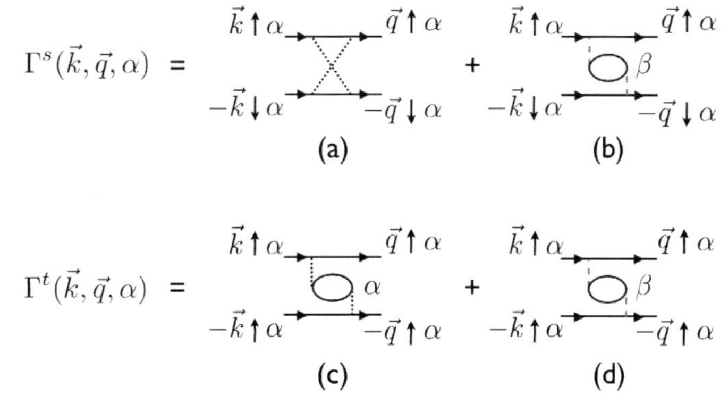

Figure 3. Lowest order Feynman diagrams which contribute to the effective interaction between electrons within the same orbital α in the Cooper channel. The dotted (black) line corresponds to the intra-orbital repulsion U and the dashed (red) line corresponds to the inter-orbital repulsion V.

In the second stage, the renormalization group flows of these effective interactions are computed and the superconducting transition temperature is related to the energy scale at which these RG flows break down. Following this basic prescription, one obtains

$$T_c \sim We^{-1/|\lambda_0^{s,t}|} \tag{5}$$

where $\lambda_0^{s,t}$ are the most negative eigenvalue of

$$g_{s,t}(\hat{k},\hat{q},\alpha) = \sqrt{\frac{\bar{v}_f}{v_f(\hat{k})}}\Gamma_{s,t}(\hat{k},\hat{q},\alpha)\sqrt{\frac{\bar{v}_f}{v_f(\hat{q})}} \tag{6}$$

The most negative eigenvalue determines the superconducting ground state and its eigenfunction is related to the pair wavefunction, as described in Ref. [9].

The central result of this section is shown in Fig. 4. When $V = 0$, the two dimensional xy band has its dominant pairing instability in the $d_{x^2-y^2}$ channel with a substantially lower pairing strength in the triplet channel. However, the xz,yz bands exhibit a close competition between singlet and triplet pairing and therefore, triplet pairing is a close contender among these degrees of freedom. A similar competition between singlet and triplet pairng is thought to occur in the quasi-one dimensional organic Bechgaard salts where similar bandstructure effects are encountered[18, 19, 20].

The triplet pair wave-function

$$\Delta_\alpha(\vec{k}) = i\left[\vec{d}_\alpha(\vec{k})\cdot\vec{\sigma}\sigma^y\right], \alpha = xz, yz \tag{7}$$

is specified by the complex vector $\vec{d}(\vec{k})$ in spin-space; in the absence of any mixing among the orbitals, its orientation is completely arbitrary. Figure 5(a) shows the sign of $\Psi_x(\vec{k})$ on the

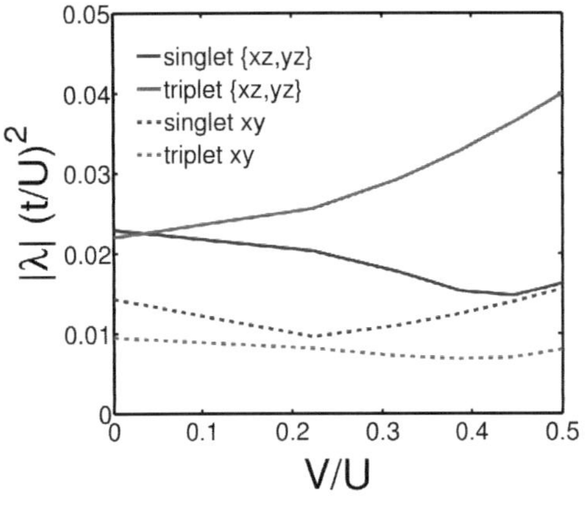

Figure 4. Pairing eigenvalues as a function of V/U for the bandstructure parameters quoted in the text. The strongest pairing strengths occur among the quasi-1D $\{\alpha, \beta\}$ bands. There is a near degeneracy of the singlet and triplet eigenvalues for $V = 0$, but with $V > 0$, the quasi-1D triplet state is the dominant superconducting configuration.

xz Fermi surface. In addition to having odd parity, the wave function has two point nodes on the Fermi surface near $k_y = \pm\pi/2$ and is well approximated by $\vec{d}_x(\vec{k}) = \sin k_x \cos k_y \hat{\Omega}$ with $\hat{\Omega}$ an arbitrary unit vector in spin-space. The nodal structure arises due to the peak in the susceptibility, and therefore the effective $\vec{k}$-dependent repulsive interaction at $\vec{k} = (2k_f^{1d}, \pi)$ which acts to enforce a sign change of the gap function on the Fermi surface. By symmetry, the pair-wave function $\vec{d}_y(\vec{k}) = e^{i\phi} \sin k_y \cos k_x \hat{\Omega}'$ on the yz Fermi surface, and both ϕ and $\hat{\Omega}'$ are completely arbitrary when the two orbitals are perfectly decoupled.

4.1. Effect of δH

The analysis above was carried through for $\delta H = 0$. Next, we consider the effect of hybridization among the different orbitals and study the Hamiltonian

$$\begin{aligned} \delta H &= \sum_{\vec{k}\sigma} \left(c^\dagger_{\vec{k},xz,\sigma} \; c^\dagger_{\vec{k},yz,\sigma} \right) \begin{pmatrix} 0 & g(\vec{k}) \\ g(\vec{k}) & 0 \end{pmatrix} \begin{pmatrix} c_{\vec{k},xz\sigma} \\ c_{\vec{k},yz,\sigma} \end{pmatrix} \\ &+ \lambda \sum_{\alpha,\beta} \sum_{\sigma\sigma'} \sum_{\vec{k}} c^\dagger_{\vec{k},\alpha,\sigma} c_{\vec{k},\beta,\sigma'} \vec{\ell}_{\alpha\beta} \cdot \vec{\sigma}_{\sigma\sigma'} \end{aligned} \tag{8}$$

where $g(\vec{k}) = -4t'' \sin k_x \sin k_y$, $t'' = 0.1t$, the second quantity above is the spin orbit coupling, and the angular momentum operators are expressed in terms of the totally anti-symmetric tensor as $\ell^a_{\alpha\beta} = i\epsilon_{a\alpha\beta}$. Recent electronic structure calculations along with angle-resolved photoemission (ARPES) studies have produced the estimate $\lambda \approx 0.1t = 0.1eV$[21, 17].

For non-zero δH, the relative phase and orientation of $\vec{d}_x$ and $\vec{d}_y$ are no longer arbitrary. A self-consistent calculation of the Bogoliubov-de Gennes (BdG) Hamiltonian taking into account δH and the same functional form of the pair wave functions obtained in the previous section reveals that two distinct types of saddle point configurations are possible for the order parameter: (i) a configuration in which $\hat{\Omega}, \hat{\Omega}' \perp \hat{z}, \hat{\Omega} \cdot \hat{\Omega}' = 0, \phi = 0$, and (ii) $\hat{\Omega} = \hat{\Omega}' = \hat{z}, \phi = \pm\pi/2$; all other configurations have a higher free energy and can therefore be neglected. While the former

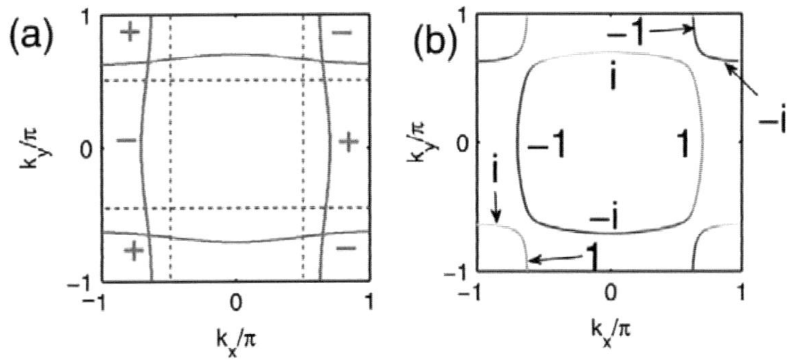

Figure 5. (a) Triplet pair-wave function with sign structure shown for the d_{xz} orbital when interorbital coupling is absent. (b) Structure of the chiral p-wave superconductor when the interorbital mixing is taken into account.

saddle point preserves time-reversal symmetry with a non-collinear order parameter in spin-space, the latter saddle point breaks time-reversal symmetry, has a collinear order parameter, and corresponds to the chiral p-wave state. Although we have found that the chiral phase is energetically favored in the presence of spin-orbit coupling, the energy difference between these two possibilities is $\mathcal{O}(\lambda/t)^2$ and is far too small for us to make any robust predictions in our theory. Therefore, we proceed with a phenomenological approach: we consider a free energy of the form

$$\mathcal{F} = \sum_{i=x,y} \left(a_z |d_i^z|^2 + a_\perp |\vec{d}_i^\perp|^2 \right) + u_1 \left(|\vec{d}_x|^2 + |\vec{d}_y|^2 \right)^2 \\ + u_3 \left(\vec{d}_x \cdot \vec{d}_y^* + \vec{d}_y \cdot \vec{d}_x^* \right)^2 + u_4 |\vec{d}_x \times \vec{d}_y^*|^2 + u_2 |\vec{d}_x|^2 |\vec{d}_y|^2 \tag{9}$$

where we have neglected the gradient terms for sake of brevity. Both $a_\perp - a_z$ and $u_3 - u_4 \sim \mathcal{O}(\lambda/t)^2$, and when $a_\perp > a_z$, the chiral state is favored but only slightly more than the non-chiral solutions with $\vec{d}$ in plane. These small energy differences imply that the d-vector is only weakly pinned along the c-axis, and is corroborated by the absence of a change in the Knight shift for magnetic fields applied along the c-axis[22] as well as for fields applied in the basal plane[1]. In our subsequent analysis, we will assume that the chiral state has a lower free energy and will study an order parameter of the form

$$\vec{d} = e^{i\theta} \left(d_x + i d_y \right) \hat{z}, \; d_x, d_y \in \mathcal{R} \tag{10}$$

Journal of Physics: Conference Series **449** (2013) 012031 doi:10.1088/1742-6596/449/1/012031

where θ is the overall phase of the two condensates. The order parameter structure on the Fermi surface is shown in Fig. 5(b).

Next, we address the issue of nodes on the Fermi surface when $\delta H \neq 0$. We have studied the BdG Hamiltonian for the chiral state with the gap functions derived in the previous section. The spectrum of such a Hamiltonian is readily obtained, and it is found that in order for the nodes to occur on the Fermi surface, the following conditions must be satisfied: $g(\vec{k}) = 0$, and $\vec{d}_x(\vec{k}) = \vec{d}_y(\vec{k}) = 0$. This in turn implies that even an infinitessimal $g(\vec{k})$ is sufficient to destroy the nodes. Although the nodes are not topologically stable, it still follows that the gap becomes parametrically small on portions of the Fermi surface so long as $g(\vec{k}) \ll t$, and the gap minima are $\sim \Delta_0 \left(t''^2 + \lambda^2\right)/t^2$, where Δ_0 is the maximum value of the gap on the Fermi surface. The energy scale of these gap minima is therefore two orders of magnitude smaller than the transition temperature and their effects are expressed in the form of power laws in the specific electronic heat and nuclear spin relaxation rate.

5. Criticism of the weak-coupling theory

Our theory can be criticized on several possible grounds. Firstly, it goes against the standard and intuitive notion that spin-triplet pairing arises when the dominant fluctuations are ferromagnetic in character. Secondly, given that the γ sheet is close to a van Hove point, the effects of electron correlations in general, and ferromagnetic fluctuations in particular, ought to be enhanced in this band. Thus, by finding weak pairing strength in the γ sheet, our theory ostensibly is missing some of the salient physics at play. We will address each of these criticisms below.

It is easy to see why ferromagnetic fluctuations can enhance the tendency towards triplet pairing. As an example, consider a system with kinetic energy $\epsilon(k) = k^2/2m$ for simplicity. In such a system, the Fermi surface is a sphere centered around $k = 0$. In the presence of short-range repulsive interactions, $U > 0$, the effective interaction in the triplet channel, shown in diagram (c) of Fig. 3 is

$$g_t(\hat{k}, \hat{q}) = -U^2\chi(\hat{k} - \hat{q}) + (O(U^4)) \simeq -U^2\rho\left[1 + a|\hat{k} - \hat{q}|^2 + \cdots\right] + (O(U^4)) \tag{11}$$

where we have made use of a long-wavelength expansion of the susceptibility, appropriate for systems with rotational invariance. Here, ρ is the density of states at the Fermi level, and $a \sim d^2\chi/dq^2$ is a number that depends on microscopic details. Since the pair wave-function has odd parity, the only component of g_t which contributes to the triplet eigenvalue in the long wavelength limit is

$$g_t(\hat{k}, \hat{q}) = V_1\hat{k} \cdot \hat{q} + (O(U^2)) \tag{12}$$

where $V_1 = 2U^2\rho a$. This is a separable interaction in the triplet channel. It leads to non-trivial solutions when $a < 0$, i.e., when the susceptibility is *peaked* at $q = 0$, implying in turn that the dominant magnetic fluctuations are ferromagnetic.

While this is a sensible argument, it does not imply a one-to-one correspondence between spin-triplet pairing and ferromagnetic fluctuations; at best, it is a helpful rule-of-thumb. As a concrete counterexample, consider a system with two Fermi pockets, which has particle-hole fluctuations that are peaked at large momentum transfer. If such a system were to exhibit superconductivity from repulsive interactions, the pair wavefunctions would change sign from one pocket to the other (see Fig. 6). Depending on the precise location of the pockets and on the lattice point group symmetry, however, different pairing symmetries will result from exactly the same pairing mechanism. For instance, on a square lattice, if the two Fermi pockets were centered at incommensurate points in the Brillouin zone, the system would exhibit p-wave pairing. In a triangular, honeycomb, or Kagome lattices, when each pocket is centered at the hexagonal Brillouin zone corner(Fig. 6), the gap function has f-wave symmetry. Our conclusions

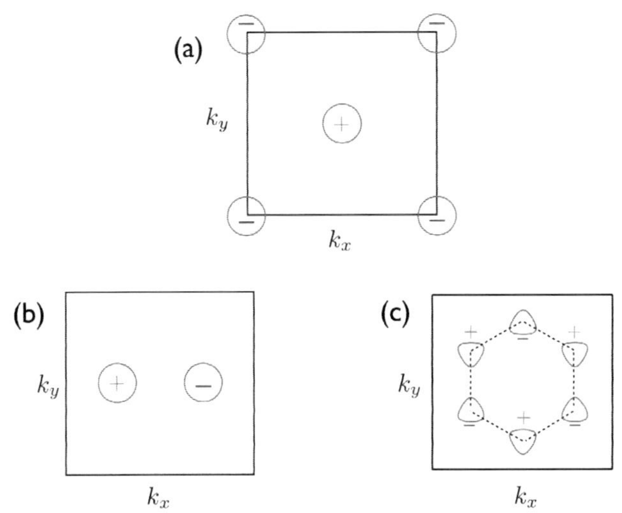

Figure 6. A system with two Fermi pockets in which large momentum transfer particle-hole fluctuations give rise to pair wavefunctions that change sign from one pocket to the other. (a) If the pockets were centered at $\boldsymbol{k} = (0, 0)$, and $\boldsymbol{k} = (\pi, \pi)$, the superconductor would be a sign-changing s-wave (i.e. extended s-wave). (b) If the pockets were centered at incommensurate momenta $\pm k_x$, the sign-changing state would have p_x-wave symmetry. (c) In hexagonal systems, the same sign-changing solution would have f-wave symmetry. Clearly all would be classified as having the same pairing "mechanism" but only the first state is characterized by a spin singlet order parameter. Ferromagnetic fluctuations are irrelevant here: yet, the latter two states exhibit spin triplet pairing.

have little to do with Ferromagnetic fluctuations.

Other examples include the quasi-one dimensional organic superconductors which, due to peaks in the susceptibility at large momentum $\boldsymbol{q} = (2k_F, \pi)$, leads to sign changing singlet and triplet order parameters that have almost identical pairing strength. While the singlet state has a higher pairing strength, application of a weak, in-plane magnetic field is believed to tilt the balance in favor of the spin-triplet state. It makes little sense to say that the field induces ferromagnetic fluctuations; instead, it plays a role of a perturbation to change the condensation energy of the triplet state relative to the singlet state.

Next, we address the role of the γ band being proximate to a van Hove singularity and the role of quasiparticle mass enhancement. While mass enhancement does suggest that the effects of electron correlations are important, the fact remains that for $T_c < T \lesssim 40K$, the normal state is a well-behaved, albeit dressed Fermi liquid. This in turn implies that the system is governed by a free fermion fixed point, and is destabilized only by the superconducting instability. Thus, while the strong correlations lead to a heavy Fermi liquid, the quasiparticles themselves are coherent and weakly coupled; this means that the mass enters mainly in the form of quasiparticle bandstructure parameters (i.e. hoppings) rather than as a bare onsite interaction, and a weak-coupling theory is nevertheless feasible. This philosophy is consistent with the experimental fact that though the system is nearly nested, and though the γ band is close to a van Hove singularity, neither antiferromagnetism nor ferromagnetism occur in the system.

Journal of Physics: Conference Series **449** (2013) 012031 doi:10.1088/1742-6596/449/1/012031

6. Phenomenological consequences of the weak-coupling theory

Throughout this article, we have stressed microscopic aspects of superconductivity in Sr_2RuO_4. However, it is important to mention that there are several phenomenological consequences of our theory that have some support from experimental observations. Firstly, as was shown in Refs. [23, 24, 25], there is an *intrinsic* Kerr response in a multi-band chiral superconductor, whereas in the single-band case, translational invariance guarantees that at least to leading order, there is zero intrinsic Kerr response (vertex corrections, which are suppressed by powers of v_f/c, provide a finite but small correction). The authors of Ref. [23, 24] found that by invoking a multi-band superconductor, the resulting Kerr angle is comparable to that found in experiments. For a particularly sharp prediction involving a nearly *resonant* Kerr response, we urge the reader to consult Ref. [24]. It would be interesting to see whether indeed such a resonance can be found when the frequency of the electromagnetic radiation is tuned to the scale characteristic of the weak inter-orbital couplings.

In the remainder of this section, we will discuss two additional phenomenological consequences of our theory. Firstly, we will discuss edge modes and their implication for edge currents. The goal is to elaborate on the discussion presented in Ref. [10]. Secondly, we will discuss the possibility of sharp, low energy collective excitations that occur in our theory, which are discussed in more detail in Ref. [11].

6.1. Topological properties and edge currents

We consider the topological properties of the quasi-1D superconductor assuming that $a_z < a_\perp$ in Eq. 9 so that the chiral state is the ground state. In this case, the z-component of the spin of the condensate remains conserved, and acts as a $\mathcal{U}(1)$ 'charge' degree of freedom. Thus, we can define the Nambu spinor in momentum space using the band indices as $\Psi_{\nu\vec{k}} = \left(d_{\vec{k},\nu,\uparrow}, d^\dagger_{-\vec{k},\nu,\downarrow}\right)^T$, $\nu = \alpha, \beta$, the BdG Hamiltonian in this basis is described in terms of two Anderson pseudospins, one for each Fermi surface:

$$H_{BdG} = \sum_{\nu=\alpha,\beta} \Psi^\dagger_{\nu\vec{k}} \left[\vec{\delta}_\nu(\vec{k}) \cdot \vec{\tau}_\nu\right] \Psi_{\nu\vec{k}}. \tag{13}$$

For a chiral p-wave state of the form $\Delta_\nu \left(s_x(\vec{k}), s_y(\vec{k})\right)$ on the ν-th Fermi surface, where $s_x(s_y)$ are functions that transform under point group operations as $k_x(k_y)$, we have the following explicit form for the Anderson pseudospin operator:

$$\delta_\nu(\vec{k}) = \left(\Delta_\nu s_x(\vec{k}), \Delta_\nu s_y(\vec{k}), \epsilon_{\nu\vec{k}} - \mu\right) \tag{14}$$

For the chiral state, the pseudospin has the form of a skyrmion in momentum space: it points along the $-\hat{z}(+\hat{z})$-direction inside(outside) the Fermi surface, and on the Fermi surface, it lies in plane, winding by 2π around the Fermi surface. The topological properties of the chiral state comes from the integer skyrmion number which is defined as

$$\mathcal{N}_\nu = \frac{1}{4\pi} \int d^2k \hat{\delta}_\nu \cdot \left(\partial_x \hat{\delta}_\nu \times \partial_y \hat{\delta}_\nu\right) \tag{15}$$

where $\hat{\delta}_\nu = \vec{\delta}_\nu / |\vec{\delta}_\nu|$. The quantity $\mathcal{N}_\nu$ is insensitive to smooth perturbations of the system, provided that the gap to quasiparticle excitations remains non-zero. Physically, the net number of chiral (i.e. the number of left-moving minus the number of right-moving) quasiparticle modes at the edge of the superconductor is given by the skyrmion number, and so long as $\mathcal{N}_\nu \neq 0$,they cannot be localized by backscattering.

The sign of $\mathcal{N}_\nu$ can change only when the skyrmion is altered to an anti-skyrmion. This can be accomplished by flipping the sign of an *odd* number of components of $\hat{\delta}$. Up to overall gauge transformations, there are two distinct ways of changing the skyrmion into an antiskyrmion: (i) a time-reversal operation under which the chirality of the order parameter is changed from $p_x + ip_y \to p_x - ip_y$, or (ii) the transformation $\epsilon_\nu - \mu \to \mu - \epsilon_\nu$ which changes the Fermi surface dispersion from electron-like to hole-like or *vice versa*. As a corollary of the latter scenario, it follows that when the Fermi surface consists of both an electron and a hole pocket - as is the case for the quasi-1D α, β surfaces - a $p_x + ip_y$ superconductor has a net skyrmion number of *zero*: i.e. it is not a topological superconductor.

As a consequence, the chiral edge modes along the boundary of the superconductor and along domain walls are no longer protected: in the presence of disorder, these modes are localized due to the existence of backscattering channels. We can understand this more explicitly by considering a semi-infinite superconductor for $x < 0$, letting $x > 0$ be the vacuum. The quasiparticle edge modes are zero energy solutions of H_{BdG}; excellent approximations for for their wave functions can be obtained by linearizing the dispersion about each Fermi surface:

$$\Psi^0_\nu(x) \propto e^{-i\text{sgn}(v^\nu_f)k_y y} e^{-\int_0^x dx'|\Delta_\nu(x')|/|v^\nu_f|} \chi_\nu \tag{16}$$

where we have defined the electron(hole) pocket to have positive(negative) Fermi velocity, and χ is a spinor which is an eigenstate of τ^y_ν. Thus, in the quasi-1D superconductor, there are two sets of quasiparticle modes, one from each Fermi pocket, which counterpropagate along the boundaries (in this case along the y direction) and along domain walls of the system. Although their propagation velocities are different, in the presence of disorder, localization will still result.

Having discussed the neutral quasiparticle edge modes, we next consider the electrical currents in a simplest chiral superconductor having a single Fermi surface. Due to the breaking of time-reversal symmetry, there are two contributions to these edge currents. Firstly, there is a bulk contribution which originates from the multi-component nature of the order parameter. In the Ginzburg Landau approximation, the contribution arises from a gradient term of the form

$$J^a_{bulk} = -i\epsilon^{ab} K \left(\vec{d}_x \partial_b \vec{d}^*_y - c.c \right) \tag{17}$$

with $K \sim \mathcal{O}(\Delta/E_f)^2$. In the continuum limit, it can be shown that this term is proportional to the curl of the total angular momentum density of the condensate. The second contribution to the electrical current comes from the chiral quasiparticle modes near the edge of the system. In a single-band system, there is no reason for these currents to be suppressed. Moreover, there are sharp predictions of mesoscopic effects in such systems, including the possibility of field-induced reentrant superconductivity[26]. The fact that edge currents have not been observed in highly sensitive scanning squid probes[27, 28, 29] leads us to question further the simple one-band descriptions that have prevailed in the field. By contrast, in our quasi-one dimensional superconductor, the magnitude of the electrical current is determined by the small interorbital couplings, since the chiral nature of the condensate is established only by these terms. Since the p_x and p_y components of the condensate are nearly decoupled, (i.e. because interorbital couplings play an important role only in regions of $\boldsymbol{k}-$space where the Fermi surfaces nearly cross), we can reasonably expect that there will be a substantial reduction of edge currents in our theory as compared to theories based on a single-band description of the chiral state. If the $\{\alpha, \beta\}$ bands were the active bands, at lower temperature when substantial pairing strength develops on the γ band, the expectation remains that observable edge currents ought to be present in the system. A more quantitative theory of edge currents shall be presented elsewhere[30].

6.2. Collective modes

A superconductor described by a multi-component order parameter (p_x and p_y in the present context), will have collective mode excitations associated with the relative phase difference,

$\phi_- \equiv \theta_x - \theta_y$, where $\theta_x(\theta_y)$ is the phase of the $p_x(p_y)$ component of the order parameter. At zero temperature, such a mode would be expected to have a frequency $\hbar\omega_0 \sim \sqrt{\mathcal{J}/\chi}$, where $\chi \sim N(0)$ is the compressibility ($N(0)$ is the density of states at the Fermi energy) and $\mathcal{J}$ is the second derivative of the condensation energy with respect to ϕ_-. Given that the condensation energy $\sim N(0)|\Delta_0|^2$ where $|\Delta_0|$ is the root mean squared gap magnitude, this means that $\hbar\omega_0 \sim |\Delta_0|$ for a single-band superconductor. Thus, in a single-band chiral superconductor, the characteristic energy scale of the collective modes is of the order of the gap, meaning that the modes are likely to be damped by the continuum. Similar considerations apply to fluctuations in the relative orientation of the spins (*i.e.* the d-vector) in a two-component triplet superconductor.

By contrast, if the two components of the order parameter are associated with different orbitals, *i.e.* the p_x component with the d_{xz} and the p_y with d_{yz} orbital respectively, and if mixing between the different orbitals were absent, then collective fluctuations of the relative phase and spin-orientation would be gapless. The orbital-mixing terms, denoted as δH above, result in a non-vanishing dependence of the condensation energy on ϕ_-. As a result, if the superconductivity arises primarily from the quasi-1D bands, then the relative phase mode is expected to have an energy $\hbar\omega_0 \approx \gamma\Delta_0$, where γ vanishes continuously as δH tends to zero. Naturally, similar considerations apply to the relative spin orientational fluctuations. The relevant question here is whether we should have $\gamma \sim 1$ or $\gamma \ll 1$ in the physical limit of $|\Delta_0| \ll \delta H \ll W$, where W is the bandwidth. The recent work by Chung *et al.*[11] showed that γ will scale with $\delta H/W$, with almost no dependence on Δ_0. This is because $\mathcal{J}$ is the cost in the pair condensation energy due to fluctuations of the relative phase and spin orientation, and fluctuations of these quantities will affect the pair condensation energy mainly in the portion of the Fermi surface that is affected by the orbital hybridization, the size of which is determined almost entirely by $\delta H/W$. Thus, if superconductivity arises from the quasi 1D bands, the collective modes should have low energies for the same reason that the edge currents are small.

7. Conclusion

We have stressed that the microscopic origin of superconductivity in Sr_2RuO_4 remains unknown. Most workers in the field believe that the physics of Sr_2RuO_4 is analogous to superfluid ^{3}He, where dominantly ferromagnetic fluctuations of a nearly isotropic Fermi surface produce triplet pairing. By contrast, we have found that the material better resembles the organics, Fe-based and cuprate superconductors in the sense that particle-hole fluctuations with large momentum transfer are responsible for spin triplet pairing in this system. From the experimental standpoint, the heat capacity measurements that have often been cited as evidence in favor of γ being the active band, do not rule out the complementary picture in which $\{\alpha, \beta\}$ are the active bands. In a well-controlled weak-coupling treatment, appropriate for the description of an unconventional superconductor that condenses from a Fermi liquid, $\{\alpha, \beta\}$ are the active bands. The resulting superconducting state has a 'hidden' quasi-one dimensionality: although it preserves the tetragonal symmetry of the underlying lattice, it consists of two nearly decoupled quasi-one dimensional condensates. In addition, the solution obtained in this limit has some phenomenological support, in that as a multi-band system, it allows for an intrinsic Kerr response[23, 24, 25]. Moreover, since the p_x and p_y components are nearly decoupled, "living" on different orbitals, the electrical currents along boundaries of this system can be substantially reduced. Another important consequence of the fact that the p_x, p_y components are nearly decoupled is that there are sharp, low energy collective modes that correspond to inter-orbital relative phase and relative spin-orientation fluctuations. These modes are "almost-Goldstone modes" and have energy well below the pairing gap. It will be vital to test some of these phenomenological predictions in the near future. It is our hope that the ideas described here will lead to further developments in the field, both along experimental and theoretical fronts.

Acknowledgments We thank D. Agterberg, J. Alicea, E. Berg, S. Brown, T. Devereaux, C. Hicks, C. Honerkamp, C. Kallin, A. Kapitulnik, J. Kirtley, S. Kivelson, A. P. Mackenzie, K. Moler, D. Scalapino, R. Thomale, and F.-C. Zhang for helpful discussions. This work is supported by the Department of Energy, Office of Basic Energy Sciences, Division of Materials Sciences and Engineering, under contract DE-AC02-76SF00515, and the Alfred P. Sloan Foundation (SR).

References

[1] Ishida K, Mukuda H, Kitaoka Y, Asayama K, Mao Z Q, Mori Y and Maeno Y 1998 *Nature* **396** 658
[2] Nelson K D, Mao Z Q, Maeno Y and Liu Y 2004 *Science* **12** 1151
[3] Kidwingira F, Strand J D, Harlingen D J V and Maeno Y 2006 *Science* **314** 1267
[4] Mackenzie A P, Julian S R, Diver A J, McMullan G J, Ray M P, Lonzarich G G, Maeno Y, Nishizaki S and Fujita T 1996 *Phys. Rev. Lett.* **76**(20) 3786–3789 URL `http://link.aps.org/doi/10.1103/PhysRevLett.76.3786`
[5] Bergemann C, Mackenzie A P, Julian S R, Forsythe D and Ohmichi E 2003 *Adv. Phys.* **52** 639
[6] Deguchi K, Mao Z Q, Yaguchi H and Maeno Y 2004 *Phys. Rev. Lett.* **92**(4) 047002 URL `http://link.aps.org/doi/10.1103/PhysRevLett.92.047002`
[7] Damascelli A, Lu D H, Shen K M, Armitage N P, Ronning F, Feng D L, Kim C, Shen Z X, Kimura T, Tokura Y, Mao Z Q and Maeno Y 2000 *Phys. Rev. Lett.* **85**(24) 5194–5197 URL `http://link.aps.org/doi/10.1103/PhysRevLett.85.5194`
[8] Mackenzie A P and Maeno Y 2003 *Rev. Mod. Phys.* **75** 657
[9] Raghu S, Kivelson S A and Scalapino D J 2010 *Phys. Rev. B* **81**(22) 224505 URL `http://link.aps.org/doi/10.1103/PhysRevB.81.224505`
[10] Raghu S, Kapitulnik A and Kivelson S A 2010 *Phys. Rev. Lett.* **105**(13) 136401 URL `http://link.aps.org/doi/10.1103/PhysRevLett.105.136401`
[11] Chung S B, Raghu S, Kapitulnik A and Kivelson S A 2012 *Phys. Rev. B* **86**(6) 064525 URL `http://link.aps.org/doi/10.1103/PhysRevB.86.064525`
[12] Veenstra C N, Zhu Z H, Ludbrook B, Capsoni M, Levy G, Nicolaou A, Rosen J A, Comin R, Kittaka S, Maeno Y, Elfimov I S and Damascelli A 2012 *e-print arxiv: 1205.4248*
[13] Agterberg D F, Rice T M and Sigrist M 1997 *Phys. Rev. Lett.* **78** 3374
[14] Sidis Y, Braden M, Bourges P, Hennion B, NishiZaki S, Maeno Y and Mori Y 1999 *Phys. Rev. Lett.* **83** 3320
[15] Scalapino D J 2012 *e-print arxiv: 1207.4093*
[16] Kontani H, Tanaka T, Hirashima D S, Yamada K and Inoue J 2008 *Phys. Rev. Lett.* **100** 096601
[17] Liu G Q, Antonov V N, Jepsen O and Andersen O K 2008 *Phys. Rev. Lett.* **101** 026408
[18] Shinagawa J, Kurosaki Y, Zhang F, Parker C, Brown S E, Jérome D, Christensen J B and Bechgaard K 2007 *Phys. Rev. Lett.* **98**(14) 147002 URL `http://link.aps.org/doi/10.1103/PhysRevLett.98.147002`
[19] Yonezawa S, Maeno Y, Bechgaard K and Jérome D 2012 *Phys. Rev. B* **85**(14) 140502 URL `http://link.aps.org/doi/10.1103/PhysRevB.85.140502`
[20] Takigawa M, Yasuoka H and Saito G 1987 *Journal of the Physical Society of Japan* **56** 873–876 URL `http://jpsj.ipap.jp/link?JPSJ/56/873/`
[21] Haverkort M W, Elfimov I S, Tjeng L H, Sawatzky G A and Damascelli A 2008 *Phys. Rev. Lett.* **101** 026406
[22] Murakawa H, Ishida K, Kitagawa K, Mao Z Q and Maeno Y 2004 *Phys. Rev. Lett.* **93** 167004
[23] Taylor E and Kallin C 2012 *Phys. Rev. Lett.* **108**(15) 157001 URL `http://link.aps.org/doi/10.1103/PhysRevLett.108.157001`
[24] Taylor E and Kallin C 2012 *e-print arxiv: 1208.2699*
[25] Wysokiński K I, Annett J F and Györffy B L 2012 *Phys. Rev. Lett.* **108**(7) 077004 URL `http://link.aps.org/doi/10.1103/PhysRevLett.108.077004`
[26] Huo J W, Chen W Q, Raghu S and Zhang F C 2012 *Phys. Rev. Lett.* **108**(25) 257002 URL `http://link.aps.org/doi/10.1103/PhysRevLett.108.257002`
[27] Bjornsson P G, Maeno Y, Hubar M E and Moler K A 2005 *Phys. Rev. B* **72** 012504
[28] Kirtley J R, Kallin C, Hicks C W, Kim E A, Liu Y, Moler K A, Maeno Y and Nelson K D 2007 *Phys. Rev. B* **76**(1) 014526 URL `http://link.aps.org/doi/10.1103/PhysRevB.76.014526`
[29] Hicks C W, Kirtley J R, Lippman T M, Koshnick N C, Huber M E, Maeno Y, Yuhasz W M, Maple M B and Moler K A 2010 *Phys. Rev. B* **81**(21) 214501 URL `http://link.aps.org/doi/10.1103/PhysRevB.81.214501`
[30] Lederer S and Raghu S in preparation

Journal of Physics: Conference Series **449** (2013) 012032 doi:10.1088/1742-6596/449/1/012032

Extended analysis of the field-angle-dependent heat capacity of $(TMTSF)_2ClO_4$ toward identification of the superconducting gap structure

Shingo Yonezawa[1], **Yoshiteru Maeno**[1], **and Denis Jérome**[2]

[1] Department of Physics, Graduate School of Science, Kyoto University, Japan
[2] Laboratoire de Physique des Solides, Université Paris-Sud, Orsay, France

E-mail: yonezawa@scphys.kyoto-u.ac.jp

Abstract. In this paper, we present detailed analyses of the field-angle dependence of the heat capacity of the quasi-one-dimensional superconductor $(TMTSF)_2ClO_4$ with various models of superconducting gap structure. We clarify that the superconducting gap structure with line nodes at $k_y = \pm 0.25b^*$ is the only structure that is consistent with our experiment, irrespective of the value of the anion gap. The observed field-angle dependence of the heat capacity indicates that two of the four nodes mainly contribute to the quasiparticle excitation.

1. Introduction

In modern researches of superconductivity, investigation of unconventional superconductors, which have anisotropic gaps with sign reversals, is one of the most important subjects. The "Bechgaard salts", consisting of tetramethyl-tetraselena-fulvalene (TMTSF) molecules and anions, are the first organic superconductors reported in 1980 [1, 2] and are among the longest-studied unconventional superconductors. The TMTSF family is also unique because they have archetypal quasi-one-dimensional (Q1D) electronic bands [3, 4, 5, 6]. Interestingly, the superconducting (SC) phase is located next to the spin-density-wave (SDW) magnetic insulating phase in the pressure-temperature phase diagram. Thus, the possibility of spin-fluctuation-mediated superconductivity has been proposed. Indeed, the existence of spin-fluctuation in the vicinity of the SC state has been recently investigated [7, 8, 9, 10].

We recently performed field-angle-resolved calorimetry of the ambient-pressure SC member of the TMTSF family: $(TMTSF)_2ClO_4$ [11]. We observed a peculiar modulation in the in-plane field-angle ϕ dependence of the heat capacity C (see Figs. 2(a) and (b)) with clear kinks at $\phi = \pm 10°$. This modulation is attributed to the quasiparticle excitation with wavenumber $\boldsymbol{k}$ near SC gap nodes/zeros [12, 13]. The position of the kinks gives us a crucial information on its SC gap structure: The SC gap nodes or zeros are located at $\boldsymbol{k}$-points where the Fermi velocity $v_F(\boldsymbol{k})$ points $\pm 10°$ away from the crystalline a axis. From this information, we proposed that the "d-wave-like" [14] gap structure with nodes at $k_y = \pm 0.25b^*$ (structure D in Fig. 1) is most consistent with the experiment. The identification of the gap structure should provide a solid bases to understand interesting SC phenomena in the TMTSF family, such as a possible Fulde-Ferrel-Larkin-Ovchinnikov (FFLO)-like state in high fields [15, 16, 17, 18, 19, 20, 21, 22].

In this paper, we present detailed analyses of the heat capacity data assuming various models of superconducting gap structures. We clarify that the superconducting gap structure with line nodes at

Journal of Physics: Conference Series **449** (2013) 012032 doi:10.1088/1742-6596/449/1/012032

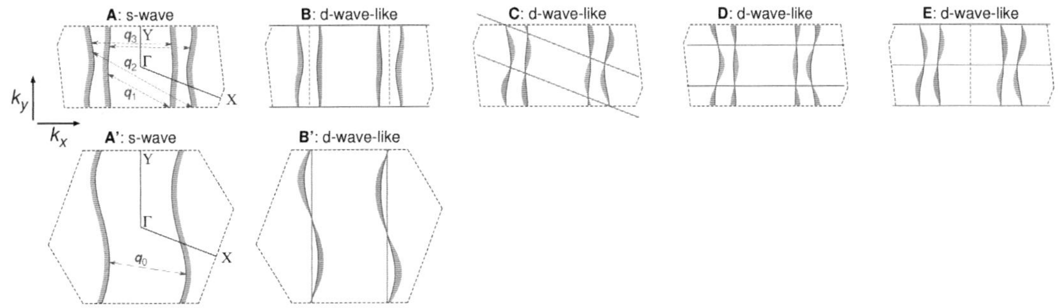

Figure 1. Schematic drawings of s-wave and d-wave-like superconducting gap structures. The orange and green colors indicate + and − signs of the superconducting gap, respectively. The red solid lines indicate nodal positions in the SC gap whereas the red broken lines indicate nodes that do not cut the Fermi surface. The purple lines in B indicate the positions of small gap minima. The first row exhibits gap structures for the split Fermi surface for $(TMTSF)_2ClO_4$. The structures in the second row are corresponding gap structures for the non-split Fermi surface for $(TMTSF)_2PF_6$. We here define k_x and k_y as a set of orthogonal basis in the reciprocal space as shown in the panel A. A and A′: s-wave state. The three nesting vectors [23] $\boldsymbol{q}_1$, $\boldsymbol{q}_2$, and $\boldsymbol{q}_3$ for the folded FS and the nesting vector $\boldsymbol{q}_0$ for the unfolded FS are also shown with blue arrows. Except the inter-band nesting vector $\boldsymbol{q}_3$, all the others are intra-band nesting vectors. B: d-wave-like state with small gap minima at $k_y = \pm 0.5b^*$ but without nodes. [24] B′: d-wave-like state, which turns into the structure B when the Fermi surface is folded due to the anion order. C: d-wave-like state with gap nodes running parallel to the Γ-X line. [25] D: d-wave-like state with gap nodes at $k_y = \pm 0.25b^*$. E: d-wave-like state with gap nodes at $k_y = 0$ and $\pm 0.5b^*$. For the structures C–E, no corresponding unfolded SC gap structures have been proposed.

$k_y = \pm 0.25b^*$ is the only structure that is consistent with our experiment, irrespective of the choice of the value of the anion gap Δ_{AO}, which causes the splitting of the quasi-one-dimensional Fermi surface into two pairs. The observed behavior indicates that two of the four nodes mainly determines the field-angle ϕ dependence of the heat capacity.

2. Possible Nodal Structures of the Superconducting Gap

Generally speaking, when the Fermi surface (FS) is strongly nested with a nesting vector $\boldsymbol{Q}_{\mathrm{nest}}$ and spin-singlet Cooper pairs are formed by the spin fluctuation with $\boldsymbol{Q}_{\mathrm{nest}}$, the SC gap $\Delta(\boldsymbol{k})$ and the pairing interaction $V(\boldsymbol{q})$ should satisfy the relation [4]

$$V(\boldsymbol{Q}_{\mathrm{nest}})\Delta(\boldsymbol{k})\Delta(\boldsymbol{k}+\boldsymbol{Q}_{\mathrm{nest}}) < 0 \quad (\boldsymbol{k}, \boldsymbol{k}+\boldsymbol{Q}_{\mathrm{nest}} \in \text{Fermi Surface}). \tag{1}$$

This is a direct consequence of the gap equation with assumption that $|V(\boldsymbol{q})|$ has a large peak at $\boldsymbol{q} = \boldsymbol{Q}_{\mathrm{nest}}$. For superconductivity near spin-density wave states, the value of $V(\boldsymbol{Q}_{\mathrm{nest}})$ should be positive. Thus, Eq. (1) can be simplified as

$$\Delta(\boldsymbol{k})\Delta(\boldsymbol{k}+\boldsymbol{Q}_{\mathrm{nest}}) < 0 \quad (\boldsymbol{k}, \boldsymbol{k}+\boldsymbol{Q}_{\mathrm{nest}} \in \text{Fermi Surface}). \tag{2}$$

The Brillouin zone (BZ) and FS of $(TMTSF)_2ClO_4$ is folded along the b^* direction due to the orientational order of the tetrahedral ClO_4 anions. Thus, the FS consists of two pairs of warped sheets as shown in Fig. 1(a). For this folded FS, three nesting vectors $\boldsymbol{q}_1$, $\boldsymbol{q}_2$, and $\boldsymbol{q}_3$ exists [23]; whereas for the unfolded FS shown in A′, $\boldsymbol{q}_0$ is the only nesting vector.

Journal of Physics: Conference Series **449** (2013) 012032 doi:10.1088/1742-6596/449/1/012032

For $(TMTSF)_2ClO_4$, there are several candidates for its spin-singlet SC gap structure as shown in Fig. 1. For example, the *d*-wave-like structure B′ with line nodes, which has been theoretically predicted for the unfolded FS [26, 27, 28, 4], is folded resulting in the structure B [24]. The difference is that the nodes in B′ disappear and are replaced by small gap minima at $k_y = \pm 0.5b^*$ due to the split of the FS. Note that both structures B and B′ satisfy the relation (2) with $\boldsymbol{Q}_{\mathrm{nest}} = \boldsymbol{q}_3$ and $\boldsymbol{Q}_{\mathrm{nest}} = \boldsymbol{q}_0$, respectively. The structure C is recently predicted by Mizuno *et al* [25] for a certain situation where charge fluctuation contributes to the Cooper pairing [29]. This structure satisfies Eq. (2) with $\boldsymbol{Q}_{\mathrm{nest}} = \boldsymbol{q}_3$. We also list other possible structures D and E. For the structure D, Eq. (2) is satisfied with $\boldsymbol{Q}_{\mathrm{nest}} = \boldsymbol{q}_1$ and $\boldsymbol{q}_2$. Thus the intra-band nesting drives the pairing for this state. The structure E is similar to the structure C and the inter-band nesting $\boldsymbol{q}_3$ is important. We should note that the structures C, D, and E are essentially different from the structure B′, in the sense that b′ is not converted to C, D, and E by the FS folding.

For simplicity we only list *s*-wave and *d*-wave-like structures in Fig. 1. However, more complicated structure such as *g*-wave-like states [30] might also be possible. The present experiment is not phase-sensitive; thus we cannot distinguish structures with the same nodal positions but with different signs.

3. Anisotropy in the Doppler shift

Within the Doppler shift model, the heat capacity C of a Q1D superconductor with line nodes/zeros on its SC gap should depend on the field-direction ϕ with respect to the a axis as

$$C(\phi)/T \propto N(\phi) \propto (\Gamma^2 \sin^2\phi + \cos^2\phi)^{1/4} \sqrt{\frac{H}{H_{c2}(0^\circ)}} \sum_{\mathrm{n:nodes}} A_{\mathrm{n}} |\sin(\phi - \phi_{\mathrm{n}})|, \tag{3}$$

as explained in Ref. [11]. Here, N is the quasiparticle density of states, ϕ_{n} is the angle between the a axis and the Fermi velocity at a node, and $\Gamma \equiv H_{c2}(\phi = 0^\circ)/H_{c2}(\phi = 90^\circ)$ is the in-plane anisotropy ratio of the upper critical field H_{c2}. The coefficient A_{n} represents the contribution of each node to the field-angle-dependent heat capacity. For superconductors with higher crystalline symmetry, all nodes should have the same value of A_{n}. In contrast, because $(TMTSF)_2ClO_4$ has a triclinic crystal as well as band structures, the values of A_{n} may be different among different nodes; A_{n} should depend on the details of the band structure as well as of the SC gap structure near the gap node, e.g. the linear slope $\tilde{\Delta}$ of the gap around the node: $\Delta(\boldsymbol{k}) \sim \tilde{\Delta}|\boldsymbol{k} - \boldsymbol{k}_{\mathrm{node}}|$.

In Ref. [11], we chose $A_{\mathrm{n2}}/A_{\mathrm{n1}} = 0.3$ in order to demonstrate that the model captures important features of the experimental result, namely the asymmetry and the kink structures in the C/T vs. ϕ curve. However, as shown in Fig. 2, this choice of the value of $A_{\mathrm{n2}}/A_{\mathrm{n1}}$ is not very essential: as long as the condition $A_{\mathrm{n2}}/A_{\mathrm{n1}} \lesssim 0.7$ is satisfied, the observed behavior is reproduced.

The experimental result is hardly explained without the anisotropy in A_{n}. This fact strongly suggest that the triclinic anisotropy in the band structure and the SC gap structure is important in the superconductivity of $(TMTSF)_2ClO_4$. Investigation of the origin of this anisotropy in A_{n}, however, requires detailed theoretical calculations and is far beyond the scope of this paper.

We should note that recent theory [31] based on the Kramer-Pesch approximation also revealed that the observed behavior is indeed attributable to the existence of the gap node.

4. Anion-gap-energy dependence of the Fermi surface geometry

In the anion ordered state, there should be an anion gap energy Δ_{AO}, which corresponds to the difference in the on-site energy of the neighboring TMTSF chains driven by the anion order. In Ref. [11], we used the value $\Delta_{\mathrm{AO}} = 100$ meV deduced from the X-ray diffraction [23]. However, several transport studies have suggested smaller values of Δ_{AO} such as 4.5 meV and 25 meV [32, 33]. In addition, a recent first-principle band calculation claimed that Δ_{AO} is nearly zero [31]. Within the tight-binding model [23], when Δ_{AO} becomes larger, the FS becomes flatter and the angle between the Fermi velocity and the a axis, $\phi_{v_{\mathrm{F}}}$, tends to become smaller, as shown in Fig. 3.

This change in the Fermi surface geometry is important when we discuss the SC gap structure based on the experimental results.

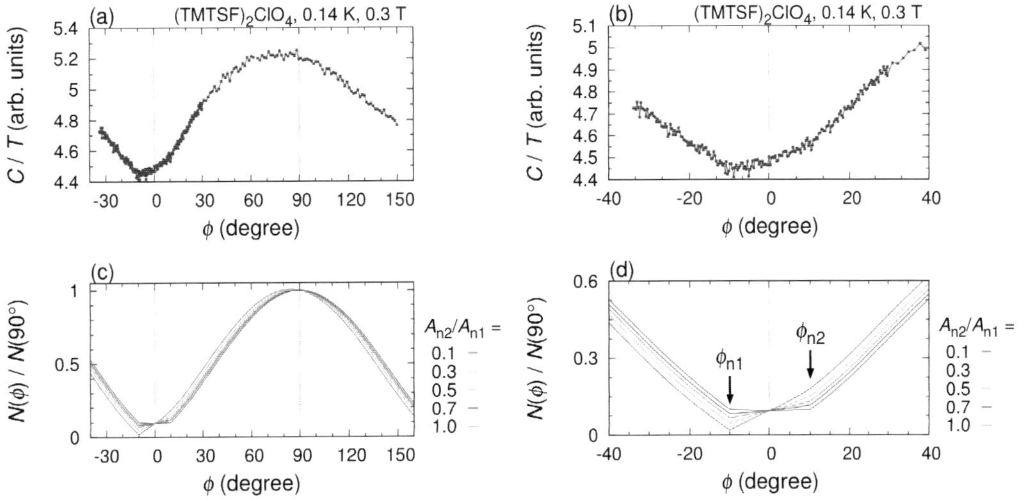

Figure 2. (a) In-plane field-angle ϕ dependence of the heat capacity of a $(TMTSF)_2ClO_4$ single crystal, measured at 0.14 K and 0.3 T [11]. The panel (b) is an enlarged view near $\phi = 0°$. (c) Normalized quasiparticle density of states (QDOS) $N(\phi)/N(90°)$ for $\Gamma = 3.5$, $(\phi_{n1}, \phi_{n2}) = (-10°, +10°)$, and $A_{n2}/A_{n1} = 0.1, 0.3, 0.5, 0.7$, and 1.0. The panel (d) is an enlarged view near $\phi = 0°$.

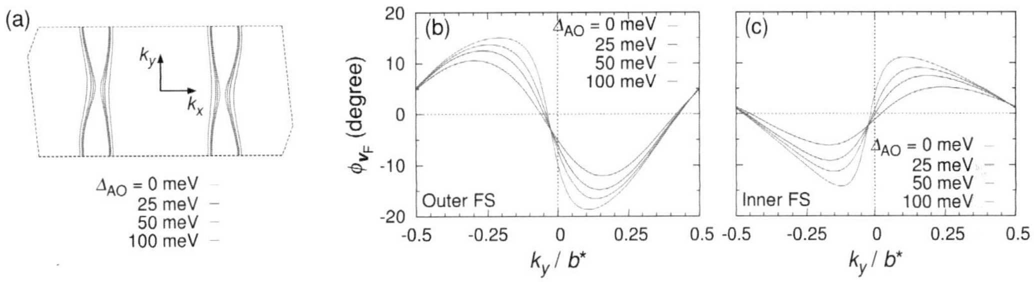

Figure 3. Dependence of the Fermi surface geometry on the anion gap energy Δ_{AO} calculated based on the tight-binding model. [23] (a) Fermi surfaces for different values of Δ_{AO}. (b) Anion-gap-energy dependence of $\phi_{v_F} = \arctan(v_y/v_x)$ for the outer Fermi surface of the $k_x > 0$ side. (c) Anion-gap-energy dependence of ϕ_{v_F} for the inner Fermi surface of the $k_x > 0$ side.

5. Detailed analysis of the gap structure

Here, we deduce the SC gap structure from the experimental results by comparing nodal structures shown in Fig. 1. It is also required to consider the dependence of the Fermi velocity on Δ_{AO} because there is some ambiguity in the value of Δ_{AO}. In Table 1, we list values of $\phi_{v_F}^{\rm node}$ for several superconducting gap structures and for different values of Δ_{AO}.

Our experimental results require that nodes of $\phi_{v_F}^{\rm node} \simeq +10°$ and of $\phi_{v_F}^{\rm node} \simeq -10°$ must *both* exist. In Table 1, if we allow error bars of 1°, there are only two possibilities that satisfy this condition: i.e. the structure D with $\Delta_{AO} = 0$ meV and with $\Delta_{AO} = 100$ meV; in case of $\Delta_{AO} = 0$ meV the inner FS has nodes of $\phi_{v_F}^{\rm node} \simeq \pm 10°$ whereas in case of $\Delta_{AO} = 100$ meV the outer FS has such nodes. Thus, irrespective of the value of Δ_{AO}, we can conclude that the structure with nodes at $k_y = \pm 0.25b^*$ is the only possible nodal structure that is consistent with our experiment.

Table 1. Nodal direction in the velocity space (i.e. the angle between the Fermi velocity at the node and the a axis, $\phi_{v_F}^{\rm node}$) for different gap structures and the anion gap energy. The values of $\phi_{v_F}^{\rm node}$ for the outer FS (O-FS) and those for the inner FS (I-FS) are listed separately. Pairs of $\phi_{v_F}^{\rm node}$ that are consistent with our experiment within an error bar of $\pm1°$ are shown with bold letters. The non-folded d-wave like structure B′ is also examined in the last column.

	$\phi_{v_F}^{\rm node}$ for nodal structures listed in Fig. 1				
$\Delta_{\rm AO}$ (meV)	structure B† O-FS / I-FS	structure C O-FS / I-FS	structure D O-FS / I-FS	structure E O-FS / I-FS	structure B′ ‡
0	+5° / +1°	+5°, −12° / +9°, −13°	+15°, −14° / **+9°, −10°**	+5°, −10° / +1°, +2°	+11°, −18°
25	+5° / +1°	+5°, −9° / +5°, −11°	+13°, −13° / +8°, −9°	+5°, −8° / +1°, 0°	-
50	+5° / +1°	+5°, −7° / +3°, −9°	+12°, −12° / +7°, −7°	+5°, −7° / +1°, 0°	-
100	+5° / +1°	+5°, −6° / +1°, −6°	**+10°, −10°** / +5°, −5°	+5°, −5° / +1°, −1°	-

†For the structure B, which is a fully-gapped state, we list here ϕ_{v_F} at the gap minima at $k_y = \pm0.5b^*$.
‡Values for the non-folded structure B′ is calculated based on the high-pressure band structure in Ref. [23].

6. Contribution from all nodes

We for simplicity calculated $N(\phi)$ by taking into account of only the two nodes with $\phi_{v_F}^{\rm node} = \pm10°$ in the preceding discussion. However, as shown in Fig. 1, the structure D has *four* nodes on its SC gap. Below, we will examine the contribution from the other two nodes.

In Fig. 4, we plot $N(\phi)/N(90°)$ for the gap structure D with $\Delta_{\rm AO} = 100$ meV (i.e. $(\phi_{\rm n1}, \phi_{\rm n2}, \phi_{\rm n3}, \phi_{\rm n4}) = (-10°, +10°, -5°, +5°)$) and $\Delta_{\rm AO} = 0$ meV (i.e. $(\phi_{\rm n1}, \phi_{\rm n2}, \phi_{\rm n3}, \phi_{\rm n4}) = (-10°, +9°, -15°, +14°)$) with several sets of $(A_{\rm n1}, A_{\rm n2}, A_{\rm n3}, A_{\rm n4})$, the strength of nodal contribution to the heat-capacity oscillation. It is clear that the additional contributions of the nodes n3 and n4 do not alter the qualitative behavior of $N(\phi)$, unless $A_{\rm n3}$ and $A_{\rm n4}$ are as large as $A_{\rm n1}$. Thus, the structure D is still consistent with the experiment. It is rather difficult to detect the contributions of all nodes even with the present high-resolution calorimeter, considering the signal-to-noise ratio of the data in Fig. 2(b).

7. Summary

In order to fully evaluate the superconducting gap structure from our calorimetry, we have examined several superconducting gap structures. We clarified that the superconducting gap structure with line nodes at $k_y = 0.25b^*$ is the structure that is consistent with our field-angle ϕ dependence of the heat capacity C. The simplest example is shown in Fig. 1D, but possibility of a structure with the same nodal positions but with different signs (e.g. g-wave-like state) is not excluded. The conclusion on the nodal position is not altered irrespective of the value of the anion gap $\Delta_{\rm AO}$. We also demonstrate that the shape of the C/T-ϕ curve is mainly determined by nodes with large contributions to the heat capacity. This analysis guarantees the validity of the model in Ref. [11] taking into account of contributions from only two of the nodes.

Acknowledgments
We gratefully acknowledge H. Aizawa, Y. Suzumura, and A. Kobayashi for valuable discussion. This work is supported by a Grant-in-Aid for the Global COE "The Next Generation of Physics, Spun from Universality and Emergence" and by Grants-in-Aids for Scientific Research (KAKENHI 21110516, 21740253, 23540407, and 23110715) from MEXT and JSPS.

References

[1] Jérome D, Mazaud A, Ribault M and Bechgaard K 1980 *J. Phys. Lett.* **41** L95
[2] Bechgaard K, Carneiro K, Olsen M, Rasmussen F B and Jacobsen C S 1981 *Phys. Rev. Lett.* **46** 852
[3] Ishiguro T, Yamaji K and Saito G 1998 *Organic Superconductors Second Edition* (Heidelberg: Springer-Verlag)

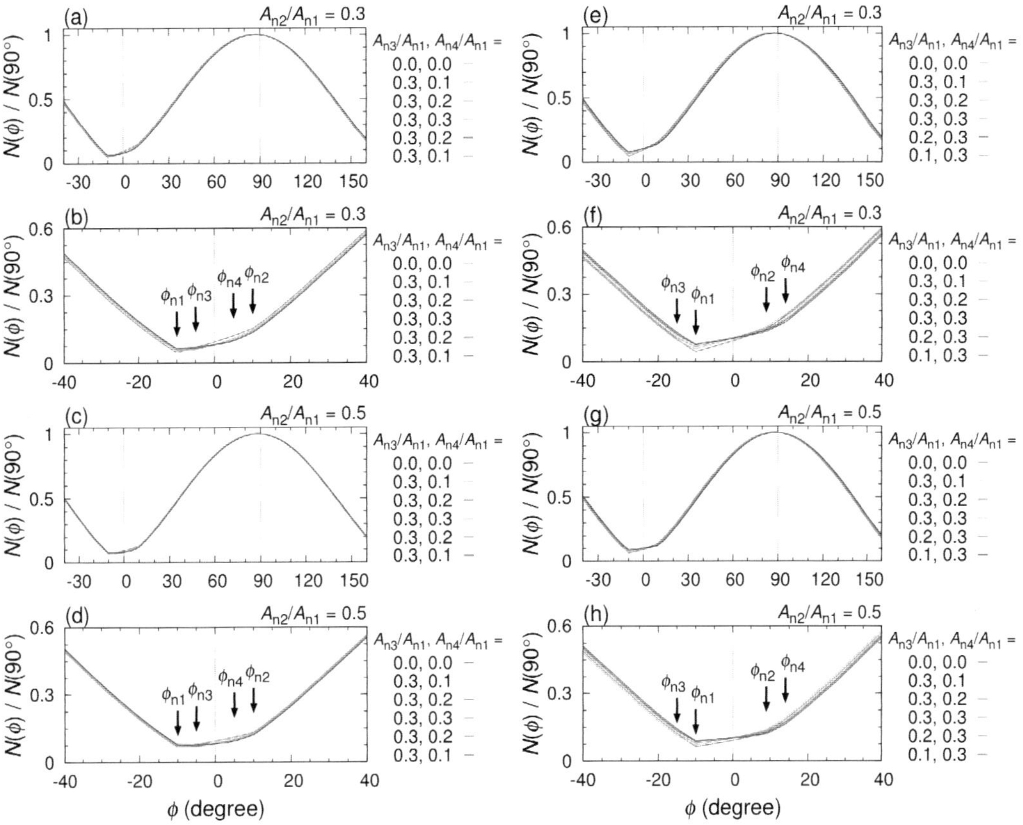

Figure 4. Normalized QDOS $N(\phi)/N(90°)$ for the gap structure D and for several sets of $(A_{n1}, A_{n2}, A_{n3}, A_{n4})$, the strength of nodal contribution to the heat-capacity oscillation. Here we used the in-plane anisotropy $\Gamma = 3.5$. (a)–(d) $N(\phi)/N(90°)$ for $(\phi_{n1}, \phi_{n2}, \phi_{n3}, \phi_{n4}) = (-10°, +10°, -5°, +5°)$. This set of $\{\phi_n\}$ corresponds to the anion gap Δ_{AO} = 100 meV (See Table 1). (e)–(h) $N(\phi)/N(90°)$ for $(\phi_{n1}, \phi_{n2}, \phi_{n3}, \phi_{n4}) = (-10°, +9°, -15°, +14°)$. This set of $\{\phi_n\}$ corresponds to Δ_{AO} = 0 meV.

[4] Kuroki K 2006 *J. Phys. Soc. Jpn.* **75** 051013
[5] Lee I J, Brown S E and Naughton M J 2006 *J. Phys. Soc. Jpn.* **75** 051011
[6] Jérome D 2012 *J. Supercond. Nov. Magn.* **25** 633
[7] Wu W, Chaikin P M, Kang W, Shinagawa J, Yu W and Brown S E 2005 *Phys. Rev. Lett.* **94** 097004
[8] Doiron-Leyraud N, Auban-Senzier P, de Cotret S R, Bourbonnais C, Jérome D, Bechgaard K and Taillefer L 2009 *Phys. Rev. B* **80** 214531
[9] Bourbonnais C and Sedeki A 2009 *Phys. Rev. B* **80** 085105
[10] Kimura Y, Misawa M and Kawamoto A 2011 *Phys. Rev. B* **84** 045123
[11] Yonezawa S, Maeno Y, Bechgaard K and Jérome D 2012 *Phys. Rev. B* **85** 140502(R)
[12] Volovik G E 1993 *JETP Lett.* **58** 469
[13] Vekhter I, Hirschfeld P J, Carbotte J P and Nicol E J 1999 *Phys. Rev. B* **59** R9023
[14] We note that the symmetries of these *d*-wave-like states are not lower than the crystalline symmetry. Thus, strictly speaking, they should not be called as the "*d*-wave" state in terms of the symmetry group theory. This is because we use the term "*d*-wave-like" here.
[15] Lebed A G 1986 *JETP Lett.* **44** 114
[16] Dupuis N and Montambaux G 1994 *Phys. Rev. B* **49** 8993
[17] Yonezawa S, Kusaba S, Maeno Y, Auban-Senzier P, Pasquier C, Bechgaard K and Jérome D 2008 *Phys. Rev. Lett.* **100** 117002
[18] Yonezawa S, Kusaba S, Maeno Y, Auban-Senzier P, Pasquier C and Jérome D 2008 *J. Phys. Soc. Jpn.* **77** 054712

[19] Aizawa H, Kuroki K, Yokoyama T and Tanaka Y 2009 *Phys. Rev. Lett.* **102** 016403
[20] Aizawa H, Kuroki K and Tanaka Y 2009 *J. Phys. Soc. Jpn.* **78** 124711
[21] Lebed A G 2011 *Phys. Rev. Lett.* **107** 087004
[22] Croitoru M D, Houzet M and Buzdin A I 2012 *Phys. Rev. Lett.* **108** 207005
[23] Pevelen D L, Gaultier J, Barrans Y, Chasseau D, Castet F and Ducasse L 2001 *Eur. Phys. J. B* **19** 363
[24] Shimahara H 2000 *Phys. Rev. B* **61** R14936
[25] Mizuno Y, Kobayashi A and Suzumura Y 2011 *Physica C* **471** 49
[26] Hasegawa Y and Fukuyama H 1987 *J. Phys. Soc. Jpn.* **56** 887
[27] Shimahara H 1989 *J. Phys. Soc. Jpn.* **58** 1735
[28] Duprat R and Bourbonnais C 2001 *Eur. Phys. J. B* **21** 219
[29] At first glance, the gap structure in Fig. 7 of Ref. [25] resembles the structure D. However, this is merely because an orthorhombic configuration is used for this figure; thus these nodes are actually parallel to the Γ-X line in the triclinic configuration (A. Kobayashi and Y. Suzumura, private communication.)
[30] Fuseya Y and Suzumura Y 2005 *J. Phys. Soc. Jpn.* **74** 1263
[31] Nagai Y, Nakamura H and Machida M 2011 *Phys. Rev. B* **83** 104523
[32] Uji S, Terashima T, Aoki H, Brooks J S, Tokumoto M, Takasaki S, Yamada J, and Anzai H 1996 *Phys. Rev. B* **53** 14399
[33] Yoshino H, Oda A, Sasaki T, Hanajiri T, Yamada J, Nakatsuji S, Anzai H and Murata K 1999 *J. Phys. Soc. Jpn.* **68** 3142

Experimental studies of the topological superconductor $Cu_xBi_2Se_3$

Yoichi Ando, Kouji Segawa, Satoshi Sasaki and M. Kriener

Institute of Scientific and Industrial Research, Ibaraki, Osaka University
Osaka 567-0047, Japan

E-mail: y_ando@sanken.osaka-u.ac.jp

Abstract. $Cu_xBi_2Se_3$ was recently elucidated to be the first example of a time-reversal-invariant topological superconductor characterized by the gapless surface quasiparticle states consisting of helical Majorana fermions. The evidence for topological superconductivity in $Cu_xBi_2Se_3$ was obtained by the point-contact spectroscopy that indirectly detected the itinerant massless Majorana state on the surface. In this paper, we show new point-contact spectroscopy data which supplement the previous conclusion. We also present the anisotropic normal-state resistivity data for superconducting samples of $Cu_xBi_2Se_3$ which strongly suggest that not only the effective mass but also the scattering rate is anisotropic in this material by a factor of $\sim$3.

1. Introduction

Topological superconductors (TSCs) are characterized by a nontrivial topological structure of the Hilbert space spanned by the bulk wave functions [1, 2, 3]. Due to the bulk-edge correspondence [4], a topologically nontrivial bulk state of a superconductor dictates the appearance of a surface quasiparticle state consisting of Majorana fermions [1, 2, 3], which are their own antiparticles and are of significant current interest [5, 6]. The Majorana nature of the quasiparticles in the surface states stems essentially from the particle-hole symmetry that is an inherent property of the low-energy physics of a superconductor described by the Bogoliubov-de Gennes equation [6].

Among the various kinds of TSCs classified by their symmetries [1], of particular interest is the new topological state preserving time-reversal symmetry [7], which dictates the Majorana fermions to have helical spin structure [2]. Recently, it was theoretically predicted [8] that the superconducting doped topological-insulator material $Cu_xBi_2Se_3$ [9] may be such a time-reversal-invariant TSC because of its peculiar band structure. This prediction motivated us to perform point-contact spectroscopy experiments on this material, and we found a pronounced zero-bias conductance peak (ZBCP) [10]; careful examinations of its magnetic-field dependence as well as the contact dependence allowed us to conclude [10] that the observed ZBCP is a signature of surface Andreev bound states, which gives direct evidence for unconventional superconductivity [11]. Knowing that only four types of superconducting gap functions [8] are allowed by the symmetry of this material [12] and that all possible unconventional states are topological [10], it was possible to further conclude that $Cu_xBi_2Se_3$ is a TSC [10].

Although $Cu_xBi_2Se_3$ has been identified to be the first concrete example of a time-reversal-invariant TSC, there are many remaining issues that need to be experimentally clarified. For example, the exact pairing symmetry should be determined, possibly by π-junction type experiments; also, existence of a Majorana bound state in the vortex core should be elucidated, possibly by STM experiments. However, before conducting such advanced experiments, one should understand this material better, because $Cu_xBi_2Se_3$ samples have a problem of intrinsic inhomogeneity and strong disorder [13, 14, 15].

One issue that remained unresolved in our previous point-contact experiment [10] was the onset temperature of the ZBCP; namely, the ZBCP was found to grow below 1.2 K, while the T_c of the measured sample was 3.2 K. One possibility for this difference was that the ZBCP is prone to thermal smearing and needs a low temperature to become observable. Another possibility was that the T_c was locally 1.2 K at the position beneath the point contact, since the temperature dependence of the diamagnetic signal suggested a broad distribution of local T_c. In this paper, we show new point-contact spectroscopy data in which the ZBCP starts to grow below 3.2 K, suggesting that the second possibility mentioned above was actually the case.

Another issue is the robustness of the odd-parity paring state in $Cu_xBi_2Se_3$ against disorder. According to the common wisdom [16, 17], the odd-parity pairing state should be strongly suppressed by impurity scattering [18]. Given the strong disorder in $Cu_xBi_2Se_3$ [14, 15], one might expect that the odd-parity pairing state cannot survive in real samples. In this context, a recent theory by Michaeli and Fu [19] showed that odd-parity superconductivity in strongly spin-orbit coupled semiconductors like $Cu_xBi_2Se_3$ are much more robust against the pair-breaking effect induced by impurity scattering than in more ordinary odd-parity superconductors. Nevertheless, this theory assumed a scalar impurity potential, so the information of the scattering anisotropy in $Cu_xBi_2Se_3$ is important for better understanding the protection of the odd-parity state from impurity scattering. In this paper, we show the anisotropic normal-state resistivity data which strongly suggest that not only the effective mass but also the scattering rate is anisotropic, but the scattering rate anisotropy is only a factor of $\sim$3.

2. $Cu_xBi_2Se_3$ superconductor

The prototypical topological insulator Bi_2Se_3 consists of basic crystallographic units of Se–Bi–Se–Bi–Se quintuple layers, which are weakly bonded by the van der Waals force. When Cu is intercalated into the van der Waals gap, superconductivity appears below the critical temperature T_c of up to $\sim$ 3.8 K [9]. However, $Cu_xBi_2Se_3$ is difficult to synthesize and samples with a large superconducting volume fraction are rarely obtained with the usual melt-growth method [9, 20]. This problem was recently ameliorated by the development of electrochemical synthesis technique [14] which allowed the synthesis of superconducting samples with the shielding fraction exceeding 50% [13]. In this work, we used $Cu_xBi_2Se_3$ single crystals of slab-like geometry prepared by the electrochemical technique described in detail in Ref. [14].

3. Measurements

The point contacts were prepared on a freshly cleaved surface with the so-called "soft point-contact" technique described in Refs. [10] and [21]. The dI/dV spectra were measured with a lock-in technique by sweeping a dc current that is superimposed with a small-amplitude ac current. The anisotropic resistivity was measured with the Montgomery method [22] by putting four line-shaped contacts along the four parallel edges of a parallelepiped; for this measurement, we employed the dc four-probe technique with a scanner switching the combinations of the current and voltage contacts. All the samples were characterized by a commercial SQUID magnetometer to check for the T_c and the shielding fraction.

4. Zero-bias conductance peak

Figure 1 shows the dI/dV spectra measured at various temperatures on a new $Cu_xBi_2Se_3$ sample with $x = 0.31$. One can see that the conductance near zero bias is slightly enhanced at 3.2 K compared to that at 4.0 K, and a clear peak is formed at 2.4 K and below. The inset of Fig.

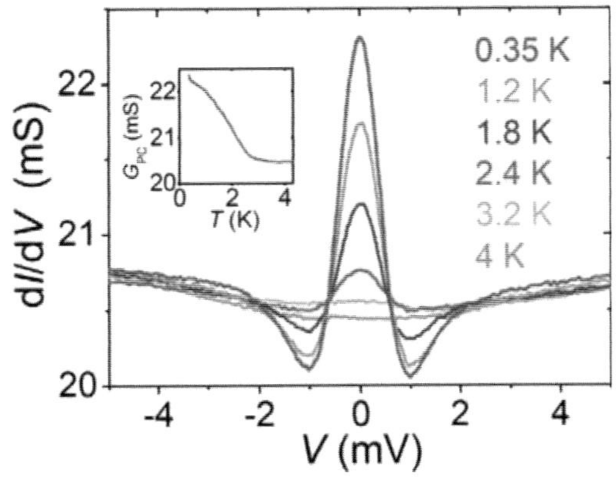

Figure 1. dI/dV spectra at various temperatures measured on a cleaved surface of a $Cu_xBi_2Se_3$ single crystal ($x = 0.31$) using the soft point-contact technique. Inset shows the temperature dependence of the conductance at $V = 0$.

1 shows the temperature dependence of the conductance at zero bias, G_{PC}, which presents a weak enhancement below ~3.8 K and a stronger enhancement below ~2.7 K. Since the T_c of this sample measured by the magnetization was 3.3 K, which is typical for $x = 0.31$, the present result strongly suggest that the zero-bias conductance can start to grow immediately below T_c if the point contact hits a good spot. (The shielding fraction of this sample at 1.8 K was 38%.) Therefore, one may conclude that thermal smearing plays little role in determining the observability of the ZBCO in $Cu_xBi_2Se_3$.

5. Resistivity anisotropy

Since the Montgomery method can sometimes be inaccurate depending on the aspect ratio and the inhomogeneity of the sample, we measured two samples with different aspect ratios to confirm the reproducibility of the result. Figures 2(a) and 2(b) show the temperature dependences of the in-plane resistivity ρ_{ab} and the c-axis resistivity ρ_c measured on two different samples cut from the same single crystal ($x = 0.35$), which showed the T_c of 3.2 K and the shielding fraction of 48% at 1.8 K. Figure 2(c) shows the resistivity anisotropy ratio ρ_c/ρ_{ab} for the two samples. One can see that ρ_c/ρ_{ab} is around 10, and the possible uncertainty (inferred from the difference

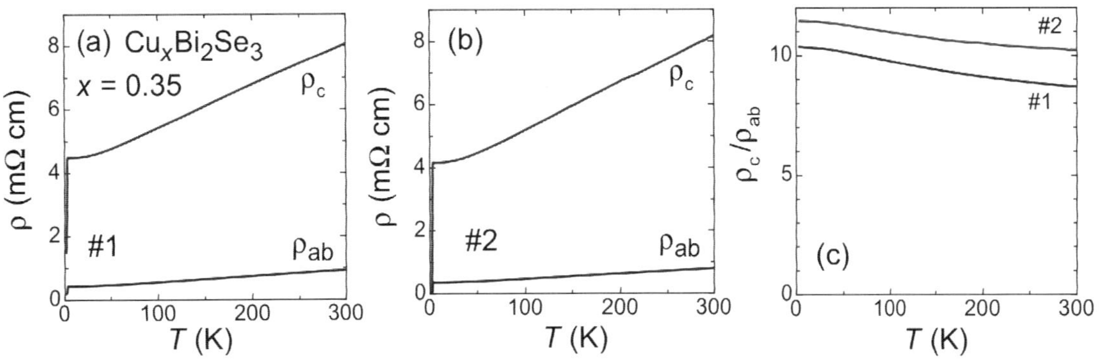

Figure 2. (a),(b) Temperature dependences of ρ_{ab} and ρ_c measured on two samples with different aspect ratios using the Montgomery method. Sample #1 had the thickness $d = 0.334$ mm, the width $w = 0.649$ mm, and the length $L = 1.33$ mm, while the dimensions of sample #2 were $d = 0.340$ mm, $w = 1.64$ mm, and $L = 2.14$ mm. (c) Resistivity anisotropy ratio calculated from the data shown in panels (a) and (b).

between the two data sets) is about 20%.

6. Scattering-rate anisotropy

One should remember that the source of the resistivity anisotropy is usually an effective-mass anisotropy, m_c/m_{ab}. In anisotropic superconductors [23], the anisotropy ratio γ is defined as $\gamma = \xi_{ab}/\xi_c$, where ξ_{ab} and ξ_c are the coherence lengths in the in-plane and the c-axis directions, respectively. By using the anisotropic Ginzburg-Landau theory, one obtains $\gamma = \sqrt{m_c/m_{ab}}$ [23]. Since the mass anisotropy also determines the resistivity anisotropy via $\rho_c/\rho_{ab} = m_c/m_{ab}$ when the scattering rate is isotropic, the resistivity anisotropy in the normal-state resistivity and the anisotropy ratio in the superconducting properties are often related by $\rho_c/\rho_{ab} = \gamma^2$. In fact, in a high-$T_c$ cuprate $YBa_2Cu_3O_y$ near optimum doping, $\gamma \simeq 5$ and $\rho_c/\rho_{ab} \simeq 25$ [23].

In the present case of $Cu_xBi_2Se_3$, we observed that $\gamma \simeq 1.8$ and it is essentially independent of x [15]. This implies that the effective-mass anisotropy in $Cu_xBi_2Se_3$ is $m_c/m_{ab} = \gamma^2 \simeq 3.2$, which is smaller than the observed ρ_c/ρ_{ab}. Therefore, it is most likely that the scattering rate is *not* isotropic in $Cu_xBi_2Se_3$ and it adds additional anisotropy in the normal-state transport properties. One can estimate the scattering-rate anisotropy from $(\rho_c/\rho_{ab})/(m_c/m_{ab})$ to be ~3.

7. Summary

We presented new point-contact spectroscopy data as well as the anisotropic normal-state resistivity data to better understand the topological superconductor $Cu_xBi_2Se_3$. The new point-contact spectra showed that the zero-bias conductance peak starts to grow immediately below T_c, removing the ambiguity about the possible role of thermal smearing. The resistivity anisotropy was found to be ~10, which, in combination with the effective-mass anisotropy of 3.2 determined from the anisotropic superconducting properties [15], suggests that the scattering rate is anisotropic by a factor of ~3. Hence, the scattering potential in $Cu_xBi_2Se_3$ is not strongly anisotropic, which is important for theoretically elucidating the robustness of the odd-parity pairing against impurity scattering.

Acknowledgments

We thank L. Fu, Y. Tanaka, and M. Sato for helpful discussions. This work was supported by JSPS (NEXT Program), MEXT (Innovative Area "Topological Quantum Phenomena" KAKENHI), and AFOSR (AOARD 124038).

References

[1] Schnyder A P, Ryu S, Furusaki A and Ludwig W W A. 2008 *Phys. Rev. B* **78** 195125
[2] Qi X L and Zhang S C 2011 *Rev. Mod. Phys.* **83** 1057
[3] Sato M 2010 *Phys. Rev. B* **81** 220504(R)
[4] Wen X G 1990 *Int. J. Mod. Phys.* B **4** 239
[5] Wilczek F 2009 *Nature Phys.* **5** 614
[6] Alicea J 2012 *Rep. Prog. Phys.* **75** 076501
[7] Qi X L, Hughes T L, Raghu S and Zhang S C 2009 *Phys. Rev. Lett.* **102** 187001
[8] Fu L and Berg E 2010 *Phys. Rev. Lett.* **105** 097001
[9] Hor Y S *et al* 2010 *Phys. Rev. Lett.* **104** 057001
[10] Sasaki S *et al* 2011 *Phys. Rev. Lett.* **107** 217001
[11] Kashiwaya S and Tanaka Y 2000 *Rep. Prog. Phys.* **63**, 1641
[12] Liu C X *et al* 2010 *Phys. Rev. B* **82** 045122
[13] Kriener M, Segawa K, Ren Z, Sasaki S and Ando Y 2011 *Phys. Rev. Lett.* **106** 127004
[14] Kriener M *et al* 2011 *Phys. Rev. B* **84** 054513
[15] Kriener M 2012 arXiv:1206.6260.
[16] Balian R and Werthamer N R 1963 *Phys. Rev.* **131** 1553
[17] Larkin A I 1965 *Sov. Phys. JETP Lett.* **92** 130
[18] Mackenzie A P and Maeno Y 2003 *Rev. Mod. Phys.* **75** 657

[19] Michaeli K and Fu L 2012 arXiv:1203.1055
[20] Wray L A *et al* 2010 *Nature Phys.* **6** 855
[21] Daghero D and Gonnelli R S 2010 *Supercond. Sci. Technol.* **23** 043001
[22] Montgomery H C 1971 *J. Appl. Phys.* **42** 2971
[23] Clem J R 1989 *Physica C* **162-164** 1137

NMR study of the impurity effects in noncentrosymmetric superconductors $Li_2(Pd,Pt)_3B$

K Arima[1], K Matano[1], S Harada[1], G Z Bao[1], Y Inada[2], G Q Zheng[1,3]

[1]*Department of Physics, Okayama University, Okayama 700-8530, Japan*

[2]*Faculty of Education, Okayama Univertsity, Okayama 700-8530, Japan*

[3]*Institute of Physics, Chinese Academy of Sciences, and Beijing National Laboratory for Condensed Matter Physics, Beijing 100190,China*

zheng@psun.phys.okayama-u.ac.jp

Abstract. We report ^{11}B nuclear magnetic resonance (NMR) studies of the impurity effect in noncentrosymmetric superconductors Li_2Pt_3B and Li_2Pd_3B. We substitute B with Al, and find that the full width of half maximum (FWHM) of the NMR spectrum increases with increasing Al content. The superconducting transition temperature (T_c) of $Li_2Pt_3\ B_{1-x}Al_x$ decreases drastically with increasing non-magnetic impurity (Al) content, while it is hardly affected in $Li_2Pd_3\ B_{1-x}Al_x$. These results provide new evidence for superconducting gap with line nodes in Li_2Pt_3B.

1. Introduction

In most superconductors, there is an inversion center in the crystal structure. In such superconductors, the Cooper pairs are either in the spin-singlet state or spin-triplet state because of parity conservation law. If a crystal lacks an inversion centre, an electric field is induced which leads to an asymmetric spin-orbit coupling (ASOC). This allows a mixing of the spin-singlet and spin-triplet superconducting states [1, 2].

Li_2Pd_3B and Li_2Pt_3B are non-centrosymmetric superconductors with T_c =7 K and 2.7 K, respectively [3,4]. Fig. 1 shows the crystal structure of Li_2M_3B (space group; $P4_332$). It consists of distorted octahedra of BM_6. The electron-correlation is weak in these compounds and they are most suitable for studying the effects of the inversion symmetry breaking. Previous works have found that the superconducting state is dominantly in a spin-triplet state with line nodes in the gap function in Li_2Pt_3 B [5], while Li_2Pd_3B is a conventional superconductor [6]. The ^{11}B spin-lattice relaxation rate ($1/T_1$) in Li_2Pd_3B shows a coherence peak just below T_c and decreases exponentially with decreasing temperature, and the Knight shift changes below T_c , which indicate a spin-singlet state with an *s*-wave gap [6]. In the case of Li_2Pt_3B, however, $1/T_1$ shows no coherence peak and decrease with T^3, and the Knight shift does not change across T_c. Recently, it has been found that the contrasting behaviour of the superconducting state of the two compounds stem from a difference in the local crystal structure that is responsible for the ASOC [7].

In order to shed new light on the superconducting gap, here we report a study of the impurity effect on these compounds by substituting Boron with Aluminum and characterizing with nuclear magnetic resonance (NMR). We replace B with Al as a non-magnetic impurity with concentrations up to 5%.

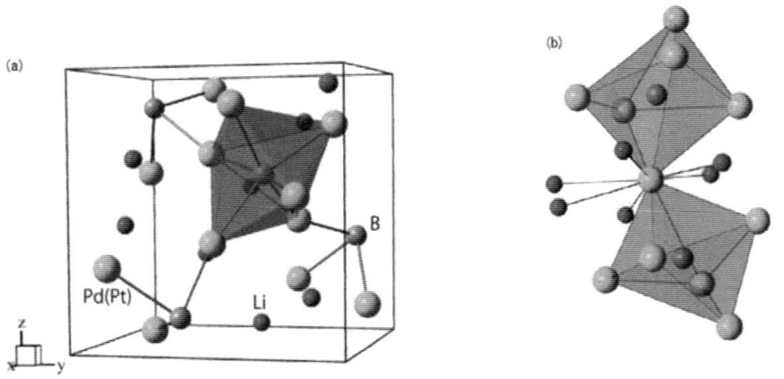

Figure 1. (a) Crystal structure of Li_2M_3B (M=Pd, Pt). In a unit cell, there are 8 Li, 12 M and 4 B. (b) Neighbourhood near Pt atom. The octahedral of BM_6 is distorted.

2. Experimental results

Poly-crystal samples of $Li_2(Pd, Pt)_3B_{1-x}Al_x$ (x=0, 0.02, 0.05) were prepared by arc melting method with starting materials of Li (99.9% purity), Pt (99.999%), Pd (99.9%), B (99.8%), Al (99.99%). The two-step arc-melting process was used [4, 5]. For NMR measurements, the polycrystalline samples were crushed into powder. The T_c at zero magnetic field was determined by measuring the ac susceptibility using the NMR coil. The T_c of Li_2Pd_3B and Li_2Pt_3B is 7.2 K and 2.9 K, respectively. The NMR spectra were obtained by fast Fourier transform (FFT) of the spin echo taken at H=5 T. The full width of half maximum (FWHM) of the Li_2Pd_3B and Li_2Pt_3B are around 6 kHz and 5.5 kHz, respectively.

Fig. 2(a) shows the ^{11}B NMR spectrum of $Li_2Pd_3B_{1-x}Al_x$ and Fig. 2(b) shows that of $Li_2Pt_3B_{1-x}Al_x$. In Fig. 3(c), we plotted the FWHM of ^{11}B NMR spectrum vs. Al content. One can see that the FWHM for both compounds increases with increasing Al content up to x=0.05, which indicates that Al indeed replaces B in the crystal, and act as impurity.

Fig. 3 shows the dependence of T_c on the impurity content x. The vertical axis is normalized by T_c of Li_2Pd_3B or Li_2Pt_3B. The T_c of $Li_2Pd_3B_{1-x}Al_x$ is 7.4, 7.3 and 7.3K for x= 0, 0.02 and 0.05, and the T_c of $Li_2Pt_3B_{1-x}Al_x$ is 2.9, 2.6 and 2.2K for x=0, 0.02 and 0.05, respectively. In Li_2Pd_3B, the Al impurity up to x=0.05 has little effect on T_c. By contrast, in Li_2Pt_3B, the Al impurity suppresses T_c drastically.

These results indicate that Li_2Pt_3B is sensitive to impurity scattering. The behavior in Li_2Pt_3B is similar to that of spin-triplet superconductor Sr_2RuO_4 [8]. Generally, non-magnetic act as strong pair breakers and suppress T_c drastically in superconductors with nodes in the gap function. Thus, our results provide new evidence for nodal superconducting gap in Li_2Pt_3B.

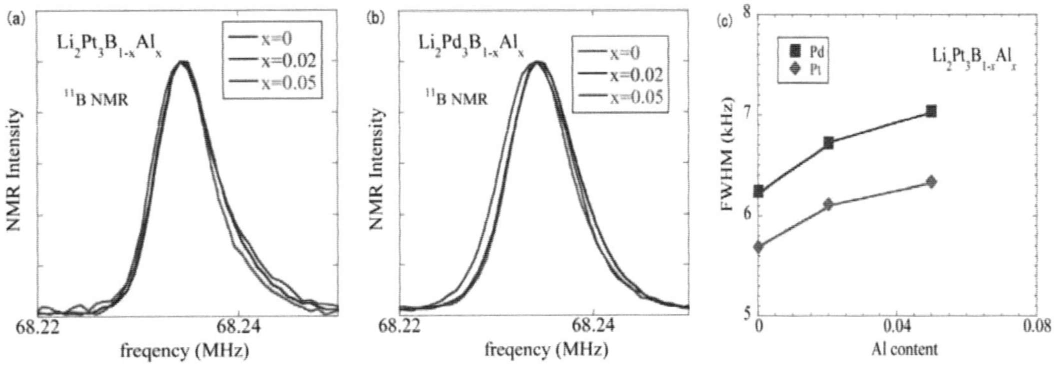

Figure 2. (a) ,(b) ^{11}B NMR spectra of $Li_2(Pd, Pt)_3B_{1-x}Al_x$ (x=0, 0.02, 0.05)
(c) the Al content dependence of the FWHM of ^{11}B NMR spectra. FWHM increases with increasing Al content.

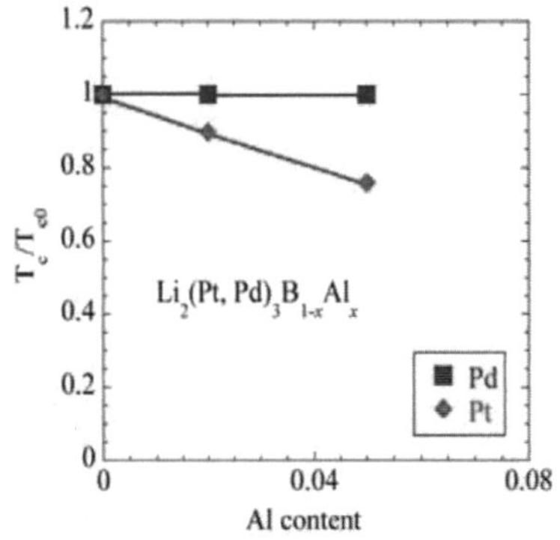

Figure 3. The Al impurity content dependence of T_c in $Li_2(Pd, Pt)_3B_{1-x}Al_x$.

3. Conclusion

In conclusion, we have presented a study of the impurity effect on the superconducting transition temperature in noncentrosymmetric superconductors Li_2Pt_3B and Li_2Pd_3B. We found that the FWHM of ^{11}B NMR spectrum increases with increasing x in both $Li_2Pt_3B_{1-x}Al_x$ and $Li_2Pt_3B_{1-x}Al_x$. However, the Al impurity suppresses the T_c drastically in $Li_2Pt_3B_{1-x}Al_x$, while it has little effect in $Li_2Pd_3B_{1-x}Al_x$. Our results indicate that Li_2Pt_3B is sensitive to impurity scattering and provide new evidence for the superconducting gap with nodes in Li_2Pt_3B.

References

[1] Golkov L P and Rashba E I 2001 *Phys. Rev. Lett.* **87** 037004
[2] Frigeri P A, Agterberg D F, Koga A and Siglist M 2004 *Phys. Rev. Lett.* **92** 097001
[3] Togano K, Badica P, Nakamori Y, Orimo S, Takeya H and Hirata K, 2004 *Phys. Rev. Lett.* **93** 247004
[4] Badica P, Kondo T and Togano K, 2005 *J. Phys. Soc. Jpn.* **74** 1014
[5] Nishiyama M, Inada Y and Zheng G Q 2005 *Phys. Rev. B.* **71** 220505(R)
[6] Nishiyama M, Inada Y and Zheng G Q 2007 *Phys. Rev. Lett.* **98** 047002
[7] Harada S, Zhou JJ, Yao YG, Inada Y and Zheng G Q, to be published
[8] Kikugawa N, Mackenzie A P and Maeno Y 2003 *J. Phys. Soc. Jpn.* **72** 237

Magnetic field induced transition in superconducting $LaTiO_3/SrTiO_3$ interfaces

J Biscaras[1], N Bergeal[1], S Hurand[1], C Feuillet-Palma[1], A Rastogi[2], R C Budhani[2,3], M Grilli[4], S Caprara[4], J Lesueur[1]

[1]LPEM- UMR8213/CNRS - ESPCI ParisTech - UPMC, 10 rue Vauquelin - 75005 Paris, France
[2]Condensed Matter - Low Dimensional Systems Laboratory, Department of Physics, Indian Institute of Technology Kanpur, Kanpur 208016, India
[3]National Physical Laboratory, New Delhi - 110012, India
[4]CNR-ISC, CNISM, and Dipartimento di Fisica, Università di Roma "La Sapienza", P.[le] Aldo Moro 5, 00185 Roma, Italy

E-mail: `jerome.lesueur@espci.fr`

Abstract.
Superconductivity at the $LaTiO_3/SrTiO_3$ interface is studied by low temperature and high magnetic field measurements as a function of a back-gate voltage. We show that it is intimately related to the appearance of a low density (a few 10^{12} cm^{-2}) of high mobility carriers, in addition to low mobility ones always present in the system. These carriers form superconducting puddles coupled by a metallic two-dimensional electron gas, as revealed by the analysis of the phase transition driven by a perpendicular magnetic field. Two critical fields are evidenced, and a quantitative comparison with a recent theoretical model is made.

1. Introduction

Superconductivity at the interface between insulating oxides[1] has raised considerable interest, all the more so the superfluid density can be tuned by a gate voltage[2, 3, 4]. This discovery opens the way to fundamental investigations on superconductivity in a two-dimensional electron gas (2DEG), as for instance the exploration of the specific role of quantum confinement, disorder, Coulomb repulsion and screening in a true two-dimensional superconductor. Specific features of quantum wells, such as Rashba spin orbit coupling[5], further enhance the interest for these systems by enabling tunable pairing symmetry [6], or exotic spin dependent behaviours[7, 8]. However, the electrostatic control of the 2DEG properties both in the normal and the superconducting state is not fully understood yet. In this article, we show that two types of carriers are present in the 2DEG, and that superconductivity is intimately related to the occurrence of high-mobility ones which set at the edge of the quantum well. By applying a perpendicular magnetic field, we show that the superconducting behaviour is characteristic of a disordered array of superconducting puddles coupled by a tunable normal metal. This intrinsic behaviour has to be taken into account when studying basic properties of the superconducting 2DEG[4].

2. Two types of carriers

Samples were grown by pulsed laser deposition of a 15 unit cells thick layer of $LaTiO_3$ on TiO_2-terminated (001)$SrTiO_3$ substrates (see Biscaras *et al.*[9] for details). A metallic back-gate was deposited at the rear of the 500 μm thick substrate and connected to a voltage source (V_G). Standard four-probe resistance measurements were made with a current sufficiently low to avoid any heating of the electron at the lowest temperatures. After the cool-down, the gate voltage V_G is ramped up to +200 V. This polarisation procedure insures that all voltage sweeps made subsequently are reversible and reproducible [10].

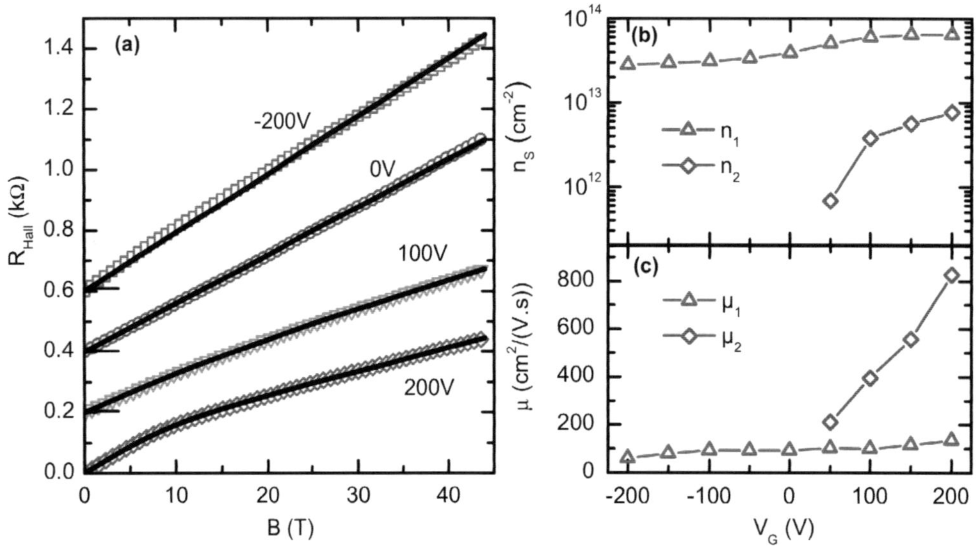

Figure 1. (a) Hall resistivity R_H as a function of the magnetic field B measured at 4.2 K for different V_G. R_H is linear in B for negative V_G, and non linear for positive V_G. (b) Carrier density n_S as a function of V_G, for the two types of carriers : n_1 majority carriers, n_2 minority carriers. (c) Mobility μ as a function of V_G, for the two types of carriers : μ_1 low mobility carriers (LMC), μ_2 high mobility carriers (HMC).

We performed Hall effect measurements up to 45 T in pulsed magnetic field. Results at a temperature of 4.2 K are presented in Fig. 1 (a). For negative V_G, the Hall effect is linear in magnetic field as expected, whereas it displays two distinct slopes (low field and high field) for positive V_G. The latter situation is characteristic of a multi-carriers electronic transport. We fitted the experimental curves using a two-carriers model whose parameters are the densities n_1 (majority carriers) and n_2 (minority carriers) and the corresponding mobilities μ_1 and μ_2[3]. Their variations with V_G are displayed in Fig. 1 (b) and (c). Majority carriers (n_1) are always present with a low and almost constant mobility (μ_1). For positive V_G, minority carriers appear (n_2), with a high and rapidly increasing mobility (μ_2). In the following, theses two types of carriers will be referred to as Low Mobility Carriers (LMC) and High Mobility Carriers (HMC).

To understand the origin of these two types of carriers we modelled the quantum well at the interface using a semi-conductor type model [3]. We assumed that the 2DEG is confined within the $SrTiO_3$ by an infinite potential barrier corresponding to the $LaTiO_3$ Mott gap. Carriers are therefore free to move within the interface plane, but are confined in the perpendicular direction, leading to a discrete 2D character of the energy spectrum. We solved the Schrödinger and the

Poisson equations self-consistently, introducing the appropriate boundary conditions imposed by the gate voltage V_G and taking into account the non-linear dielectric constant of $SrTiO_3$. The shape of the quantum well given by the bending of the $SrTiO_3$ conduction band and the envelops of the wave-functions for different sub-bands are computed (figure 2 inset). Details of the calculations are given in reference [3]. For negative V_G, a few sub-bands are occupied, corresponding to wave-functions confined close to the interface. These carriers are submitted to surface scattering and their mobility is low. For positive voltage, more sub-bands are filled, and the wave functions of the upper energy levels extent deeper in $SrTiO_3$ substrate (inset Fig. 2), where the scattering is weaker and the mobility higher. According to our simulations, these carriers appear only for positive gate voltage, when the conduction band is sufficiently bent. As seen in Fig. 2, their extension increases with V_G, so does the HMC density ($n_{HMC} = n_2$). Therefore, it is clear that LMC correspond to carriers deep in the well and HMC to those which set at the edge of the quantum well.

When adding on the same graph T_C as a function of V_G (Fig. 2), we observe that superconductivity is intimately related to the presence of HMC in the well ($n_{HMC} \neq 0$). Unfortunately, as no superfluid density measurement is available on the $LaTiO_3/SrTiO_3$ interface, it is not possible to directly linked the HMC to the superconducting carriers at low temperature. However, we notice that our observation is consistent with a recent measurement of the superfluid density in the similar superconducting $LaAlO_3/SrTiO_3$ interface. [11]. Indeed, they found that only a small fraction of the total carrier density is superconducting (typically a few 10^{12} cm^{-2}). This value is of the same order as the HMC density we measure in the normal state of the $LaTiO_3/SrTiO_3$ interface. They also showed that the superfluid density increases roughly linearly with V_G for positive gate voltage as we observed for HMC (Fig. 2). As a conclusion, we think that only the HMC condense in the superconducting state while the LMC remain in a metallic state. It is worth noticing that Bert *et al.* found that the superfluid density is spatially inhomogeneous, especially in the low doping region. It seems that superconductivity is observed in some regions of the sample, connected by non-superconducting ones. We will come back to this point in the second part of this article.

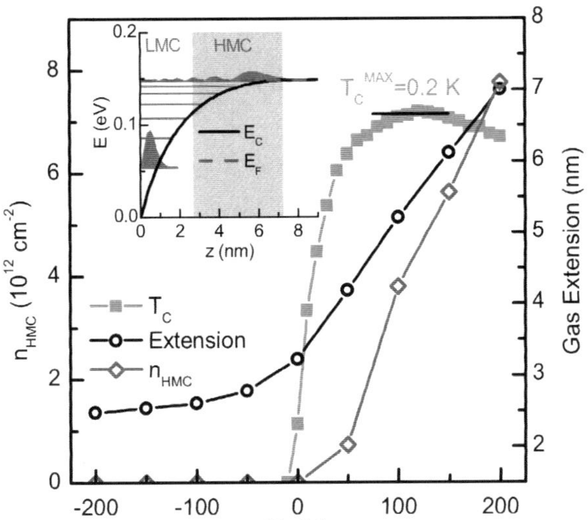

Figure 2. HMC sheet density n_{HMC} (left scale), calculated gas extension (right scale) and T_C as a function of V_G. Inset : Shape of the quantum well at the interface for positive V_G. The sub-bands energy levels and the corresponding wave-function envelops are shown in red. The dashed green line is the Fermi level E_F. Majority carriers (yellow area) lie very close to the interface, whereas minority carriers (blue area) set deeper in the bulk.

3. Magnetic field induced transition

As can be seen in Fig. 3 (a) for $V_G = -15$ V, a perpendicular magnetic field drives the system towards a non superconducting state. A zoom on the low temperature part of the graph reveals that for a critical field $B_\times = 0.1$ T, a plateau corresponding to at the sheet resistance $R_\times = 2176\ \Omega/\square$ separates the superconducting behaviour from a weakly localizing one (Fig. 3 (b)). When $V_G = +80$ V, the transition occurs in two steps, as can be seen in Fig. 3 (c) and (d). A first plateau in R_S is observed between 0.1 and 0.2 K for a characteristic field $B_\times = 0.185$ T ($R_\times = 372.4\ \Omega/\square$), and a second one at lower temperature for a critical field $B_C = 0.235$ T, ($R_C = 376.6\ \Omega/\square$).

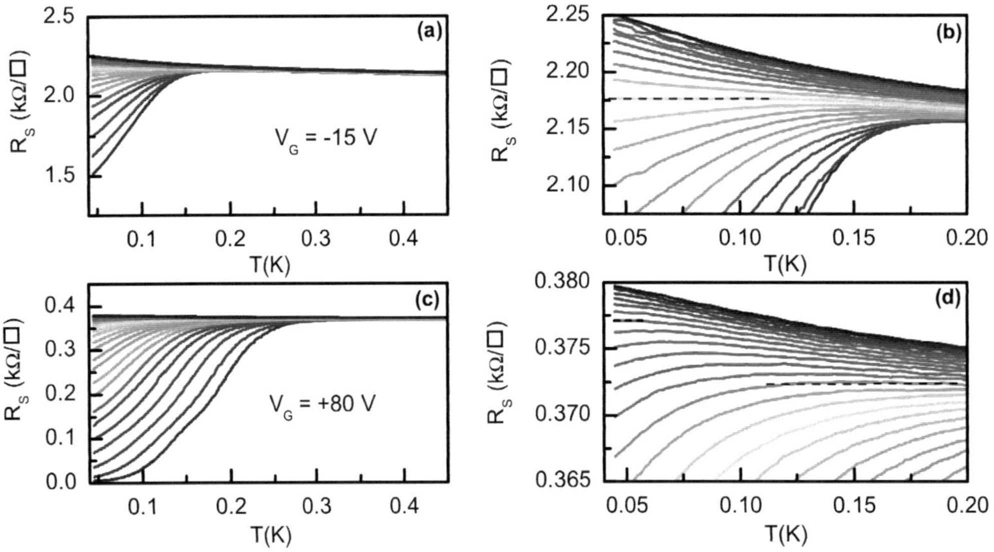

Figure 3. (a) Sheet resistance R_S as a function of temperature T for magnetic field ranging from 0 to 0.3 T ($V_G = -15$ V). (b) Zoom on the same data : a plateau in R_S is seen below 0.12 K (dashed line). (c) Sheet resistance R_S as a function of temperature T for magnetic field ranging from 0 to 0.3 T ($V_G = +80$ V). (d) Zoom on the same data : two plateaus (dashed lines) in R_S are evidenced (see text).

In Fig. 4 (a), we report the characteristic fields as a function of V_G, and distinguish two regions. In region I, a single critical field is observed, whereas in region II, $B_\times$ and B_C are distinct. In this region, we can see that $B_\times$ is roughly constant at a value B_d. In the whole phase diagram, T_C clearly follows B_C, which is therefore the field for which superconductivity disappears. To account for these observations, we propose the following scenario based on the XY model where superconductivity is destroyed by phase fluctuations in a two-dimensional superconductor. We suppose that the system consists in superconducting islands coupled by non-superconducting metallic regions (see Fig. 4(b) for a sketch). Taking into account the two types of carriers mentioned above, we assume that the HMC form the superfluid with an intrinsic inhomogeneity due to its very low average density and the associated density fluctuations, and that the LMC form the metal which provides a long range coupling. The system is therefore described as a disordered array of superconducting puddles (HMC) of size L_d coupled by a metallic 2DEG (LMC) (see sketch in Fig. 4(b)).

Such a situation has been theoretically studied by Spivak, Oreto and Kivelson (SOK)[12]. Superconducting phase coherence of the array is governed by the transport properties of the metallic part through the proximity effect, and depends on the 2DEG conductance G_{2DEG}. If the coupling is strong enough (high G_{2DEG}), puddles can develop full local superconductivity and two transitions can be observed corresponding respectively to the one of the puddles themselves and to the one of the disordered array of puddles. In the opposite case (low G_{2DEG}), decoupled puddles are always in a fluctuating regime, and only the full array transition can be seen at low temperature. These two situations are observed in our $LaTiO_3/SrTiO_3$ interfaces, corresponding to regions II and I respectively. The gate voltage V_G is the tuning parameter to switch from one to the other as it controls the sheet conductance of the 2DEG $G_S = 1/R_S$ [3, 4], and therefore the coupling between puddles. Here, G_S is taken to be the inverse of the sheet resistance measured in the normal state. In region I, G_S takes small values and the coupling is too weak to insure full local superconductivity on the puddles. There is therefore only one critical field that corresponds to the transition of the whole array. In region II, G_S takes larger value and we can observe separately the transition of the puddles ($B_\times$) at high temperature and the one of the array (B_C) at lower temperature. In that case, the first transition occurs when the magnetic flux threading a puddle of typical size L_d reaches the flux quantum Φ_0. Experimentally this takes place for a constant dephasing field $B_d \sim \Phi_0/L_d^2$ (Fig. 4 (a)). The typical value of L_d is of the order of 100 nm. The SOK model [12] predicts that B_C is related to the coupling parameter G_{2DEG} (in our case G_S) in a specific way. At low conductance, $B_C \propto G_{2DEG}$, whereas as high conductance, $B_C \propto [G_{2DEG}]^{1/4}$. To test the model in more details, we plotted the critical field B_C as a function of the conductance G_S on a log-scale in Fig. 4 (c). The data are in good agreement with the theory. Not only the two regimes can be clearly identified, but also the values of the slopes correspond to the calculated one. This is a strong indication that the SOK model is a good representation of the physics involved in these experiments.

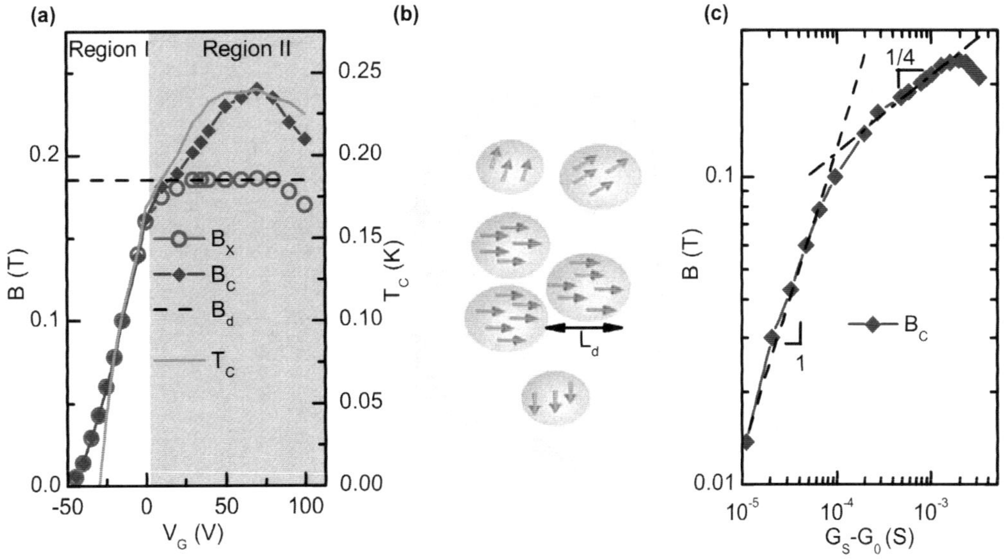

Figure 4. (a) $B_\times$, B_C (left scale) and T_C (right scale) as a function of V_G. The dashed line corresponds to B_d (see text). Regions I and II refer to the low, and respectively high, coupling regimes (see text). (b) Sketch of a piece of material. In this drawing, superconducting puddles (in blue) are coupled through a 2DEG (in yellow). The arrows symbolize the local phase of the superconductor. (c) B_C as a function of G_S on a log scale. $G_0 = 356$ μS is the conductance at V_G =-45V where BC goes to zero. Dashed lines have slopes 1 and 1/4 respectively as expected from the SOK model.

4. Conclusion

As a conclusion, superconductivity in $LaTiO_3/SrTiO_3$ interfaces is related to the appearance of Highly Mobile Carriers which set at the edge of the quantum well upon positive back-gate biasing. Their density is very low, in the 10^{12} cm^{-2} range, as compared to the majority Low Mobility Carriers one ($10^{13} - 10^{14}$ cm^{-2}). The transition to the non-superconducting state in a perpendicular magnetic field reveals that this low density superconductivity is inhomogeneous, and that the superfluid is organized in puddles (HMC) coupled by a metallic 2DEG (LMC). The back-gate voltage controls the coupling parameter (namely the 2DEG conductance), and drives the system from a single transition situation to a double one, where the puddles themselves and the disordered array of puddles display separate transitions.

References

[1] Reyren N, Thiel S, Caviglia A D, Kourkoutis L F, Hammerl G, Richter C, Schneider C W, Kopp T, Ruetschi A S, Jaccard D, Gabay M, Muller D A, Triscone J M and Mannhart J 2007 *Science* **317** 1196–1199

[2] Caviglia A, Gariglio S, Reyren N, Jaccard D, Schneider T, Gabay M, Thiel S, Hammerl G and Triscone J M 2008 *Nature* **456** 624

[3] Biscaras J, Bergeal N, Hurand S, Grossetete C, Rastogi A, Budhani R C, LeBoeuf D, Proust C and Lesueur J 2012 *Phys. Rev. Lett.* **108** 247004

[4] Biscaras J, Bergeal N, Hurand S, Feuillet-Palma C, Rastogi A, Budhani R C, Grilli M, Caprara S and Lesueur J 2012 *arXiv.org* 1209.6464

[5] Caviglia A D, Gabay M, Gariglio S, Reyren N, Cancellieri C and Triscone J M 2010 *Phys. Rev. Lett.* **104** 126803

[6] Yada K, Onari S, Tanaka Y and Inoue J I 2009 *Phys. Rev. B* **80** 140509

[7] Tanaka Y, Yokoyama T, Balatsky A and Nagaosa N 2009 *Phys. Rev. B* **79** 060505

[8] Liu X, Borunda M F, Liu X J and Sinova J 2009 *Phys. Rev. B* **80** 174524

[9] Biscaras J, Bergeal N, Kushwaha A, Wolf T, Rastogi A, Budhani R C and Lesueur J 2010 *Nature Commun.* **1** 89

[10] Biscaras J, Hurand S, Feuillet-Palma C, Rastogi A, Budhani R C, LeBoeuf D, Proust C, Lesueur J and Bergeal N 2012 *arXiv* 1206.1198

[11] Bert J, Nowack K, Kalisky B, Noad H, Kirtley J, Bell C, Sato H, Hosoda M, Hikita Y, Hwang H and Moler K 2012 *Phys. Rev. B* **86** 060503

[12] Spivak B, Oreto P and Kivelson S 2008 *Phys. Rev. B* **77** 214523

Journal of Physics: Conference Series **449** (2013) 012036 doi:10.1088/1742-6596/449/1/012036

Anomalous Hall conductivity of clean Sr_2RuO_4 at finite temperatures

Edward Taylor and Catherine Kallin

Department of Physics and Astronomy, McMaster University, Hamilton, Ontario, L8S 4M1, Canada

E-mail: kallin@mcmaster.ca

Abstract. Building on previous work, we calculate the temperature- and frequency-dependent *anomalous* Hall conductivity for the putative multiband chiral superconductor Sr_2RuO_4 using a simple microscopic two-orbital model without impurities. A Hall effect arises in this system without the application of an external magnetic field due to the time-reversal-symmetry breaking chiral superconducting state. The anomalous Hall conductivity is nonzero only when there is more than one superconducting order parameter, involving inter- as well as intra-band Cooper pairing. We find that such a multiband superconducting state gives rise to a distinctive resonance in the frequency-dependence of the Hall conductivity at a frequency close to the inter-orbital hopping energy scale that describes hopping between Ru d_{xz} and d_{yz} orbitals. The detection of this feature, robust to temperature and impurity effects in the superconducting phase, would thus constitute compelling evidence in favour of a multiband origin of superconductivity in Sr_2RuO_4, with strong superconductivity on the α and β bands. The temperature dependence of the Hall conductivity and Kerr rotation angle are studied within this model at the one-loop approximation.

1. Introduction

Since superconductivity was discovered in the layered perovskite Sr_2RuO_4 in 1994 [1], there has been much speculation as to the nature of the superconducting order parameter. Almost immediately, it was realized that Sr_2RuO_4 might be chiral p-wave [2, 3], an electronic analogue of the A-phase of ^{3}He. However, after more than a decade and a half of intense experimental and theoretical work, basic questions—is the order parameter chiral and if so, where on the three active bands of Sr_2RuO_4 does it live?—remain controversial. (For recent reviews, see Refs. [4, 5], as well as the comprehensive earlier review in Ref. [6].)

If Sr_2RuO_4 turns out to be chiral p-wave, it would be the only known example so far of this remarkable state in electronic systems. Chiral superconductivity breaks time-reversal symmetry with order parameters describing Cooper pairs with nonzero magnetic moments. For chiral p-wave superconductivity, the time-reversal symmetry breaking triplet order parameter is

$$\Delta_{\mathbf{k}} \propto \langle c^\dagger_{\mathbf{k}} c^\dagger_{-\mathbf{k}} \rangle \sim k_x \pm i k_y, \tag{1}$$

corresponding to a nonzero angular momentum $\mathbf{L}_z = \pm\hbar$ in the z-direction per Cooper pair. Apart from the intrinsic interest in observing and understanding such an unusual quantum state, the quest to find chiral p-wave superconductors has generated enormous interest since they have been predicted to harbour Majorana bound states [7], quasiparticles which are their own antiparticles and which may play an important role in quantum computation.

One important feature of Sr_2RuO_4 that can impact the nature of the possibly chiral order parameter is the fact that it is multiband [8]. Three atomic orbitals dominate the electronic structure of Sr_2RuO_4 close to the Fermi surface: Ru d_{xy}, d_{xz}, and d_{yz} orbitals (strongly hybridized with O p orbitals). The d_{xz} and d_{yz} orbitals hybridize to form the quasi-one-dimensional α and β bands, while the d_{xy} orbital predominates the γ band, which has an approximately isotropic two-dimensional dispersion. Spin-orbit coupling leads to some hybridization between the d_{xz}/d_{yz} and d_{xy} orbitals, which otherwise would not hybridize [9]. Importantly for Sr_2RuO_4, the Fermi surface crosses all three of the bands resulting from this orbital structure.

Where does superconductivity arise amongst these three bands? The most widely-held viewpoint holds that superconductivity arises primarily on the γ band (see, for instance, Refs. [10, 11, 12, 13]). A persistent contrarian position, however, maintains that superconductivity instead arises primarily on the α and β bands [14, 15, 16, 17]. The latter model has one very appealing feature insofar as it may give rise to negligible spontaneous supercurrents at sample edges [17]. Such edge currents have been predicted to arise in one-band chiral p-wave superconductors [18], but bafflingly, have never been observed [19]. (For a recent review of experimental probes of the order parameter in Sr_2RuO_4, see Ref. [20].) Non-topological chiral superconductivity on the α and β bands could explain this.

Amongst experiments purporting to show evidence for time-reversal symmetry breaking in Sr_2RuO_4, one of the most direct probes is the Kerr rotation experiment carried out in the group of Kapitulnik at Stanford [21, 22]. In this experiment, a near-infrared beam ($\hbar\omega = 0.8$eV) is reflected off a sample of Sr_2RuO_4, the beam's axis of polarization rotating in the process. The angle of rotation is called the *Kerr angle*. It is related to the anomalous Hall conductivity, σ_H, by [23]

$$\theta_K(\omega) = (4\pi/\omega d)\mathrm{Im}[\sigma_H(\omega)\alpha(\omega)]. \tag{2}$$

Here, d is the interlayer spacing and $\alpha \equiv 1/(n(n^2-1)$ with $n(\omega)$ the complex, frequency-dependent index of refraction. As we discuss in detail in Sec. 2, a nonzero Hall conductivity indicates that time-reversal symmetry is broken.

If time-reversal symmetry is broken by chiral superconductivity, spectroscopic knowledge of the frequency-dependent Kerr angle can tell us a lot about the chiral superconducting order parameter in Sr_2RuO_4 – in particular, where it lies in relation to the three active bands of Sr_2RuO_4. (Spectroscopy of Leggett-like collective modes [24] would also provide considerable insight into this matter.) It is the point of this paper to expand on this idea and suggest ways for future experiments to make further progress in answering questions related to the nature of the order parameter.

On its own, time-reversal symmetry breaking is not sufficient to have a nonzero Hall conductivity in the absence of an applied magnetic field and hence, a non-zero Kerr angle. As pointed out by Read and Green [7], for instance, the Hall conductivity of a translationally-invariant chiral p-wave superconductor is identically zero [25]. Since the publication of Kapitulnik's group's data in 2006 [21], two credible mechanisms have emerged in which chiral p-wave superconductivity in Sr_2RuO_4 can give rise to a nonzero anomalous Hall conductivity. First, a purely extrinsic mechanism due to impurity scattering was suggested by Goryo [26] and Lutchyn *et al.* [27]. Subsequently, an *intrinsic* mechanism was identified by us [28] as well as Wysokiński *et al.* [29]. (A previously identified intrinsic contribution from chiral p-wave collective modes [30] is estimated to be several orders of magnitude too small to explain the experiments [21] and will not be discussed here.)

Very usefully for the questions identified above, the intrinsic mechanism gives rise to a nonzero anomalous Hall conductivity if and only if there are multiple chiral order parameters spanning multiple optically active bands. For Sr_2RuO_4, this means that there *must* be superconductivity on the α and β bands in order for the intrinsic σ_H to be nonzero. In contrast, the extrinsic mechanism contributes as well when chiral superconductivity occurs on the γ band. Thus, if it was found that the intrinsic mechanism dominated the Hall conductivity, one could reliably conclude that the origin of superconductivity in Sr_2RuO_4 lay in the α and β bands and not the γ band.

In this paper, we suggest two ways in which experiments can deduce the relative importance of the intrinsic and extrinsic mechanisms. First, the intrinsic mechanism produces a very strong spectral feature in the frequency-dependent anomalous Hall conductivity (and Kerr angle) close to twice the inter-orbital $d_{xz} - d_{yz}$ hopping energy, near 0.1eV. The detection of this feature, robust to temperature and impurity effects in clean samples of Sr_2RuO_4, would provide nearly unequivocal evidence in our opinion for strong superconductivity on the α and β bands. Second, by measuring the dependence of σ_H on the impurity concentration, knowing the dependence of the superconducting gap on this quantity as well, one could estimate the relative roles of the intrinsic and extrinsic contributions.

2. Chiral two-band superconductor

Following our earlier work [28], we analyze a two-orbital model in order to make clear the physics of the anomalous Hall conductivity in a multiband superconductor at finite temperatures. We will use this model later to describe the physics of Sr_2RuO_4 since, although three orbitals are needed to account for the Fermi surface properties [6], only two orbitals–the Ru d_{xz} and d_{yz} ones–are directly relevant for the Hall conductivity.

We consider the following BCS pairing Hamiltonian describing superconductivity in a two-orbital system:

$$H=\sum_{\mathbf{k}}\begin{pmatrix} c^{\dagger}_{\mathbf{k}1} & c^{\dagger}_{\mathbf{k}2}\end{pmatrix}\begin{pmatrix}\xi_1(\mathbf{k}) & \epsilon_{12}(\mathbf{k})\\ \epsilon_{12}(\mathbf{k}) & \xi_2(\mathbf{k})\end{pmatrix}\begin{pmatrix} c_{\mathbf{k}1}\\ c_{\mathbf{k}2}\end{pmatrix}+\sum_{\alpha,\beta}\sum_{\mathbf{k},\mathbf{k}'}V_{\alpha,\beta}(\mathbf{k},\mathbf{k}')c^{\dagger}_{-\mathbf{k}\alpha}c^{\dagger}_{\mathbf{k}\beta}c_{\mathbf{k}'\beta}c_{-\mathbf{k}'\alpha}. \tag{3}$$

Here, $\xi_{1(2)} \equiv \epsilon_{1(2)} - \mu_{1(2)}$ is the dispersion for the Bloch states constructed from the 1(2) orbital, ϵ_{12} is the inter-orbital coupling, and $V_{\alpha,\beta}(\mathbf{k},\mathbf{k}')$ describes the pairing interaction, allowing for the possibility of inter-orbital ($\alpha = 1, \beta = 2$ and $\alpha = 2, \beta = 1$) as well as intra-orbital ($\alpha = \beta = 1, 2$) pairing. In the discussion to follow, we use Δ_{11} and Δ_{22} to denote the two intra-orbital order parameters while $\Delta_{12} = \Delta_{21}$ is the inter-orbital order parameter. Spin labels have been suppressed.

In the basis set by the spinor, $\hat{\Psi}^{\dagger}_{\mathbf{k}} = (c^{\dagger}_{\mathbf{k}1}, c_{-\mathbf{k}1}, c^{\dagger}_{\mathbf{k}2}, c_{-\mathbf{k}2})$, the inverse mean-field 4×4 Green's function for this model is

$$\mathbf{G}_0^{-1}(\mathbf{k},\omega_n) = \begin{pmatrix} i\omega_n - \xi_1\hat{\tau}_3 + \Delta'_{11}\hat{\tau}_1 - \Delta''_{11}\hat{\tau}_2 & -\epsilon_{12}\hat{\tau}_3 + \Delta'_{12}\hat{\tau}_1 - \Delta''_{12}\hat{\tau}_2 \\ -\epsilon_{12}\hat{\tau}_3 + \Delta'_{12}\hat{\tau}_1 - \Delta''_{12}\hat{\tau}_2 & i\omega_n - \xi_2\hat{\tau}_3 + \Delta'_{22}\hat{\tau}_1 - \Delta''_{22}\hat{\tau}_2 \end{pmatrix}. \tag{4}$$

Here, $\hat{\tau}_l$ are the usual 2×2 Pauli matrices, Δ'_a (Δ''_a) is the real (imaginary) part of the intra- and inter-orbital order parameters, and ω_n is a Fermi Matsubara frequency. The two branches of the BCS quasiparticle spectrum, E_- and E_+, found from the solution of $\det\mathbf{G}_0^{-1}(\mathbf{k},\omega_n) = (\omega_n^2 + E_-^2)(\omega_n^2 + E_+^2)$, are

$$\begin{aligned} E_{\pm}^2 = \frac{1}{2}\Big\{&|\Delta_{11}|^2 + |\Delta_{22}|^2 + 2|\Delta_{12}|^2 + 2\epsilon_{12}^2 + \xi_1^2 + \xi_2^2 \pm \big[(|\Delta_{11}|^2 + |\Delta_{22}|^2 + 2|\Delta_{12}|^2 + 2\epsilon_{12}^2 + \xi_1^2 \\ &+ \xi_2^2)^2 - 4\big(|\Delta_{22}|^2\xi_1^2 + |\Delta_{11}|^2\xi_2^2 + |\Delta_{11}|^2|\Delta_{22}|^2 - 4\epsilon_{12}\xi_1\mathrm{Re}[\Delta_{12}\Delta_{22}^*] - 4\epsilon_{12}\xi_2\mathrm{Re}[\Delta_{12}\Delta_{11}^*] \\ &+ 2\epsilon_{12}^2\mathrm{Re}[\Delta_{11}\Delta_{22}^*] - 2\mathrm{Re}[\Delta_{11}\Delta_{22}(\Delta_{12}^*)^2] + (\epsilon_{12}^2 - \xi_1\xi_2)^2\big)\big]^{1/2}\Big\}. \end{aligned} \tag{5}$$

In the situation (relevant for Sr_2RuO_4) that the Fermi surface crosses the two sheets formed by the bare 1 and 2 orbital dispersions and $\mu_1 \simeq \mu_2 \equiv \mu \gg |\Delta_{ij}|\ \forall i,j$, the minimum in the *sum* of the two Bogoliubov quasiparticle bands, $E_- + E_+$, is close to twice the inter-orbital hopping energy:

$$\min(E_- + E_+) \sim 2\epsilon_{12}(k_x \simeq k_y \simeq |\mathbf{k}_0|). \tag{6}$$

Here, $\mathbf{k}_0$ are the wavevectors at which the bare dispersions ξ_1, ξ_2 both vanish; i.e., where the Fermi surface crosses the 1 and 2 sheets. In this case, as we will see in the following, the inter-orbital hopping energy scale plays a crucial role in determining the frequency dependence of the anomalous, intrinsic Hall conductivity in a multiband chiral superconductor, with multiple order parameters spanning the bands.

3. Intrinsic and anomalous Hall conductivity

3.1. Formalism

The optical Hall conductivity $\sigma_H(\omega)$ is defined in terms of the antisymmetric part of the $\hat{J}_x$-$\hat{J}_y$ current correlator $\pi_{xy}(\mathbf{q},\omega)$ by

$$\sigma_H(\omega) \equiv -\frac{1}{2i\omega}\lim_{\mathbf{q}\to 0}\left[\pi_{xy}(\mathbf{q},\omega) - \pi_{yx}(\mathbf{q},\omega)\right]. \tag{7}$$

The total current operator in the i direction is given by [31] $\hat{J}_i = e\sum_{\mathbf{k}} \mathrm{tr}\hat{\Psi}^\dagger_{\mathbf{k}}\hat{\mathbf{v}}(\mathbf{k})_i\hat{\Psi}_{\mathbf{k}}$, where

$$\hat{\mathbf{v}}_i(\mathbf{k}) = \begin{pmatrix} v_{i,11}(\mathbf{k})\hat{1}_2 & v_{i,12}(\mathbf{k})\hat{1}_2 \\ v_{i,12}(\mathbf{k})\hat{1}_2 & v_{i,22}(\mathbf{k})\hat{1}_2 \end{pmatrix} \tag{8}$$

is the 4×4 bare current vertex ($\hat{1}_2$ is the 2×2 identity matrix). In the orbital basis, $v_{i,aa}(\mathbf{k}) = \partial_{k_i}\epsilon_a$ and $v_{i,12}(\mathbf{k}) = \partial_{k_i}\epsilon_{12}$.

When there are multiple orbitals present, it is possible to have a nonzero anomalous Hall conductivity evaluating the current correlator at the one-loop level, and so in the following we ignore vertex corrections:

$$\pi_{xy}(\mathbf{0},\nu_m) = e^2T\sum_{\mathbf{k},\omega_n}\mathrm{tr}[\hat{\mathbf{v}}_x(\mathbf{k})\mathbf{G}_0(\mathbf{k},\omega_n)\hat{\mathbf{v}}_y(\mathbf{k})\mathbf{G}_0(\mathbf{k},\omega_n+\nu_m)]. \tag{9}$$

Here, ν_m is a Bose Matsubara frequency. In the case of a single orbital (or multiple uncoupled orbitals), $\hat{\mathbf{v}}_\sigma$ is purely diagonal and commutes with $\mathbf{G}_0$. Consequently, π_{xy} equals π_{yx}, and the one-loop value for the Hall conductivity is zero, irrespective of details such as band anisotropy, pairing symmetry, or self-energy corrections [27, 28]. Broken time-reversal and translational symmetries are necessary but not sufficient conditions for a nonzero anomalous Hall conductivity at this level. It follows that for superconductivity on a single orbital, vertex corrections are crucial to having a nonzero anomalous Hall conductivity. Goryo [26] and Lutchyn *et al.* [27] considered impurity-scattering vertex corrections for a model of superconductivity in Sr_2RuO_4 assuming superconductivity takes place predominantly on the γ band (or d_{xy} orbital).

For a multiorbital superconductor, on the other hand, the one-loop contribution (9) can be nonzero, and if so, should provide a major contribution to the anomalous Hall effect in a clean superconductor. This contribution is straightforwardly evaluated by analytically continuing to real frequencies, $i\nu_m \to \omega + i0^+$, to obtain the real and imaginary parts of the anomalous Hall conductivity (7). The complex conductivity is given by

$$\sigma_H(\omega) = 2e^2\sum_{\mathbf{k}}(\delta\mathbf{v}_{21}\times\mathbf{v}_{12})_z\left[\epsilon_{12}\mathrm{Im}(\Delta^*_{11}\Delta_{22}) + \xi_1\mathrm{Im}(\Delta^*_{22}\Delta_{12}) - \xi_2\mathrm{Im}(\Delta^*_{11}\Delta_{12})\right]\times$$
$$\frac{1}{E_+E_-}\left\{\frac{1-f(E_+)-f(E_-)}{(E_-+E_+)[(E_-+E_+)^2-(\omega+i\eta)^2]} + \frac{f(E_+)-f(E_-)}{(E_+-E_-)[(E_+-E_-)^2-(\omega+i\eta)^2]}\right\}. \tag{10}$$

Here, $\mathbf{v}_{ab} \equiv (v_{x,ab}, v_{y,ab})$, and $\delta\mathbf{v}_{21} \equiv \mathbf{v}_{22} - \mathbf{v}_{11}$. $\eta = 0^+$ is an infinitesimal in the clean limit. Otherwise, we can make a crude estimate of the effects of impurity scattering by using a finite $\eta \equiv 1/\tau$, where $1/\tau$ is the impurity scattering rate. For unconventional p-wave superconductivity, superconductivity is easily destroyed by impurity scattering and one expects to be in the clean limit, $|\Delta_{ij}|\tau \gg 1 \Rightarrow \eta \ll |\Delta_{ij}|$, for samples with the highest T_c.

3.2. Generic features of the Hall conductivity in a multiband chiral superconductor

Before discussing the implications of Eq. (10) for Sr_2RuO_4, we point out some generic features relevant for any multiorbital chiral superconductor. At the one-loop level, due to the cancellation between the two terms in Eq. (7), the only contributions to a nonzero σ_H arise from time-reversal symmetry-breaking inter-orbital transitions. The factor $(\delta\mathbf{v}_{21} \times \mathbf{v}_{12})_z$ in Eq. (10) indicates that one photon vertex involves an electron transition between different orbitals, while the other vertex involves an intra-orbital transition. The two terms in the curly brackets of Eq. (10) indicate that, in the quasiparticle basis, only processes which create or destroy two quasiparticles on *different* branches (first term) or which scatter a quasiparticle from one branch to the other (second term) contribute. The rest of Eq. (10), the terms in square brackets together with energy denominators, are the form factors, or coherence factors, that relate the quasiparticle states to the electron orbital states. In passing from Eq. (9) to Eq. (10), almost all such terms cancel, leaving only two very specific terms which carry the signature of the broken time reversal symmetry of the superconducting state. The relevant terms contain a coupling between the orbitals, either through the hopping, ϵ_{12}, or through intra-orbital pairing, Δ_{12}, that is essential to connect an inter-orbital vertex with an intra-orbital vertex. The signature of broken time reversal symmetry comes into these terms through a product of different intra- and inter-orbital superconducting order parameters. A nonzero *relative* phase of these order parameters,

$$\Delta\phi_{11,22} \equiv \phi_{11} - \phi_{22} \equiv \mathrm{Im}(\Delta_{11}^*\Delta_{22})/|\Delta_{11}||\Delta_{22}|, \tag{11}$$

for example, then gives rise to a nonzero anomalous Hall conductivity. Consequently, the existence of multiple order parameters with different complex phases spanning multiple orbitals is a necessary condition to have a nonzero anomalous Hall conductivity.

Beyond the fact that multiple orbitals are needed in order to have a nonzero anomalous Hall conductivity at the one-loop level (i.e., no vertex corrections), Eq. (10) shows that inter-*band* Cooper pairing is also a necessary condition. Although this expression shows the Hall conductivity with respect to orbital degrees of freedom, it is straightforward to apply a unitary transformation to represent this in terms of band degrees of freedom. The second line of Eq. (10) remains the same since it is expressed in terms of the Bogoliubov quasiparticle basis. On the other hand, the velocities $\delta\mathbf{v}_{21}$ and $\mathbf{v}_{12}$ will be different and the term in square brackets involving the coupling ϵ_{12} will vanish since the bands are, by definition, uncoupled. This leaves the remaining terms in square brackets which, in the band basis, describe inter-band pairing, $\Delta_{\alpha\beta} \neq 0$ where $\alpha \neq \beta$ are the band indices. Thus, inter-band pairing is needed to have a nonzero anomalous Hall conductivity in a chiral superconductor at the one loop level. Moreover, as in the orbital picture, the inter-band order parameter cannot have the same phase as at least one of the intra-band order parameters.

At zero temperature, only the first term in curly brackets in Eq. (10) contributes to the Hall conductivity; it corresponds to a process wherein a photon destroys a Cooper pair and creates two quasiparticles. The imaginary part of σ_H, which is the absorptive part, turns on at the minimum of $E_- + E_+$. As discussed above, this minimum is set by the inter-orbital coupling, ϵ_{12}, and, in general, is expected to be a much larger energy scale than the BCS quasiparticle gap $|\Delta_{ij}|$. The zero temperature anomalous Hall conductivity was studied in Ref. [28] and the absorptive part exhibits sharp features due to the van Hove singularites in the quasiparticle density of states near the minimum of $E_- + E_+$.

There are two effects of finite temperature: (1) to modify the contribution from the first term in curly brackets and (2) the new contribution represented by the second term in the curly brackets. If $\Delta_{ij} \ll \epsilon_{12}(\mathbf{k}_0)$, the primary effect of (1) is to replace the zero temperature values of the Δ's with their finite temperature values. In particular, the sharp features emerging due to the van Hove singularities near the minimum of $E_- + E_+$ (see Figs. 2 and 3) will not be appreciably smoothed out at nonzero temperatures. This is because, relative to $\epsilon_{12}(\mathbf{k}_0)$, the temperature, $T \leq T_c \sim \Delta$, is always very small in the superconducting phase where the anomalous Hall effect arises. For this same reason, contribution (2), due to quasiparticle scattering, is also very small. This contribution vanishes at $T = 0$, as well as at $T = T_c$, and is largest at temperatures comparable to but less than T_c. For most frequencies, this term is smaller than contribution (1) by a factor of order $\Delta/\epsilon_{12}(\mathbf{k}_0)$ because the Fermi factors restrict the momentum integral to a very narrow region where $E_1 \sim \Delta$. Consequently, at most frequencies, and in particular at the high frequency relevant to current experiments, the anomalous Hall conductivity is dominated by the first term in the curly brackets and the main effect of finite temperature is simply to replace the zero temperature Δ with $\Delta(T)$.

As noted above, the intrinsic Hall conductivity arises from inter-orbital transitions when there are order parameters of different phases living on the orbitals. The van Hove singularities represent a resonance condition for these transitions. In this way, the detection of a strong increase in the Hall conductivity at frequencies on the order of the inter-orbital hopping energy would provide strong evidence in favour of multiband/multiorbital Cooper pairing in a chiral superconductor, such as has been proposed in Sr_2RuO_4 [14, 15, 16, 17]. In contrast, features in the Hall conductivity arising at frequencies on the order of the gap, such as impurity scattering contributions which yield a nonzero Hall conductivity in a single band system [26, 27], are comparatively insensitive to whether or not superconductivity is a multiband phenomenon.

4. Optical Hall conductivity in Sr_2RuO_4

Having used a two-orbital model to elucidate the basic physics of the anomalous intrinsic Hall effect in a multiorbital chiral superconductor, we now discuss the implications for Sr_2RuO_4. Although Sr_2RuO_4 is a three-band system, in calculating the intrinsic anomalous Hall conductivity, σ_H, an important simplification arises because, to a very good approximation, the x and y components of the current operator do not couple d_{xy} to the other two orbitals and, hence, only $d_{xz} - d_{yz}$ inter-orbital transitions contribute to its σ_H. Consequently, the conclusions reached from our two orbital model are still valid. In particular, the intrinsic, anomalous Hall effect in Sr_2RuO_4 is only nonzero when there is complex inter-band pairing and this pairing must involve the d_{xz} and d_{yz} orbitals to a significant extent. Higher order spin-orbit coupling, of order λ^2 (where λ is the spin-orbit coupling parameter; see e.g. Ref. [32]), and interlayer hopping involving d_{xy} orbitals, will bring in the d_{xy} orbital, but these are are sufficiently small that they can be ignored. Consequently, the d_{xy} orbital only plays a passive role in determining the Hall conductivity. In particular, the terms outside the curly brackets in Eq. (10) would be unchanged if we used a three-orbital model of Sr_2RuO_4, but neglected interlayer hopping and higher order spin-orbit effects. Only the quasiparticle dispersions would be modified and would renormalize our estimate of the Hall conductivity somewhat. Thus, we simply ignore the d_{xy} orbital and use our two-orbital result for the Hall conductivity, (10), as this gives a reasonable order of magnitude estimate.

Building on our previous work [28] where we neglected next-nearest neighbour intra-orbital hopping, we now include this effect, taking $\epsilon_1 = -2t\cos(k_x) - 2t''\cos(k_y)$, $\epsilon_2 = -2t\cos(k_y) - 2t''\cos(k_x)$, and $\epsilon_{12} = 2t'\sin(k_x)\sin(k_y)$. We use parameters from a recent LDA study of the band structure of Sr_2RuO_4 [32]: $t = 0.4$eV, $t' = 0.1t$, and $t'' = 0.125t$. We also fix the chemical potential to be $\mu = t$, consistent with this study. Following Ref. [17], we assume purely intra-

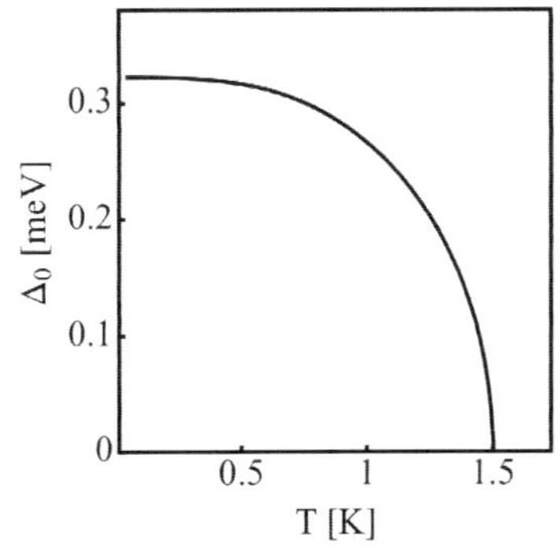

Figure 1. Temperature dependence of the order parameter amplitude, $\Delta_0(T)$, defined in Eq. (18).

orbital pairing on the d_{xz} and d_{yz} orbitals. As a simple ansatz, we use

$$\Delta_{11}(\mathbf{k},T) = \Delta_0(T)\sin(k_x)\cos(k_y),\ \Delta_{22}(\mathbf{k},T) = i\Delta_0(T)\sin(k_y)\cos(k_x),\ \Delta_{12}(\mathbf{k},T) = 0. \quad (12)$$

Here, $\Delta_0(T)$ is the purely real temperature-dependent amplitude of the order parameters, assumed to be equal for both. Crucially in order to obtain a nonzero Hall conductivity, the two order parameters have different phases, meaning that $\mathrm{Im}(\Delta_{11}^*\Delta_{22})$ and hence, σ_H, is nonzero.

Mineev [33] has argued that a model of the form shown in Eq. (12), based on a quasi-1D approach, is not energetically stable in the case relevant to Sr_2RuO_4, where $t' \gg \Delta_0$. However, since $t' \ll t$, t' only affects a very small region of the Fermi surface, while Δ_0 operates over the entire Fermi surface. As explained by Chung *et al.* [24], this allows the quasi-1D approach to remain valid even for $t' \gg \Delta_0$. While we use a somewhat simplified model, more realistic details, such as allowing different pairing amplitudes on the α and β bands would not change our results in any significant way. *The key property of the model is that it includes interband pairing.* As stressed in Ref. [28], this is a necessary condition for a non-zero intrinsic Hall effect, a point also emphasized by Mineev [33].

To see the structure of the purely intraorbital order parameter, Eq. (12) in the band basis, we diagonalize the single-particle contribution to the Hamiltonian, Eq. (3), into the α and β bands. In this basis, there are three order parameters: two intra-band,

$$\Delta_{\alpha\alpha} = \frac{1}{(\xi_+ - \xi_1)^2 + \epsilon_{12}^2}\left[\epsilon_{12}^2\Delta_{11} + (\xi_+ - \xi_1)^2\Delta_{22}\right], \quad (13)$$

$$\Delta_{\beta\beta} = \frac{1}{(\xi_- - \xi_1)^2 + \epsilon_{12}^2}\left[\epsilon_{12}^2\Delta_{11} + (\xi_- - \xi_1)^2\Delta_{22}\right], \quad (14)$$

and one inter-band,

$$\Delta_{\alpha\beta} = \Delta_{\beta\alpha} = \frac{1}{\sqrt{(\xi_+ - \xi_1)^2 + \epsilon_{12}^2}\sqrt{(\xi_- - \xi_1)^2 + \epsilon_{12}^2}}\left[\epsilon_{12}^2\Delta_{11} + (\xi_+ - \xi_1)(\xi_- - \xi_1)\Delta_{22}\right], \quad (15)$$

order parameters.

As expected, the inter- and intra-band order parameters have different phases and hence, the Hall conductivity will be nonzero, as we already inferred from the fact that the phases of the order parameters in the orbital basis were also different. It is interesting to note that in the "hot zone" $\xi_1 \simeq \xi_2$ (with $k_x \simeq k_y \simeq |\mathbf{k}_0|$) which delineates the regions of the Brillouin zone that dominate the conductivity integral in Eq. (10), these order parameters have approximately opposite chirality:

$$\Delta_{\alpha\alpha} \simeq \Delta_{\beta\beta} \simeq [\Delta_{11} + \Delta_{22}]/2 \tag{16}$$

and

$$\Delta_{\alpha\beta} \simeq [\Delta_{11} - \Delta_{22}]/2. \tag{17}$$

Turning now to the task of evaluating numerically the Hall conductivity, we assume a separable potential, $V_{\alpha\alpha}(\mathbf{k}, \mathbf{k}') = g f_{\mathbf{k}}^{(\alpha)} f_{\mathbf{k}'}^{(\alpha)}$ where $f_{\mathbf{k}}^{(1/2)} = \sin(k_{x/y})\cos(k_{y/x})$, and use this to determine the amplitude $\Delta_0(T)$ of the order parameter. The resulting BCS gap equation is

$$\frac{\Delta_0(T)}{g} = T \sum_{\mathbf{k},\omega_n} \sin k_x \cos k_y G_{0,12}(\mathbf{k}, \omega_n). \tag{18}$$

Here, $G_{0,ij}$ is the (i, j) element of the 4×4 matrix Green's function, the inverse of which is given by Eq. (4). We choose the dimensionless coupling constant (restoring the crystalline spacing a but keeping $k_B = \hbar = 1$) $t\pi^2 a^2/g$ to be $\simeq 5.59$ in order to have a critical temperature $T_c = 1.5K$ [21], appropriate for ultraclean Sr_2RuO_4 samples. We plot $\Delta_0(T)$ in Fig. 1. Note that the $T = 0$ value, $\Delta_0(0) \simeq 0.32$meV, is larger than the canonical weak-coupling s-wave BCS-limiting value, $\Delta_{0,BCS}(0) \simeq 1.76T_c \simeq 0.23$meV, because of anisotropy.

In Figs. 2 and 3, we make use of the calculated $\Delta_0(T)$ in Eq. (10) to plot the frequency-dependent Hall conductivity at $T = 0$ and $T = 1.4K$, respectively in the ultraclean limit, $\eta = 10^{-6}t$. There is very little difference between the two plots apart from an overall scale factor: $\sigma_H(\omega, T) \simeq (\Delta_0(T)/\Delta_0(0))^2 \sigma_H(\omega, 0)$. This can be seen explicitly in the inset of Figs. 4 and 5 which show the temperature dependence of the real and imaginary parts of the Hall conductivity (at $\hbar\omega = 0.8$eV), respectively, divided $(\Delta_0(T)/t)^2$. One sees that this quantity is essentially independent of temperature. This result,

$$\sigma_H(\omega, T) \underset{\sim}{\propto} \Delta_0^2(T) \text{ for } \omega \not\sim 2t', \tag{19}$$

where the proportionality factor is independent of temperature, holds for all frequencies except those very close to the van Hove singularity in the vicinity of $\omega \sim 2t'$. In the main panels of Figs. 4 and 5, we plot the temperature dependence of the real and imaginary parts of the conductivity.

The fact that the Hall conductivity is approximately proportional to the square of the temperature-dependent order parameter (at least away from the van Hove singularities) accounts for its smallness. Dividing the calculated Hall conductivity by the relevant dimensionless energy scale $x \equiv (\Delta_0(0)/t)^2 \sim 6 \times 10^{-7}$, we see that the zero-frequency and temperature Hall conductivity $\sigma_H(0)/x \sim e^2/\hbar$ is approximately unity (in natural units of $e^2/\hbar$). It is only the smallness of the order parameter in Sr_2RuO_4 that makes the anomalous Hall conductivity as small as it is.

Equation (19) also suggests that useful information about the origin of the anomalous Hall effect in Sr_2RuO_4 can come from looking at the impurity concentration dependence of the Hall conductivity in Sr_2RuO_4. T_c in non-s-wave superconductors is strongly suppressed by non-magnetic impurities [34]. In Sr_2RuO_4, this dependence has been studied in the group of Mackenzie [35, 36]. Assuming that $\Delta_0(T \sim 0) \propto T_c$, measurement of the impurity concentration dependence of the Hall conductivity (or Kerr angle, as discussed below) in conjunction with the impurity concentration dependence of T_c would help reveal the relative importance of the intrinsic and extrinsic contributions to σ_H.

Figures 2 and 3 also clearly exhibit the sharp features anticipated in Sec. III, arising from the van Hove singularities near $\min[E_-(\mathbf{k}) + E_+(-\mathbf{k})] \sim 2t' \gg \Delta_0$. As discussed there, these are

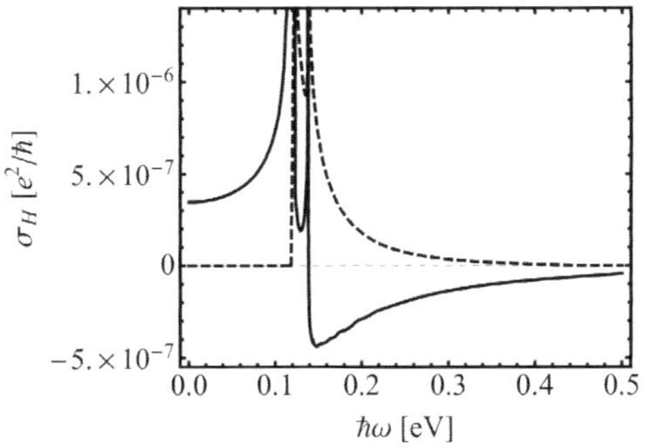

Figure 2. Frequency dependency of the real (solid line) and imaginary (dashed line) parts of the Hall conductivity at $T = 0$. In all figures, the model parameters are as given in the text surrounding Eqs. (12) and (18).

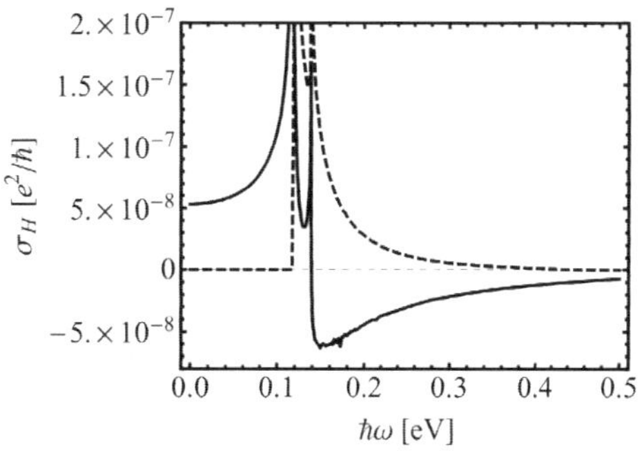

Figure 3. Frequency dependence of the real and imaginary parts of the Hall conductivity at $T = 1.4K(\sim 0.93T_c)$. Note the different scale as compared to Fig. 2.

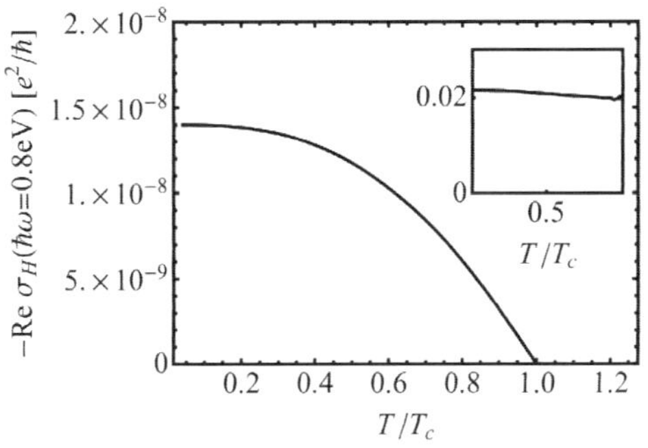

Figure 4. Temperature dependence of the real part of the Hall conductivity at $\hbar\omega = 0.8$eV. Inset: Temperature dependence of the real part of the Hall conductivity at $\hbar\omega = 0.8$eV divided by $(\Delta_0(T)/t)^2$.

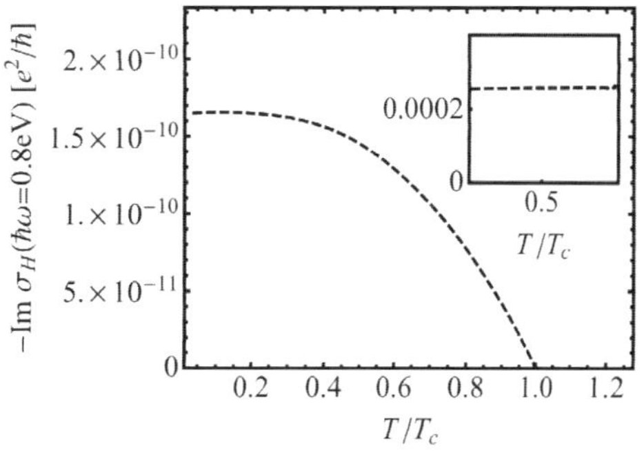

Figure 5. Temperature dependence of the imaginary part of the Hall conductivity at $\hbar\omega = 0.8$eV. Inset: Temperature dependence of the imaginary part of the Hall conductivity at $\hbar\omega = 0.8$eV divided by $(\Delta_0(T)/t)^2$.

not at all washed out at finite temperatures since, relative to this energy scale ($t' = 40$meV), the temperature is very small in the superconducting phase ($T_c \sim 0.13$meV). We have also checked the effect of adding a small amount of impurity scattering, taking $\eta = 10^{-4}t$ (corresponding to $\Delta_0\tau \sim 10$) and, unsurprisingly given this small broadening (recall that $\Delta_0 \ll t'$), find very little difference from the results shown in Figs. 2 and 3. Notably, the resonance features remain intact. If measured, these strong features, arising as they do from inter-orbital transitions involving the creation of BCS quasiparticles from multiple complex order parameters spanning these orbitals, would constitute strong evidence for multiband chiral superconductivity.

The second term in Eq. (10) adds very little to the Hall conductivity. For our parameters, we find this contribution ($\equiv \sigma_H^T$) to be always much smaller than the contribution coming from the first term and do not find the coherence features found by Wysokiński, Annett, and Györffy at low frequencies, $\omega \sim \Delta_0$ [29]. Such features only arise in our calculations by artificially inflating the order parameter (and hence T_c and crucially, the temperature) to energies on the order of several meVs. Due to the thermal Fermi factor, $f(E_-)$, which heavily suppresses all momenta

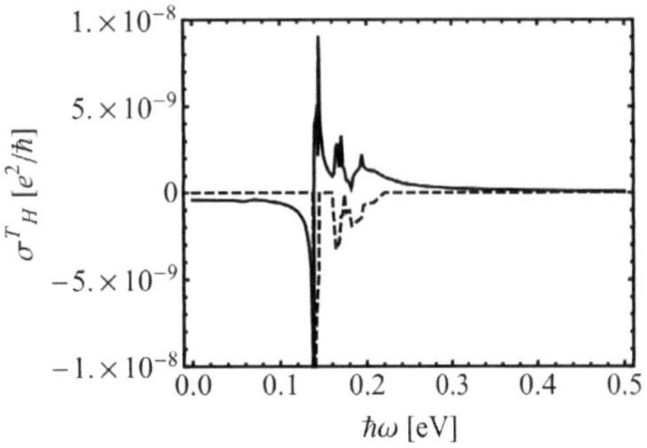

Figure 6. Frequency dependence of only the quasiparticle scattering contribution, σ^T_H, to the Hall conductivity, coming from the second term in curly brackets in Eq. (10), at $T = 1.4K(\sim 0.93T_c)$. The solid (dashed) line shows the real (imaginary) part of the Hall conductivity. Compare with the full Hall conductivity at the same temperature in Fig. 3.

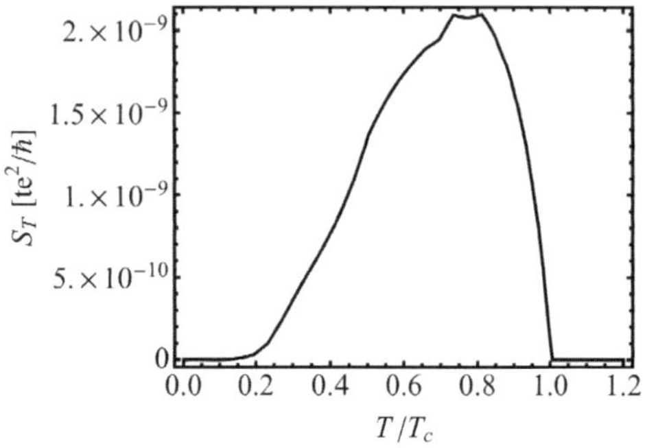

Figure 7. Temperature dependence of the integrated spectral weight, $S_T \equiv -\int_0^\infty d\omega \text{Im}\sigma^T_H(\omega)$, of the imaginary part of the contribution, σ^T_H, to the Hall conductivity from quasiparticle scattering (the second term in curly brackets in Eq. (10)).

except where $E_- \gtrsim \Delta_0$ is small, $\sigma^T_H(\omega)$ will always be much smaller than $\sigma_H(\omega)$ unless T/t' and hence, Δ_0/t', is appreciable. (Note, $E_+ \gtrsim 2t'$ and $f(E_+) \approx 0$.)

Because of the Fermi factor, the largest contribution to $\sigma^T_H(\omega)$ will come when $\omega \sim E_+ - E_- \sim 2t'$, with $E_- \sim \Delta_0$ and $E_+ \sim 2t'$. In Fig. 6, we plot the frequency-dependence of $\sigma^T_H(\omega)$ at $T \simeq 0.93T_c$ showing this strong frequency dependence [37]. To better understand the temperature dependence of this contribution, in Fig. 7, we also plot the integrated spectral weight (defined to be positive)

$$S_T \equiv -\int_0^\infty d\omega \text{Im}\sigma^T_H(\omega) = -\pi e^2 \sum_{\mathbf{k}} \frac{\epsilon_{12}\text{Im}(\Delta^*_{11}\Delta_{22})\,(\delta\mathbf{v}_{21} \times \mathbf{v}_{12})_z}{E_+E_-(E_+-E_-)^2}\,[f(E_+)-f(E_-)] \tag{20}$$

of $\text{Im}\sigma^T_H(\omega)$ in the clean limit ($\eta \to 0^+$) as a function of temperature. As expected, the low-temperature spectral weight from scattering E_- BCS quasiparticles to E_+ is exponentially suppressed.

5. Kerr angle at finite temperatures

We now turn our attention to the Kerr angle, the quantity that has been measured in experiments [21, 22] and which provides an indirect measure of the Hall conductivity. The frequency-dependent Kerr angle θ_K is given by Eq. (2). (See, for instance, Ref. [27].) In addition to depending on the frequency-dependent Hall conductivity, $\sigma_H(\omega)$, the Kerr angle also depends on the diagonal element of the optical conductivity tensor, $\sigma(\omega)$, through its dependence on the complex index of refraction $n(\omega)$:

$$\alpha(\omega) = \frac{1}{n(n^2-1)}, \tag{21}$$

where $n(\omega) = \sqrt{\varepsilon_{ab}(\omega)}$ is equal to the square root of the component of the permeability tensor,

$$\varepsilon_{ab} = \varepsilon_\infty + (4\pi i/\omega)\sigma(\omega), \tag{22}$$

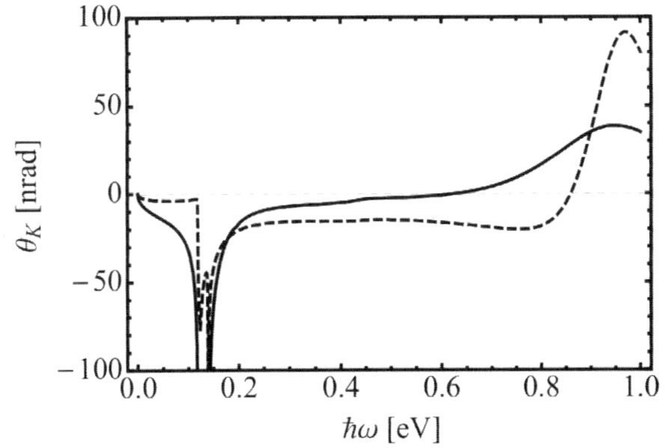

Figure 8. Frequency dependence of the Kerr angle at $T = 0$. The solid line shows the Kerr angle for an inverse transport lifetime 0.4eV; the dashed line corresponds to an inverse transport lifetime 0.2eV. The large difference between the two results shows the sensitivity of the Kerr angle to material parameters beyond the Hall conductivity. Note the large resonant feature at $\omega \gtrsim 0.1$eV arising from the van-Hove singularity in the Hall conductivity (see Fig. 2).

in the ab plane. Here, ε_∞ is the background dielectric tensor.

Having calculated the Hall conductivity, we can combine this calculation with experimental conductivity data to determine the frequency-dependent Kerr angle. The conductivity in Sr_2RuO_4 seems to be well approximated by a generalized Drude expression with a frequency-dependent effective mass and transport scattering rate [38]. For simplicity, however, we use a Drude model with constant transport scattering rate γ:

$$\sigma(\omega) = -\frac{\omega_{pl}^2}{4\pi i(\omega + i\gamma)}. \tag{23}$$

This is a reasonable approximation since the transport scattering rate varies relatively slowly over a large frequency range (see Fig. (4b) in Ref. [38]), from approximately 2.5 eV/$\hbar$ at $\hbar\omega = 0.1$eV to 4.5 eV/$\hbar$ at $\hbar\omega = 1$eV at low temperature. This variation is small in comparison with that exhibited by the Hall conductivity over the same range and we simply use the value at 0.8eV (the frequency at which the Kerr angle is measured in Ref. [21]) to compute the frequency-dependent Kerr angle: $\gamma(0.8\text{eV}) \sim 0.4\text{eV}/\hbar$.

The other parameters needed to calculate the Kerr angle are also extracted from experiments on Sr_2RuO_4. Following Ref. [27] we take the interlayer spacing $d = 6.8\mathring{A}$, $\varepsilon_\infty = 10$, and $\omega_{pl} = 2.9$eV. Using these, in Fig. 8 we plot the $T = 0$ Kerr angle θ_K as a function of frequency using our calculated $T = 0$ Hall conductivity (see Fig. 2). The results are shown for two values of the transport scattering rate, the "experimental value" of 0.4eV/$\hbar$ (extracted at $\omega = 0.8$eV) and a somewhat smaller value of 0.2eV/$\hbar$. Using the former value, we find $\theta_K \simeq 17$nrads at $\hbar\omega = 0.8$eV, while the latter value gives $\theta_K \simeq -20$nrads. As one sees from the plot, in the vicinity of the plasma edge, the precise value of the Kerr angle is very sensitive to material parameters such as the transport scattering rate. Setting the scattering rate to zero, as is sometimes done, is not a good approximation because of the significant inelastic scattering at the frequencies of interest.

Over the frequency range of interest, the temperature dependence of the Kerr angle will track that of the Hall conductivity (see Figs. 4 and 5), which in turn is approximately proportional to the square of the order parameter.

6. Discussion

In this paper, we have calculated the anomalous Hall conductivity in a simple two-orbital model of a chiral superconductor, with an eye to understanding the Kerr effect in Sr_2RuO_4 [21]. Although Sr_2RuO_4 is a three-orbital system, we have argued that the essential physics responsible for the anomalous Hall effect arises from only the Ru d_{xz} and d_{yz} orbitals, since these are the only orbitals that are optically coupled. This fact gives an important experimental handle in understanding *where* superconductivity arises in Sr_2RuO_4: In contrast to the impurity-scattering extrinsic contribution to the Hall conductivity [26, 27], there must be significant pairing on the α and β bands in order to have an intrinsic anomalous Hall conductivity. The "smoking gun" for such superconductivity would be a large resonant feature in the Hall conductivity at a frequency close to the twice the inter-orbital hopping scale $t' \sim 0.04$eV, as clearly seen in Figs. 2 and 3 (also in Fig. 8 for the Kerr angle). This feature is robust both to nonzero temperature as well as impurity broadening effects since both these energy scales are on the order of Δ_0, much smaller than $2t'$.

To make contact with the Kerr rotation experiment [21], using our two-orbital model in conjunction with parameters appropriate for Sr_2RuO_4, we have arrived at the result $\theta_K(0.8\text{eV}) \simeq$ 17nrads for the Kerr angle at $T = 0$. This is of the same order of magnitude—but smaller than—the experimental value of $\sim$ 65nrads [21].

Although the inclusion of the d_{xy} orbital would not lead to any qualitatively new physics, it can lead to a quantitative change to this prediction. We also note that the d_{xz}/d_{yz} hopping parameters used in our calculations were obtained from a microscopic *three*-orbital model [32] including spin-orbit coupling (SOC), which we have also excluded. As we showed in Ref. [28], SOC does not lead to any Hall effect unless there is a Zeeman splitting $\mu_\uparrow \neq \mu_\downarrow$ between the two spin species. Its only effect otherwise is to renormalize the dispersions that enter Eq. (10).

In comparison to our earlier work [28], where we predicted a $T = 0$ Kerr angle of 50nrads using a two-orbital model without next-nearest neighbour hopping, the present work produces a smaller Kerr angle, resulting from a smaller prediction for the Hall conductivity. This difference is due almost entirely to the form for the order parameter given in Eq. (12), appropriate for next-nearest neighbour hopping, an effect not included in our earlier calculations. The additional cosine factors in the order parameters lead to a significant reduction in the Hall conductivity since the integrand in Eq. (10) is dominated by a small region in the vicinity of $k_x \sim k_y \lesssim |\mathbf{k}_0| \equiv \cos^{-1}(-\mu/2t)$ (see Eq. (6)). This reduction is offset somewhat by the larger value of $\Delta_0(T = 0) \simeq 2.45T_c$ (where $T_c = 1.5$K) used in the present work as a result of solving the BCS gap equation appropriate for our model, instead of using the s-wave BCS result $\Delta_0(T = 0) \simeq 1.76T_c$

Although the Kerr angle is additionally very sensitive to the low-temperature optical properties of Sr_2RuO_4, and more information about these are needed to reliably make contact with experiment, there are a few ways in which experiments should be able to distinguish between the two contributions – extrinsic [26, 27] and intrinsic – to the Hall conductivity identified by theory. First, as we have noted in this paper, the intrinsic mechanism will exhibit a strong frequency dependence in the vicinity of twice the inter-orbital hopping, around 40 meVs. Unfortunately, this seems to be outside the range of the current experimental setup at Stanford [21, 22]. Secondly, experiments could probe the dependence of the Kerr angle on the impurity concentration.

To a good approximation, the intrinsic mechanism discussed here leads to a Hall conductivity that depends on the impurity concentration only through its approximately quadratic dependence on the order parameter, Eq. (19). In turn, the impurity dependence of T_c, and hence, Δ_0, has been studied [35, 36]. In contrast, the extrinsic contribution to the Hall conductivity [26, 27] will exhibit an additional, stronger dependence on the impurity concentration.

In summary, a more detailed understanding of the anomalous Hall conductivity in Sr_2RuO_4, either by probing its frequency dependence or its dependence on the concentration of non-magnetic impurities, would contribute significantly to understanding this material. In particular, the Hall conductivity is sensitive to the way in which the different bands crossing the Fermi energy of Sr_2RuO_4 enter into the chiral p-wave order parameter. We hope that the present work stimulates further experimental studies of the anomalous Hall effect in Sr_2RuO_4 as well as in any other materials that might exhibit chiral superconductivity [39].

Acknowledgments
We thank Steve Kivelson, Sri Raghu, and Ronny Thomale for useful discussions. This work was supported by NSERC and the Canadian Institute for Advanced Research.

References

[1] Maeno Y, Hashimolo H, Yoshida K, Nishizaki S. Fujita T, Bednorz I G and Lichtenberg F 1994 *Nature* **372** 532.
[2] Rice T and Sigrist M 1995 *J. Phys.: Condens. Matter* **7** L643.
[3] Baskaran G 1996 *Physica B* **224** 490.
[4] Kallin C 2012 *Rep. Prog. Phys.* **75**, 042501.
[5] Maeno Y, Kittaka S, Nomura T, Yonezawa S and Ishida K 2011 *J. Phys. Soc. Jpn.* **81** 011009.
[6] Mackenzie A and Maeno Y 2003 *Rev. Mod. Phys.* **75** 657.
[7] Read N and Green D 2000 *Phys. Rev. B* **61** 10267.
[8] Puetter C M and Kee H-Y 2012 *Eur. Phys. Lett.* **98** 27010.
[9] Haverkort M W, Elfimov I S, Tjeng L H, Sawatzky G A, and Damascelli A 2008 *Phys. Rev. Lett.* **101** 026406.
[10] Agterberg D F, Rice T M, and Sigrist M 1997 *Phys. Rev. Lett.* **78** 3374.
[11] Zhitomirsky M E and Rice T M 2001 *Phys. Rev. Lett.* **87** 057001.
[12] Nomura T and Yamada K 2002 *J. Phys. Soc. Jpn.* **71** 404.
[13] Deguchi K, Mao Z Q and Maeno Y 2004 *J. Phys. Soc. Jpn.* **73** 1313.
[14] Takimoto T 2000 *Phys. Rev. B* **62** R14641.
[15] Kuroki K, Ogata M, Arita R and Aoki H 2001 *Phys. Rev. B* **63** 060506(R).
[16] Annett J F, Litak G, Gyórffy B L and Wysokiński K I 2002 *Phys. Rev. B* **66** 134514.
[17] Raghu S, Kapitulnik A and Kivelson S A 2010 *Phys. Rev. Lett.* **105** 136401.
[18] Stone M and Roy R 2004 *Phys. Rev. B* **69** 184511.
[19] Kirtley J R, Kallin C, Hicks C W, Kim E-A, Liu Y, Moler K A, Maeno Y and Nelson K D 2007 *Phys. Rev. B* **76** 014526.
[20] Kallin C and Berlinsky A J 2009 *J. Phys.: Condens. Matter* **21** 164210.
[21] Xia J, Maeno Y, Beyersdorf PT, Fejer M M and Kapitulnik A 2006 *Phys. Rev. Lett.* **97** 167002.
[22] Kapitulnik A, Xia J, Schemm E and Palevski A 2009 *New J. Phys.* **11** 055060.
[23] Argyres P N 1955 *Phys. Rev.* **97** 334.
[24] Chung S B, Raghu S, Kapitulnik A and Kivelson S A 2012 *Phys. Rev. B* **86** 064525.
[25] Roy R and Kallin C 2008 *Phys. Rev. B* **77** 174513.
[26] Goryo J 2008 *Phys. Rev. B* **78** 060501(R).
[27] Lutchyn R M, Nagornykh P and Yakovenko V M 2009 *Phys. Rev. B* **80** 104508.
[28] Taylor E and Kallin C 2012 *Phys. Rev. Lett.* **108** 157001.
[29] Wysokiński K I, Annett J F and Györffy B L 2012 *Phys. Rev. Lett.* **108** 077004.
[30] Yip S K and Sauls J A 1992 *J. Low T. Phys.* **86** 257.
[31] Although we suppress spin in our model, the Hall conductivity receives equal contributions from both spins. The current operator includes contributions from both particle and hole contributions, mimicking spin counting. Hence, the Hall conductivity calculated from this spin operator correctly accounts for two spins.
[32] Rozbicki E J, Annett J F, Souquet J R and Mackenzie A P 2011 *J. Phys.: Condens. Matter* **23** 094201. Note that our factor of t' is equivalent to twice $f_{xz/yz}$ in this reference.
[33] Mineev V I 2012 *J. Phys. Soc. Jpn.* **81** 093703.
[34] Balian R and Werthamer N R 1963 *Phys. Rev.* **131** 1553.
[35] Mackenzie A P, Haselwimmer R K W, Tyler A W, Lonzarich G G, Mori Y, Nishizaki S and Maeno Y 1998 *Phys. Rev. Lett.* **80** 161.
[36] Mackenzie A P, Haselwimmer R K W, Tyler A W, Lonzarich G G, Mori Y, Nishizaki S and Maeno Y 1998 *Phys. Rev. Lett.* **80** 3890.

Journal of Physics: Conference Series **449** (2013) 012036 doi:10.1088/1742-6596/449/1/012036

[37] For $t' \gg \Delta_0$, the $\sigma_H^T(\omega)$ integrand is *very* strongly peaked about the Fermi surface and it is difficult to carry out the integration to high accuracy. Although the primary feature visible at $\sim 2t'$ in Fig. 6 is robust, the details of the other features above this are only approximate.
[38] Katsufuji T, Kasai M and Tokura Y 1996 *Phys. Rev. Lett.* **76** 126.
[39] Nandkishore R, Levitov L S and Chubukov A V 2012 *Nat. Phys.* **8** 158.

Journal of Physics: Conference Series **449** (2013) 012037 doi:10.1088/1742-6596/449/1/012037

Chemically gated electronic structure of a superconducting doped topological insulator system

L A Wray[1,2], S Xu[2], M Neupane[2], A V Fedorov[1], Y S Hor[3,4], R J Cava[4] and M Z Hasan[2]

[1]Advanced Light Source, Lawrence Berkeley National Laboratory, Berkeley, California 94305, USA

[2]Department of Physics, Joseph Henry Laboratories, Princeton University, Princeton, NJ 08544, USA

[3]Department of Physics, Missouri University of Science and Technology, Rolla, MO 65409

[4]Department of Chemistry, Princeton University, Princeton, NJ 08544, USA

Abstract.

Angle resolved photoemission spectroscopy is used to observe changes in the electronic structure of bulk-doped topological insulator $Cu_xBi_2Se_3$ as additional copper atoms are deposited onto the cleaved crystal surface. Carrier density and surface-normal electrical field strength near the crystal surface are estimated to consider the effect of chemical surface gating on atypical superconducting properties associated with topological insulator order, such as the dynamics of theoretically predicted Majorana Fermion vortices.

Three-dimensional topological insulators (TIs) are a class of materials distinguished by topological parameters of bulk band structure, and by the consequent appearance of distinctive spin-helical surface states [1, 2, 3, 4, 5, 6, 7]. When combined with superconductivity, the surface environment of TIs is expected to host effective Majorana Fermion particles with collective non-Abelian statistics. Electron doped Bi_2Se_3 has been identified as one model system in which superconducting phase coherence can be induced in TI surface states [8, 9, 10, 11]. In this study, copper atoms are deposited on the cleaved surface of N-type $Cu_xBi_2Se_3$ crystals to better understand how the surface energetics of Bi_2Se_3 can be modified, and to explore the material as a potential platform for the manipulation of Majorana vortex modes. Angle resolved photemission spectroscopy is used to observe surface energetics and the formation of new surface-localized electronic states.

Surface-normal charge density distributions are estimated using the modified Thomas Fermi approximation (MTFA) [12], an approach that addresses important effects of surface electrostatics [13]. Because it is our goal to address implications of this electrostatic environment, we will not discuss other important chemical effects such as atomic diffusion and bonding near the TI surface.

Bismuth selenide has a bulk band gap of $\Delta\sim0.3$eV, which is spanned by a single surface state Dirac cone at the 2D reciprocal space $\bar{\Gamma}$-point (Fig. 1(a)). Light copper doping does not significantly change the bulk band structure of $Cu_xBi_2Se_3$ [8, 9]. A highly variable renormalization of surface electron velocities is found as a function of carrier doping, and has been suggested to be linked to the chemical potential rather than deriving from hybridization with chemical impurities [14]. Electrostatic repulsion and quantum exclusion from the topological

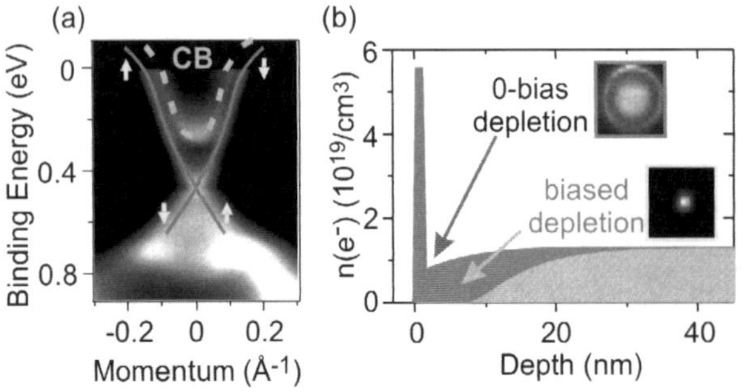

Figure 1. Intrinsic charge inhomogeneity at a doped TI surface: (a) (blue) Bulk and (green) surface bands intersecting the Fermi level of $Cu_xBi_2Se_3$ with x=0.12 superconducting composition are traced above ARPES measurements. A diagram above the ARPES image shows topological quantum numbers and the location of the topological surface states. (b) The electron density distribution is modeled for (gray) as-grown Bi_2Se_3 and (gold) the same sample with chemical potential set at the Dirac point by NO_2 deposition. Even with no surface deposition, large electron density in the surface state (top nanometer) is expected to create a minor depletion dip in charge density. Insets show the Fermi surface, from which the surface state charge density is determined using the Luttinger theorem.

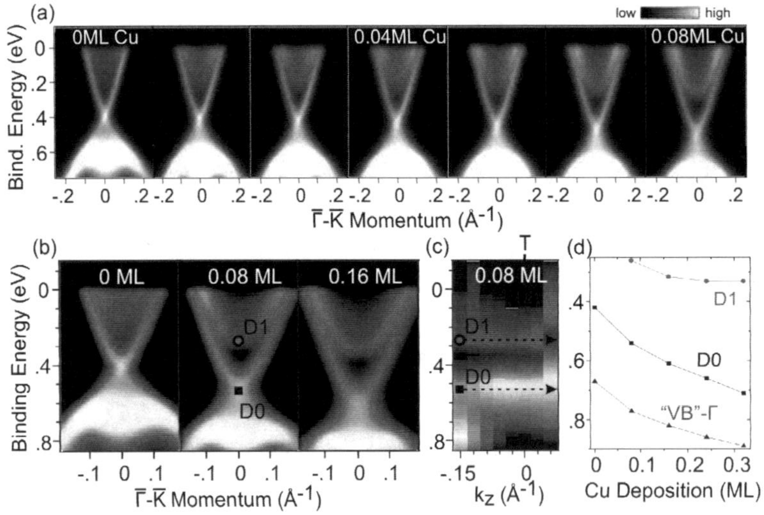

Figure 2. Cu deposition on $Cu_xBi_2Se_3$: (a) Photoemission measurements show the incremental formation of a 2DEG surface state as an estimated 0.08ML of Cu is added to the surface of bulk-doped $Cu_xBi_2Se_3$. (b) Deposition is shown over a wider range. (c) The intensity maximum of the new band that appears after 0.08 ML Cu deposition, labeled 'D1' at the BZ center, has no momentum dispersion along the z-axis. The original surface state Dirac point is labeled 'D0'. (d) The doping evolution of band energies at the Γ-point is shown, including a local maximum of ARPES intensity associated loosely with the valence band ("VB"-Γ).

surface state shift the energies of bulk-derived electronic states near the surface [15, 9, 14], and are expected to create a region with slightly depressed carrier density even when no external bias is applied (Fig. 1(b)). As we will show later, this "0-bias depletion" effect can potentially create a unique surface environment for interacting with superconducting vortices.

Bulk electron charge density is calculated using MTFA from the measured band dispersions and surface carrier density for pre-deposition samples, matching the approximations adopted to calculate band bending energies in Ref. [15]. After deposition, surface charge is estimated from a combination of the ARPES Luttinger count and the deposited density of Cu atoms. These calculations are intended as qualitative illustrations only, and therefore make use of the nominal dielectric constant of ϵ_r=113 [16] as opposed to attempting to treat ϵ_r more accurately as a function of depth inside the crystal. Additionally, the Dirac cone surface state of freshly cleaved Bi_2Se_3 is assumed to be localized in the top nanometer of the crystal, consistent with DFT simulations and experimentally base estimates [9]. Similar ARPES-based methodologies for exploring the effect of electrostatic perturbations on TI surfaces have been established in previous studies [6, 7, 15, 19, 17, 18, 14].

In-situ copper deposition and concurrent ARPES measurements were performed at Advanced Light Source beamline 12, at temperatures lower than 20K to minimize migration of Cu atoms into the sample. ARPES energy resolution was better than 20 meV, and the vacuum was maintained below 8×10^{-11} Torr. Deposition depth is presented in units of monolayers (ML) to approximately represent the number of deposited atoms per hexagonal unit cell of the surface, based on rough linear assumptions outlined in Ref. [14]. Data in Fig. 2(a-b) are obtained using 61eV photons. Incident energy dependence in Fig. 2(c) was obtained by incrementally raising the photon energy, and converting to z-axis momentum using an inner potential of 9.5eV [8, 9].

Intensity resembling a new surface state appears near the Fermi level after deposition of just ~0.01ML Cu on the surface of N-type $Cu_{0.02}Bi_2Se_3$ (Fig. 2(a)). After 0.08ML deposition, the new surface band is clearly seen at all momenta beneath the Fermi level, and shows no z-axis momentum dispersion (Fig. 2(c)). Excess surface-deposited Cu adatoms cause the electron energies near the surface of the TI to bend towards higher binding energy, as they would in an N-N Schottky diode (Fig. 2(d)).

Because TI surface states cross the bulk band gap and have different energetics than bulk electrons [3, 5], when interfacing non-TI materials with a topological insulator, the surface states are expected trap charge carriers and attract or repel bulk electrons even when the bulk carrier doping of each material is identical. The surface charge density evaluated from the Luttinger count suggests that prior to depositing additional copper on the vacuum-cleaved $Cu_{0.02}Bi_2Se_3$ surface, a negative charge is trapped by the surface state, and will be screened over a range of more than 10 nanometers as simulated by MTFA (Fig. 3(a)). Approximating +1 valence for Cu at low coverage, the simulation suggests that depositing 0.08ML of copper causes the net charge at the crystal surface to undergo a significant change from negative to positive. The approximation of nearly full +1 ionization of surface Cu atoms is known to break down at higher coverage [14]. Between 0 and 0.08ML (shaded in Fig. 3(a)) this picture suggests that the electric field penetrating beyond the first few nanometers of the crystal surface is very small, allowing electrons to retain their bulk-derived energetics and symmetries as they approach the surface from deeper in the crystal. The existence of a sign change in the surface charge is supported by the slower change in surface energies when adding Cu from 0.08ML to 0.16ML surface coverage, relative to the first 0.08ML of deposition. Titrating with electropositive adatoms or a gating voltage (V_G~0.1V from our data) to achieve an electrically neutral surface may therefore provide a method to enhance bulk interactions with the surface states, such as magnetic and superconducting proximity effects.

When superconducting phase coherence is introduced at the surface of a TI, effective Majorana Fermion modes are expected to manifest where magnetic vortex lines intersect the surface states [20]. These Majorana states can give rise to non-Abelian statistics, and may be possible to manipulate adiabatically to store quantum information, given the right combination of superconducting order parameter and chemical potential [8, 9, 21]. There is strong evidence that $Cu_xBi_2Se_3$ ($x\geq0.1$) is a so-called topological superconductor [8, 9, 22, 23], which would

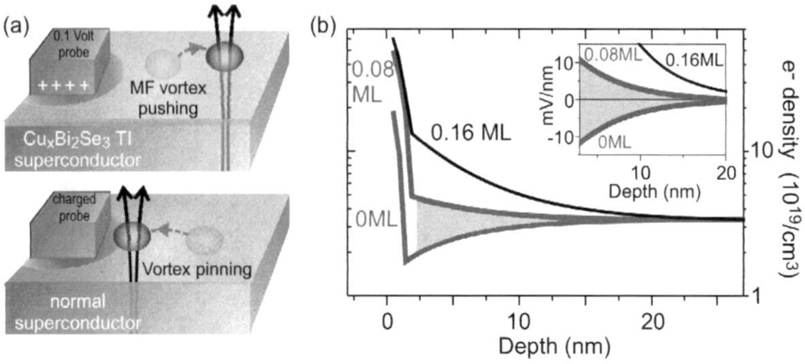

Figure 3. Majorana vortices and surface bias: (a) Probes interfacing with normal superconductors suppress the superfluid density, attracting ("pinning") vortices. On a topological insulator, a gentle probe with bias voltage similar to 0.1eV can potentially repel vortices by enhancing the superfluid density. Regions with suppressed superfluid density are indicated with green shading. (b) The z-axis electron density distribution is modeled for different degrees of surface doping. (shaded region) Deposition levels between 0ML and 0.08 ML are expected to reduce bulk screening and strengthen superconductivity near the surface. An inset shows the MTFA estimation of electric field strength inside the crystal.

create gapless states that preclude adiabatic manipulation of the surface. However, bismuth selenide thin films can also allow standard s-wave superconducting phase coherence to propagate into surface states from a niobium diselenide substrate [11], satisfying the known prerequisites for adiabatic manipulation of Majorana Fermions.

The MTFA calculation in Fig. 3(b, inset) provides an estimate that the electric field from the negatively charged surface state will have an energy scale per nanometer that is large relative to the $2\Delta \sim 1$meV superconducting gap function in the top ~20nm of the crystal [8, 9, 10, 22]. This is expected to cause phase fluctuations that suppress superconductivity if the surface charge is not neutralized. Moreover, because vortices are repelled by regions of high superfluid density, this surface environment provides a new mechanism to control the dynamics of Majorana fermions and potentially observe their non-Abelian properties. For normal superconductors, surface perturbations suppress superfluid density, generating an attractive force that pins vortices in place [24, 25]. For a topological insulator however, our data show that adjusting the surface chemical energy to cancel out the negative surface state charge can likely enhance superfluid density under a probe, causing surface vortices to be repelled by the probe rather than pinned to it (Fig. 3(b)).

It should be emphasized that the MTFA calculations we have discussed are provided only as a basis for qualitative discussion. We have not gone to great lengths to make these calculations quantitatively accurate, as that would be very difficult for such a complex surface environment. Error relating to the z-axis distribution of dielectric permeability and z-axis localization of the newly bound 2DEG-like surface states is also difficult to estimate. We have assumed that the newly bound surface state is highly localized in the top 2 nanometers of the crystal, which provides a sufficient minimum basis for describing the states in a k.p theory derived model [14]. This approximation is consistent with the strong manifestation of this band within the shallow ~1nm ARPES penetration depth [9], and with simulations that assume a sharp z-axis gradient for the surface dielectric constant [14]. While there is strong evidence that dielectric permeability near the surface is renormalized after electropositive surface deposition, a more shallow gradient would cause the surface states to be less localized [17].

In summary, we have presented measurements of surface electronic structure as a function of

Cu deposition on the electron doped topological insulator $Cu_xBi_2Se_3$. These measurements are interpreted with the MTFA to present a broader discussion of how chemical and electrical gating of TI surface states may interact with surface electron density. We propose that carefully tuned surface bias potentials within the explored perturbation energy range could make it possible to enhance the proximity effects of bulk electronic ordering at the surface. The unusual charge inhomogeneity at the surface of topological insulators may also allow a method for creating tunable repulsive forces that could be useful to control the motion of superconducting vortices. For intrinsically homogeneous surfaces, the same types of external perturbation would likely suppress bulk-derived properties at the crystal surface and pin vortices in place.

Acknowledgements:

The synchrotron X-ray-based measurements are supported by the Basic Energy Sciences of the US DOE (DE-FG-02-05ER46200, AC03-76SF00098 and DE-FG02-07ER46352).

[1] Moore J E 2009 *Nature Physics* **5** 378
[2] Hasan M Z and Kane C L 2010 *Rev. Mod. Phys.* **82** 3045; Qi X-L and Zhang S-C 2011 *Rev. Mod. Phys.* **83** 1057
[3] Fu L and Kane C L, 2007 *Phys. Rev. B* **76** 045302
[4] Hsieh D *et al* 2008 *Nature* **452** 970
[5] Hsieh D *et al* 2009 *Science* **323** 919
[6] Hsieh D *et al* 2009 *Nature* **460** 1101
[7] Xia Y *et al* 2009 *Nature Physics* **5** 398
[8] Wray L A *et al* 2010 *Nature Physics* **6** 855
[9] Wray L A *et al* 2011 *Phys. Rev. B* **83** 224516
[10] Kriener M *et al* 2011 *Phys. Rev. B* **84** 054513
[11] Wang M-X *et al* 2012 *Science* **336** 52
[12] Trott S, Trott M and Nakov V 1993 *Phys. Stat. Sol.* **177**, 389
[13] M. Bianchi *et al* 2012 *Semicond. Sci. Technol.* **27** 124001
[14] Wray L A *et al* 2012 Preprint at ⟨http://arxiv.org/abs/1206.1087⟩
[15] Analytis J G *et al* 2010 *Phys. Rev. B* **81** 205407
[16] Madelung U R O and Schulz M, in Non-Tetrahedrally Bonded Elements and Binary Compounds I (1998), pp. 112
[17] King P D C *et al Phys. Rev. Lett.* **107** 096802
[18] Pan Z-H *et al* 2012 *Phys. Rev. Lett.* **108** 117601
[19] Wray L A *et al* 2011 *Nature Physics* **7** 32
[20] Fu L and Kane C L 2008 *Phys. Rev. Lett.* **100** 096407
[21] Hosur P, Ghaemi P, Mong R S K and Vishwanath A 2011 *Phys. Rev. Lett.* **107** 097001
[22] Sasaki S *et al* 2011 *Phys. Rev. Lett.* **07** 217001
[23] Hsieh T H and Fu L 2012 *Phys. Rev. Lett.* **108** 107005
[24] Campbell A M and Evetts, J E 1972 *Adv. Phys.* **21** 199
[25] Villegas J E *et al* 2003 *Science* **302** 1188

AUTHOR INDEX

AUTHOR INDEX